Lysosomes

Eric Holtzman

Columbia University
New York, New York

Plenum Press • New York and London

Library of Congress Cataloging in Publication Data

Holtzman, Eric, 1939–
  Lysosomes.

  (Cellular organelles)
  Includes bibliographies and index.
  1. Lysosomes. I. Title. II. Series. [DNLM: 1. Lysosomes. QH 603.L9 H758La]
  QH603.L9H63   1989   574.87′4   88-32085
  ISBN 0-306-42966-7
  ISBN 0-306-43126-2 (pbk.)

© 1989 Plenum Press, New York
A Division of Plenum Publishing Corporation
233 Spring Street, New York, N.Y. 10013

Printed in the United States of America

Preface

In 1976 I wrote a monograph on lysosomes (*Lysosomes: A Survey*, Springer-Verlag, Vienna) that was intended as an up-to-date, comprehensive survey. Whatever success I may have achieved then in fulfilling that intention, even the effort now would be foolhardy. The literature has grown so rapidly in the past decade that I certainly could not even read all of the essential papers, let alone understand and analyze them. My goal here, therefore, is simply to introduce the major features of lysosomes at a level I hope will be useful both to advanced students and to researchers interested in obtaining a broad background. This is in keeping with the design of the Cellular Organelles series: the series is more a set of advanced texts than of review monographs.

This design carries with it the decision not to support each point by references to the original literature. I apologize for the injustice involved in such a decision but feel that in any event it would be impossibly unwieldy to cite, adequately and in a balanced manner, the contributions of the vast network of researchers responsible for the information upon which I draw. I have tried, in the selection of illustrations and in the Acknowledgments and Further Reading sections at the end of each chapter, to identify as many as possible of those researchers who have constructed the central views and provided the information with which I deal, and to point out short routes to the heart of the original literature.

Inevitably, the organization of the text mixes logic, taste, and arbitrariness. The first chapter sets the stage, introducing fundamental concepts and terminology, and sketching some history to help explain the present state of the field. The next three chapters focus on **heterophagy**—lysosomal digestion of material taken up by the cell from its surroundings. This is the lysosomal function that has been studied most intensively. In Chapter 2 the basic phenomena of heterophagic uptake and digestion are described. Next, the mechanisms of uptake and degradation are examined in greater detail (Chapter 3) and examples are provided of the diverse biological roles of heterophagy and related processes (Chapter 4).

Chapter 5 takes up **autophagy**—lysosomal digestion of the cell's own constituents. Considered here also are several topics whose relations to autophagy are controversial, especially intracellular turnover and facets of the recycling and reutilization of the cell's membranes.

The normal and the pathological are handled side-by-side throughout the book, reflecting an interplay that has characterized the history of work on lysosomes. But Chapter 6, which focuses on "exaggerations" of release and storage of lysosome-related materials, devotes particular attention to the pathological. For example, it presents perspectives on the "storage" diseases whose study has been of cardinal significance for "lysosomology."

The final chapter, Chapter 7, concerns lysosome biogenesis. This topic is dealt with last because the present state of knowledge is difficult to handle without the information in the previous chapters. Section 7.6 presents important information that became available at a late stage in the volume's production.

My belief in the fundamental unity of lysosome-related functions has led me to intermingle discussions of animals, plants, and microorganisms rather than to treat the different kingdoms in separate lumps. Also, by considering or illustrating selected experiments and observations in detail and by intercalating sections on methodology and terminology, I have tried to convey a critical sense of the state of the evidence underlying current concepts. Some of this is done through extended footnotes, to avoid excessive digression within the text.

My interest in lysosomes was first sparked by the period I spent in the laboratory of the late Dr. A. B. Novikoff, to whom I am grateful on many levels. Brian Storrie read most of the manuscript and made valuable comments and suggestions. Additional commentators are acknowledged at the end of each chapter. I thank Philip Siekevitz, the series editor, for having invited me to write the book, the editors and staff of Plenum Press, especially Shuli Traub, for their considerable assistance, and Kathleen Kehoe and the other librarians of the Columbia University Department of Biological Sciences Library for their help in obtaining material. I appreciate the permission granted to me by The Rockefeller University Press to reprint the many illustrations taken from the *Journal of Cell Biology*; original publishers of other illustrations are acknowledged in the individual figure legends.

The research in my laboratory, described or illustrated at various points in the book, was supported by NIH grants from NINCDS and NEI.

Eric Holtzman

New York

Contents

Chapter 3
Acidification; Membrane Properties; Permeability and Transport

Chapter 4

Uses and Abuses of Endocytotic and Heterophagic Pathways

Chapter 5
Autophagy and Related Phenomena

Chapter 6

Extensive Release. Excessive Storage

Chapter 7

Genesis

Historical Fragments; Methods; Some Terminology

1.1. Introduction

Lysosomes are cytoplasmic organelles, each delimited by a single lipoprotein membrane. Typically, they contain several dozen different enzymes (Table 1.1) mostly of the variety called **acid hydrolases;** these catalyze reactions of the type A-B + $H_2O \rightarrow$ A-H + B-OH and have acid pH optima. Collectively, the enzymes are capable of degrading virtually all large cellular molecules—nucleic acids, proteins, polysaccharides, and lipids—to low-molecular-weight products (Fig. 1.1).

The lysosomes of most cells function principally in intracellular digestion. Materials to be digested are incorporated within the same membrane-bounded compartments as the lysosomal enzymes. In this way the cell deploys its acid hydrolases economically (concentrations of some of the proteolytic enzymes in lysosomes approach the millimolar range) and it fosters the cooperative activities of a voracious group of enzymes whose actions would likely prove fatal, or at least severely injurious, were the enzymes turned loose to roam the cytoplasm (see Section 6.3). The compartmentation also facilitates the establishment of the low-pH environment needed for efficient activity of the hydrolases.

1.2. Discovery; Ubiquity

The occurrence of intracellular digestive activity is an "ancient" discovery of cell biology and pathology. The studies of phagocytosis by Metchnikoff and his followers in the late 19th century were seminal. These studies, of feeding by protozoa and of the uptake of particles by ameboid defensive cells in multicellular organisms, led to the articulation of the view that digestion of phagocytosed material takes place in acidified, enzyme-containing intracellular structures. These structures are now known to be lysosomes.

In large measure, the modern concept of the lysosome emerged from the effort to fractionate rat liver cells into their component organelles by homogenization and centrifugation. As this work evolved in the 1950s in the laboratory headed by C. de Duve, it became clear that what the pioneer cell fractionators had designated as the "mitochondrial fraction," actually contains an assort-

TABLE 1.1. Enzymes of the Lysosomal Metabolic Pathways[a]

Proteolytic pathway	Glycanolytic pathway	Nuclease pathway	Lipolytic pathway
Cathepsin D	Hyaluronidase	Ribonuclease II	Triacylglycerol lipase
Cathepsin B	Heparin endoglucuronidase	Deoxyribonuclease II	Phospholipase A_1
Cathepsin H	Heparan sulfate endoglycosidase		Phospholipase A_2
Cathepsin L	Lysozyme[b]		Phosphatidate phosphatase
(Glycosidases and phospho-protein phosphatase also act on intact proteins)		Exonuclease (5'-terminal)	Acylsphingosine deacylase
	α-L-Fucosidase	Acid phosphatase	Sphingomyelin phosphodiesterase
	α-Galactosidase	(There are additional enzymes with phosphodiesterase, pyrophosphatase, nucleoside triphosphatase, and phosphoamidase activities in lysosomes)	(Other enzymes including glycosidases and sulphatases also make important contributions to the lipolytic pathway. Moreover, there are necessary 'activator' proteins)
	β-Galactosidase		
Tripeptidyl peptidase	α-Glucosidase		
Dipeptidyl peptidase I	β-Glucosidase		
Dipeptidyl peptidase II	α-N-Acetylgalactosaminidase		
Arginyl aminopeptidase (cathepsin H)	α-N-Acetylglucosaminidase		
Peptidyl dipeptidase C (cathepsin B)	β-N-Acetylglucosaminidase		
	β-Glucuronidase		
	α-L-Iduronidase		
Carboxypeptidase A	α-Mannosidase		
Carboxypeptidase B	β-Mannosidase		
Prolyl carboxypeptidase	Neuraminidase		
Tyrosine carboxypeptidase	β-Aspartylglucosylaminase		
Dipeptidase I	Chondroitin 6-sulfatase		
Dipeptidase II	Heparin sulfamatase		
	Iduronosulfatase		
	Sulfatases A and B		

[a]Major lysosomal enzymes are grouped here according to the principal classes of natural substrates on which they act. Enzymes restricted to lyscsomes of one or a very few cell types have mostly not been included here. As indicated for selected examples, enzymes listed in a given pathway participate in other degradation as well. Certain of the enzymes have broader specificities than their names imply (e.g., the triacylglycerol lipase probably hydrolyzes cholesterol esters as well as triglycerides). For IUPAC–IUB Enzyme Commission numbers, consult Barrett, A. J., and Heath M. F. (1977) In *Lysosomes: A Laboratory Handbook* (2nd ed.) (J. T. Dingle, ed.), Elsevier: North-Holland, Amsterdam, pp. 22 ff. Table from Barrett, A. J. (1984) *Trans. Biochem. Soc.* **12**: 899. (Copyright: The Biochemical Society, London.)

[b]Not widespread among cell types; present principally in phagocytes.

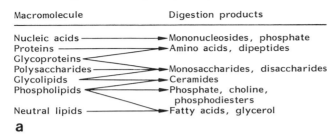

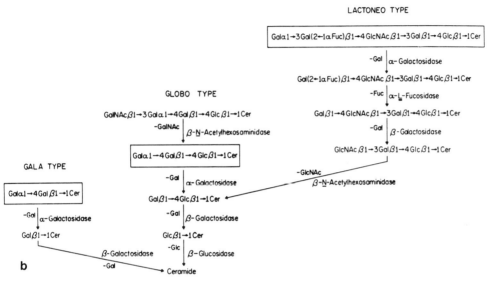

a

b

FIG. 1.1. (a) Probable products of lysosomal digestion as evaluated from experiments in the 1960s and 1970s with isolated lysosomes, and from the known enzymatic capacities of the lysosomes. The importance of the larger products listed (dipeptides, phosphodiesters, disaccharides) is unclear for ordinary circumstances (but see also Chapter 6): Lysosomes have machinery for further breakdown of various of these molecules, but some do seem to arise in detectable amounts *in vitro* and some (e.g., certain dipeptides) may escape from the lysosomes *in vivo* to be further degraded in the extralysosomal cytoplasm (Chapter 3). Courtesy of B. Poole and C. de Duve.

(b) Degradation of large, complex substrates requires sequential action of many hydrolases. Shown here are the likely pathways for digestion of three major types of neutral glycosphingolipids (a class of ceramide-based glycolipids widespread in tissues of animals). The enzymes involved, and the unit removed by each enzyme, are indicated next to each vertical arrow. The structures of the lipids and of the intermediates in their degradation are indicated by the horizontal sequences of units. (Cer, ceramide; Fuc, fucose; Gal, galactose; Glc, glucose; GalNAc, N-acetylgalactosamine; GlcNac, N-acetylglucosamine. Positions and orientations of the bonds linking the units are given by the conventional numbers, Greek letters, and horizontal arrows.) Figures 6.8, 6.9, and 7.2 illustrate other pathways. From Desnick, R. J., and Sweeley, C. C. (1983) In *Metabolic Basis of Inherited Disease* (5th ed.) (J. B. Stanbury, J. B. Wyngaarden, D. S. Frederickson, J. L. Goldstein, and M. S. Brown, eds.), McGraw-Hill, New York.

ment of enzymes unlikely to coexist in a single particle. Along with the expected mitochondrial enzymes of cellular respiration, acid hydrolases are present, as well as enzymes that produce or decompose hydrogen peroxide. The acid hydrolases exhibit the property of **latency.** That is, their activity toward substrates added to the suspension medium increases dramatically with treat-

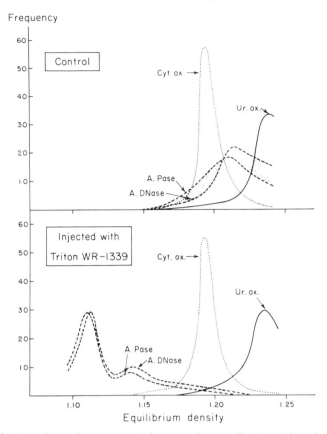

FIG. 1.2. Distribution of "marker enzymes" for several organelles in rat liver homogenates centrifuged on sucrose density gradients. The lower preparation is from a rat injected with Triton 4 days previously; the upper preparation is from a noninjected rat. The use of Triton WR-1339 brings about a selective shift in the location of lysosomes (detected by their content of acid phosphatase and acid DNase) to lower density positions on the gradient. Mitochondria (cytochrome oxidase) and peroxisomes (urate oxidase) are not affected by the detergent. The slight difference in sedimentation behavior between the two lysosomal hydrolases may reflect heterogeneity of the lysosomes in liver (see Section 7.5.1.1). From de Duve, C. (1965) *Harvey Lect.* **59**:49. (Copyright: Alan R Liss Co.)

ments that disrupt membrane structure: prolonged standing at room temperature; freezing and thawing; exposure to detergents or other membrane-perturbing agents. Disruption allows hydrolases and substrates to intermix.

Eventually, these observations were interpreted as indicating the existence, in the "mitochondrial" fraction, of hydrolase-containing particles, distinct from the mitochondria and also distinct from the **peroxisomes,** a third type of particle present in the fraction. de Duve and his co-workers developed differential and density gradient centrifugation procedures for resolving the "mitochondrial" fraction into subfractions (e.g., Fig. 1.2). When certain of the cell fractions in which lysosomal enzymes from normal liver are concentrated were examined in the electron microscope, they proved to be enriched in "pericanalicular dense bodies," a distinctive category of structure a few tenths of a micrometer in diameter, with a single delimiting membrane surrounding a heterogeneous electron-dense content (Fig. 1.3). These bodies come from the liver's most abundant cell type, the parenchymal cells known as **hepatocytes,** which are responsible for most familiar liver functions.

Lysosomes have since been found in virtually all organisms and cell types

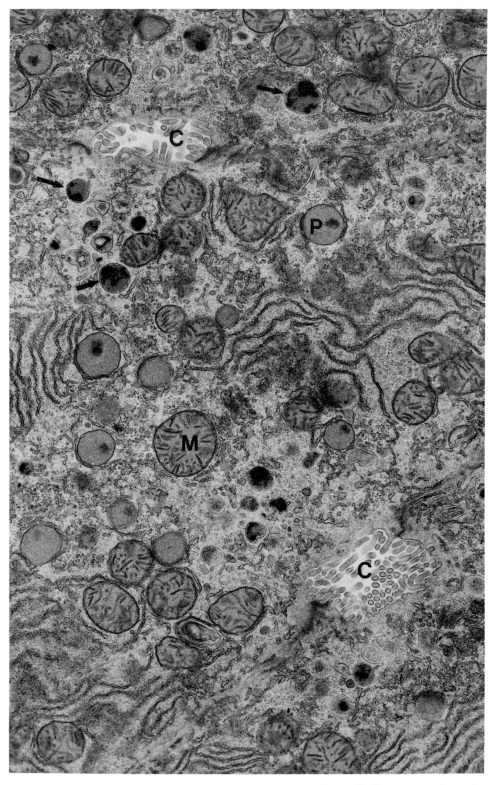

FIG. 1.3. Hepatocyte cytoplasm near bile canaliculi (C) in rat liver. The hepatocytes show the "pericanalicular" dense bodies (arrows indicates three of them) described in the text. Other organelles seen include mitochondria (M) and peroxisomes (P). (See also Fig. 1.7a.) Courtesy of P. M. Novikoff and A. B. Novikoff. × 19,000.

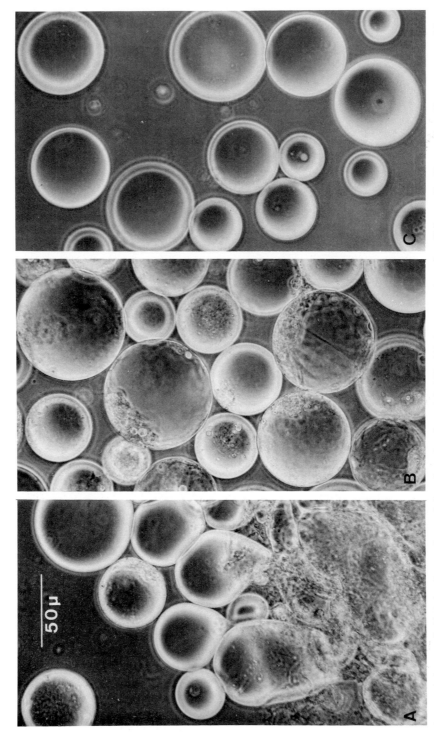

FIG. 1.4. Isolation of vacuoles from tubers of the plant *Stachys sieboldii*. **(A)** A suspension of wall-free cells ("protoplasts") was prepared by enzymatic digestion of the cell walls and the protoplasts were purified by filtration through cheesecloth and centrifugation **(B)**. **(C)** Intact vacuoles were then isolated by disruption of the cells (using low osmotic pressure, the chelating agent EDTA, and gentle shearing) followed by centrifugation. Table 5.3 summarizes biochemical properties of the preparations. From Keller, F., and Matile, P. (1985) *J. Plant Physiol.* **119**:369. (Copyright: Gustav Fischer Verlag.)

TABLE 1.2. Distribution of Enzyme Activity between Protein Body Fractions and Supernatant ("Load") Fractions on a Discontinuous Ficoll Gradient[a,b]

Enzyme	% in fraction	
	Supernatant	Protein bodies
α-Mannosidase	21.6	78.4
Carboxypeptidase	17.9	82.1
N-Acetyl-β-glucosaminidase	24.7	75.3
Acid phosphatase	21.3	78.7
Phosphodiesterase	22.9	77.1
Ribonuclease	24.5	75.5
Phospholipase D	21.5	78.5
Leucine aminopeptidase	95.3	4.7
β-Amylase	>99.9	<0.1
NADH-cytochrome c reductase	>99.9	<0.1

[a]Some of the major hydrolases of protein bodies (a form of vacuole; Section 5.4.2) isolated from germinating mung bean (*Vigna*) seedlings. From Van der Wilden, W., Herman, E. M., and Chrispeels, M. J. (1980) *Proc. Natl. Acad. Sci. USA* **77**:428.
[b]Protoplasts were prepared from cotyledons of seeds 36 hr after the start of imbibition and the ruptured protoplasts were layered on a 5/25% discontinuous Ficoll gradient that was centrifuged at 9.0g for 20 min.

studied. The most familiar exception is the mammalian erythrocyte (red blood cell), which also lacks the other types of intracellular organelles. Whether plant cells have compartments that can reasonably be viewed as homologous to the lysosomes of animal cells was debated for some time. Cell fractionation (Fig. 1.4) and other approaches have now shown clearly that the compartment known as the **vacuole,** prominent in many plant cells, maintains a low pH and contains hydrolytic enzymes, among other components (Fig. 1.4, Table 1.2, Section 3.1.4). There are interesting differences in the functional emphases of the lysosomes of plant cells versus those of animal cells (e.g., see Sections 3.4.2 and 5.4.1), but the lysosomes of both kingdoms represent variations on common themes. The homologies extend as well to the vacuoles of yeast and fungi (e.g., Sections 4.7, 6.1.4.3, and 7.3.3.2), some of whose hydrolases exhibit familial resemblances to lysosomal enzymes of plant and animal cells.

1.2.1. Consumer Alert I: What Is a Lysosome?

Most readers will have encountered the term "vacuole" in uses quite different from the designation of lysosome-related compartments in plants (see Section 1.4.1). As we cover more ground it will become apparent that this is just one of a myriad of terminological and definitional problems encountered in dealing with lysosomes (e.g., see Section 1.5.1). These problems arise partly from history: Study of lysosomes has progressed very rapidly and there has not yet been enough time for universal agreement on the terminology to be used. But many of the problems stem from the difficulties inherent in coping with biological dynamics. We will repeatedly confront situations in which structures arise, undergo transformations in appearance and in biochemical and physiological properties, fuse with other structures, and sometimes even disappear, all in the course of a relatively few minutes. In such situations, it will often not be clear just when, in this sequence of events, a given body should be called a lysosome. I understand lysosomes to be membrane-delimited bodies that contain a characteristic set of acid hydrolases and are capable of participating in intracellular digestion, but even these simple criteria do not apply in straightforward fashion to all the phenomena I will discuss. Although the resulting confusion is a nuisance for the reader (and for the writer) it is important to get beyond the semantic muddle and stay focused on the underlying biological diversity and the dramatic interactions and rapid transformations of cellular organelles in which lysosomes are involved.

1.3. Methods

The study of lysosomes nicely exemplifies the symbiosis between biochemical approaches and microscope-based approaches that characterizes much of modern cell biology. Biochemical methods based on cell fractionation have provided the intimate details about the mechanisms of degradation within lysosomes. Microscopic studies, spearheaded from early on by A. B. Novikoff, have contributed much of what we know about lysosomal diversity and about how lysosomes acquire their contents.

1.3.1. Cell Fractionation

In most cell types, lysosomes make only a small contribution to the cell's mass, a factor complicating the preparation of purified cell fractions. In mammalian hepatocytes or fibroblasts for example, lysosomes recognizable in the "morphometric" microscopic procedures used to evaluate organelle abundance and size (see Section 1.5.3), contribute 0.5 to 5%, or less, of the cell's volume. Rat hepatocytes are usually estimated to contain on the order of 200 recognizable lysosomes per cell, though this excludes small vesicles that transport lysosomal hydrolases (Section 2.1.3.1) and ignores the likelihood that some lysosomes appearing separate in a given electron micrograph actually are connected to one another (Sections 1.5.3 and 2.4.1). In cell types specializing in phagocytosis (Section 1.4.1) lysosomes take up more of the cytoplasm; an active mammalian **macrophage** (Section 1.4.1.1) can contain many hundreds of lysosomes that collectively occupy 2 to 10%—sometimes, during intense phagocytosis, considerably more—of its volume. But until recently, methods for obtaining phagocytes in quantity and for isolating their lysosomes have lagged behind the procedures developed for liver. At its heart, therefore, the biochemical portrait of lysosomes still is derived from studies on rat liver.

Though matters have improved since the days of the first "mitochondrial fractions," traditional differential and density-gradient centrifugation procedures using sucrose media still generally yield impure lysosomal fractions, a situation that strongly colors the lysosome literature. An important advance was the discovery that when rats are injected with the nonionic detergent Triton WR-1339, the detergent accumulates in the lysosomes via "heterophagic" routes like those discussed in Chapters 2 to 4, reducing lysosomal density drastically. The resulting **tritosomes** can be purified on sucrose gradients to the point where less than 5% of the proteins in the fractions come from nonlysosomal sources (Fig. 1.2). Such preparations were used to confirm many of the inferences about lysosomal enzymatic constitution and digestive capacities drawn from work on less pure fractions. However, it remains unclear whether in subtler respects such as membrane properties, lysosomes loaded with indigestible detergents adequately represent their normal counterparts.

In some contexts, naturally occurring lipoproteins or indigestible agents other than Triton WR-1339 (latex beads, paraffin oil, dextrans, metal compounds) can be used to load lysosomes for fractionation. But even so, density-alteration procedures have proved practical only for a handful of tissues.

Recent improvements in purification of unaltered lysosomes have come through the use of density gradient media better than sucrose for controlling osmotically induced alterations in organelle properties during centrifugation. Percoll (Fig. 1.5) or related colloidal silica gradients, and metrizamide gra-

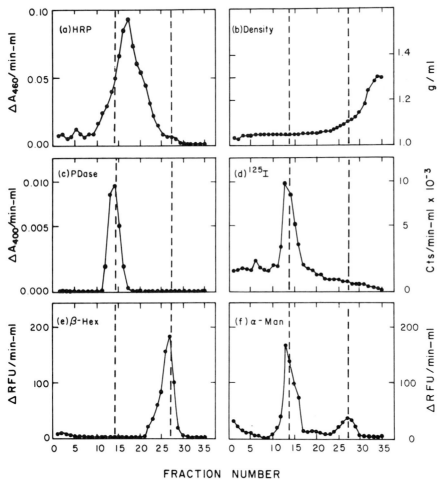

FRACTION NUMBER

FIG. 1.5. Percoll gradient fractionation of Chinese hamster ovary (CHO) cells illustrating the distribution of "markers" for cellular structures of central interest in this book. Before fractionation, the cells were exposed to the tracer protein horseradish peroxidase (HRP; Section 1.4.1.2), which the cells take up and eventually deliver to lysosomes, but the exposure here was short enough that most of the tracer is in "prelysosomal," endocytotic compartments (Section 2.3.1; the compartments presumably include "endosomes"). Panel **e** shows the distribution of β-hexosaminidase A, a lysosomal enzyme. Panels **c** and **d** show the distribution of a plasma membrane-associated alkaline phosphodiesterase (PDase) and of the ^{125}I-labeled proteins produced by iodinating intact cells under conditions in which only the cell surface is labeled. Panel **f** shows the distribution of α-mannosidase activity, which is present both in the Golgi apparatus (left peak) and in lysosomes (right peak). From Pool, R. R., Maurey, L. M., and Storrie, B. (1983) *Cell Biol. Int. Rep.* **7**:361. (Copyright: Academic Press.)

dients have been utilized with considerable success to separate lysosomes from common contaminants, and to characterize different stages in the life history of lysosomes and prelysosomal structures. Yields tend to be low, and even with these improved methods it frequently is difficult to accomplish complete purification (e.g., Percoll gradients often fail to separate lysosomes from mitochondria). However, with care, and by using successive centrifugations in a sequence of gradients, enrichments of better than 100-fold and separation from most other structures can be accomplished.

In free-flow electrophoresis, lysosomes and certain related structures (Section 2.3.1.6) move more rapidly toward the anode than do the other major organelles, affording a useful approach for improving fraction purity.

Vacuoles of some mature plant cells are of large size and relatively low density (see Sections 3.1.4 and 3.4.2), which facilitated their isolation, once adequate methods for disrupting plant cells were devised. Most often, as in Fig. 1.4, "protoplasts" are prepared by enzymatic digestion of the walls and then are disrupted to liberate the vacuoles or to obtain vesicles derived from fragmented vacuoles. With certain plant roots, however, vacuoles can be freed directly by slicing the tissue with sharp blades.

1.3.2. Microscopy; Cytochemistry; Immunocytochemistry

Biochemical analysis of cell fractions provides detailed information about the average properties of organelles. Microscopic methods are needed to investigate the individual structures that contribute to the averages. This is important both because few cell fractions are reliably "pure" and because most tissues from which cell fractions are obtained contain more than one cell type. In liver for instance, lysosomal roles and properties differ appreciably between the hepatocytes and the phagocytotic Kupffer cells, both of which are cell types with quite prominent populations of lysosomes (e.g., Sections 4.4.1, 4.5.2.3, and 4.5.2.4). The cells in an established cultured cell line are less diverse than those in a tissue but the cells do vary. for example at different stages of the division cycle, in ways that might well affect the lysosomes (e.g., in the distribution of organelles in the cytoplasm and in the intensity of lysosomal accumulation of materials from the cells' surroundings). Moreover, even the lysosomes and related structures within a given cell are heterogeneous, owing to the differences in appearance and properties of lysosomes that are sequestering and degrading different types of substrates or are at different stages in their functioning (e.g., Sections 2.1.2.2 and 7.5.1).

1.3.2.1. Dense Bodies

There are morphological criteria to help identify lysosomes and to discriminate among different lysosomal types and stages (e.g., Section 5.1) but by themselves these are not adequate to resolve all of the many ambiguities. For instance, like the first ones studied by electron microscopy, lysosomes often can be described as "dense bodies"—structures delimited by a single membrane and having an electron-dense content (Fig. 1.3). Some such bodies are unique-looking but many are not and can be confused with other structures. The pericanalicular dense bodies are recognizable in hepatocytes by virtue of their accumulations of distinctive materials, mostly poorly digestible intralysosomal deposits rich in metals (Section 3.5.2.2) or lipids (Section 6.4) that persist within the organelles. In contrast, other types of lysosomal dense bodies look very much like secretion granules. For example, those in Fig. 2.2 include bodies rich in hydrolases stored for later use (Section 2.1.2.1); these bodies can be recognized in leukocytes because they are quite abundant and there are no other abundant granules like them in the cells. But they are morphologically "nondescript" in the sense of lacking stigmata that would allow microscopists to pick them out readily were they few in number and scattered, say, in the cytoplasm of exocrine gland cells.

1.3.2.2. Enzyme Cytochemistry

Thus, recognizing which "dense bodies" in a cell are lysosomes often requires, at a minimum, considerable experience with the cell type being studied, More generally, because lysosomes assume so many different forms additional to dense bodies, it frequently is impossible to decide, from morphology alone, whether a particular structure seen in the microscope is or is not a lysosome. This situation is quite unlike that with mitochondria and chloroplasts, which almost always can be identified from their unique ultrastructure. A consequence is that microscopic methods for demonstrating the localization of enzymes *in situ* within cells and subcellular structures, have played a particularly important part in the history of study of lysosomes. Such methods bridge the gap between morphology and biochemistry, providing crucial, albeit rudimentary, information about the biochemical properties of individual structures visualizable in the microscope. They enable the (tentative; see Section 1.3.2.3) discrimination of hydrolase-containing bodies from similar-looking ones that lack lysosomal enzymes. And the methods give access to the many cells and tissues that cannot be approached adequately through cell fractionation because the available quantities are too small, or because the material is too heterogeneous in cell content or is refractory to homogenization and centrifugation procedures.

A handful of the lysosomal enzymes can be localized reliably through light microscopic cytochemical procedures employing substrates that the enzymes alter into products that can be precipitated as visible deposits by trapping agents included in the incubation media (Figs. 1.6, 1.7, 2.15, 4.4, and 6.2). Such methods have proved appropriate for many studies, including the demonstration that the phagocytic compartments of interest to Metchnikoff and his followers do contain lysosomal enzymes. For electron microscopy, one lysosomal hydrolase, an acid phosphatase that hydrolyzes substrates such as β-glycerophosphate (2-glycerophosphate), can be demonstrated routinely (Figs. 1.7, 5.18, and 7.4). Most other lysosomal enzymes, even those that can be studied light microscopically, either do not survive the fixation necessary for adequate preservation of ultrastructure, or cannot be made to yield an electron-dense product that reliably remains at its site of formation.

1.3.2.3. Acid Phosphatase as a "Marker" Enzyme for Lysosomes

Electron microscopists frequently make the assumption that membrane-bounded structures containing acid phosphatase also contain other acid hydrolases, and thus qualify as lysosome-related. This assumption, especially when coupled with evidence for the participation of the acid phosphatase-containing bodies in digestive processes, has been a valuable operational one that propelled crucial progress in the early work on lysosomes. The presumption that acid phosphatase is accompanied by additional acid hydrolases has been ver-

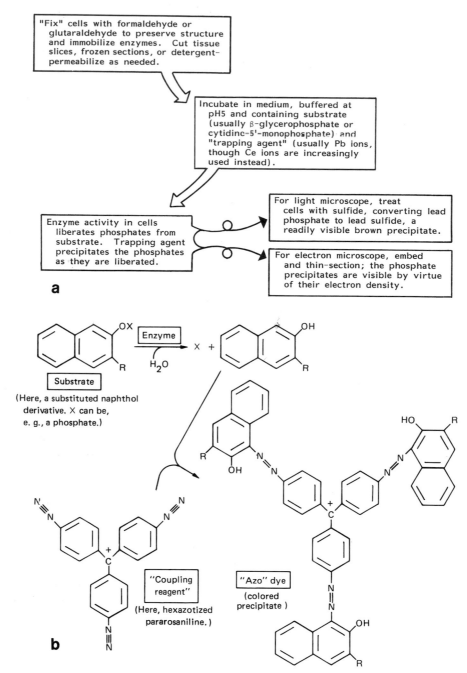

"Fix" cells with formaldehyde or glutaraldehyde to preserve structure and immobilize enzymes. Cut tissue slices, frozen sections, or detergent-permeabilize as needed.

Incubate in medium, buffered at pH5 and containing substrate (usually β-glycerophosphate or cytidine-5'-monophosphate) and "trapping agent" (usually Pb ions, though Ce ions are increasingly used instead).

Enzyme activity in cells liberates phosphates from substrate. Trapping agent precipitates the phosphates as they are liberated.

For light microscope, treat cells with sulfide, converting lead phosphate to lead sulfide, a readily visible brown precipitate.

For electron microscope, embed and thin-section; the phosphate precipitates are visible by virtue of their electron density.

a

Enzyme
H_2O
Substrate
(Here, a substituted naphthol derivative. X can be, e. g., a phosphate.)

"Coupling reagent"
(Here, hexazotized pararosaniline.)

"Azo" dye
(colored precipitate)

b

FIG. 1.6. Principles of enzyme cytochemistry. The methods aim at producing a visible deposit at the sites of enzyme activity while preserving cell structure as intact as possible. **(a)** "Metal-salt" methods for demonstration of acid phosphatase. **(b)** Principles of "azo-dye" methods for light microscopic localization of enzymes. In practice, a variety of substituted derivates of naphthol (or indoxyl) compounds are used as substrates. Diazonium compounds often are employed as coupling reagents. (The representation of the central carbon in hexazotized pararosaniline by a C+ with only three bonds is one of several alternative conventions to indicate the resonance structure of the molecule.) The "metal-salt" methods, especially as pioneered by G. Gomori, have formed the basis for the most effective cytochemical studies of lysosomes, partly because the metal-containing precipitates can be seen by electron microscopy (e.g., see Figs. 1.7a,b, 2.20b, 5.2b, 5.18, 7.3, and 7.4), but light microscopic methods like those in b have also proved very useful (see Figs. 1.7d and 4.4). b based on Lojda, Z., Grossrau, R., and Schiebler, T. H. (1979) *Histochemistry*, Springer, Berlin, and Burstone, M. S. (1962) *Enzyme Histochemistry*, Academic Press, New York.

ified in some cases by correlation of cytochemical findings with findings on cell fractions. In other cases, verification has come from use of electron microscopic methods for localizing aryl sulfatase activity (Figs. 1.7 and 7.3) or for demonstrating an acid esterase (Fig. 5.2; the enzyme demonstrated is often called a "nonspecific" esterase in part because its major natural substrates are still uncertain). At present, immunocytochemical procedures are the most important means for establishing that bodies suspected to be lysosomes from their acid phosphatase activity contain other lysosomal proteins—enzymes and characteristic membrane antigens (Figs. 1.8, 3.10, 7.5, 7.6, and 7.10). As well as helping identify lysosomes in many cell types, these several approaches have also supported the conclusion that the different lysosomal hydrolases coexist within the same bodies rather than being spread among distinct bodies, each containing one or a very few species of enzymes (Section 7.5.1.2).

Such successes, however, do not imply that acid phosphatase is an infallible marker for lysosomes. Phosphatases active at low pH are known to exist outside the lysosomes; one is even present in erythrocytes. Indeed, the early biochemical literature on lysosomes is greatly complicated by an overreliance on p-nitrophenyl phosphate as the substrate for phosphatase assays. This substrate is convenient for spectrophotometric assays but it is as readily split by nonlysosomal phosphatases as by lysosomal enzymes. Biochemists and cytochemists have tried to minimize such problems by establishing criteria for distinguishing among phosphatases, based on careful choice of substrates and pH and on the use of inhibitors. For instance, the lysosomal acid phosphatase differs from some major nonlysosomal enzymes, like the erythrocytic acid phosphatase, in being inhibitable by tartarate and by fluoride ions. But there still is no unambiguous way of being sure, from acid phosphatase assays alone, that a structure being scrutinized biochemically or cytochemically is lysosomal.

Two other cautions merit mention. First, problems like those just outlined for acid phosphatase exist for other lysosomal enzymes as well. Sections 6.1.4 and 7.3.3.1 will present examples of acid hydrolases found in nonlysosomal locales. Second, negative results in cytochemistry, immunocytochemistry, and biochemistry must always be interpreted conservatively. Even when the enzymes being sought actually are present, cytochemical reaction product or immunocytochemical label may be absent as the result of technical inadequacies such as the limited sensitivities of the methods used or the effects of preparative procedures. (For example, even when cytochemical methods reliably show where an enzyme activity in the fixed cell is located, only a small fraction of the enzyme activity generally survives the fixations and other procedures used to preserve cells and to prepare them for cytochemical incubation.)

Biochemists now tend to rely much more on enzymes like the glycosidases than on acid phosphatase, as diagnostic markers for lysosomes. Microscopists still use the phosphatase extensively, but with a more critical eye than hitherto.

1.4. Intracellular Digestion; Heterophagy and Autophagy (Fig. 1.9)

1.4.1. Endocytosis and Heterophagy: Phagocytosis and Pinocytosis

Phagocytosis is one of the two major varieties of **endocytosis,** the uptake of material from the extracellular space directly into membrane-delimited com-

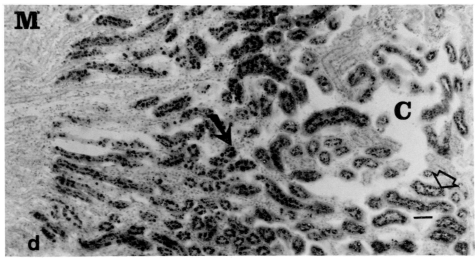

partments that form at the cell surface and separate off into the cytoplasm. Classically, **phagocytosis** was defined from light microscopy, as the uptake of visible particles. The other major form of endocytosis, **pinocytosis,** was taken to be the uptake of fluid. These definitions still serve as useful starting points, but as we will see, the distinctions between phagocytosis and pinocytosis are now more complex.

The terms "vacuole" and "vesicle" are used to refer to the membrane-bounded compartments generated by endocytosis. "Vacuole" denotes the larger compartments and "vesicle" the smaller ones, but there is no hard-and-fast distinction between the two and particularly for structures with diameters of a few tenths of a micrometer the terms are often used interchangeably. More unfortunate is that no terminological convention distinguishes plant vacuoles (Section 1.2), which are mostly not endocytotic, from the endocytotic vacuoles of animal cells. This terminological ambiguity was of little concern in the good old days when cell biologists working on plants and those working on animals had little concourse but is becoming more of an impediment with the appreciation of the homologies between the acid hydrolase-containing structures of plants and animals.

The delivery of the contents of endocytotic vesicles and vacuoles to lysosomes is the basis of **heterophagy,** the digestion, in lysosomes, of **exogenous** material—material originating outside the particular cell under consideration.

FIG. 1.7. Enzyme cytochemistry of lysosomes. **(a)** Light micrograph of a section of rat liver incubated to demonstrate the sites of acid phosphatase activity by "classical" lead methods (as in Fig. 1.6a). The numerous stained "granules" in the regions like those indicated by arrows include the "dense bodies" shown in Fig. 1.3. The heavily reacted larger structures at K are examples of Kupffer cells, the macrophage-related cells found in the lining of the modified capillaries (sinusoids) of the hepatic parenchyma. (The reactive structures within the Kupffer cells are discrete lysosomes, but they are difficult to make out at low magnification when the reaction product is very abundant, as in this illustration.) Courtesy of P. M. Novikoff and A. B. Novikoff. × 600 (approx.).

(b, c) Lysosomal aryl sulfatase activity demonstrated by a lead salt method modified by the author from a procedure originated by S. Goldfischer. (b) Light micrograph showing reactive lysosomes scattered throughout the cell body of a neuron in a mouse ganglion. The prominent sphere in the center of the cell is a nucleolus within the nucleus; it is seen by virtue of its intrinsic refractility and does not contain heavy reaction product. From Holtzman, E. (1969) In *Lysosomes in Biology and Pathology* (J. T. Dingle and H. B. Fell, eds.), Vol. 1, North-Holland, Amsterdam, p. 192. × 1000. (c) Electron micrograph of a lysosome from an epinephrine cell of rat adrenal medulla showing sulfatase reaction product as a dense deposit. (As in this case, deposits of cytochemical reaction product often are restricted to limited zones in a reactive organelle but because the formation of cytochemical precipitates is a complex and poorly understood process, it cannot be concluded that the responsible enzyme is similarly restricted.) The membrane bounding the body (arrow) and the moderately electron-dense intrinsic content of the body can readily be seen. Note the narrow zone of low electron density ("halo") between the membrane and content (see Fig. 3.8). From Holtzman, E., and Dominitz, R. (1968) *J. Histochem. Cytochem.* **16:**320. × 50,000.

(d) Light micrograph of a section of mouse kidney stained to demonstrate the sites of β-glucuronidase activity using a Naphthol AS-BI glucuronide (Fig. 1.6b) as substrate. The tissue is from a mouse of the "beige" line (see Section 4.4.2.4 and Fig. 7.7) in which lysosomes are often so enormous there is room for only one or two per cell [the solid arrow indicates such a body: the bar at the lower right corresponds to 50 μm; the more ordinary-sized lysosomes of the outer cortex of the kidney (C) are indicated by the clear arrow]. M indicates the medullary region of the kidney. From Swank, R. T., and Brandt, E. J. (1978) *Am. J. Pathol.* **92:**755. × 90. (Copyright: Harper and Row.)

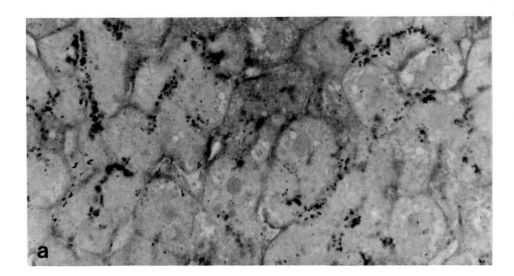

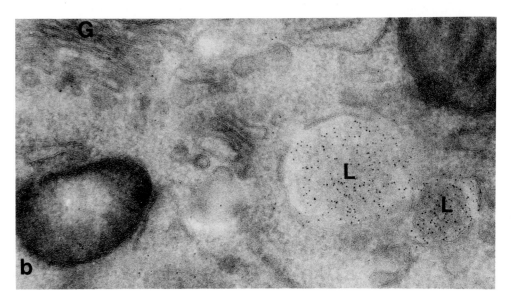

FIG. 1.8. Localization of hydrolases by immunocytochemistry. **(a)** Light micrograph of a section of rat liver in which the lysosomal α-mannosidase has been demonstrated immunocytochemically, using antibodies conjugated to HRP (Section 7.3.2.3): The sites of antibody binding to the tissue are visualized as HRP enzymatic reaction product. This mannosidase has a similar distribution, within hepatocytes, to the acid phosphatase shown in Fig. 1.7a. Courtesy of P. M. Novikoff and A. B. Novikoff. × 600.

    (b) Electron microscope immunohistochemical localization of cathepsin D in cultured human hepatoma cells (G2 line) using an immunogold procedure (see Section 7.3.2.3 and footnote on p. 373) in which the sites of antibody binding are visualized by the presence of electron-dense gold particles. The two bodies with heavy deposits of gold particles, marked L, are lysosomes or cathepsin-containing "prelysosomal" structures (Section 2.4.4.1). A few gold particles are also seen over sacs of the Golgi apparatus (G). From Van Dongen, H. J., Barneveld, R. A., Geuze, H. J., and Galjaard, H. (1984) *Histochem. J.* **16:**941. × 80,000. (Copyright: Chapman and Hall.)

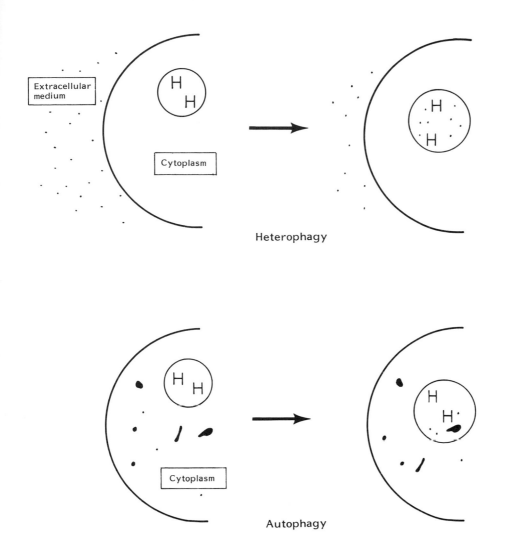

FIG. 1.9. The two major classes of lysosomal functions. In heterophagy, material from outside the cell is degraded as a consequence of being taken into membrane-delimited intracellular compartments that come also to contain lysosomal hydrolases (H). In autophagy, the degraded material originates within the cell.

1.4.1.1. Professional Phagocytes

Phagocytosis serves nutritional roles for unicellular organisms and is quite widespread among common protozoa; it is observed also in other protists, including some organisms related to algae (Section 4.7.2). When they are crawling in culture, fibroblasts, the tips of growing axons, and several other cell types of multicellular animals can take up small numbers of particles commonly used to study phagocytosis (e.g., polystyrene latex spheres with diameters of tenths of micrometers). But in "higher" multicellular organisms, phagocytosis by most cell types probably is rare under usual *in situ* circumstances;

rather, it is a function chiefly of special populations of "professional phagocytes." In higher vertebrates, the principal phagocyte populations are the **polymorphonuclear leukocytes** circulating in the blood, the **macrophages** and other related "mononuclear" phagocytes found in a variety of tissue sites,* and certain other cell types with specialized roles such as the pigment epithelial cells of the retina (Section 4.6.2).

Under favorable conditions in culture, professional phagocytes can readily engulf particles ranging in diameter from a fraction of a micrometer to 5–10 μm or, for some cell types, even more. The most widely used test particles are latex beads, erythrocytes, and microorganisms or cell walls therefrom. A given phagocyte can take up many dozens of the smaller of these particles in the course of a few minutes to an hour.

As Chapter 4 will detail, the phagocytes of multicellular organisms are responsible for defensive activities such as the disposal of invading pathogens; for the clearing of the dead cells and other debris that arise in pathology or in the remodeling phenomena of normal development; and for the normal, steady-state turnover of cell populations such as the erythrocytes of higher animals.

1.4.1.2. Pinocytotic Tracers I: Ferritin, HRP, and Gold

Pinocytosis originally was described from light microscopy as the uptake of droplets of fluid into small intracellular vacuoles by unicellular organisms and cultured cells. It was initially conceived as a sort of cell "drinking" and was of uncertain importance and distribution among cell types. The advent of electron microscopic tracers with macromolecular dimensions (diameters of 5–20 nm or so) has both clarified and complicated the picture. Some of these tracers can be seen in the microscope by virtue of intrinsic electron density; examples are colloidal particles of thorium dioxide ("Thorotrast"; Fig. 2.4), colloidal gold particles coated with proteins (e.g., Figs. 4.6 and 5.14), and molecules of the iron-storage protein **ferritin** (Figs. 1.10, 4.9, and 6.4). Other

---

*The names of mammalian phagocytes: The long history of study of the "white blood cells" (leukocytes) and their relatives and derivatives has left us with a complex terminology, still employed somewhat differently by different authors. In this book, **polymorphonuclear leukocyte** refers to the category of leukocyte possessing a multilobed or bilobed nucleus (hence "polymorphonuclear") and prominent cytoplasmic granules visible in the light microscope (hence these are among the "granulocytic" leukocytes). In the human circulation, 75% of the polymorphonuclear leukocytes are of the **neutrophil** variety, whose name and identification derives from work with "classical" blood strains. ("Heterophil," a term originating in studies of nonhuman mammals, is occasionally still used in place of "neutrophil.") Most of our discussions of granulocytic leukocytes will concern the neutrophils. **Eosinophils** and **basophils** constitute the other two major classes of granulocytic, polymorphonuclear leukocytes; the names of these cells too come from classical stains.

Macrophages and related cells are collectively categorized as the "mononuclear phagocytes." **Monocytes** are the circulating progenitors of these cell types. At sites of inflammation and under other abnormal or experimental circumstances (Section 4.4), monocytes accumulate from the circulation and differentiate into macrophages. But populations of macrophages and closely related cell types also are resident in many tissues under normal conditions. Examples are the Kupffer cells lining the modified capillaries—sinusoids—of the liver, the macrophages of the peritoneal cavity and connective tissue spaces, the alveolar macrophages of the lungs, and perhaps some of the glial cells of nervous tissue. Many current studies of macrophage lysosomes utilize macrophage cell lines (more properly, lines of "macrophage-like" cells) generally derived from spontaneously transformed cancer cells. The murine (mouse) cell lines J774 and P388D are most widely employed.

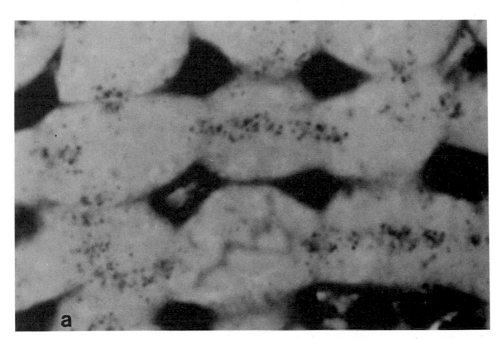

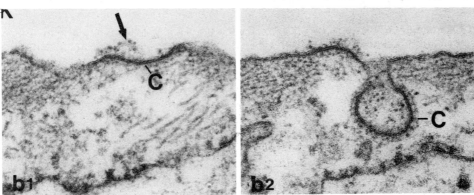

FIG. 1.10. Tracers for studying endocytosis. **(a)** Light micrograph of a section of liver incubated to demonstrate sites of peroxidase activity, from a rat injected intravenously with HRP. The tracer is visualized, by its cytochemical reaction product, in numerous small granules, mostly lysosomes, which are aligned near bile canaliculi (see Figs. 1.3, 1.7a, 1.8a). The tracer also is present in the circulatory spaces, seen as large, dark areas. Courtesy of P. M. Novikoff and A. B. Novikoff. × 1600.

(b) Cultured human fibroblasts exposed to low-density lipoprotein (LDL) particles linked to molecules of the protein **ferritin.** Each ferritin molecule is seen, in the electron microscope, as an electron-dense dot (as at the arrow) because the polypeptide portion of the protein molecule surrounds a core of many iron atoms. In b1, LDL–ferritin conjugates have clustered along the extracellular surface of the cell at a region ("coated pit") whose cytoplasmic surface (C) shows the "coating" described in the text. In b2, a similar cell surface region has invaginated into the cell, preparatory to separating off as an endocytotic structure; the coating (C) is more readily seen than in b1. From Orci, L., Carpentier, J.-L., Perrelet, A., Anderson, R. W., Goldstein, J. L., and Brown, M. S. (1978) *Exp. Cell Res.* **113:**1. × 80,000. (Copyright: Academic Press.)

tracers can be visualized through their enzyme activity, as with horseradish peroxidase (HRP), which oxidizes the cytochemical substrate **di-aminobenzidine** to yield an insoluble polymeric product; this product is brown and so is easily visible in the light microscope, and its affinity for osmium tetroxide facilitates its visualization in the electron microscope (Figs. 1.10, 2.15, 4.9, and 4.10; see also Fig. 1.5).

By including such tracers in the medium surrounding cells *in situ* or in culture, it has been shown that virtually all animal cell types studied can take up droplets of the medium into intracellular vesicles with diameters of a few tens to a few hundred nanometers (e.g., Figs. 1.10 and 4.9). Work with these tracers also gave impetus to a shift in interest from the **fluid** taken up from the medium, to the **molecules** suspended or dissolved in the fluid. Studies on material like kidney tubules, whose lysosomes are prominent (Fig. 1.7d) and rapidly accumulate proteins pinocytosed from the tubule lumen, led quickly to the realization that much of the material pinocytosed by cells eventually winds up in lysosomes. Later, quantitative studies of pinocytosis and related phenomena forced attention to the fact that endocytosis often involves a bidirectional (or multidirectional) movement of membranes—the use of membrane to carry material into the cell is linked to passage of membrane back to the cell surface. This type of circulation is now understood to be a major feature of many cellular processes (Chapters 2–4).

1.4.2. Autolysis versus Autophagy

Early speculation was that one role of lysosomes is in the self-destruction of dying or dead cells. This suggestion was echoed in proposals that the release of lysosomal enzymes into the surrounding cytoplasm is a widespread **causal** mechanism of cell death, both pathological and developmentally programmed. Proposals that lysosomal rupture **causes** cell death are still much in dispute (Section 6.3.1) but it is very widely held that once cells have died, hydrolases released from the lysosomes do participate in disposing of the remains. Biologically important **autolytic** phenomena of this latter sort are prominent in plant development (Section 6.3.2.1).

Lysosomes are also responsible for nonlethal cellular self-digestion. Most cell types can incorporate portions of their own cytoplasm into membrane-delimited lysosomal vacuoles, and digest the material so sequestered (Fig. 1.9). Such **autophagy**—the lysosomal breakdown of material of origin within (**endogenous** to) the particular cell being studied—is regularly seen under normal, nonpathological circumstances as well as in sick cells or cells experiencing other extreme stresses. It is therefore often conjectured that lysosomes are big-time players in the degradative arm of the normal turnover through which cells continually destroy and replace their own macromolecules. Testing this surmise about lysosomal roles has, however, proved harder than might have been anticipated (Chapter 5).

1.4.3. Genetically Determined Failures of Digestion

Two truisms of modern biology are: that study of the abnormal often clarifies normal processes, and that abnormalities of genetic origin are among the most interesting or useful disorders. For lysosomes these truisms are nicely illustrated by consideration of a set of rare human diseases. Each lysosomal "storage disease" is characterized by the intracellular accumulation of large quantities of a particular spectrum of molecules, as a consequence of genetically engendered defects in the lysosomal machinery. Analysis of these failures to digest (Section 6.4.1) has provided pivotal information about the sequences of digestion of complex molecules in lysosomes and about the synthesis and transport of lysosomal enzymes.

1.5. Alert II: Cautions on Terminology and Inhibitors

1.5.1. More Terminology

1.5.1.1. The "Cytosol"

At many points in subsequent chapters, reference will be made to proteins and other molecules whose intracellular locations are thought to be in the "unstructured" spaces surrounding the microscopically visible cytoplasmic organelles and cytoskeletal arrays. Controversies of ancient origins continue about these molecules.

What should their locations be called? Microscopists have tended to refer to them as components of the "cell sap," "ground substance," or "hyaloplasm." Biochemists have generally used terms like "cytosol" or "soluble fraction." "Cytosol" is the term I have chosen for this book, completely on arbitrary grounds.

Where do they really come from? The cytosol was originally conceived as a pervasive aqueous medium of truly soluble or suspended materials occupying the spaces between the cell's formed elements. Now it is appreciated that some of the materials biochemists find in the corresponding cell fractions (the supernatants left after extensive centrifugations) actually have leaked out of cellular organelles during fractionation or have been detached from loose in vivo associations with intracellular structures. The extent of such artifacts generally is difficult to evaluate, creating unresolved ambiguities in almost all discussions of the behavior of cytosolic materials (see Section 6.3.1). Microscopists and cell physiologists too have begun to think that some of the intracellular materials that seem unstructured at first glance may actually be bound in labile or subtle arrays, difficult to visualize or detect with classical methods. Even intracellular water is regarded as structured by hydrogen bonding to surfaces and macromolecules.

1.5.1.2. Proteases, Proteinases, etc.

What to call enzymes that hydrolyze proteins is also unsettled. Again, on arbitrary grounds I have adopted one of the schemes now in wide use: "Protease" and "peptidase" will be employed as overall terms to include all enzymes that degrade polypeptide chains or small peptides. There is no sharp boundary between "proteases" and "peptidases." "Exopeptidase" will refer to enzymes that act only near the ends of proteins or peptides. "Proteinase" and "endopeptidase" will be used to designate hydrolases that can cleave peptide bonds near the middle of peptides and polypeptide chains.

The major lysosomal proteinases are called "cathepsins" (see Table 1.1). Of these, several (prominently, cathepsins B, H, and L) are of the sort known as "cysteine" (or "thiol") proteinases because of the presence of essential SH groups at their active sites. Cathepsin D is the principal "aspartyl" ("acid"; "aspartic"; "carboxyl") proteinase, so designated because carboxyl groups of aspartic acid residues are essential components of the active sites of the enzymes.

1.5.1.3. "Receptors" and "Ligands"

The term "receptor" has provided terrain for merry semantic struggles among pharmacologists, biochemists, virologists, neurobiologists, and many others. The term will be used here in its present "loose" connotation, to refer to cellular molecules, usually proteins but sometimes lipids or saccharide chains, that form specific complexes with other molecules (the receptor's "ligands") that reach the cell via extracellular media. The receptors help mediate the biological effects or behavior of the ligands that bind to them. Such rough-and-ready definitions are in wide current use though their limitations must not be forgotten. For instance, though the cell surface molecules to which a virus binds before entering a cell are conventionally denoted "receptors" for that virus, the teleological resonances are askew: It is not likely that evolution has provided the cell with receptors **intended** for these ligands. Rather the virus has "adopted" cellular molecules as its targets.

1.5.2. Inhibitors

Cytochemists and biochemists have made extensive use of enzyme inhibitors in working out the details of lysosomal properties and function. Section 1.3.2.3 mentioned that the susceptibility of the major lysosomal acid phosphatase activities to inhibition by fluoride or tartarate has helped distinguish lysosomal phosphatases from others. Widely used inhibitors of proteolysis (see Section 5.5) include **pepstatin,** a pentapeptide of microbial origin, which blocks the action of cathepsin D. Several inhibitors are used for cysteine proteinases, especially the microbial peptide-aldehyde **leupeptin,** and synthetic epoxysuccinyl peptides such as **E-64. Bestatin** has been employed to inhibit exopeptidases.

Helpful as such inhibitors have been, the results so obtained need careful examination (Section 5.5.3.2). For example, the inhibitors' effects on different lysosomal proteinases vary and sometimes assumptions of complete inhibition of proteolysis have turned out to be unfounded. Moreover, most of the inhibitors are not intrinsically limited to acting only on lysosomal enzymes. When inhibitors of proteolysis are used on living cells, it often is assumed that the lysosomes and related compartments are the principal loci of effects. This is because the inhibitors seem to be compounds of types that can gain access to the lysosomes through endocytosis, but cannot readily penetrate membranes and thus "should" not enter the cytosol or other loci of nonlysosomal enzymes. From time to time, doubts are raised about this reasoning—cell fractionation studies sometimes seem to locate inhibitor molecules like leupeptin in the cytosol as well as in lysosomes (but see Section 1.5.1.1), and certain of the biological effects of the inhibitors could signal impacts on nonlysosomal enzymes (see Section 5.5.5).

1.5.3. Reconstruction: Time and Space

Because conventional techniques mostly depend upon preserving cells with microscopic fixatives that kill them, or on disruption of cells for fractionation, microscopic and biochemical data usually provide only snapshots of processes. Temporal sequences must be inferred from these snapshots and from the experimental design. When complex, rapidly changing structures are involved, as will often be true in the ensuing discussions, such reconstruction can be difficult. Wherever possible therefore, recourse must be had to the evolving sets of methods—light microscopic, cell physiological, and the like— that examine processes directly in living cells.

Interpreting data in terms of cell structure can also prove problematical. Extrapolating from cell fractions to the intact cell is not always easy; reconstitution of processes in cell-free mixtures of fractions is one of the more powerful approaches now being used to overcome some of the difficulties. Electron microscopists rely very heavily on thin sections of tissues—two-dimensional slices in which, for example, structures that actually are continuous with one another in three dimensions often appear to be separate. Improved microscopic reconstructions are coming through computer-assisted processing of images from sequential series of sections ("serial sections") or by examining much thicker sections than has been customary. Of particular value for microscopists has been the development of **morphometric** ("stereological") approaches— sampling and statistical techniques that greatly facilitate the use of thin sections for accurate estimation of the volumes and surface areas of cell structures with irregular or complex shape.

Acknowledgments

Discussions over the past twenty years with Drs. A. B. Novikoff, C. de Duve, Z. Cohn, S. Silverstein, S. Goldfischer, and D. Bainton were of particular utility in providing the background information put forth in this chapter. Dr. A. J. Barrett provided useful references and information on enzymes and inhibitors and Drs. F. Marty, D. Branton, M. Chrispeels, and P. Matile helped orient me in the literature on plants.

Further Reading

Bainton, D. F. (1981) The discovery of lysosomes, *J. Cell Biol.* **91**:66S–76S.

Barrett, A. J. (1984) Proteolytic and other metabolic pathways in lysosomes, *Biochem. Soc. Trans.* **12**:899–902.

Bond, J. S., and Butler, P. E. (1987) Intracellular proteases, *Annu. Rev. Biochem.* **56**:333–364.

Dean, R. T. (1978) *Cellular Degradative Processes*, Wiley (Halsted Press), New York.

de Duve, C. (1975) Exploring cells with a centrifuge, *Science* **189**:186–194.

de Duve, C. (1983) Lysosomes revisited, *Eur. J. Biochem.* **137**:391–397.

de Reuck, A. V. S., and Cameron, M. P. (eds) (1963) *Lysosomes*. Little, Brown, Boston.

Dingle, J. T. (ed.) (1977) *Lysosomes: A Laboratory Handbook* (2nd ed.), North-Holland, Amsterdam.

Dingle, J. T., Fell, H. B., Dean, R. T., and Sly, W. S. (eds.) (1969–1984) *Lysosomes in Biology and Pathology*, North-Holland (currently Elsevier), Amsterdam. [Seven volumes of this series were published between 1969 and 1984. The articles cover both the basics of lysosome form and function and the specifics of lysosome properties in many cell types and tissues, both vertebrate and invertebrate.]

Gaudreault, P. R., and Beevers, L. (1984) Protein bodies and vacuoles as lysosomes, *Plant Physiol.* **76**:228–232.

Holtzman, E. (1976) *Lysosomes: A Survey*, Springer-Verlag, Berlin.

Marty, F., Branton, D., and Leigh, R. A. (1980) Plant vacuoles, in *The Biochemistry of Plants* (P. K. Stumpf and E. E. Conn, eds.), Vol. 1, Academic Press, New York, pp. 625–658.

Matile, P. (1975) *The Lytic Compartment of Plant Cells*, Springer-Verlag, Berlin.

Novikoff, A. B. (1973) Lysosomes: A personal account, in *Lysosomes and Storage Diseases* (H. G. Hers and F. Van Hoof, eds.), Academic Press, New York, pp. 1–41.

Wattiaux, R., Wattiaux-de Coninck, S., Ronveaux-Dupal, M.-F., and Dubois, F. (1978) Isolation of rat liver lysosomes by isopycnic centrifugation in a metrizamide gradient, *J. Cell Biol.* **78:**349–364.

Wiemken, A., Schellenberg, M., and Urech, K. (1979) Vacuoles: The sole compartment of digestive enzymes in yeast (*Saccharomyces cerevisiae*)? *Arch. Microbiol.* **123:**23–35.

Wolff, D. A. (1975) The separation of cells and subcellular particles by colloidal silica density gradient centrifugation, *Methods Cell Biol.* **10:**85–104.

Endocytosis and Heterophagy

Lysosomes have been studied most extensively in the context of heterophagy (Fig. 2.1). This is so partly because there are readily obtainable cell types, like the mammalian phagocytes or protozoa, that show elaborate heterophagic apparatus, and use their lysosomes for easily identifiable, major cellular activities. Important phases of these activities can be studied in living cells. Moreover, heterophagic structures are accessible to many materials introduced into the extracellular environment, and this has greatly facilitated experimentation.

Keep in mind, however, that present views of heterophagy are dominated by findings on a small number of cell types, especially neutrophils, macrophages, cultured fibroblasts, and a few strains of protozoa. Each of these cell types has specialized features that compromise its appropriateness as a "model" system representing cells in general. Put another way: though the outlines of endocytosis and heterophagy are closely similar in diverse organisms and cell types, significant variations in detail are already known and more are likely to appear with intensive scrutiny of a broader variety of cells (Chapter 4).

2.1. Basics plus Tastes of Diversity

The most obvious mechanism for mixing of acid hydrolases with endocytosed materials is through the fusion of preexisting lysosomes with the compartments in which endocytosed materials accumulate upon their entry into the cell.

2.1.1. Phagosomes and Pinosomes

Material destined for heterophagic degradation enters the cell in phagocytotic vacuoles or pinocytotic vesicles arising from the plasma membrane. Special terms like **phagosome** and **pinosome** are used for these vehicles of endocytotic entry to emphasize that some time—at least a minute or two and often much more—generally elapses before endocytosed materials enter compartments in which extensive degradation is evident.* During this interval the

---

*This labored formulation ("in which extensive degradation is evident") is necessitated by present uncertainties as to when endocytosed materials first encounter acid hydrolases (Section 2.4.4.1). The point being made is simply that the initial vehicles of endocytotic entry lack much of the vigorous hydrolase activity demonstrable in lysosomes by conventional techniques (see Fig. 1.5).

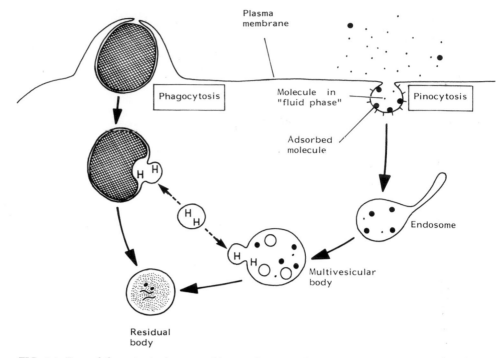

FIG. 2.1. Two of the principal routes of heterophagy. In **phagocytosis** particles are taken up into vacuoles that subsequently acquire hydrolases (H) by fusion with preexisting lysosomes. In the form of **pinocytosis** illustrated here, nascent vesicles invaginating into the cytoplasm carry both molecules adsorbed from the surrounding medium, and fluid droplets of the medium itself. Precisely how the pinocytosed materials are brought together with hydrolases is still uncertain (see text), but part of the story involves entry of the materials into intermediate compartments ("endosomes"), and their eventual accumulation in bodies that show small internal vesicles— "multivesicular bodies." Hydrolases become evident at the multivesicular body stage, though conceivably some of the enzymes are present earlier. Eventually, both for phagocytosis and for pinocytosis, the degradative structures evolve into residual bodies.

entry vehicles can undergo obvious, marked changes. For example, phagosomes of ciliated protozoa alter substantially in volume prior to their fusion with recognizable lysosomes: these alterations reflect both the addition of components through fusions with nonlysosomal cytoplasmic structures, and the withdrawal of membrane through budding of vesicles (see Fig. 2.5). Pinosomes, through fusions and transformations, can contribute their components to **endosomes,** compartments thought of as "intermediate" or "prelysosomal" in the sense that they probably do not possess the full range of lysosomal degradative activities (Sections 2.3.1.4, 2.4.4.1, and 7.3.2.2).

Terms like "endocytotic vesicle," "phagocytotic vacuole," "food vacuole" are used in generic senses to designate vacuoles and vesicles of known endocytotic origin, both at the pinosome or phagosome stage and also later in their development. "Digestive" vacuoles are ones in which digestion is taking place. These terms are needed to cover cases where structures have yet to be more precisely classified.

2.1.2. Admixture of Hydrolases and Materials to Be Degraded

2.1.2.1. Primary and Secondary Lysosomes in Neutrophils

In neutrophils, the delivery of enzymes to phagosomes is observed microscopically as a process of "degranulation." This process resembles the exocytotic release of secretions from gland cells. That is, the resting cell contains numerous cytoplasmic granules and when phagocytosis begins, some of these granules fuse with the new phagosomes. The granule's delimiting membrane merges with the membrane bounding the phagosome so that the contents of the two bodies mix but the interior of the hybrid body resulting from the fusion remains always separated from the remainder of the cytoplasm by a continuous membrane.

Among the neutrophil's principal granules one class is lysosomal. These granules (Fig. 2.2) are about 0.75–1 μm in diameter and are known classically as **azurophilic** granules from their staining properties (see footnote on p. 18). They are packages of lysosomal enzymes, generated by the cell's Golgi apparatus (Fig. 2.2 and Section 7.2.1) and stored for use when the cell engages in phagocytosis. Such packages of lysosomal hydrolases that have yet to be used for intracellular digestion are referred to as **primary lysosomes.** The phagocytotic "digestive vacuoles" ("phagolysosomes") formed when the azurophilic granules fuse with phagosomes exemplify **secondary lysosomes,** defined as bodies containing both the lysosomal hydrolases and materials that are undergoing digestion or that already have been digested to the extent possible under the prevailing conditions.

2.1.2.2. Lysosomal Heterogeneity I

Both the primary and the secondary lysosomes of the neutrophilic leukocytes contain enzymes not present in lysosomes of most other cell types. Along with the usual acid hydrolases, the azurophilic granules—primary lysosomes—have a peroxidase known as myeloperoxidase, as well as some lysozyme. The phagocytotic digestive vacuoles—secondary lysosomes—receive these enzymes along with the acid hydrolases by fusions with the azurophilic granules. The digestive vacuoles also obtain additional components that are atypical for secondary lysosomes, through fusion with **specific granules;** these are a set of specialized neutrophil storage structures that, in rabbits,* are somewhat more numerous and somewhat smaller (diameters up to 0.5 μm) than the azurophilic granules. The specific granules contribute an alkaline phosphatase, lysozyme, and other materials (Section 4.4.2.1) to the digestive vacuoles.

---

*The description of the two major types of neutrophil granules presented here is based chiefly on accounts of rabbit cells, which have been studied especially carefully both by cytochemistry and by biochemical analysis of isolated granules. The neutrophils of other mammalian species contain an armory of granule-associated enzymes and other proteins fundamentally similar to those of the rabbit. There seem, however, to be major species-to-species variations in the relative abundance of different components and in the details of distribution of different proteins among granule types. Few of these variations have been thoroughly studied and investigators have not always distinguished carefully between storage structures, like those under consideration, and derivative bodies resulting from digestive activities. But, for example, it is claimed that the neutrophils of cows have three main storage granule classes: lysosomal azurophilic granules; granules containing a vitamin B_{12} binding protein (as do the specific granules of other species; Section 4.4.2); and a third class of granule characterized by "antibacterial" proteins—lactoferrin, a class of small cationic proteins (Section 4.4.2), and possibly others. (In rabbits, lactoferrin is found in the specific granules and a set of cationic proteins is present in the azurophilic granules.) For humans, some reports suggest that alkaline phosphatase is segregated to a class of granules separate from the specific granules, in contradistinction to the situation in rabbits.

Particularly in the earlier literature, azurophilic granules often are called "primary" granules, and specific granules, "secondary" granules. Additional granule types are sometimes called "tertiary" granules.

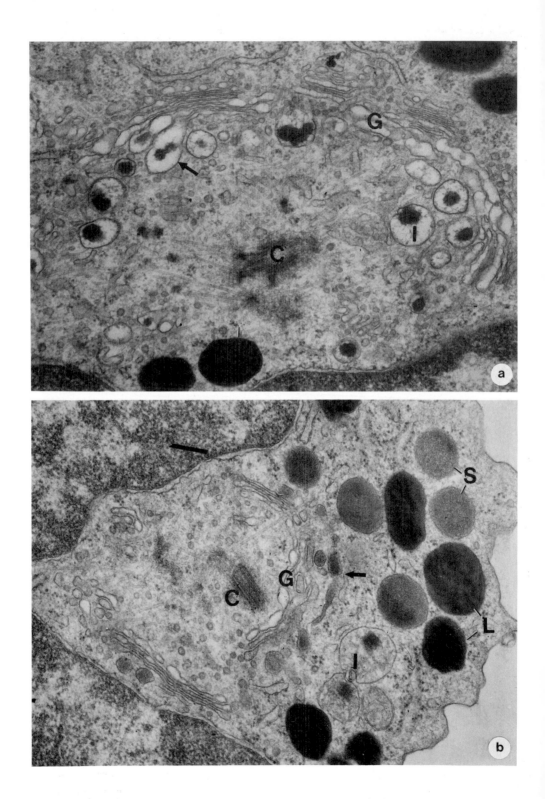

The point of immediate note is that, despite their common enzymatic machinery, the lysosomes of different cell types differ from one another. The presence of enzymes like myeloperoxidase in the neutrophil's lysosomes provides a clear case of qualitative difference from other cell types; the distinctive components are mostly involved in the microbicidal and inflammatory functions in which the cells specialize (Section 4.4). The characteristic major cytoplasmic granules of the **eosinophil** class of granulocytic leukocyte are lysosomes that contain a specialized peroxidase different from the myeloperoxidase of neutrophils. (Eosinophil granules have also been claimed to possess oxidative enzymes more usually found in peroxisomes, but this claim is based on immunocytochemical evidence and needs strengthening from other types of data.) For lysosomes of most other cell types, such qualitative distinctiveness is not known, but quantitative differences in the relative levels of different lysosomal enzyme activities have been detected (Section 7.5.1).

Peculiarities in lysosome properties arising from the nature of the materials digested by specific cell types are also to be expected (see Section 7.5.1.1).

2.1.3. The Recycling of Lysosomes; Their Fate

2.1.3.1. Who Are the Primary Lysosomes?

In addition to the ones seen in polymorphonuclear leukocytes, primary lysosomes are evident as prominent cytoplasmic granules originating near the Golgi apparatus in the monocyte precursors of macrophages, in platelets, and in a few other cell types. But most of the recognizable lysosomes in most cells are secondary lysosomes. In the large majority of cell types, primary lysosomes that can be identified morphologically have yet to be found. Even functioning macrophages lack a morphologically distinctive set of storage organelles for yet-to-be-used hydrolases. The few protozoa that have been studied in this connection possess microscopically recognizable structures that could be primary lysosomes but have yet to be definitively identified as such, largely because the cells contain a confusing diversity of cytoplasmic structures involved in heterophagy. In numerous cases, structures have been designated primary lysosomes with little evidence beyond, sometimes, a superficial resemblance to the granules of neutrophils, ignoring the fact that secondary lysosomes often have this appearance.

Despite all this, the prevailing belief is that digestive structures in most cell types can receive charges of recently made hydrolases by transport from the Golgi apparatus in vesicular vehicles. From cytochemical studies of the distribution of acid phosphatase in cells lacking distinctive primary lysosomes, this transport has been provisionally ascribed to vesicles, on the order of 50–

FIG. 2.2. Golgi regions of rabbit neutrophils developing in bone marrow. Panel **a** is of a progranulocyte, the stage at which azurophilic granules form. Panel **b** shows a myelocyte at the time of specific granule formation. L indicates mature lysosomes (azurophilic granules), S specific granules, I immature granules, and G Golgi sacs. The mature azurophilic granules are larger and more electron-dense than are the specific granules.

The arrows indicate apparent formation of azurophilic granules (in panel a) and specific granules (panel b) from Golgi-associated sacs; note that if one uses the positions of the centrioles (C) as a landmark, the two granule types appear to form from opposite surfaces of the Golgi stack (see Chapter 7). From Bainton, D., and Farquhar, M. G. (1966) *J. Cell Biol.* **28:**277. × 30,000 (approx.).

100 nm in diameter, arising from the Golgi apparatus or from closely associated systems. Propinquities of acid phosphatase-containing Golgi vesicles of this size to endocytotic (and autophagic) structures are frequently observed in the vicinity of the Golgi apparatus and elsewhere (Fig. 2.15) and images suggesting fusion of the vesicles with digestive structures have also been obtained. [There are inevitable ambiguities in interpreting these images from the electron micrographs: When a vesicle is seen in continuity with a larger body, is it fusing with or budding from the larger body? Is the image actually one of a vesicle, or really a thin section through a tubule (Section 2.4.2.1)?] During at least a portion of their life history, the presumptive transport vesicles are of the variety referred to as "coated vesicles," because of the layer of organized material that lines their cytoplasmic aspects (Fig. 2.15b; see Fig. 1.10, Sections 2.3.1.1, and 7.2.1.1). But this coating is not diagnostic for primary lysosomes—there are nonlysosomal vesicles of similar appearance in the Golgi region and at present we lack definitive morphological criteria for distinguishing the hydrolase transport vesicles from the others.

2.1.3.2. Mannose-6-Phosphate Receptors as Markers for Modern "Primary Lysosomes"

A more detailed perspective on primary lysosomes is currently crystallizing from work on the molecular mechanisms by which lysosomal hydrolases are synthesized and sorted from other cellular proteins. Full consideration of this work has been deferred until Chapter 7 because it is still very much in progress and is difficult to discuss smoothly without the background information that will be provided in the intervening chapters. But a central, provisional conclusion useful here, is that many recently made lysosomal hydrolase molecules pass, via the Golgi apparatus, to their sites of eventual use, in concert with specific transport or sorting proteins. These proteins are known as **mannose-6-phosphate** (M6P) **receptors** because they recognize and bind to phosphorylated mannose residues ("phosphomannoses") present in the oligosaccharides of newly made lysosomal hydrolases.

Through immunocytochemistry, M6P receptor molecules have been found in presumptive primary lysosomal vesicles like those thought to be primary lysosomes from the considerations already summarized (see also Fig. 7.5). Receptors are found as well at some of the sites to which the primary lysosomes are believed to deliver hydrolases (Section 7.3.2: Figs. 7.5 and 7.10). But studies of the behavior of the M6P receptors suggest that the compartments in which recently made hydrolases are carried to sites of intracellular digestion are not one-way transport devices, like the granules of leukocytes. Rather, the M6P receptors seem to be reusable membrane-associated "shuttles" associated with membranes that move back and forth between the Golgi apparatus and the sites to which hydrolases are sent. The receptors also may spend some time at the cell surface (see Section 6.4.1.4 and Chapter 7).

2.1.3.3. Residual Bodies; Lysosome Reuse

Neutrophils are short-lived cell types that usually survive only a few days after maturing (Section 4.4.1). The depletion of their supply of granules ends their ability to participate in the microbicidal and phagocytotic processes for which they are specialized. They cannot make more granules and in many

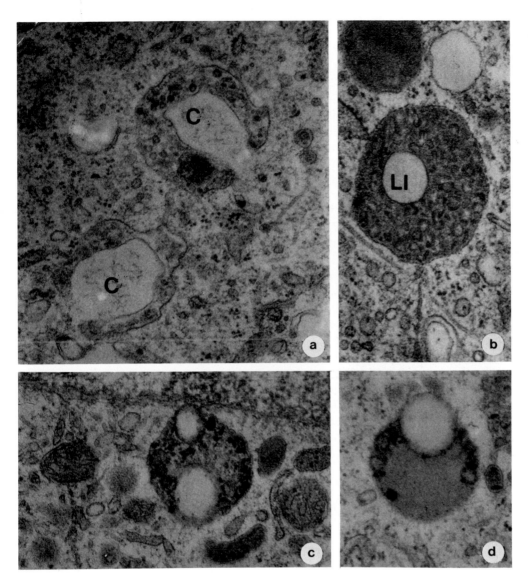

FIG. 2.3. Residual bodies: Probable sequence of development of lysosomes in rat adrenal medulla cells during the period of membrane degradation that follows insulin-induced depletion of secretory stores. The membranes degraded seem to be ones that initially delimited secretory granules, were added to the cell surface during exocytotic release of secretion, and are then retrieved, endocytotically, from the cell surface (Section 5.3.3). In **a,** cuplike stages in the development of multivesicular bodies (Fig. 2.14) are present (C). These appear to evolve into multivesicular bodies with electron-dense content, as in **b;** bodies of this sort contain cytochemically demonstrable hydrolases (as do those in **c** and **d**). Later, the multivesicular bodies transform into residual bodies like those in c and d—internal vesicles no longer are seen, the bodies instead containing lipid globules (LI) and heterogeneous electron-dense material. It may be that the lipid globules sometimes are released into the cytoplasm and that d is a stage in this release. From work by S. Abrahams and E. Holtzman. Derived from a figure in Holtzman, E. (1976) *Lysosomes: A Survey,* Springer-Verlag, Vienna. × 40,000 (approx.).

circumstances they die within hours after functioning. Most other cell types, including macrophages, live much longer and many of them engage in endocytosis repeatedly, or even continuously (Section (2.2.2.6). To maintain this capacity without massive production of primary lysosomes, the cells rely upon reuse of lysosomes.

As digestion proceeds, endocytotic digestive structures typically evolve into **residual bodies** (Figs. 2.3 and 7.7), a form of dense body containing the acid hydrolases along with content that often has a heterogeneous, electron-dense appearance. This content—materials resistant to degradation and degradation products that cannot escape rapidly from lysosomes (Section 3.4)—varies in appearance, depending principally upon the nature of the materials being digested. Often, as in the pericanalicular dense bodies of hepatocytes (Section 1.3.2.1), residual bodies contain globular or lamellar lipid-rich inclusions, probably because lipids, especially when present in organized arrays, tend to be digested relatively slowly and perhaps sometimes also because products of lipid breakdown aggregate within the lysosome and so cannot readily exit to the cytoplasm. Deposits rich in metals also are common within residual bodies (Section 3.5.2.2). Occasionally, paracrystalline structures are seen.

That secondary lysosomes fuse with new endocytotic structures, both pinocytotic and phagocytotic, is evident from observations like the one in Fig. 2.4. Because the constituent enzymes of lysosomes are resistant to lysosomal digestion—they can retain activity for many hours, or even days, within secondary lysosomes (Section 7.1)—this fusion of "old" secondary lysosomes with newly formed endocytotic bodies permits use of the same hydrolase molecules for repeated bouts of digestion. Along with patent connections among lysosomes (Section 2.4.1), such fusions also help account for the finding that when cells take up indigestible or slowly digestible materials, these materials eventually spread throughout the population of secondary lysosomes in a given cell, rather than being confined to a subset of lysosomes. This sort of "homogenization" of the contents of mature lysosomes is evident within a few hours of the uptake of suitable tracers. It is regularly observed in normal cells, but has also been demonstrated in heterokaryons formed between cultured cells whose lysosomes were marked, before cell fusion, with different recognizable tags. (See also Sections 3.5.2.4 and 7.5.1.2.)

2.1.3.4. Defecation

As guardians of multicellular organisms, macrophages are among the cells responsible for destroying potentially dangerous materials and for sequestering materials that cannot be destroyed. In carrying out these functions, the macrophages tend to retain residues of digestion intracellularly, within the lysosomal system, for prolonged periods extending even to the several weeks or months the cells survive. In contrast, protozoa, as free-living single-celled organisms, are not constrained to retain digestive residues. They dispose of undigested materials by expulsion (**defecation**): The residual bodies' **contents** are released by fusion of the bodies with the cell surface through a process like the exocytotic release of secretions. The residual bodies' **membranes** are retained and reused (Section 3.3.1). In protozoa with well-defined shapes, such as the ciliates, defecation often occurs at a specialized "cytoproct" region of the cell surface.

From disconcertingly scanty data, different cell types of multicellular orga-

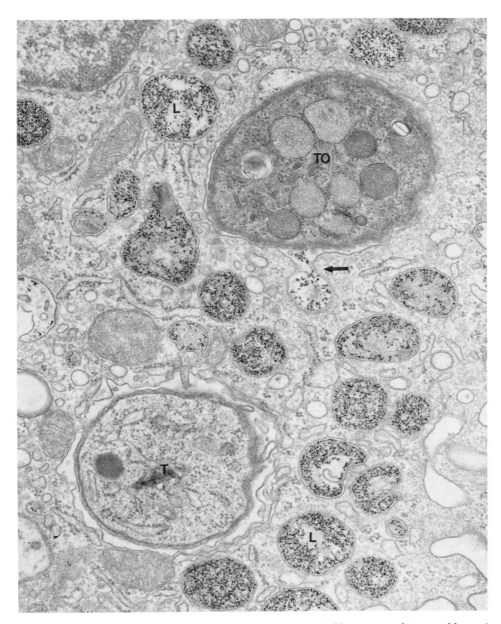

FIG. 2.4. Region of cytoplasm from a mouse macrophage that had been grown for several hours in the presence of a colloidal thorium endocytotic tracer (Thorotrast) and then was exposed to *Toxoplasma gondii*, a protozoan that the macrophages also endocytose (see Fig. 4.14). Electron-dense Thorotrast granules are seen in many secondary lysosomes (L) scattered in the cytoplasm. At the arrow, one of these bodies seemingly has just fused with a phagocytotic vacuole containing a *Toxoplasma* (TO); as the text discusses, this sort of image suggests that "old" lysosomes can be reused for repeated bouts of digestion. The protozoan in the vacuole with which the lysosome is fusing has an electron-dense cytoplasm and other signs of disruption suggesting that it is dying or dead. Nearby, another vacuole (T) contains a healthy-looking *Toxoplasma* and lacks Thorotrast. Evidently some of the protozoa can effectively inhibit fusion of lysosomes with the vacuoles in which they enter cells (Section 4.4.3.3) but this inhibition operates on a local level so the same host macrophage can fuse lysosomes with some vacuoles and not with neighboring vacuoles. From Jones, T. C., and Hirsch, J. G. (1972) *J. Exp. Med.* **136**:1173. × 30,000 (approx.). (Copyright: Rockefeller University Press.)

nisms are thought to differ considerably in the extent to which they retain or defecate the contents of secondary lysosomes. Circumstantial evidence suggests that neurons and certain other cell types additional to macrophages, retain digestive residues for prolonged periods (Section 6.4.2.1). If so, defecation probably is avoided because it could lead to persistent dangerous accumulations of enzymes and digestive residues: for example, the extracellular spaces and chambers of the brain have few if any outlets through which to dispose of the released materials and may have few scavenger cells that could take up lysosomal debris (Section 4.4.1). In contrast, some of the cell types held to engage in extensive defecation, are able to focus release at surfaces facing spaces specialized to handle "wastes" or problematic materials. Hepatocytes, for example, probably release lysosomal contents largely into the bile canaliculi (Section 3.5.2) that communicate with the lumen of the gut. Kidney tubule cells release lysosomal enzymes into the urine.

Shortly before digestive residues are defecated from protozoa such as *Paramecium*, vesicles containing cytochemically demonstrable acid phosphatase bud from the digestive vacuole surface and move away into the adjacent cytoplasm (Fig. 2.5). This has been interpreted as a conservation device whereby lysosomal hydrolases are retrieved for reuse in digestion.

Defecation accounts for some of the normal presence of acid hydrolases in extracellular media of multicellular organisms (Section 6.2) and for some of the "secretion" of lysosomal enzymes by unicellular organisms. But it would be incorrect to assume that most release of lysosomal enzymes from cells is simply a by-product of mechanisms for ridding cells of digestive residues. In fact, a still viable, though controversial, hypothesis suggests that during their passage from ribosomes to lysosomes, most newly made lysosomal hydrolase mole-

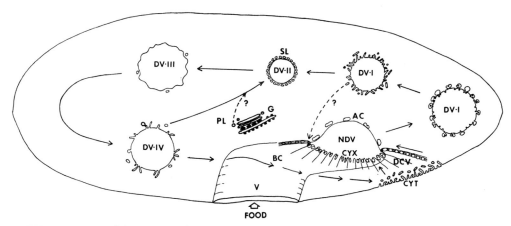

FIG. 2.5. Proposed digestive cycle in *Paramecium*. NDV ("nascent digestive vacuole") and DVs I–IV are successive stages in the development of a food vacuole forming at the cytopharynx (CYX) and eventually defecated at the cytoproct (CYT). The vacuoles migrate through the cytoplasm, moving posteriorly, then anteriorly, and then again posteriorly. As they move, "acidosomes" (AC; Section 3.1.5.5) fuse with them and their internal pH drops; subsequently, membranous vesicles and tubules bud off, some of which may migrate back to the cytopharynx to supply membrane for further endocytosis (Chapter 3). Next, lysosomes fuse with the vacuoles [the secondary lysosomes (SL) are most readily identified but Golgi-derived primary lysosomes (PL) may participate as well]. Before defecation, membranous vesicles again bud from the surface of the vacuole, probably carrying acid hydrolases, which can be reused for subsequent digestion. The membrane added to the cell surface at the cytoproct during defecation is retrieved by the cell, endocytotically, and circulates back to the cytopharynx in the form of discoidal vesicles (DCV; see Fig. 3.11). V and BC are parts of the feeding apparatus. Summarizing work by R. D. Allen, A. Fok, and their colleagues. From Allen, R. D. (1984) *J. Protozool.* **31**:1.

cules experience a brief interlude of exposure to the extracellular world (Section 7.2.3). Chapter 6 will consider additional facets of hydrolase secretion and Section 3.5.2 will continue discussion of the fate of lysosomal residues.

2.1.4. Condensation; Matrices; Nonenzymatic Components

The contents of the primary lysosomes of leukocytes, and those of residual bodies and other secondary lysosomes, appear highly concentrated in that they are distinctly electron-dense. By analogy to secretion granules, this "condensation" of content is thought to serve efficient storage in minimal space. *A priori*, one might think that the lysosomes or secretion granules are dehydrated through the active expulsion of inorganic ions or other small solutes, with water following the ions osmotically. Transport systems that control the internal ionic milieu or carry low-molecular-weight organic solutes across the membrane do exist in the membranes bounding lysosomes and secretion granules (Section 3.1.5). But there is little evidence for direct roles of these systems in condensation, beyond the general assumption that lysosomes shrink as digestive products move out into the cytoplasm (Section 3.4).

For secretion granules, a popular alternative proposal is that as the granules form, their components aggregrate spontaneously into large, multimolecular complexes, and that this leads to the osmotic efflux of water. The macromolecular contents of the granules—typically proteins and saccharide polymers such as glycosaminoglycans—are thought to associate with one another by weak bonds. The associations and the insolubility of the complexes are supposedly fostered by high Ca^{2+} levels, low pH (Section 6.1.5), the abundance of charged nucleotides, or other conditions known or thought to exist within various nascent secretion granules. Perhaps the contents of lysosomes behave similarly. The neutrophil primary lysosomes, for example, are rich in cytochemically demonstrable "acid mucosubstances" (probably anionic glycoconjugates) and also contain cationic proteins. These large anions and cations would be expected to associate with each other by electrostatic interactions. Many other types of lysosomes contain carbohydrate-rich materials, which probably include charged glycoconjugates (Fig. 2.6). But like most other phenomena related to the control of lysosomal volume and to changes in the concentration of lysosomal contents (Sections 2.2.2.5 and 2.4.1), condensation of primary or secondary lysosomes has so far been subjected only to casual, cursory exploration. How, for instance, condensation is coordinated with the coming and going of membranes at the lysosome surface (Section 3.3) or how it influences the state of soluble components within the lysosome (Section 2.2.2.5) are not known. Presumably the process is among those that account for the greater buoyant density of "mature" lysosomes as compared with the structures involved in early phases of heterophagic digestion.

From the fact that many of the hydrolases are released into the supernatant when the membranes bounding lysosomes are disrupted, it is concluded that once primary or secondary lysosomes have fused with endocytotic structures, most of the enzymes are substantially free to diffuse within the resultant digestive "chamber."* Were this not the case, it is difficult to see how effective

---

*The release of enzymes from disrupted lysosomes in cell fractions, however, is reportedly slow at low pH (pH 4.8 was used for the experiments). And electron-dense contents of secondary lysosomes often remain visibly aggregated for some time after the lysosomes have fused with other structures (Figs. 2.4 and 2.21) or when they enter extracellular spaces, as sometimes happens when cells die and their cytoplasm disperses (see Chapter 6).

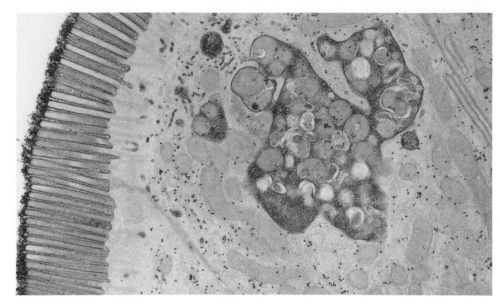

FIG. 2.6. Part of a cell in the intestinal epithelium from a preparation exposed to chloroquine (10^{-4} M; see Section 3.1.2) for 24 hr. The micrograph shows several lysosome-related bodies containing numerous small vesicles along with larger membranous and lipidlike inclusions; one of these bodies is a very large, multilobed structure, whose presence is due to the treatment with chloroquine. The electron density of the contents of these bodies, and that of the glycocalycal extracellular fuzz atop the microvilli protruding from the cell at the left, is due to staining with a cytochemical method (silver proteinate) for glycoconjugates. From Blok, J., Mulder-Stapel, A. A., Daems, W. T., and Ginsel, L. A. (1981) *Histochemistry* **73:**428. (Copyright: Springer-Verlag, Berlin.)

digestion could take place rapidly. [Indeed, access to substrates is not yet explained for the few hydrolases that are not readily liberated by lysosomal disruption, and especially for the enzymes that show evidence of intimate association with the lysosome's membrane (Section 7.3.3.5).] But is it possible that when lysosomes are awaiting use, or in the intervals between use, condensation into structured or insoluble aggregates helps keep lysosomal enzymes from digesting one another? Though the idea is appealing, the answer is not known. Lysosomal enzymes are relatively long-lived despite the fact that the proteases in lysosomes are in molecular forms ready for immediate action and thus should pose threats to their intralysosomal neighbors. [The proteases do seem to require an initial activation but this takes place early in their life history (Section 7.4.1.1).]

Most often the ability of lysosomal proteins to coexist is explained by postulating that the molecules have evolved with protease-resistant conformations or other protective features (Section 3.2.1). It might also be that the digestive cycle has evolved so as to provide conditions fully conducive to intralysosomal enzyme mobility and activity only when appropriate substrates are present or when preliminary events, such as fusions with endocytotic compartments, have appropriately diluted or otherwise altered the intralysosomal milieu. Considerations of this sort could, for example, help explain the fact that the storage granules of mature neutrophils show very little cytochemical reaction for some of the enzymes they are known to contain. And restrictions on the diffusibility of lysosomal contents have been invoked to explain apparent delays in the transfer of materials from lysosomes to structures with which they fuse (Section 2.4.2).

2.1.4.1. Is There an Intralysosomal Matrix?

One slant on these matters is to suppose that lysosomes contain special components whose principal evolutionary *raison d'être* is to provide a matrix that controls the diffusibility and activities of the hydrolases and the state of the materials to be digested. This idea has a long history but until recently it was supported almost exclusively by questionable evidence. Ostensible matrix substances—highly charged lipids, glycoconjugates, and other macromolecules—have been isolated from lysosomes. The sources, however, have mostly been secondary lysosomes with quite complex contents—hydrolases, residues of digestion, and materials that were undergoing degradation at the time of organelle isolation. And because most lysosomal hydrolases and many lysosomal membrane proteins are glycoproteins (Sections 3.2.2 and 3.2.3.3, Chapter 7), some with quite extensive oligosaccharides, it is difficult to establish whether, for example, neutrophil granules truly do have additional, separate saccharide polymers responsible chiefly for controlling the state of other granule constituents.

Few "lysosomologists," therefore, are convinced of the wide existence of **special** matrix components. But whether or not distinctive lysosomal matrices exist, many interactions are likely to occur in the lysosome interior that could affect the mobilities and activities of the enzymes and of the other materials present. Even lysosomes actively engaged in digestion would be expected often to behave as though their interiors were semiorganized. Most obviously, secondary lysosomes rarely are completely devoid of visible internal structure: For secondary lysosomes digesting bacteria, dead cells, or cellular organelles, lysosomes bearing indigestible "junk," or even many lysosomes degrading soluble proteins (Sections 2.3.1.4 and 2.3.1.5), the content is hardly to be portrayed as a simple suspension of enzymes mixed with molecules being digested. Membranes, cell walls, or insoluble polymers are present, creating impediments to free diffusion and affording surfaces for differential adsorption (Sections 2.4.2.2 and 2.4.3).

2.1.4.2. Nonenzymatic Contents

Some of the macromolecules intrinsic to lysosomes have no known or evident enzymatic activities. Examples already mentioned are the "antimicrobial" proteins contributed by the neutrophil's granules to the phagocytotic digestive vacuoles, including the set of cationic proteins and the iron-sequestering protein, lactoferrin (see footnote on p. 27; see also Section 4.4.2). More widespread (Section 6.4.1.3) are lysosomal proteins whose roles are to "assist" certain hydrolases in attacking lipids, and proteins that complex with particular hydrolases helping to activate the enzymes or to stabilize them against intralysosomal degradation. These proteins lack independent enzymatic activity but do condition the intralysosomal environment and in these regards at least, do conform to expectations for "matrix" substances.

2.2. Heterophagic Selectivity

We will see below that ameboid cells engulf occasional particles or droplets of medium "accidentally," as an accompaniment to cellular motility. In these and some other cell types, a certain amount of endocytosis may take place constitutively, whether or not there is anything of obvious interest to take up.

But under normal conditions, endocytosis rarely if ever is really random in the sense of cells using any old part of their plasma membrane in unprovokedly taking up indiscriminate gulps of whatever is nearby. Rates of endocytosis vary strikingly in different cell types and with varying conditions, and the membranes the cell invests in taking up exogenous macromolecules or particles can show special features that dramatically affect uptake.

2.2.1. Phagocytosis

The conclusion that phagocytosis usually must be induced rather than occurring spontaneously rests on observations that cells capable of phagocytosis form few phagocytotic vacuoles unless presented with appropriate particles or other stimuli. Each phagocytotic cell type responds to a characteristic spectrum of particles.

2.2.1.1. Unicellular Organisms: "Filter Feeders" and Others

Protozoa and comparable cells that feed by phagocytosis show fascinating devices for acquiring their fare, though most of the underlying mechanisms are only sketchily known. For example, many ciliated protozoa have highly differentiated feeding apparatus located at specific zones along the cell surface; the cells pass large volumes of water through this apparatus and filter out particles to eat. Cilia and ciliary membranelles are deployed along "gullet" ("pharyngeal") regions and other extracellular pathways in patterns that create currents and establish sieving pathways whose effect is to increase the rate of retention of particles in desired size ranges at sites where phagocytosis takes place. By such means, ciliates that feed on bacteria efficiently retain particles down to 0.2 μm in diameter. Most other ciliates tend to trap somewhat larger particles.

Carnivorous protozoa frequently immobilize prey by secreting mucus or other products to foul cilia, or by releasing threadlike trichocysts that generally are described as penetrating or entangling their targets (though their modes of action need to be better understood).

Many protozoa show distinct preferences for particular foods and some have quite restricted ranges, feeding only on one or a very few types of material. The discriminatory mechanisms probably include chemotactic responses. But in at least some cases the protozoa adhere to their food before engulfing it; thus, direct surface recognition may be important both in selecting prey and in initiating uptake. A few types of "receptors" reminiscent of the plasma membrane proteins that permit cells of higher organisms to select particular materials for uptake (Fig. 1.10, Sections 2.2.1.2 and 2.2.2) have been tentatively identified in unicellular organisms. For example, amebas of the slime mold *Dictyostelium discoideum* possess cell surface molecules that may enable them to enhance uptake of materials bearing specific saccharides or having suitable hydrophobic surfaces (Section 2.2.1.2).

A number of familiar protozoa are opportunistic feeders, phagocytosing "almost anything" to see if they can digest it. They take up many sorts of suitably sized particles presented to them, even if they prefer particular types. Commonly, such organisms will readily engulf inert, indigestible particles such as polystyrene latex beads, which move into lysosomal digestive vacuoles and subsequently are defecated. Still, even those ciliates that seem to feed by such relatively nonspecific filtering form markedly fewer endocytotic vacuoles in particle-free media than they do when particles are present. And amebas, *Paramecium,* and other organisms with nearly omnivorous appetites do show chemotactic responses, thought to guide them to desirable foods. They sometimes also exhibit notable differences in the rates of uptake of particles with different surface charges or other chemical properties. *Amoeba proteus,* for example, takes up the ciliated protozoan *Tetrahymena* markedly more avidly than it does agarose or latex beads. Slime mold amebas when feeding on other strains of slime mold amebas selectively spare their own close genetic relatives.

Though mechanical factors such as the size of particles can be decisive in determining what is taken up, certain protozoa engulf surprisingly large bodies intact (Fig. 2.7) or by "biting" off pieces. Elaborate intake systems based on basketlike arrays of cytoskeletal and membranous elements, and probably involving also the secretion of digestive enzymes (Section 6.1.4), equip organisms to break up material, such as elongate multicellular filaments of cyanobacteria (blue-green algae), that at first glance would appear far too large and asymmetrical to be manageable. Uptake of large organisms often is preceded by adhesion of a predatory protozoan to the surface of its prey; sometimes Ca^{2+} ions are needed for this reaction, or changes in surface charge markedly affect it.

A number of unicellular organisms are suspected to sort out endocytosed materials intracellularly, hinting at the existence of "sophisticated" recognition devices and membrane phenomena. The amebas of slime molds, for example, reputedly can take up complex mixtures of different particles into a large common vacuole, and then segregate different types of particles into separate, smaller vacuoles. Figure 2.8 illustrates an experiment with *Acanthamoeba* in which the cell was found to discharge indigestible particles it had previously taken up, in exchange for digestible particles. Such capacities are imagined to compensate for indiscriminate feeding behavior, by permitting routine retention of those particles that can serve as food, and expulsion of those that cannot.

2.2.1.2. Mammalian Phagocytes; Immune Recognition; Opsonins; Frustrated Phagocytosis

The professional phagocytes of multicellular organisms face problems not encountered by unicellular predators. They must be versatile enough to handle a variety of invasive microorganisms and other foreign particles, and to deal with dead or abnormal cells and extracellular debris from their own organism. But they must do so while sparing most of the healthy cells of their own organism. Different types of professional phagocyte face different requirements; for instance, because of their location, the macrophages of the lung (alveolar macrophages) confront inorganic particles more often than do many of the other phagocytes.

The most highly refined—or at least the most intensively studied—of the mechanisms used by mammalian phagocytes to select their targets depend on the cells and circulating proteins of the immune system. During infections or inflammatory responses to other injuries, chemotactic and motility-influencing molecules are generated that guide the phagocytes to locales where they are needed (see Section 4.4.1.1). There, selective uptake of foreign particles by the phagocytes is promoted by the presence of antibodies bound to corresponding

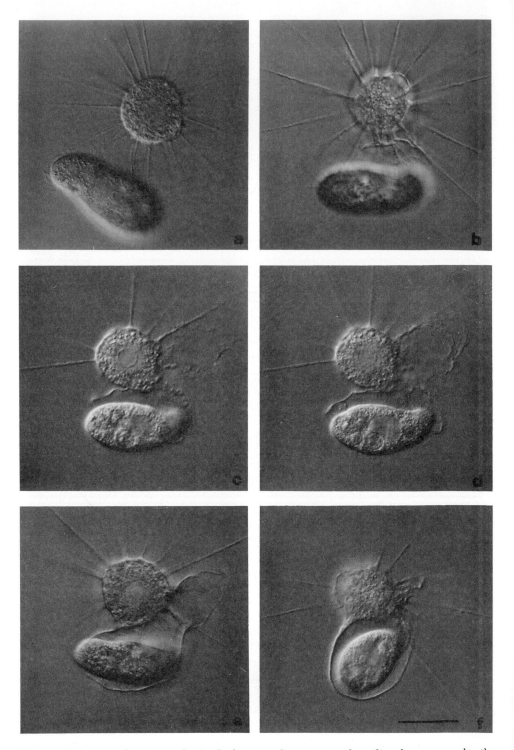

FIG. 2.7. Sequence of micrographs **(a–f)** showing phagocytosis of a ciliated protozoan by the heliozoan *Actinophrys sol*. Bar in **f** = 50 μm. From Hausmann, K., and Patterson, D. J. (1982) *Cell Motility* **2**:9.

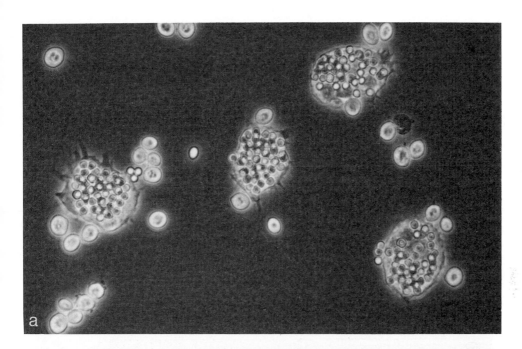

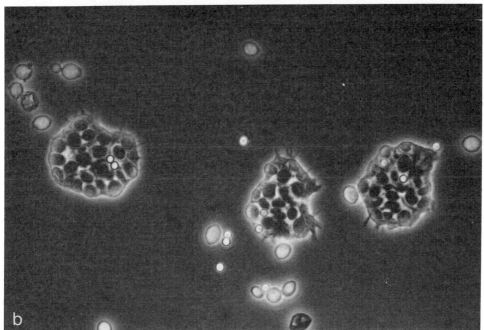

FIG. 2.8. (a) *Acanthamoeba castellanii* that had phagocytosed latex beads (2-μm diameter) for 60 min. **(b)** Similar cells that, after taking up latex, were washed and incubated with yeast for 60 min. The cells have lost most of the latex beads and in their stead have accumulated the larger, darker-looking yeast. Courtesy of B. Bowers. (Experiments as in *J. Cell Biol.* **97:**317, 1983.)

antigens of the particles' surfaces (see also Sections 4.4 and 4.6.1.2). Macrophages and neutrophils possess cell surface **receptor** proteins (Section 1.5.1.3) that selectively recognize "immune complexes" (antibodies complexed with antigens) and bind to them with high affinity. The best known of the receptors are the "Fc receptors." These recognize and associate with the Fc portions of IgG-type immunoglobulins; in dispute is whether tight association is promoted by conformational changes undergone by the IgG molecules when they complex with antigens. The binding of immune complexes to cell surface Fc receptors can trigger phagocytosis of the complexes (Section 2.3.2.10) and of particles coated with the complexes.

When suitably "activated" (Section 4.4.1.3), macrophages also initiate uptake of particles through receptors specific for **C3b,** a protein generated in the responses to several sorts of insult or injury, from the set of serum proteins known as **complement** (Section 4.4.1.1).

Immunoglobulins and the C3b complement protein are examples of **opsonins,** soluble molecules that complex with and promote the uptake of other materials. In addition to the receptors for opsonins, mammalian phagocytes have receptors that foster uptake of particles without the intervention of opsonins. For example, in some of the physiological states they can assume (Chapter 4), macrophages actively take up **zymosan** particles (preparations of yeast cell walls) because the saccharide polymers that abound in the particles are rich in mannose and therefore are recognized by macrophage receptors specific for mannose and N-acetylglucosamine ("Man/GlcNAc" receptors). Macrophage or fibroblast engulfment of particles like polystyrene latex spheres is frequently referred to as "nonspecific" but the term is intended more to connote versatility rather than a very broad nonselectivity. One possibility is that cells adhere to and engulf materials like the polystyrene beads because the cell surface can establish many "weak bonds" (hydrophobic interactions?) with the particle surface.

When mammalian professional phagocytes encounter structures that are too large to be taken up, but show the presence of opsonins or other surface properties that tend to trigger phagocytosis, the cells spread out on the structures as if attempting to surround them (see Fig. 2.16). Such **frustrated phagocytosis** probably relates to the capacities of macrophages to initiate formation of multicellular "capsules" that surround and sequester large particles. Sometimes, however, there are pathological consequences (Section 6.2.2).

2.2.2. Pinocytosis and Receptor-Mediated Endocytosis

2.2.2.1. Fluid Phase versus Adsorptive Uptake; "Macropinocytosis"

Many cell types can incorporate tracers dissolved in the medium around them into fluid-filled vacuoles large enough to be seen readily in the light microscope (diameters well above the 0.2-μm limit of resolution). For most cell types *in situ*, pinocytotic mechanisms have been studied chiefly with the electron microscope, and the findings have led to the conclusion that fluid enters the cells in pinocytotic vesicles with diameters on the order of 100–200 nm. In such cases, the formation of fluid-filled vacuoles large enough to be seen readily in the light microscope is thought to be chiefly by fusion of these smaller vesicles. But under certain culture conditions, amebas and macrophages or other cultured mammalian cells seem, from light microscopy of living cells, to form fairly large fluid-filled vacuoles directly from their surfaces. In ameboid protozoa, relatively large vacuoles can arise by budding from elongate channels that protrude into the cytoplasm from the cell surface (see Fig. 2.10). Cells of the thyroid gland *in situ* take up the extracellular "colloid," which serves as a storage depot for the precursors of thyroid hormone, in small vesicles and in larger vacuoles, both of which can be produced directly at the cell surface (Section 4.2.4.1).

Macropinocytosis is a term used, loosely, for direct formation of a large fluid-filled vacuole or tubule from the cell surface. Particularly in the past, **micropinocytosis** was used for uptake in small vesicles.

In strains of the amebas *Amoeba* and *Chaos*, inorganic cations such as sodium ions (at 10–100 mM), and cationic proteins (at 1 mM or less) are particularly effective inducers of pinocytosis; they markedly increase the abundance of pinocytotic channels. How significant these inductive responses are for the normal physiology of the cells is not certain. One line of argument maintains that the cations are experimental surrogates for yet-to-be-identified natural inducers. Another view asserts that it is advantageous for protozoa living in fresh water to be able to gulp up proteins and salts whenever the organisms encounter them. Irrespective of this uncertainty, the observations on induction of pinocytosis in amebas were of particular importance because the fact that the inducing cations bind to anions in the ameba's cell surface helped focus attention on adsorption of material to the cell surface as a frequent prelude to pinocytosis. Subsequently, such adsorption was demonstrated directly, in amebas, by using fluorescent derivatives of pinocytosable proteins, or electron-dense tracer particles (Section 2.2.2.3). The particles could be seen to bind to the saccharide-rich cell surface coat, which, in some amebas, is a distinctive filamentous extracellular layer 100 nm or more in thickness. The pinocytosed particles enter intracellular compartments while still bound to these filaments.

Though the details vary from case to case, one type or another of adsorption is now understood to be an extremely widespread and important prelude to most forms of pinocytosis. Its effect is to increase the effective concentration of materials undergoing endocytosis, by factors of as much as a thousandfold or even more as compared to the bulk solution. Moreover, unlike the bulk uptake of material in the "fluid phase," which was originally conceived as central to pinocytosis, uptake involving adsorption will vary differentially for specific components of a solution or suspension. The overall rates of pinocytotic uptake of a given substance will depend both upon the sizes and rates of formation of pinocytotic vesicles and upon the affinity of the substance for pertinent regions of the cell surface (e.g., on the availability and distribution of cell surface sites to which the substance can bind). (A third factor—the degree of retention within the cell—will be considered in the next chapter.)

Is adsorption at the heart of pinocytosis? In most well-analyzed situations it seemingly is. Still, amebas and other protozoa, and some cultured vertebrate cell types do continue to form fluid-filled vacuoles, at basal rates, in the absence of known inducers of pinocytosis. In cells that crawl by one or another form of ameboid motion, some of this uptake may be an "accidental" accompaniment of the membrane movements and cytoskeletal reorganizations involved in motility: In amebas, fluid-filled "droplets," and also vacuoles containing particles as large as bacteria form particularly often at cell regions where motility-related reorganization is especially prominent, such as at the posterior, "uroid," region of a crawling cell. With macrophages and tissue culture cells, it could be that fluid-filled vacuoles arise directly from folds and channels present between the extensive and somewhat irregularly shaped surface protrusions that serve motility—the lamellipodia for example.

Simple considerations of surface-to-volume ratios make it evident that small vesicles, or tubules with narrow bore, carry higher proportions of membrane and membrane-adsorbed materials relative to fluid content, than do larger vacuoles or broader tubules. We will encounter several cases where these relations are likely to be of considerable importance (e.g., Section 3.3.4.4).

2.2.2.2. Receptor-Mediated Endocytosis

Like amebas and other unicellular organisms, cells of multicellular organisms also concentrate experimentally administered polycations, such as polylysine, efficiently into pinocytotic compartments because of the adsorption resulting from electrostatic interactions with anionic cell surface glycoconjugates. But for naturally occurring materials, pinocytosis by many vertebrate cell types depends on more selective adsorption, involving "receptors" in the plasma membrane (Section 1.5.1.3). These receptors recognize and bind specific molecules or classes of particles with considerable specificity and affinity and mediate the subsequent uptake of the molecules or particles into cytoplasmic vesicles. Most of the known receptors are integral membrane proteins (Section 2.3.2.9) but some may be lipids or glycosaminoglycan moieties of proteoglycans.

Selective receptors confer upon adsorptive pinocytosis capacities for efficient handling of many biologically important materials—hormones like insulin, regulatory proteins like epidermal growth factor (EGF), and materials transported in the blood, such as lipoproteins or yolk proteins (see Chapter 4). Although other forms of endocytosis, notably most types of phagocytosis in vertebrates, are also mediated by specific receptors, the term **receptor-mediated endocytosis** is increasingly used to refer specifically to those processes of the adsorptive pinocytotic type that are known to depend upon binding of ligands to specific receptors. This term has partly supplanted "pinocytosis" because the latter word still evokes fugitive echoes of the original emphasis on uptake of fluids whereas "receptor-mediated endocytosis" emphasizes the selective, adsorptive, aspect of uptake.

2.2.2.3. Tracing Pinocytosis II

The tracer particles and proteins used in the early studies of pinocytosis (Section 1.4.1.2) are now supplemented by many others more specially tailored for particular uses. For example, cationic derivatives of ferritin are now often employed because their binding to cell surface anions produces a more intimate and stable association with endocytotic membranes than is true of native ferritin (Sections 5.3.2 and 5.3.3.2). Ligands for specific receptors can be conjugated covalently to molecules that are visualizable in the electron microscope or they can be adsorbed to electron-dense colloidal gold particles: The resulting "hybrid" tracers are used to follow the distribution of ligand-binding sites at the cell surface and within endocytotic compartments. Figure 1.10 illustrates such an approach with LDL particles and Figs. 4.9 and 4.10 show the use of ferritin conjugates of IgGs, and peroxidase-conjugated "asialoglycoproteins."*

With increased tracer variety has come increased appreciation that the tracers can influence their own fate. We will see, for instance, that when single tracer molecules or particles are capable of binding simultaneously to more than one cell surface site ("bivalent" and "multivalent" ligand systems), the behavior of the complexes between tracers and cell surface components is often different from that observed when univalent ligands are bound to the same sites (Sections 3.3.4.3 and 4.1.2.3). Ligand modifications expected to be innocuous can turn out to influence events subtly. Thus, attachment of fluorescent labels ("fluorochromes") such as the commonly used FITC (fluorescein isothiocyanate) to proteins has been thought to result in increased hydrophobicity of the proteins. Perhaps for similar reasons, attachment of ^{125}I, a radioactive tag that frequently is linked either to ligands or to cell surface proteins, seems sometimes to increase rates at which the labeled molecules are degraded. Often in such situations it is not certain whether it is the modification of the protein *per se* that causes the observed change or whether some other perturbation has been inadvertently introduced by the procedures used to attach the tag. Problems of these sorts are not disabling, but they do lead most investigators to be wary of conclusions based solely upon use of a single type of tracer.

2.2.2.4. Fluid Phase Tracers

To evaluate the extent of pinocytotic uptake of fluid by cells, recourse is had to tracer molecules and particles thought not to bind appreciably to the cell surface. One of the ways to verify this presumption is to demonstrate that tracer accumulation by the cells increases linearly with concentration of the tracer in the medium; high enough concentrations must be used to avoid "spurious" near-linear uptake as is seen, for example, with adsorptive endocytosis when available receptors are much more abundant than are ligands. The more usual means to establish that a tracer is likely to be taken up chiefly with the fluid phase is to show that the tracer does not become associated with the cells under conditions, such as temperatures of 0–4°C, in which pinocytosis is inhibited (Section 2.3.2.4) but most forms of binding to the cell surface are not. As quantitative probes for the extent to which fluids enter cells, materials are used that lysosomes do not degrade; most of these are molecules that physiologists have traditionally employed to measure the distribution of water in tissues— inert molecules such as sucrose or inulin and polymers such as dextrans. The extent of their uptake by cells can be estimated, for example, by using radioactive forms of the tracers. One concern with these tracers is that unless they are used at very low concentrations, their accumulation in intracellular compart-

---

* "LDL receptors"—proteins that bind LDL particles and control their uptake by receptor-mediated endocytosis (see Section 4.1.1)—are present on many cell types; those of fibroblasts are among the most intensively studied of endocytotic receptors.

For hepatocytes, the ligands used to study receptor-mediated endocytosis include the "asialoglycoproteins"—glycoprotein derivatives produced by enzymatically cleaving off the sialic acid residues that terminate the oligosaccharide chains of many secreted glycoproteins, such as the serum protein **orosomucoid**. The receptors to which the desialylated proteins bind before endocytotic uptake are referred to as "asialoglycoprotein" (ASGP) receptors though, in fact, what the receptors recognize is not the absence of sialic acids but rather the presence of galactoses or certain other sugars exposed when the sialic acids are removed (Section 4.5.2.4).

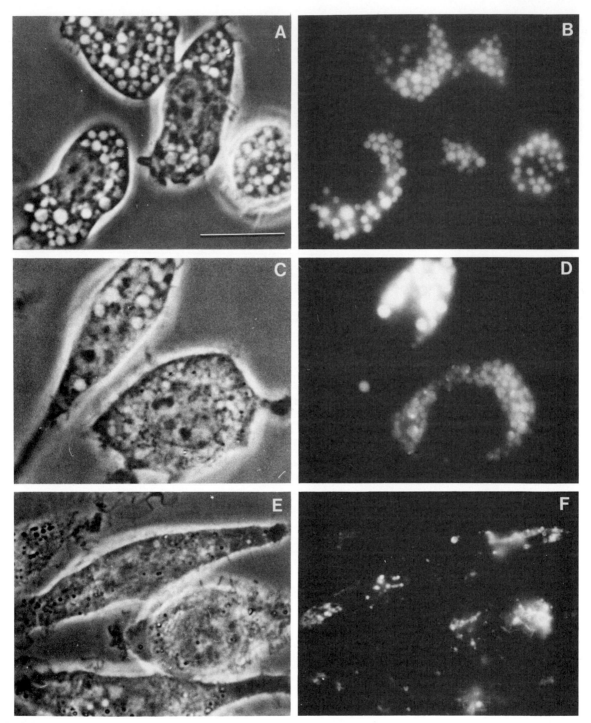

FIG. 2.9. Cultured macrophages (J777.2 mouse macrophage-like line) were grown in a medium containing sucrose and Lucifer Yellow, which were endocytosed together into the vacuoles seen in panels **A** and **B**. (The micrographs are paired so that the left column shows phase-contrast images and the right, the fluorescence of Lucifer Yellow.) Sucrose induces an increase in the average volume of heterophagic structures (A, B; most of these structures probably are "swollen" secondary lysosomes). When the cells are now grown in the presence of invertase, this enzyme is endocytosed, gains access to the sucrose-containing vacuoles, and hydrolyzes the sucrose into monosaccharides that can escape from the vacuoles; the consequence is the shrinkage of vacuoles seen in **C–F**. Bar in A = 10 μm. From Swanson, J., Yirinec, B., Burke, E., Bushnell, A., and Silverstein, S. C. (1986) *J. Cell Physiol.* **128**:195–201. (Copyright: Alan R. Liss, Inc.)

ments can alter osmotic conditions (see Section 3.4.1.3 and Fig. 2.9) and thereby perturb the normal distribution of fluid.

For a time, HRP (Section 1.4.1.2) tended to be used uncritically as a "fluid phase marker" because in many cases microscopists saw little binding of this tracer to the cell surface. HRP is particularly convenient because it is soluble, stable, and available at low cost in large quantities, and especially because the cytochemical means for demonstrating its intracellular locations, both for light microscopy and for electron microscopy, are quite sensitive. Single molecules of the enzyme generate multiple molecules of electron-dense cytochemical reaction product, an "amplification" not achievable with tracers such as metal-containing particles whose visualization is based on intrinsic electron density. However, HRP is not inert—some of the isozymes that are abundant in usual HRP preparations evoke histamine release or other "pharmacological" responses when injected into animals. And it is even conceivable that oxidation–reduction reactions catalyzed by the enzyme sometimes damage cells during experiments. Judging from control experiments—e.g., comparisons of cells exposed to different tracers—problems of these types with HRP generally are of manageable magnitude. But the peroxidase molecules also are mannose-containing glycoproteins and so can be bound by the mannose-specific endocytotic receptors present on macrophages in some physiological states, and on certain other cell types. When extensive, this binding severely compromises the suitability of HRP as a fluid phase marker.

Current favorites for light microscopic tracing of fluid uptake are particles like dextrans conjugated to FITC or other fluorescent labels, and small dyes like Lucifer Yellow, a soluble, nontoxic, low-molecular-weight (389) molecule whose anionic nature minimizes binding to the anions that predominate in cell surface coats and whose fluorescent properties are advantageous for microscopic visualization (Figs. 2.9 and 4.19). With small dyes there is, of course, always the worry that some molecules can cross membranes; with Lucifer Yellow this is known to occur under certain unusual conditions (Section 3.4.1.2) and may be more common than is generally assumed.

2.2.2.5. Nonadsorptive Solutes Can Become Concentrated

Dyes and other solutes thought to enter the lysosomes in the fluid phase but to be incapable of crossing the lysosome membrane, can accumulate in the lysosomes at concentrations severalfold higher than the concentrations in the medium. An apparent implication is that as secondary lysosomes shrink and "condense" (Section 2.1.4), soluble contents as well as insoluble materials become increasingly concentrated within them. Skeptics point to possibilities of undetected adsorption prior to endocytosis, or unanticipated transport of the solutes across membranes. But considerable concentration of solutes could be driven simply by the osmotic sequelae to transport of digestive products out of the lysosomes and to the aggregation of residues within (Section 2.1.4). Inventories of the amounts and states of the components inside lysosomes and of the movements of materials across the lysosomal membrane will be needed to sort out the possibilities.

2.2.2.6. "Constitutive" Endocytosis

Inductions of pinocytosis—increases in the rates at which vesicles bud from the plasma membrane—do occur in diverse circumstances, such as the exposure of amebas to salts as mentioned above, or the stimulation of secretory cells to exocytose their secretion (Section 5.3.2.1). Many of the ligands that cultured cells take up by receptor-mediated endocytosis, however, behave as if entering an ongoing stream of internalization that flows whether or not these particular ligands are present in the medium. When such molecules or particles are introduced into the medium, they can move into the cells rapidly and extensively, with no major increases in rates of vesicle formation. This is indicated by morphological study, by the absence of large changes in fluid–tracer accumulation by the cells upon exposure to the ligands, and by studies of the rates of cycling of receptors between intracellular and cell surface locations (Sections 2.3.2.7 and 3.3).

Cultured human fibroblasts are among the cells that accumulate LDL particles and other materials by this sort of "constitutive" process. The definition of "constitutive" here is of course a limited one: It means simply that when maintained in their normal growth media, the cells need no special stimulus to form vesicles that do pick up ligands added to the medium but are produced whether or not one has deliberately placed known ligands in the cells' surroundings. Even cells *in situ* mostly do give the impression, to microscopists, of forming at least a few endocytotic vesicles under usual, normal conditions. Not ruled out, however, is that normal tissue fluids and all the media thus far found suitable for keeping cultured cells healthy include inducers of pinocytosis among their standard ingredients, so that the cells are always being stimulated. Many of the hormones and growth factors commonly used in culture media are endocytosed by receptor-mediated endocytosis, and the endocytosis of insulin, for example, though perhaps partly constitutive, may also be partly insulin-induced.

2.3. Internalization

Pioneering-period proposals about the mechanisms by which endocytotic vesicles and vacuoles form, argued that the adsorption of endocytosable materials to the cell sufrace produces local changes in the physicochemical properties of the plasma membrane—most likely, in the surface tension. These changes supposedly expressed themselves as the "deformation" of the cell surface into vesicles or vacuoles. Proposals of this genre foundered partly on problems in accounting for the morphology of internalization and the localization of uptake to favored domains of the cell surface. And when nonadsorptive materials were taken up, it was necessary to postulate (see Section 2.2.2.6) that endocytosis was being produced by unidentified adsorptive components present in the same medium. Most persuasively negative was the realization that diverse types of "adsorption" can precede endocytosis—the binding of inorganic salts to anions, the nonspecific association of macromolecules with the cell surface via charged or hydrophobic groups, or the specific binding of ligands to receptors—and these different phenomena are unlikely all to perturb the physical chemistry of the plasma membrane directly in the same way.

Adsorption clearly is not by itself **sufficient** to bring about internalization because, even for avidly endocytosed materials, adsorption and internalization can be dissociated. As Section 2.3.2.4 will detail, most endocytosis is inhibited at 0–4°C but at these temperatures ligands continue to bind firmly to the cell surface sites with which they normally interact. This situation has been extensively used experimentally to "pulse" cells with endocytosed materials by exposing the cells to adsorptive particles or molecules at low temperature (usually 4°C) and then warming the preparations to initiate endocytosis. (Sometimes the cells are rinsed briefly before warming.) In these experiments, departure of the cohort of adsorbed molecules or particles to the cell interior can be monitored quantitatively, with living cells, by determining the rates and extents to which radioactive ligands become inaccessible to treatments that do affect the ligands when they are still outside the cell. For instance, internalized materials can no longer be dissociated from the cells by altering concentrations of ions such as Ca^{2+} or by exposure to molecules that compete for the ligands' receptors. And they cannot be reached by enzymes or antibodies added to the medium. (Suitable care must, of course, be taken during the assays to avoid internalization of the enzymes or antibodies.)

Though temperature-based pulsing protocols have proved very useful, exposure to 4°C can alter cells so that when they are warmed, the endocytosis that resumes may differ at least transiently from comparable processes in the normal steady state. For instance, some cells, during the warm-up, show enhanced "ruffling"—the extension of flattened pseudopod-like processes—which could entail significant incidental increases in macropinocytosis.

2.3.1. The Morphology of Endocytotic Internalization

Broadly considered, endocytotic structures form by one of two routes (Fig. 2.10). In "typical" phagocytosis by ameboid cells, and in macropinocytosis like that seen in the thyroid gland, folds or other elaborations **protrude** from the cell

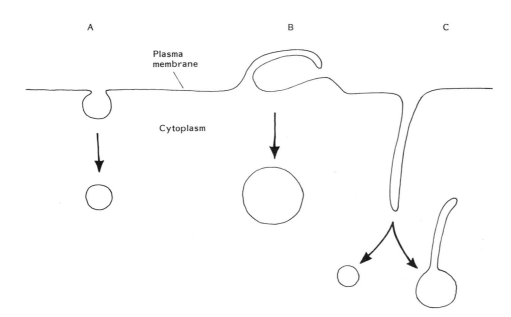

FIG. 2.10. Major internalization mechanisms of endocytosis as would be seen in different cell types or during uptake of different substances by a given cell type: Illustrated are the direct budding of vesicles from invaginations of the cell surface **(A)**, the extension of cytoplasmic processes around particles or droplets of fluid **(B)**, and the formation of tubule-like channels that bud off as tubules or give rise to vesicles **(C)**. Combinations and variants of these phenomena also are encountered.

surface and surround the material to be taken up. In contrast, small pinocytotic vesicles most often arise by **invagination** of the plasma mambrane. It is not always easy to distinguish which of these mechanisms is operating. For instance, when modest-sized particles (say, a few tenths of a micrometer in diameter) are phagocytosed by ameboid cells, electron micrographs often give the impression that the particles "sink" into the cell as if by invagination. But, in at least some such cases, direct observations on living preparations indicate that the cell surface actually first forms small folds or conical protrusions to engulf the particles. Amebas seem to use a combination of surface protrusion and invagination to establish the elongate cell-surface connected channels whose ends fragment into or otherwise generate pinocytotic vacuoles.

2.3.1.1. Coated Pits and Vesicles

In the large majority of cases studied in detail thus far, receptor-mediated endocytosis is mediated by endocytotic vehicles forming from **coated pits.** These pits are discrete patchlike, slightly indented regions of the plasma membrane (see Fig. 1.10). Their "coating" is an organized cytoplasmic layer with distinctive morphology; it looks, in conventional electron micrographs, as if it were composed of bristles, 20 nm long, protruding perpendicularly from the membrane. The protein "clathrin" is prominent in the coat (Section 2.3.2.8) and can be demonstrated at the pits by immunocytochemistry.

Coated pits take part in pinocytosis by invaginating into the cytoplasm and rounding up to generate structures usually called **coated vesicles** (but see Section 2.3.1.3). In the thin sections used for electron microscopy, the latter structures appear as circular or ovoid profiles with diameters of 100–200 nm and showing the clathrin coating along much or all of their surface. The vesicles are delimited by the membrane that belonged to the coated pit and they contain the materials that were adsorbed to this membrane, plus fluid internalized during invagination of the pits. At any given moment, 0.5–2% of the area of the plasma membrane of mammalian fibroblasts or other familiar cultured cells under ordinary culture conditions has the form of coated pits. In a fibroblast, there are one or two thousand pits at a given time and the cell invaginates them to form one or two thousand vesicles per minute. Large cells particularly active in receptor-mediated endocytosis, such as oocytes accumulating yolk (Section 4.2.1.1), produce correspondingly vast numbers of coated pits and vesicles (e.g., Fig. 4.3). Much of the surface of a chick oocyte can be occupied by coated membrane (Section 4.2.1.1); the coated pits and vesicles are somewhat larger than the more usual 100- to 200-nm structures.

In ciliated protozoa like *Tetrahymena* and *Paramecium*, pinocytotic coated pits form chiefly at the tips of specialized cell surface invaginations ("parasomal sacs") associated with the cilia. In a few cell types of vertebrates there also are persistent cell surface specializations tipped by endocytotic coated pits, such as networks of channels connected to the cell surface.

Section 2.1.3.1 pointed out that many cell types contain "putative primary lysosomes" coated in a manner similar to the vesicles described here and we will see (e.g., Section 7.2.2) that certain membranes of the Golgi apparatus exhibit comparable coats. There still is considerable confusion about the relations among the various types of coated structures in cells—only recently have adequate beginnings been made in subfractionating and analyzing the different membranes (see Section 7.4.3.4) or in immunocytochemical discriminations.

But it appears that endocytotic coated pits and vesicles tend to be somewhat larger, on the average, than the primary lysosomal coated vesicles generated in the Golgi region of the same cells. The different classes of coated structures seem also to vary in the proteins accompanying clathrin in their coats.

2.3.1.2. Endosomes

When microscopically detectable ligands destined for receptor-mediated endocytosis are provided to cells in culture or to hepatocytes and other cell types in situ, the tracers are readily seen to adsorb to the cell surface at coated pits (e.g., Fig. 1.10). Then, within one to a few minutes, the tracers begin to appear in intracellular compartments whose surfaces largely lack the pits' distinctive coating; with time, more and more of the tracer accumulates in such compartments. Even at the earliest times that tracer is seen within them, some of the endocytotic structures are two or more times larger than would be expected were they to form simply by the coated pits rounding up and pinching off to generate vesicles. Often they appear as tubules, or vacuoles, or as vacuoles with tubular extensions (Figs. 2.11, 2.12, 2.15, and 4.9). These structures, larger than coated vesicles, which serve as the principal early depots for the accumulation of materials taken up by receptor-mediated endocytosis, are designated **endosomes.***

Endosomes are of variable shape and size and they are dynamic structures that evolve rapidly through a series of states in which their morphology and content change (Section 3.3.4). The dynamics are proving difficult to grapple with. And, in many studies, endosomes have not been adequately separated, in cell fractions, from fragments of the plasma membrane or from lysosomes (Section 2.3.1.6). Therefore, a morass of uncertainties and disagreements, about the formation of endosomes and about their behavior once they have formed, has yet to drained.

2.3.1.3. Relations of Coated Pits to Endosomes

Most observers presently hold that endocytotic coated pits frequently first give rise to true coated **vesicles;** i.e., to roughly spherical, clathrin-covered structures 100 nm or more in diameter, which detach completely from the plasma membrane to lie free in the cytoplasm near the plasma membrane. Very soon thereafter, the coating on these vesicles disappears, apparently by dis-

---

*The terminology in this arena too is still unpleasantly messy. "Receptosome" has its adherents as an alternative to "endosome" (Fig. 2.12). "Pinosome" (Section 2.1.1) overlaps with "endosome" for it is used as a generic term to designate "prelysosomal" pinocytotic bodies (i.e., early endocytotic structures containing solvent and solutes with or without materials adsorbed to the membrane). "Endosome" is sometimes used for any "prelysosomal" pinocytotic intermediate compartment (Section 2.4.4.1) whether or not the process being studied is known to be receptor-mediated endocytosis; the term has even been used for phagocytosis as well as for pinocytosis.

Recently, "endosome" has acquired specific, important functional connotations related to the establishment of a low intracompartmental pH and to particular behaviors of receptors and of ligands (Section 3.3.4.2). These functional features have been studied in detail only for receptor-mediated endocytosis. Consequently, I will use "endosome" chiefly for the intermediate, endocytotic structures in which materials accumulate soon after their uptake by receptor-mediated endocytosis. For the many cases in which bodies resembling endosomes form through phenomena not shown to be receptor-mediated endocytosis, I will use "endosome-like body" or some similar formulation.

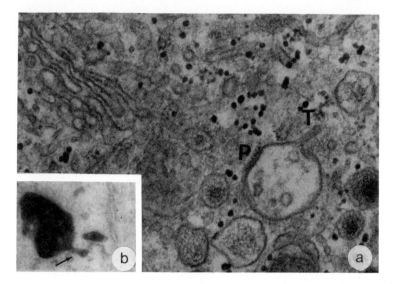

FIG. 2.11. These structures, from the rat adrenal medulla, illustrate what are thought to be common features of endosomes and multivesicular bodies at midstage in their lives. **a** shows a structure with a surface plaque (P), a tubular extension (T), and a few internal vesicles. **b** shows a similar structure containing endocytosed HRP. Note that the internal vesicles in b are pale because the electron-dense peroxidase reaction product (and presumably the peroxidase molecules that generated it) is in the space surrounding the vesicles. a from Holtzman, E., and Dominitz, R. (1968) *J. Histochem. Cytochem.* **16**:320. × 50,000. b from Holtzman, E. (1969) In *Lysosomes in Biology and Pathology* (J. T. Dingle and H. B. Fell, eds), Vol. 1, Elsevier: North-Holland, Amsterdam, p. 192. × 40,000.

sociating into its component protein complexes, which diffuse away to be reused for new coats (Section 2.3.2.8). The vesicles, now uncoated, contribute their membrane and content to larger, endosomal structures, by fusing with one another and with preexisting endosomes.

But microscopists long have known that cells taking up visible tracers into coated vesicles, often also show very rapid accumulation of some of the tracer in larger structures—tubules, or tubules attached to vacuoles—that appear as if forming directly from the cell surface. The membranes bounding these structures generally show no coating along most of the surface, though frequently

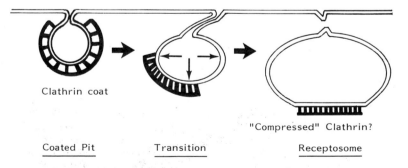

FIG. 2.12. Proposed mechanism for direct transformation of a coated pit into an endosome ("receptosome") by way of an intermediate vacuole attached to the cell surface through a narrow (closed?) channel. Also indicated is the possibility that the "plaques" on endosomes represent an altered form of clathrin. Courtesy of M. C. Willingham and I. Pastan.

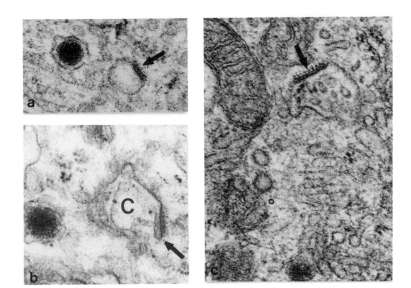

FIG. 2.13. Plaques (arrows) on structures of types generally regarded as related to endosomes and multivesicular bodies, from rat adrenal medulla (**a, b**) and a neuron in a rat ganglion (**c**). Panel a shows an "early" vacuolar body. Panel b shows a "cuplike" structure, so called because the large space within this body at C still is in communication with the outside, in the region near the arrow; this body probably was on its way to developing into a multivesicular body when the tissue was fixed. Panel c shows a multivesicular body. The plaque in c is reminiscent of the coating on coated vesicles. a and b from Holtzman, E., and Dominitz, R. (1968) *J. Histochem. Cytochem.* **16**:320. × 55,000 (approx.). c from Holtzman, E., Novikoff, A. B., and Villaverde, H. (1967) *J. Cell Biol.* **33**:419. × 45,000 (approx.).

they do have small patches ("plaques") of material about the size of coated pits and appearing somewhat like clathrin coating (Figs. 2.11 and 2.13). Some of these configurations have been regarded as macropinocytotic, some as channels from whose ends coated vesicles bud, and some as reflecting minor routes of endocytosis or as being too ambiguous to be interpreted. A band of dissenters insists, however, that vacuoles and tubules of this sort are the principal form taken by coated pits as they invaginate and that they serve directly as the chief sources of the earliest forms of endosomes—vacuoles with diameters of about 200 nm. Coated vesicles rarely if ever form as simple, free, rounded-up coated pits, these investigators assert, claiming instead that when one reconstructs the three-dimensional electron microscopic structure of the cytoplasm near the plasma membrane from serial sections, supposedly free coated vesicles usually turn out to be connected to elongate tubules. In many cases the tubules still are in close contact with the plasma membrane (Fig. 2.12), appearing as if continuous with this membrane. But these apparent continuities may no longer be avenues of entry for material from the extracellular world. At least, they are not patent to tracers like Ruthenium Red, an electron-dense molecule that, when introduced with electron microscope fixatives, generally has ready access to compartments that are in direct, structural communication with the cell surface.

The disagreements between the majority and the dissenters center on unresolved issues about the reliability of different electron microscopic techniques; on the possibility that the temperature treatments widely used to study endocytosis have pertinent distorting effects; and on possible differences among cell types (e.g., one suggestion is that coated vesicles form less often in very flat, cultured cells, than they do in cells *in situ*). Cell fractionation has been of quite limited value here because, for instance, though fractions enriched in "vesicles" bearing coats can be prepared, these fractions include at least some structures produced by fragmentation of larger, partially coated surfaces—the plasma membrane and Golgi sacs—during homogenization. Moreover, early endocytotic structures may normally be so prone to fusion with one another as to stymie efforts at sharply distinguishing a class of "entry" vehicles from bodies that have already initiated fusion.

In fundamental conceptual terms, the differences among the competing views are actually not all that substantial. At both poles of the dispute, there is agreement that coated pits first give rise to structures that retain at least some clathrin coating and whose interiors are substantially or completely isolated from the extracellular space.

2.3.1.4. Endosome Morphology; Plaques; Cuplike Bodies

Endosomes have been studied in detail in a few cell types including hepatocytes and cultured fibroblasts. Their diameters are generally reported to be in the range 0.2 to 1 μm, with the larger sizes being reached presumably by fusions among smaller bodies. Their shape varies, reflecting the dynamics of fusions and transformations of endocytotic structures. Though they often do have the appearance, in thin section, of vacuoles with one or more tubular extensions (above and Figs. 2.11 and 2.15), not infrequently, their overall morphology is more tubular than vacuolar. Three-dimensional reconstructions (Section 1.5.3) sometimes reveal that what seem to be clusters of separate endosomes, in a given thin section, actually are networks of interconnected bodies—the extent and elaborateness of these networks vary.

Though the plaques on the surface of endosomes sometimes resemble the coating on coated pits fairly closely (see Fig. 2.13), more often such similarities are minimal and typically the plaques do not show immunocytochemical binding of the anticlathrin antibodies that bind to coated pits. One view (Fig. 2.12) is that the plaques represent a modified derivative of the coat of coated pits— this notion is occasionally extended to suggest that the molecules of the coat remain with endocytotically internalized membrane throughout the membrane's migrations in the cell. But there is a near consensus that the coating that assembles at coated pits eventually dissociates from the membrane. Most investigators believe that this occurs very soon after the pits invaginate (Section 2.3.2.8) and remain dubious or agnostic about the relations of the plaques to clathrin coating of pits and vesicles. An interesting alternative now starting to be explored is that the plaques represent functionally important specializations of the endosome membrane such as clusters of the enzymes or pumps discussed in Chapter 3.

Not unusually, endosomes contain one or two small, interior membrane-bounded vesicles (Fig. 2.11). Sometimes, much larger vacuole-like inclusions are seen (Fig. 2.14) but three-dimensional reconstruction suggests that most such large inclusions represent deep infoldings of the endosome's surface that still are continuous with the membrane bounding the endosome. Infolded structures of this sort, which I and others have called "cuplike" bodies, are among the compartments showing early accumulation of endocytosed tracers in many cell types, both cells engaged in receptor-mediated endocytosis and

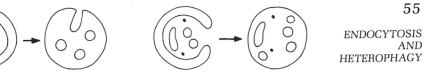

a   1   2   3

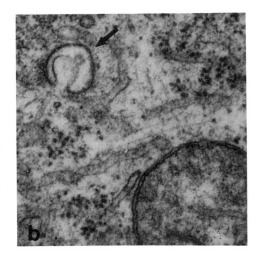

FIG. 2.14. Formation of multivesicular bodies. (a) The vesicles within heterophagic multivesicular bodies ("multivesicular endosomes") can arise by invagination of the membrane bounding an endosome: vesicles can bud off directly as in **1** or larger, cuplike invaginations can form and subsequently fragment into vesicles (**2**). **3** illustrates an alternative mode of formation of a multivesicular-type structure, through the autophagic-like engulfment of pre-existing vesicles. **b** is from a neuron in a mouse ganglion and shows a cuplike structure (arrow), apparently generating its internal vesicles. b from Holtzman, E., and Peterson, E. R. (1969) *J. Cell Biol.* **40**:863. × 55,000 (approx.).

ones carrying out other endocytosis-like activities such as the retrieval of membrane after secretion (Section 5.3.2).

2.3.1.5. Multivesicular Bodies (MVBs)

Relatively soon after endocytotic uptake commences (often within 5–30 min), increasing amounts of tracer begin to appear in bodies of types known as **multivesicular bodies** (Fig. 2.15). Like the endosome-related structures containing tracers at earlier times, MVBs tend to exhibit surface plaques (Fig. 2.13), tubular extensions (Fig. 2.15), and infoldings of the delimiting membrane. A large proportion of the MVBs are clearly "vacuolar" (circular or ovoid) in electron microscope profile, and some have a large population of the internal vesicles for which the bodies are named (the vesicles measure 30–50 nm or more in diameter).

Electron microscope images suggest that MVBs evolve from earlier endosomal structures, with the MVBs' internal vesicles arising by budding directly from the delimiting membrane or through fragmentation of larger infoldings of the MVB surface (Fig. 2.14). The cuplike morphology of many early endocytotic bodies and the presence of a few vesicles within such bodies probably reflects these processes of vesicle origin. So does the fact that endocytotic tracers within MVBs generally lie in the spaces **between** the vesicles or coat the vesicles' outer surfaces (Fig. 2.15), rather than being present within the vesicles (see Section 5.2.1). In the photoreceptors of invertebrates, where the membranes internalized during endocytosis have distinctive freeze-fracture appearance, the interior vesicles of MVBs share this appearance, consistent with their origin from membrane withdrawn from the cell surface (Section 5.2.1.1). Certain cell

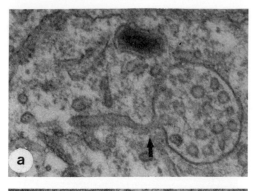

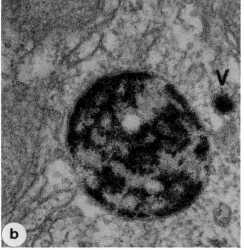

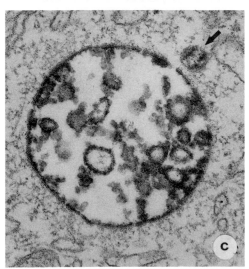

FIG. 2.15. (a) Multivesicular body (MVB) from rat adrenal medulla showing a long, tubular extension of its surface (arrow). Courtesy of S. Abrahams. From Holtzman, E. (1976) *Lysosomes A Survey*, Springer-Verlag, Vienna. × 50,000 (approx.).

(b, c) MVBs from the epithelium of rat epididymis. b is an acid phosphatase preparation showing reaction product surrounding the vesicles within the MVB, and in a coated vesicle (V) nearby. c is from a preparation exposed to HRP and shows the cytochemical reaction product for HRP surrounding the vesicles in the MVB as well as in a small vesicle, probably of endocytotic origin, near the MVB (arrow). From Friend, D. S. (1969) *J. Cell Biol.* **41:**269. × 70,000 (approx.).

surface receptors also wind up in the membranes of vesicles within MVBs (Section 5.2.1), as do some of the immunocytochemically demonstrable proteins of the lysosome or endosome surface discussed in Chapter 3. (Chapter 3 will make clear that these last considerations by no means imply that the MVBs' membranes are unmodified offspring of the plasma membranes; there is considerable selectivity in the movement of membrane components into and within the cell.)

The MVBs in which endocytotic tracers accumulate are increasingly regarded as a typical, late stage in the evolution of endosomes and therefore often are called "multivesicular endosomes" (or "late endosomes"). Sections 5.2.1 and 5.3.1.2 will, however, indicate that MVBs can arise in more than one way and that bodies looking like MVBs can participate in various autophagic phenomena as well as in heterophagic processes (see Figs. 2.14 and 5.14): a multivesicular morphology does not always signal endosomal origin. Note also that some MVBs contain acid hydrolases and therefore qualify as lysosomes (see Sections 2.4.4.1 and 7.3.2.2).

2.3.1.6. Purifying Endosomes

In part because the endosomes in a given cell are heterogeneous, most conventional cell fractionation schemes do not separate endosomes adequately from lysosomes, Golgi structures, or other organelles. Some success is being had with Percoll gradient sedimentation and free-flow electrophoresis (Section 1.3.1) of cell homogenates prepared at carefully chosen intervals after endocytosis of tracers commences: As in Fig. 1.5, endosomal densities in widely used fractionation procedures, mostly fall in the range of 1.02–1.1; the most recently formed endosomes having densities toward the lower end of this range and older ones having higher densities, approaching those of mature secondary lysosomes. In electrophoresis, endosomes move even more rapidly toward the anode than do lysosomes.

Density-altering procedures also are being tried. Figure 4.10b shows an experiment on early phases of endocytosis in the liver, using an artifical ligand—bovine serum albumin (BSA) molecules with galactoses and HRP attached to them—which is taken up via the hepatocyte's asialoglycoprotein receptor (see footnote on p. 45 and Section 4.5.2.4) and then is used to generate a reaction product formed, enzymatically, by the HRP. The reaction product markedly increases the buoyant density of the compartments in which it is formed and this facilitates the separation of endocytotic structures containing the galactose–BSA–HRP conjugates from other bodies. Unfortunately, the manipulations and the peroxidase incubation also inactivate many enzymes and otherwise alter membranes, so the endosomes obtained are of somewhat restricted utility for biochemical analyses.

Other approaches are based on immunoadsorption techniques, using beads coated with antibodies directed against naturally occurring proteins of the endosome membrane or against viral proteins introduced into the endosomes by incorporation into the cells' plasma membranes before endocytosis.

2.3.2. Mechanisms of Internalization

2.3.2.1. Phagocytosis by Zippering

The most widely accepted detailed model for phagocytosis by macrophages and similar ameboid cells proposes that the binding of particles to the cell surface not only helps trigger phagocytosis but also is an essential and integral facet of the mechanism by which the cell actually engulfs the particle. The cell surface is envisaged as moving sequentially around a particle from points of initial adhesion, progressing by virtue of the binding of plasma membrane receptors to corresponding ligands on the particle's surface (Fig. 2.16).

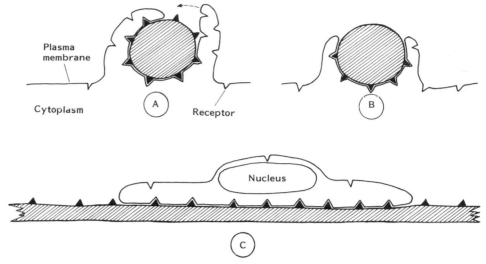

Plasma
membrane

Cytoplasm   A   Receptor   B

Nucleus

C

FIG. 2.16. Proposed mechanisms for phagocytosis-related activities of mammalian phagocytes. (**A**) During phagocytosis dependent upon interaction of cell surface receptors with ligands on a particle, the cell is thought to engulf the particle by the progressive extension of the cell surface around the particle, guided by the receptor–ligand interactions. Receptors from distant regions of the cell surface can be recruited for the process, by migration in the plane of the plasma membrane. (**B**) Artificial particles bearing ligands only along part of their surfaces are only partly engulfed. **C** schematizes "frustrated" phagocytosis in which the cell binds to an extended ligand-covered surface, and flattens out to the maximal extent possible, as if trying to engulf the surface. Based on ideas of S. C. Silverstein, F. M. Griffin, S. D. Wright, and their colleagues.

The cytoplasm and plasma membrane "flow" around the particle until only a small aperture connects the forming vacuole with the outside world; the cell surface then fuses with itself to seal off this opening.

The phagocytotic mechanism just described has been likened to a zipper, with the difference that in phagocytosis the cellular binding sites that participate are not rigidly fixed in place as the teeth of a zipper are. As large particles are progressively surrounded, the binding sites (receptors) can be recruited from distant zones of the plasma membrane, by migration in the plane of the membrane. Investigators have not yet decided whether this migration is directed in some way or whether it is essentially a diffusion in which the receptors move randomly until they are "trapped" by binding to particles.

Evidence that zipperlike receptor–ligand interaction is intrinsic to phagocytosis by macrophages includes the fact that the phagosomes in which newly engulfed particles become engulfed fit closely around the particles; there is little free space in the vacuoles. That the interaction is **necessary** for phagocytosis is inferred from experiments with particles coated with opsonizing antibodies over only a limited sector of their surface; macrophages, operating via their Fc receptors, adhere only to the coated sectors and therefore fail to complete internalization (Fig. 2.16). Contrasting with this finding, however, are observations on amebas exposed to high concentrations of a soluble ligand (the "chemotactic peptide" N-formyl-Met-Leu-Phe; see Section 4.4.1.1); the cells form large, phagosome-like endocytotic vacuoles, despite the absence of particles.

Phagocytosis by ciliates like *Paramecium* differs in morphology from that in ameboid cells: food vacuoles are generated at specialized zones of the "oral apparatus" by processes resembling invagination of the cell surface. Rather than conforming to the surface of the particles taken up, the phagosomes arise with sizes and shapes set largely by the geometry and organization of the oral apparatus. When the ciliature delivers suitable concentrations of particles such as small food microorganisms, the vacuoles that form can each contain many particles. Thus, for instance, when *Paramecium* is exposed to latex particles about 1 μm in diameter, it produces food vacuoles about 16 μm in diameter each containing dozens of the particles. (In macrophages, phagocytotic vacuoles with multiple particles most often form by fusion of phagosomes after internalization, though macrophages also can "zipper" clusters of small particles into a single new vacuole, when the particles are presented to the cells in closely bunched arrays or tightly adherent to one another.)

2.3.2.3. Experimental Perturbation of Phagocytosis; Involvements of Filaments and Energy

The internalization phase of phagocytosis by mammalian phagocytes is not drastically inhibited by exposure to colchicine or by other treatments that affect microtubules. But cytochalasins, and other agents affecting the actin-based microfilament system, do inhibit particle uptake. This is in line with observations that microfilaments abound, often to the extent of excluding most other structures, in the cytoplasm underlying the cell surfaces that are taking part in forming phagosomes. These filaments almost certainly contribute to the motility of the cytoplasmic extensions that move around the particles, bringing the cell surface into appropriate juxtaposition with potential ligands on the particles. Amebas of *Dictyostelium* mutants defective in myosin still can phagocytose bacteria, making actin the central candidate for the role of crucial filament protein.

Cytochalasin B also prevents normal separation of food vacuoles from the cell surface in *Paramecium* and other nonameboid protozoa; actin filaments are associated with the vacuoles in these organisms too. The effects of antimicrotubule agents like colchicine on phagocytosis in protozoa are complicated by the fact that these agents may affect the cytoskeletal organization upon which maintenance of specialized "oral" regions depends, and may also interfere with intracellular membrane transport phenomena upon which long-term maintenance of phagocytotic competence depends (see Section 3.3.2.1). Thus, although it has sometimes been observed that colchicine can inhibit food vacuole formation, it is not certain that this is due to direct perturbation of the internalization process *per se*.

Phagocytosis is energy dependent. For the mammalian professional phagocytes, agents inhibiting glycolysis—fluoride, iodoacetic acid, or 2-deoxyglucose (2-DOG)—are more effective inhibitors of phagocytosis than are agents affecting oxidative phosphorylation. This probably means that the phagocytes are capable of sustaining phagocytosis through anaerobic metabolism, though interpretation is clouded by evidence that the perturbing effects of some of the inhibitors are variable and may not be limited to energy metabolism. (For example, with macrophages, 2-DOG has more impact on phagocytosis of opsonized particles than it does on "nonspecific" uptake.) ATP is the likely essential carrier and direct provider of energy for phagocytosis. However, in macrophages inhibited with 2-DOG, phagocytosis continues despite reductions of ATP levels to below 25% of normal; the explanation is that macrophages maintain a pool of creatine phosphate, severalfold larger than that of ATP, which can regenerate ATP rapidly and in so doing fuel phagocytosis.

There seems to be a temperature threshold at 15–20°C, for phagocytosis by mammalian phagocytes, though how sharp this threshold is has yet to be fully explored. At lower temperatures, macrophages still are seen to bind test particles such as opsonized red blood cells avidly, but internalization of the particles is largely prevented. The basis of this inhibition could be a reduction in membrane fluidity by the low temperature, but low temperature might influence many other pertinent cellular properties or activities.

2.3.2.4. Effects of Low Temperature on Receptor-Mediated Endocytosis and Other Forms of Pinocytosis

As implied near the beginning of Section 2.3, rates of uptake of fluid phase tracers like sucrose by macrophages, and other cells capable of pinocytosis, decline to very low levels as the temperature is lowered to 0–4°C. But, unlike the virtually complete inhibition of phagocytosis in mammalian phagocytes at temperatures below 15–20°C, the uptake of HRP by cultured fibroblasts shows no sharp threshold down to 2°C. [HRP seems to be largely a fluid phase tracer (Section 2.2.2.4) for fibroblasts as it shows little evidence of adsorption to the cell surface. The curve of its uptake as a function of temperature is characterized by a logarithmic increase up to 37°C with a 2.7-fold increase in uptake for each 10° increment.]

The literature on the temperature sensitivity of receptor-mediated endocytosis is confusing, though it is clear for many cell types in culture that such uptake also decreases to negligible levels at 0–4°C. Receptor-mediated uptake of materials such as lipoprotein particles, by cultured mammalian cells is 5–10 times more rapid at 37°C than at 15–20° and for a time it was thought that receptor-mediated endocytosis by mammalian cells might have a sharp threshold at about 16°C, like that of phagocytosis. However, the methods used in the experiments underlying this belief were not always sensitive enough to monitor very low levels of uptake. Recently, continued receptor-mediated uptake at 10°C or below has been reported for internalization of viruses by cultured cells (Section 3.1.3), and for the endocytosis of desialylated glycoproteins by hepatocytes in perfused liver (Section 4.5.2.4). And in the experiments where temperature thresholds seemed to be operating, it was not always established that the effects were specifically on the internalization phase of endocytosis as opposed to some of the subsequent steps in heterophagy: Net uptake of materials by receptor-mediated endocytosis and other forms of pinocytosis can be quite responsive to interference with various of the fusions and membrane movements by which endocytotically internalized materials and membranes are handled after endocytotic entry; some of these processes are now thought to show temperature thresholds at 16–20°C (Section 2.4.7). Low temperature also affects the pH within endosomes, and related structures and these effects could influence membrane behavior (Section 3.1.3).

From life-style considerations, one would expect the optimal temperature ranges for endocytosis by cold-blooded animals and by protozoa to differ from those for mammalian cells. This expectation has been little examined, especially for populations of animals freshly obtained from natural settings. *Xenopus* oocytes continue to pinoctyose down to 4°C. Pinocytosis by laboratory strains of amebas slows considerably at temperatures much below 20°C or much above 30°C; little is seen below 15°C or, in contrast to mammalian cells, at 37°C. Paramecia still form particle-containing food vacuoles at 10°C, a temperature lower than the generally accepted threshold for particle uptake by mammalian phagocytes.

2.3.2.5. Effects of Metabolic Inhibitors and Cytoskeletal Perturbation on Pinocytosis

Many agents and conditions reduce rates of pinocytosis (see Fig. 3.13)—the list in current use includes inhibitors of energy metabolism; increases in the osmolarity of the medium or depletion of intracellular K^+ (Section 2.3.2.8); exposure to sulfhydryl reagents, to inhibitors of energy metabolism, or to agents that raise the pH of endocytotic or lysosomal compartments (Section 3.3.4.3). The effects on pinocytosis, of these and other inhibitors, and of agents that modify the cytoskeleton, are often partial and difficult to predict. Troubles in interpretation arise from problems in discriminating among different types of pinocytosis and from difficulties in distinguishing among inhibitor effects at various steps in heterophagy, like the difficulties to which I alluded above when discussing temperature. Thus, for example, though it is known that steps early in heterophagy are affected by interfering with the supply of energy, precisely where energy is needed is not well established. Some experiments suggest, for example, that the steps directly responsible for internalization can continue, for a short time at least, even when ATP levels are very low but that other processes needed for sustained pinocytosis fail under these conditions (as does the delivery of endocytosed material to lysosomes).

Treatments that affect microtubules often have minimal impact on uptake of fluid phase tracers or adsorptive ones and when such treatments do partially inhibit uptake, it is not evident that their effect is directly on the internalization process.

Cytochalasin B does not reliably inhibit fluid phase or adsorptive pinocytotic uptake. Many investigators have therefore concluded that direct participation of microfilaments is not essential for pinocytosis. But this negative conclusion should be applied conservatively. For one thing, the effects of cytochalasins are not infallible in inhibiting other types of processes known to depend on filaments (nor are the effects limited exclusively to filament systems). In addition, from time to time there are reports that cytochalasins do inhibit macropinocytosis or formation of coated vesicles, and considerable circumstantial evidence suggests that the filament system of the cell's cytoskeleton does influence certain aspects of pinocytosis. In some experiments the distribution of coated pits, of endocytotic receptors, or of other proteins in the cell surface has seemed to show striking correlations with the distribution of relevant cytoskeletal proteins (e.g., Fig. 2.17). More generally, cells of many types have an extensive meshwork of actin filaments lying just below the plasma membrane and this meshwork must likely undergo local dismantling or rearrangement if vesicles or larger structures are to bud into the cytoplasm. There have even been claims that the ends of actin filaments show intimate associations with coated pits as if the filaments could pull upon the pits thereby producing or guiding invagination. At the minimum, endocytosis would seem to require coordination and interactions between membranes and the cytoskeleton, so as to produce the appropriate movements of vesicles, endosomes, and the like.

Linked to these last considerations is speculation that endocytotic mechanisms are closely related to mechanisms for cell motility or even that cell motility itself may depend on endocytosis: One conjecture for the relatively slow crawling of mammalian cells in culture is that the cells move in part by continually adding membrane to their forward tip through exocytotic-like fusion of cytoplasmic vesicles; as the cells crawl they supposedly retrieve this membrane endocytotically from more posterior regions and circulate the endocytotic structures forward for reuse at the anterior tip. Schemes of this sort often include proposals that the cycling membranes carry receptors through which the cells adhere to substrates so as to pull themselves forward. As would be expected for some versions of these mechanisms, the ameboid protozoan *Pelomyxa* shows particularly intense endocytosis at the posterior "uroid pro-

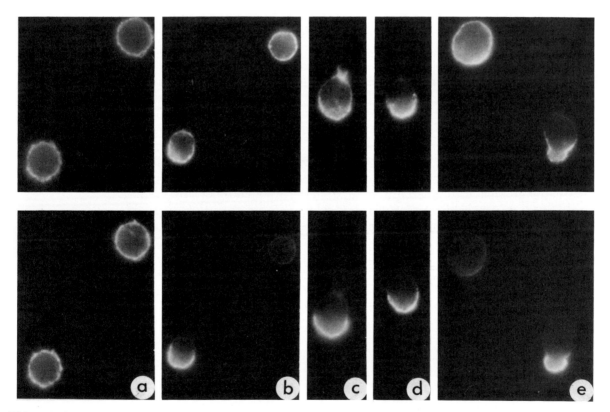

FIG. 2.17. Capping in mouse splenic lymphocytes, induced by exposure to fluorescein-labeled antibodies against immunoglobulins of the lymphocytes' plasma membranes. The lower series of panels **(a–e)** shows cells fixed at different stages in the movement of the fluorescent ligand to one pole. The upper panels show the same cells stained, immunocytochemically, for myosin (with a rhodamine-tagged antibody that fluoresces red, and thus can be distinguished from the anti-immunoglobulin, which fluoresces yellow). Myosin, in the cytoplasm underlying the plasma membrane, evidently undergoes a migration paralleling that of the receptor–ligand complexes in the cell surface. From Schreiner, G. F., Fujiwara, K., Pollard, T., and Unanue, E. R. (1977) *J. Exp. Med.* **145**:1393. × 750 (approx.). (Copyright: Rockefeller University Press.)

cess" (the prominent trailing zone, seen in cells that crawl by extending a single large pseudopod or ruffle of membrane). Moreover, cultured mammalian cells (and *Amoeba*) often do focus the exocytotic addition of membrane at their leading edge when they are moving (though endocytotic vesicles or vacuoles can form at virtually all regions of their surface).

Endocytosis is much reduced during mitosis owing presumably either to the extensive reorganizations of the cytoskeleton and cell surface taking place during cell division, or to more direct effects of the regulatory systems or ionic changes operating at this time.

2.3.2.6. Endocytotic Receptors Can Cluster in the Plane of the Plasma Membrane; Capping

A highly productive realization in the analysis of receptor-mediated endocytosis was the appreciation that coated pits not only show accumulations of clathrin and other proteins of the coat, but also show elevated concentrations of endocytotic receptors.

Given types of receptors are present in many thousands or tens of thousands of copies per cell, at the cell surfaces of hepatocytes, fibroblasts, or macrophages (see Table 3.3). At coated pits, the number of these receptors per unit area can be as much as 50- to 100-fold greater than the number elsewhere along the plasma membrane (see Section 2.3.2.7) and coated pits are able to internalize segments of the plasma membrane correspondingly enriched in ligands associated with these receptors. (Note, however, that even for receptors that are highly clustered at the coated pits, like the LDL receptors of fibroblasts, at any given moment under normal circumstances 20–30% or more of the cell surface's total population of receptors are located outside the pits.)

Initially, it was thought that receptor clustering is a response to ligand binding preparatory to endocytosis, because there are readily observable situations in which ligand binding preceding endocytosis does lead to the redistribution of receptors into concentrated patches or more extensive accumulations. The phagocytotic "zipper" involves such redistribution (Section 2.3.2.1). A particularly dramatic example is the "capping" observed at the surfaces of lymphocytes (Fig. 2.17 and Section 4.2.4.2). Capping can be triggered by multivalent ligands such as antibodies or lectins,* most likely reflecting the ligands' linking of previously separate membrane proteins to one another. The proteins might come into close enough justaposition to be cross-

---

*Receptor redistribution induced by ligands is only partly understood. The ligands may sometimes induce relevant conformational changes in receptors or changes in the interactions of membrane proteins with one another or with the cytoplasm. But more obvious features of the ligands and receptors that are germane include "valency," a term used (somewhat ambiguously) in this context to refer to the number of sites on a single ligand molecule or particle that can bind to a given type of receptor, or to the number of binding sites for ligands present on a single receptor.

Among the multivalent ligands most often used to induce experimental redistribution of membrane molecules are antibodies, and **lectins**—a heterogeneous category of proteins, each type of which binds to specific configurations of saccharides and saccharide derivatives. Antibodies of the IgG class are bivalent so that polyclonal antisera, containing mixtures of antibodies directed against different determinants on a given membrane-associated antigen, can interlink the antigens into two-dimensional aggregates analogous to the precipitates of immune complexes that form between IgGs and soluble antigens. Concanavalin A, a widely used lectin with especially high affinity for mannose residues, is tetravalent; Con A can therefore cross-link glycoproteins by binding to their oligosaccharides.

linked by ligands, simply by random diffusion in the plane of the membrane. But movements of membrane proteins undergoing capping also may be directed by interactions with other membrane molecules and with the adjacent cytoplasm. For instance, the observations in Fig. 2.17 suggest that oriented migrations of cytoskeletal components take place during capping, perhaps accounting for findings that metabolic energy is required for capping. Possibly the cytoskeleton somehow recognizes cross-linked plasma membrane proteins and controls their distribution. Alternatively, models for cell motility, like those summarized in the last section, suggest that the cells able to form caps are types that continually transport membrane molecules toward one cell end as a facet of the mechanisms by which the cells themselves can move; normally, diffusion in the plane of the membrane largely counteracts the resulting tendency for motile membrane molecules to pile up toward one end of the cell, but when the molecules are interlinked so that they no longer can diffuse rapidly enough, they accumulate in larger and larger rafts at that end.

Whatever the mechanism of capping, many of the complexes of ligand with membrane proteins that pile up in the caps eventually are internalized by endocytosis. Others are shed from the cell.

2.3.2.7. Ligand-Independent Clustering of Receptors

Capping, however, turns out to be a misleading model for the more modest normal accumulation of receptors seen at coated pits. True, certain of the receptors participating in ordinary receptor-mediated endocytosis—those binding epidermal growth factor (EGF) on cultured cells for example—do seem to cluster into pits as a response to ligand binding (but see Section 7.6). But several important types of receptors are found to aggregate at the pits whether or not their ligands are present. Thus, the coated pits of cultured fibroblasts are enriched in binding sites for LDL particles whether or not the medium contains LDL, and even if the cells are maintained at 4°C to suppress endocytosis. This distribution of receptors can be demonstrated by immobilizing the receptors with the preservatives ("fixatives") used for microscopy and then visualizing the sites of LDL binding by exposing the fixed cells to visible LDL derivatives.

Techniques like this, and "bookkeeping" considerations (estimates of the number per cell of receptor molecules, coated pits, and coated vesicles), also indicate that receptors for several different ligands are simultaneously present in a given coated pit and in the intracellular vesicle it generates; the limits of such versatility are not yet known. A typical pit or vesicle has a surface area on the order of 0.05 to 0.5 μm^2 and given receptors are present in numbers ranging from a half-dozen to several dozen or more. It is often presumed that receptor types that cluster in response to ligands enter the same coated pits as those that cluster independently of ligands, but the extent of such overlap has not been studied in detail.

Immunocytochemical and cell fractionation data on receptors that accumulate in coated pits, whether or not they are occupied by their ligands, show that unoccupied receptors are internalized endocytotically by the cell at rates of the same order as those seen when the receptors are occupied. Here, as in Section 2.2.2.6, the picture that emerges is of a constitutive endocytotic "conveyor belt." But as before, it is premature to assume that the belt keeps running even if *none* of the pertinent ligands are present; conceivably, internalization of unoccupied receptors is always dependent on the uptake of ligands bound to other types of receptors present in the same pits.

With neutrophils, the receptor that mediates phagocytosis of particles opsonized with the C3b component of complement (Section 2.2.1.2), can be clustered in the absence of C3b by pretreating the cells with phorbol esters.

2.3.2.8. Cycling of Clathrin

Although multivalent ligands can induce some clustering of receptors without the obvious involvement of clathrin, most proposals about normal mechanisms of receptor-mediated endocytosis suppose that the receptors clustered at coated pits are held there by the material that forms the cytoplasmic coat. In part because clustering of receptors like those for LDL, is abolished when assembly of the coat is blocked (see below), recruitment of receptors into the clusters is usually modeled as the trapping of diffusible receptors by the pits. Alternative views envisage the recruitment of coating by previously established clusters of receptors, or propose mutual cooperation between coat proteins and receptors. (See also the recent findings in Section 7.6.)

For receptors that cluster only after binding ligands, it remains to be clearly established whether the receptors, having bound to ligands, group together before or after the corresponding coated pits acquire their coats. Nor is it understood why ligand binding induces assembly of clathrin coats along the plasma membrane of phagocytes spread on opsonized surfaces too large to be engulfed ("frustrated phagocytosis"; Section 2.2.1.2).

Clarification of the mechanisms for accumulation of receptors and of coats should come soon from further exploitation of recently developed experimental and genetic means for perturbing the formation of clathrin coats. With cultured fibroblasts and other cells, manipulation of ionic or osmotic properties of the culture medium so as, for example, to deplete intracellular K^+, reduces the abundance of the pits probably by inhibiting assembly of new coats. Lowering the cytosolic pH by a few tenths of a unit can also inhibit formation of coated pits. Strains of yeast have been developed that lack operational genes for clathrin (Section 4.7.1) and thus are presumed incapable of forming coated pits. Also promising is the rapid progress being made in analyzing the structure and biochemistry of coated structures and in assembling and disassembling coats *in vitro*. The major coat proteins—clathrins—are now known to be organized into tripartite structures called **triskelions** each of which contains three "heavy" (180 kD) clathrin chains and three "light" (about 35 kD) chains. The triskelions are the basic elements of the basketwork of pentagons and hexagons (Fig. 2.18) that is the framework of the coat on pits and vesicles. The basketwork readily assembles *in vitro* whether or not membranes are present: Coats forming *in vitro* do have affinity for the membranes of vesicles that have been experimentally stripped of their coats, but the triskelions can also spontaneously assemble into membrane-free "baskets" looking much like the ones on coated vesicles. Assembly of such baskets can be promoted, in the cell, by low cytosolic pH. It may also eventually prove of considerable interest for understanding the involvement of coated membranes in lysosome functioning, that clathrin, plus additional ("accessory") proteins obtained by disassembling coats, can selectively interact with purified M6P receptors (see Section 2.1.3.2 and Chapter 7) to generate assembled baskets highly enriched in the receptors. Other endocytotic receptor molecules also have been rumored to bind clathrin *in vitro* but work on these interactions has not been extensive.

The presence of coating impedes membranes from fusing with one another *in vitro*. Control of coat assembly and of the uncoating of coated structures (or

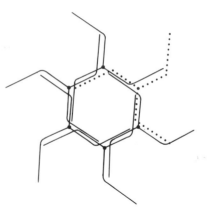

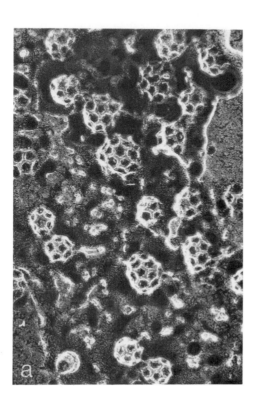

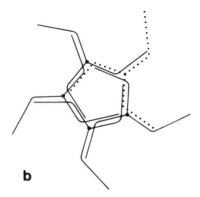

FIG. 2.18. (a) Coated vesicles (diameters about 100 nm) isolated from brain and prepared for microscopy by the freeze–deep etch procedure. The "geodesic dome"-like structure of the coating is readily seen as a lattice, chiefly of hexagonal and pentagonal elements. From Heuser, J. (1981) *Trends Biochem. Sci.* **6**:64. (Copyright: Elsevier, Cambridge.)

(b) Schematic illustration of how clathrin molecules, arranged in the "triskelions" that are the building blocks of clathrin coats, assemble to form a hexagonal unit and a pentagonal unit of the coating. One triskelion is shown in dotted lines. After work by Branton, Pearse, Ungewickell, Crowther, and many others.

drastic reorganization of the coat) in the first minutes of endocytosis thus must be of considerable importance in governing interactions of new endocytotic structures with other bodies. Uncoating seems not to be a spontaneous process. Instead, studies on isolated coated vesicles suggest the process requires "uncoating enzymes" that hydrolyze ATP in the course of dismantling the coat into its constituent triskelions. The majority view (see Section 2.3.1.3) is that these triskelions enter a soluble pool before being used to make new coats.

The capacities of clathrin triskelions to self-associate into baskets is the basis of suggestions that the assembly of coating along a region of membrane exerts mechanical forces on the membrane. These forces could produce the indented form of coated pits and contribute to the pinching off of vesicles from the cell surface. Other proposals for roles of the coat are that clathrin anchors directly to receptors and other proteins of the membrane, thereby maintaining receptor clustering in coated pits, or that the baskets link coated membranes to the cytoskeleton so as to transmit or harness forces generated by the cell's motility apparatus.

Proteins thought to aid in assembly of the baskets, regulatory agents such as calmodulin, substrates for phosphorylation by protein kinases (and kinases themselves), cytoskeletal proteins such as tubulin, and other interesting molecules have been reported to be present in cell fractions highly enriched in coated membranes or in supposedly purified extracts of the coats. Though many of these proteins probably do pertain to coated membranes, some are present because of fraction impurity, adsorptions during homogenization, or other types of artifact. Thus, although it is clear that clathrin is accompanied by other important proteins in the coat, confident inventorying of these proteins is not yet possible; nor are the differences among the coats of different types of clathrin-coated vesicles well analyzed or understood.

2.3.2.9. Ligand Binding

Several of the receptor proteins that mediate endocytosis have been purified and their molecular structures analyzed, and a few of the corresponding genes have also been studied (see also Section 7.3.1). Most of the receptors identified thus far are transmembrane glycoproteins. In the human LDL receptor (Fig. 2.19), which is a single glycoprotein chain of 839 amino acids, the domain protruding at the extracellular surface of the plasma membrane accounts for most of the mass of the protein. This domain includes the NH_2 terminus of the protein. Near this terminus there is a stretch of almost 300 amino acids extensively organized by internal disulfide bonds and rich in negative charges. The disulfides probably stabilize the domain against the vicissitudes of life in the outside world and in endocytotic compartments, such as the varied pHs encountered during receptor functioning (see Section 3.3.4.2). The charges are thought to participate in the receptor's recognizing and binding to positively charged residues in the protein components of the lipoprotein ligands that the LDL receptor handles.

The cytoplasmic end of the LDL receptor is a 50-amino-acid domain. This cytoplasmic "tail" seems to be implicated in the clustering of receptors in that when this region is altered as a consequence of genetic change (Fig. 2.19b), cells fail to aggregate LDL receptors at coated pits and consequently internalize LDL at much reduced rates. One could imagine that the cytoplasmic domain interacts with the cytoskeleton or serves in the associations of LDL receptors with one another. Recent findings suggest, in fact, that LDL receptor molecules in solution, as well as in membranes, can cluster selectively ("self-associate") with one another and that this capacity depends upon the presence of the cytoplasmic "tail" portion of the receptor.

The polypeptide chains of several other receptors that cluster at coated pits resemble the LDL receptor in having a single membrane-spanning region 20–30 amino acids long and a large, extensively organized extracellular domain. Their cytoplasmic domains vary in extent from a few dozen amino acids to several hundred—the larger tails characterize receptors that catalyze enzymatic activities in the cytoplasm (see below). Other details of the organization of the polypeptides also vary: Some of the receptors have their COOH terminus protruding into the cytoplasm and some their NH_2 terminus; some are dimers, some are single chains, and one or two have been claimed to exist in the plasma membrane as multimeric arrays. (For example, mammalian asialoglycoprotein

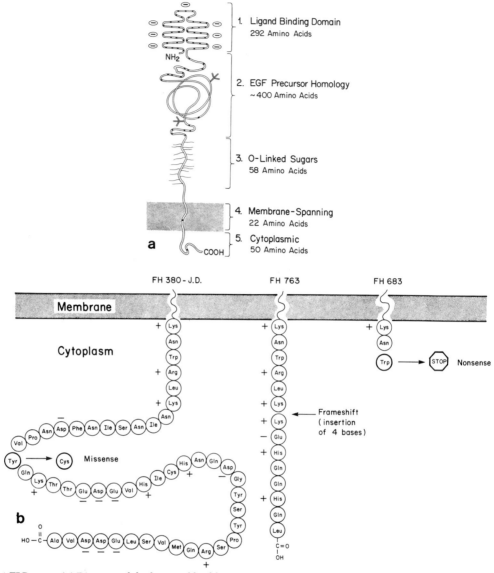

FIG. 2.19. (a) Diagram of the human fibroblast LDL receptor indicating its major domains. (b) Three of the mutations in the cytoplasmic domain of the LDL receptor that result in failure of the receptor molecules to cluster into coated pits. Diagrams courtesy of M. S. Brown and J. Goldstein. From *Science* **232**:34, 1986. (Copyright: American Association for the Advancement of Science, and The Nobel Foundation.)

receptors may be hexamers involving products of at least two distinct, though similar, genes; such as multimer could presumably bind several ligand molecules.) As yet, few generalizations or functional insights have emerged from comparative work on the different receptors although, for example, one major type of macrophage Fc receptor resembles the LDL receptor in that it seems, from genetic transfection experiment, to require its cytoplasmic tail in order to cluster in the plane of the membrane and to deliver its ligands to lysosomes. Molecular engineering and genetic studies of receptors are underway, but they are at too early a stage to have yielded definitive information about the molecular events responsible for endocytosis. One of the possibilities currently being examined is that specific tyrosine residues are key components in the cytoplasmic tails of endocytotic receptors: loss of certain tyrosines (e.g., by substituting non-aromatic amino acids) can depress uptake of some receptors, while inserting tyrosines into the cytoplasmic domains of viral hemagglutinins can result in enhanced uptake of the hemagglutinins via coated pits.

Binding of ligands to most endocytotic receptors is noncovalent. The affinity of many of the receptors for their ligands is markedly pH-sensitive, a property that governs fundamental features of the intracellular behavior of internalized receptor–ligand complexes (Section 3.3.4.2). Other ionic influences on binding vary among the receptors. For example, Ca^{2+} is necessary for effective ligand binding to the hepatocyte's asialoglycoprotein receptor, but divalent cations seem not as important for some of the receptors that bind hormones or growth factors.* Several receptors—those for EGF for example—behave as though two or more subpopulations or states, with different affinities, coexist in the same cell surface; the molecular basis for this is not known. Cases are also encountered of receptors showing unexpectedly high apparent affinities or anomalous binding curves: Such properties sometimes have been attributed to simultaneous association of the same ligand particle or molecule with two or more receptors; alternatively, the behavior has been explained as grounded in the very high local concentrations of receptors in coated pits—a ligand dissociating from one receptor molecule might tend to be recaptured by a closely adjacent one before it can diffuse away from the pit.

2.3.2.10. Signaling

The binding of ligands to endocytotic receptors can trigger such a diversity of physiological and biochemical changes that sorting out the ones crucial to endocytosis has not yet been feasible. The initial effects presumably include conformational changes in the receptors and alterations in their interactions with immediately neighboring molecules in the membrane, the cell coat, and the adjacent cytoplasm. Local rearrangements can take place resulting, many believe, in the ligand-driven assembly of multiprotein systems within the membrane, such as the arrays that generate cyclic AMP. In a few cases—notably that of the insulin receptor—the binding of bivalent antireceptor antibodies to the cell surface can mimic certain of the physiological effects of the natural ligand whereas univalent derivatives of the same antibodies have little effect. The customary interpretation is that in these systems, local redistribution of the

---

*Generalizations have been advanced to the effect that receptors mediating uptake of hormones, growth factors, and other agents with direct biological activity differ, in their endocytotic behavior, from receptors responsible for internalizing "nutrients" like LDL or for clearing materials from the circulation (see Chapter 4). Supposedly the hormone and growth factor receptors, as a class, show less dependency on pH or Ca^{2+} in binding their ligands, are less "prone" to ligand-independent clustering into coated pits, and are less likely to recycle repeatedly between intracellular compartments and the cell surface (Section 3.3.4.1). This viewpoint has stimulated useful thought, though there are exceptions to the "rules" and not all receptors are readily classifiable in terms of the scheme. Still, differences along such lines could relate, for example, to the endocytotic mechanisms that help cells regulate the abundance of receptors in their plasma membranes under varying physiological conditions (Section 5.2.1.2).

receptors into small aggregates is an activating mechanism that normally is triggered by the corresponding ligand.

Some models envisage stabilization of ligand-induced clusters of receptors by the formation of covalent links among the receptors or between the receptors and other proteins in the membrane or the cytoplasm. For a time, transglutaminase enzymes were thought likely to catalyze such linkage. But the data underlying these models are ambiguous, and enthusiasm for the proposals has waned of late.

During endocytosis, cells often show changes in intracellular levels of Ca^{2+} or cyclic nucleotides or exhibit activation of familiar enzyme-driven regulatory cascades, such as those generating arachidonic acid derivatives (Section 4.4.1.2) or the ones responsible for production of diacylglycerol and phosphoinositides from phospholipids. Binding of EGF to its receptors can lead to changes in cytoplasmic pH. Kinase-mediated phosphorylations have been detected in a growing number of cases—both the insulin receptor (see Sections 4.5.2 and 5.2.1.2) and the EGF receptor are protein kinases that can phosphorylate tyrosine residues, including ones in their own amino acid sequences ("autophosphorylation"). No doubt some of these responses bear importantly on endocytosis. But none is known to be universal and for none is the causal relationship with endocytosis known (see also Section 5.2.1.2). Most of the changes have been detected at the level of the whole cell; to understand their relevance to endocytosis, more work is needed on how the changes impact at local levels. Macrophages, for instance, discriminate among different particles that are simultaneously adherent to the cell's surface, engulfing particles coated with opsonins while leaving outside adjacent particles that lack this coating. Responses of this sort seem likely to be controlled by discrete, transient, localized alterations in the cell surface and in the immediately subjacent cytoplasm; these changes might not be well reflected in the overall levels of cellular pH, Ca^{2+} ions, or cyclic nucleotides.

Local alterations in membrane potential or in ion concentration mediated by transmembrane channels controlled by endocytotic receptors could serve to communicate "inductive" stimuli for phagocytosis from the cell surface to the underlying cytoplasm. Phagocytosing macrophages, for instance, are known to show changes in potential that begin very early in the initiation of phagocytosis. (For a time, the principal Fc receptor of macrophages was thought to be capable of functioning directly as an ion channel that changed its permeability to cations such as Na^+ upon binding ligands, but this idea has lost ground of late.) Work now in progress is aimed at determining whether the observed changes in the permeability and membrane potential in macrophages impact directly on concentrations of key cytoplasmic ions or whether they initiate more complicated chains. If, for example, Na^+ fluxes into the cell are increased, this might in turn alter local levels of Ca^{2+} or pH due, say, to stimulation of the plasma membrane Na^+/H^+ antiport system that pumps H^+ out of many animal cell types. With amebas, changes in ionic permeability and in related electrical properties of the cell surface, seen in cation-induced pinocytosis, could have analogous implications. For amebas, a speculative model surviving from early studies suggests that the binding of cations to the ameba's surface displaces Ca^{2+} locally, and that this triggers subsequent events. But though Ca^{2+} has long been a favored second messenger for all sorts of phenomena, experimental results on its importance for endocytosis have been inconsistent.

2.3.2.11. Phagocytosis versus Pinocytosis; Coated Vesicles versus Noncoated Vesicles; More on Fluid Phase versus Adsorptive Uptake

71

*ENDOCYTOSIS
AND
HETEROPHAGY*

Over the decades, opinion has swung back and forth as to whether phagocytosis and pinocytosis are fundamentally different processes or whether both are manifestations of the same underlying mechanisms whose expressions vary with the size and physical consistency of the material engulfed. The observations that pinocytosis is less obviously affected by cytochalasin B than is phagocytosis (Sections 2.3.2.3 and 2.3.2.4), and the findings of apparent differences in the responsiveness of these two processes to moderately low temperatures (Section 2.3.2.4) have lent new life to this debate but the question remains unresolved.

Similarly, views have shifted repeatedly as to whether certain of the endocytotic processes by which fluids are taken up differ fundamentally from receptor-mediated phenomena. Aspects of fluid uptake do seem separable from receptor-mediated endocytosis—accumulation of endocytosed fluid phase tracers by neutrophils in culture, for example, is only minimally affected by changes in the tonicity of the medium that substantially suppress receptor-mediated uptake of chemotactic peptides (Section 4.4.1.1) by the same cells. Conversely, with macrophages certain of the biologically active proteins of the "interferon" class can inhibit pinocytotic uptake of supposed fluid phase tracers (e.g., fluorescent dextrans) whereas the same cells show enhanced phagocytosis of opsonized erythrocytes. In the ileum of suckling rats, tracers thought to discriminate between adsorptive and fluid phase uptake are endocytosed into the same initial set of structures but soon separate, with the fluid phase tracers moving into vacuoles and the others into tubules and MVBs; later, both sets of tracers enter common (quite large) lysosomes (Chapter 4).

For a while after coated vesicles were discovered, it appeared that different types of endocytosis might be classifiable simply in terms of the involvement or noninvolvement of the coating. In contrast to the coated structures that participate obviously in receptor-mediated endocytosis and take part as well in other endocytotic processes (Sections 2.3.1 and 5.3.2), mature phagosomes of mammalian cells and food vacuoles of ciliated and ameboid protozoa generally lack coating; so do the large, fluid-containing macropinocytotic vacuoles that arise directly from the cell surface, and the large pinosomes of amebas (though these last structures have not been fully studied from this viewpoint).

But clathrin coats are only transiently associated with endocytotic membranes. Moreover, they are not always easily seen unless suitable methods have been used to prepare the cells for microscopy; without careful staining they can be almost invisible in thin sections, and "deep-etching methods" like those used for Fig. 2.18 have only recently been extensively employed to achieve face-on views of coated surfaces. In a number of cases, where noncoated structures had been thought to be responsible for endocytosis on the basis of microscopy of cells soon after tracer administration, systematic reexamination has revealed coating on the earliest endocytotic structures. Even mammalian phagosomes sometimes show extensive coated regions during their formation from the cell surface.

There are, however, a few cell types—the endothelial cells that line blood vessels, the **adipocytes** that store fat in connective tissues, and some muscles—whose surfaces show numerous noncoated invaginations resembling vesicles in the process of budding off. "Caveolae" is among the terms used to designate these structures. For the cell types just mentioned, it once was thought that the invaginations served simply in extensive "smooth-vesicle" endocytosis but views have modified. In the endothelial cells, for example, the invaginations seem not to be much involved in delivering material to intracellular depots. Rather, they participate in a transport system for moving molecules across the cell from one extracellular medium to another (see Section 4.3.3).

The two thousand or so coated vesicles formed by a cultured mammalian fibroblast every minute carry a sufficient volume to account for virtually all of the cell's endocytotic fluid uptake so no appreciable direct formation of non-coated endocytotic bodies by these cells need be postulated. For most other familiar cells of higher organisms there simply is insufficient evidence as to whether pinocytosis initiated by noncoated structures is quantitatively significant or even if it ordinarily occurs at all. Nonetheless, careful and reputable microscopists continue to report cases in which endocytosis, even of adsorbed ligands such as cholera toxin (Section 4.4.3.2), depends on entry vehicles that lack evident clathrin coats; in several such cases the endocytosed materials wind up in the same types of compartments as do materials taken up by coated vesicles though this may not always be the case. Along with the microscopy there is a handful of carefully done cell physiological or biochemical studies suggesting that the uptake of various molecules, including insulin, by a number of cell types is different from the uptake of molecules known to be endocytosed by these same cells via coated vesicles (e.g., Section 5.2.1.2). For example, endocytotic uptake of certain molecules—both fluid phase and adsorptive—can continue under conditions in which formation of clathrin coats is largely or completely suppressed (e.g., Section 4.4.3.2).

2.4. Movements and Fusions

Tracers internalized into small vacuoles or MVBs in nerve terminals move up the entire length of the axon to the nerve cell body where, eventually, the tracers accumulate in mature lysosomes. This is an extreme example of the general observation that, in most cells, heterophagy involves considerable migration of recently formed endocytotic structures. The movement often centers on cell regions where lysosomes abound, such as the vicinity of the Golgi apparatus or the neighborhood of the nucleus of cultured cells. (The Golgi apparatus of cultured cells often has a juxtanuclear location that may partly account for the abundance of lysosomes near the nucleus.) Such localization should, however, not be overemphasized even though it is frequently striking. Lysosomes rarely are strictly confined to particular cell regions; they move about and in many cell types are widely distributed in the cytoplasm.

The movements of, and the fusions among, membrane-delimited heterophagic structures, such as between lysosomes and endocytotic bodies, sometimes seem not to be rigidly programmed. When, for example, neutrophils or macrophages are suitably stimulated, the cells can be induced to fuse their lysosomes (and the specific granules of neutrophils) with the plasma membrane, releasing enzymes to the outside world. This behavior is seen especially when phagocytotic internalization is selectively impeded but phagocytotic or secretory stimuli are present, as in frustrated phagocytosis or when neutrophils are exposed to zymosan after treatment with cytochalasin B to inhibit the formation of phagosomes from the cell surface. Still, in many cases, including the normal functioning of the mammalian phagocytes, heterophagic sequences of fusions and movements are highly ordered. In *Paramecium* for example (Fig. 2.5), a special set of vesicles move to and fuse with the plasma membrane of the "cytopharynx" region during food vacuole formation, contributing to the growth of new food vacuoles (Section 2.3.2.2); once they separate from the cell surface, the food vacuoles move toward the cell posterior and as they do, they fuse with another set of membrane-delimited bodies, ones thought to be responsible for lowering the pH within the food vacuoles (Section 3.1.5.5); only then, as the food vacuoles move in an anterior direction, do lysosomes fuse with the vacuoles, converting them into digestive structures. Later, residual bodies undergo defecation at the cytoproct located near the cell's posterior. (This contrasts with the situation in amebas, which seem able to defecate at any region along their surface.)

In the ciliated protozoan *Pseudomicrothorax dubius*, which feeds by pinching off lengths from filamentous cyanobacteria (Section 2.2.1.1), acid phosphatase-containing bodies believed to be lysosomes fuse directly with newly forming food vacuoles before the vacuoles have completely separated from the cell surface. This difference in timing from *Paramecium* may be due to the cell's need to use the enzymes or other contents of lysosomes to fragment its food into bite-size pieces.

Endocytotic structures fuse with one another, with lysosomes, and perhaps with certain Golgi-associated compartments (Section 7.2.2) but as far as is known, they never fuse with mitochondria, endoplasmic reticulum, the nuclear envelope, or other membrane-bounded cellular compartments. So little is known of the membranes of endocytotic bodies and of lysosomes (Section 3.2) that only general and obvious hypotheses about the origins of these fusion specificities can be advanced: There is little to go on in choosing among proposals that proteins or lipids target the movements or control the fusions, or that there are special interactions of the participating bodies with the cytoskeleton or with popular regulatory systems such as those involving G-proteins and GTP.

On the theory that the phagosome membrane is essentially the same as the plasma membrane from which it arises (see Section 3.2.3.1) and that lysosomal fusions are akin to the exocytosis of secretions from gland cells, neutrophils stimulated experimentally to secrete enzymes to the extracellular space (see above) have been used as a model system for studying lysosomal fusion phenomena. Ca^{2+} ions are required for secretion of lysosomal hydrolases by neutrophils and, correspondingly, the Ca^{2+} ionophore A23187 stimulates secretion. Energy, available from glycolysis, seems to be required for the secretion, and cyclic nucleotides can modulate its rates. But there are obvious limits to the confidence with which these findings can be extrapolated to normal, intracellular fusions.

2.4.1. Size Must Be Regulated

From the available information—gained largely through casual inspections—the overall volume of the lysosome population and the sizes and shapes of the individual lysosomes in given cell types seem not to change much ordinarily, except for the transient changes that take place during bouts of digestion as secondary lysosomes fuse with new digestive structures and the products of these fusions evolve into residual structures. This apparent "homeostasis" implies the existence of balances among the volume of individual lysosomes, the production of new lysosomes, the fusions of old lysosomes with one another or with new digestive structures, and defecatory fusions with the plasma membrane. The balances are such that cell types very active in lysosomal digestion generally give the impression of maintaining a larger volume of lysosomes than do cell types less active in digestion. The mechanisms controlling the sizes, volumes, and behaviors of heterophagic compartments are also flexible enough to produce yolky oocytes and other cell types stuffed with very large (5–10 μm or more in diameter) structures that arise through endocytosis and persist for long times, as if protected from too much fusion with lysosomes (Section 4.2). In several other cell types, seemingly separate vacuole-like lysosomes turn out, in three-dimensional reconstructions, to be connected to one another by saclike or tubular channels.

In certain diseases the volumes of the individual lysosomes and/or the volume of the overall lysosomal compartment can increase drastically. This is true in the genetic "storage diseases," where defects in degradative capacities lead the lysosomes to accumulate larger-than-normal stores of digestive residues (Section 6.4). Another set of genetic disorders exemplified by Chediak–Higashi disease of humans (Section 4.4.2.4; Fig. 1.7d) is characterized by abnormalities of unknown basis that apparently affect lysosomal fusions and can lead cells to contain one or a few giant lysosomes in place of their normal content of many small bodies.

2.4.2. Assaying Fusions; Monitoring Movements

One way to determine whether vehicles carrying endocytosed materials have encountered lysosomes is by tracking the intracellular digestion of the materials. Often, degradation products can be detected in growth media within a few minutes after endocytosis begins, providing an upper-limit indication as to when the first admixtures with active hydrolases take place. Another approach, used in cell fractionation studies of the degradation of EGF and some other molecules, is to expose cells to inhibitors of lysosomal enzymes, like leupeptin (Section 1.5.2), to encourage the accumulation in lysosomes of detectable levels of endocytosed materials that would otherwise disappear rapidly through degradation. Sometimes cell fractionation is used to show that endocytosed materials have passed from the structures of relatively low buoyant density, in which they enter the cell (Section 2.3.1.6), into structures with the higher buoyant densities thought to be typical of lysosomes (Section 2.4.3). Most often, however, fusion of lysosomes with endocytotic bodies is inferred from observations with the usual microscopic methods: images of bodies associated in the appropriate configurations; the demonstration that endocytosed materials are in compartments with cytochemically visualizable acid hydrolases; or the finding that old digestive residues have been admixed with newly endocytosed materials (Fig. 2.4).

Studies of fusions have chiefly concerned secondary lysosomes. With the usual exception of the neutrophils, fusions of presumptive primary lysosomes with digestive structures have been studied very little beyond the occasional demonstration that small acid phosphatase-containing vesicles lie near recently formed endocytotic structures such as MVBs, at times when transfer of hydrolases to the endocytotic structures seems to be occurring (Fig. 2.15). This situation is changing, however, with the availability of immunocytochemical probes to detect M6P receptors and other proteins characteristic of the membranes of lysosomes, endosomes, or related compartments. Recent findings, for example, may indicate that newly synthesized lysosomal enzymes are delivered preferentially to recently formed endocytotic bodies such as endosomes, rather than to older secondary lysosomes (Section 7.3.2.2).

2.4.2.1. Living Cells; Vital Dyes

Phase-contrast and interference microscopy, or vital staining procedures, have been employed to watch lysosomes and recently formed endocytotic structures move and fuse in living cells. These approaches have been particularly powerful when coupled with slow-motion microcinematography and video-processing technology; it is now possible to watch and record movements, in cultured cells, of tracers such as individual gold particles 10–50 nm in diameter.

For a decade or so, fusions were frequently monitored by administering two tracers, one to mark new endocytotic structures and a different one to identify preexisting lysosomes. Mixing of the two dyes in common compartments was taken as an indication of fusions of compartments. In many of these experiments, vital dyes like Neutral Red or Acridine Orange (a fluorescent dye) were used to label presumptive lysosomes (Section 3.1.2). Endocytotic structures forming in the same cells were loaded with materials distinguishable from these dyes: Large or distinctive-looking particles were used to mark phagosomes, and fluorescent compounds such as Lucifer Yellow (Section 2.2.2.4) or macromolecules labeled with fluorescein were used for pinosomes.

Unfortunately, interpretation of findings with these approaches is not always straightforward. For one thing, it is now understood that the principal basis for lysosomal accumulation of dyes like Neutral Red or Acridine Orange is the low intralysosomal pH (Section 3.1.2; polyanions within the lysosomes may also contribute by electrostatically binding the dyes, which are cationic). This introduces uncertainties both because supposedly "prelysosomal" compartments—notably endosomes—also maintain a low pH and thus may stain (Section 3.1.3), and because changes in staining intensity can result from changes in intracompartmental pH. In a few cases, electron microscope tracers (see Fig. 2.4) have revealed fusions not reliably detectable with the vital dyes. And sometimes the dyes have seemed to alter the movement or other behavior of the structures they enter. Another complication is the bidirectional nature of the traffic in which lysosomes and endocytotic structures are involved; compartments come together by fusion but membrane-bounded vehicles also carry materials away from lysosomes and endosomes (see below and Chapter 3).

Nevertheless, observations on living cells have been important in establishing that endocytotic bodies do begin to fuse with one another from early in their life history. The studies also have helped clarify aspects of the three-dimensional structure of lysosomes and endocytotic bodies. A case in point is the set of tubular structures illustrated in Fig. 2.20a. These clearly are elongate tubules in the living cell, but they can appear as chains of small, separate bodies in electron micrographs, due either to artifactual fragmentation during preparation of the material or to the misleading impressions conveyed by the thin sections. On the other hand, studies of living cells have confirmed that many lysosomes and endocytotic structures are discrete bodies that can move rapidly, as individual entities, over long distances in the cell and a few observa-

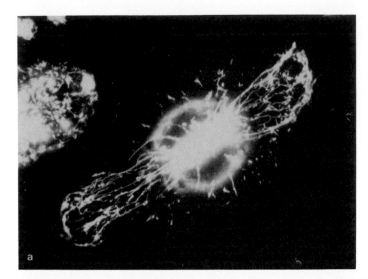

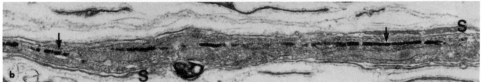

FIG. 2.20. Tubular lysosomes. (a) Fluorescence micrograph of a macrophage (J774.2 mouse cell line) that had taken up Lucifer Yellow under conditions of exposure and timing such that most of the dye should be in lysosomes. Numerous elongate tubules contain the dye. Their identity as lysosomes has been confirmed by enzyme cytochemistry and immunocytochemistry. From Swanson, J., Bushnell, A., and Silverstein, S. C. (1987) *Proc. Natl. Acad. Sci. USA* **84**:1921. × 1000 (approx.).

(b) Portion of an axon from an X-irradiated cultured rat ganglion showing electron microscopically demonstrated acid phosphatase in a tubular structure that extends a considerable distance along the axon. Segments of this structure are indicated by the arrows, and S indicates Schwann cells associated with the axon. From Holtzman, E., Teichberg, S. T., Abrahams, S. J., Citkowitz, E., Crain, S. N., Kawai, N., and Peterson, E. R. (1973) *J. Histochem. Cytochem.* **21**:349. × 10,000 (approx.).

tions suggest that endosomes can undergo "fission" into smaller bodies that then move independently of one another. These findings serve as counterweights to the evidence for continuities among seemingly separate lysosomes or endosomes (see Section 2.3.1.4); extensive connections do sometimes exist, but the bodies are not all tied into one large continuous network, as occasionally has been proposed.

Overall, the studies on living cells have made it evident that the intracellular endocytotic compartments of the early phases of heterophagy are quite dynamic, undergoing rapid fusions, fissions, and shape changes as they move within the cytoplasm.

2.4.2.2. Is Fusion All-or-None?

When macrophages or fibroblasts form pinosomes larger than the cell's preexisting lysosomes, microcinematography and videotape analyses of living cells seem to show that bodies of the size and appearance of lysosomes occasionally interact with the vacuoles as if nibbling away small chunks, by repeated transient "attacks." The images have been interpreted as showing that lysosomes fuse transiently with the larger vacuoles and then back off, carrying a portion of the vacuolar content (and membrane?). This would be a quite different sort of process from the "one-shot," almost "explosive" degranulations seen in cells such as neutrophils in which the granule contents are transferred to phagosomes rapidly and, apparently, completely (Fig. 2.21).

Transient, reversible fusions could permit somewhat selective transfers between the lysosomes and the endocytotic structures; small, rapidly diffusing molecules would be expected to cross temporary bridges more efficiently than larger molecules or molecules retarded by interactions, say, with the semi-organized contents of secondary lysosomes (Section 2.1.4.1). Several observations have been taken as showing that lysosomes do engage in differential transfers of some sort. For instance, lysosomes of cultured cells seem to transfer small dyes like Lucifer Yellow and Sulforhodamine 101 to endocytotic bodies, to one another, or eventually to the extracellular space, more rapidly than they transfer larger tracers like HRP and dextrans. And Sections 3.2.3 and 3.3 will emphasize that each of several compartments thought to interact by fusion

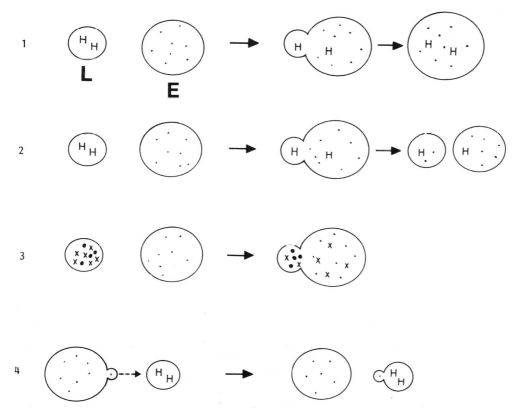

FIG. 2.21. Some proposed relations between endocytotic structures (E) and lysosomes (L). (1) The two types of structures may simply fuse, mixing the lysosomal hydrolases (H) with the materials (dots) carried within the endocytotic bodies. (2) Fusions might sometimes be transient: the two types of bodies separate shortly after fusing and only partial transfer of contents takes place during the encounter. (3) The bodies may fuse permanently but different components of their content may intermingle at different rates. (4) The bodies themselves might remain separate but transfer contents via vesicles; the vesicles might shuttle repeatedly back and forth, or merge with the recipient organelle for an extended time.

during heterophagy—the plasma membrane, endosomes, and lysosomes—maintains a distinctive membrane composition.

But only a few observers are yet persuaded that transient fusions are the rule or even occur widely. In certain of the cases where lysosomes seem to be transferring smaller and larger markers to endocytotic bodies at different rates, it may be that lysosomal fusion is persistent rather than transient but that transfer of the more readily diffusible of the lysosomal contents seems to take place first because these components mingle more rapidly with the contents of the endocytotic body (Figs. 2.4 and 2.21; see also Fig. 5.15 for the image of a secretion granule that seems to have retained its compact morphology after fusion with a lysosome-related body). Findings with dyes like Lucifer Yellow must be dealt with cautiously because, as already mentioned, the assumption that the dyes cannot cross membranes is not always valid (Section 3.4.1.2). If, as is very likely (see Section 2.4.3 and Chapter 3), small vesicles and tubules are continually budding from and fusing with the several compartments that participate in heterophagy, these vesicles and tubules could be vehicles both for partial exchanges of contents among the compartments and for selective movements of membranes. Similarly, impressions of transient fusions could come from situations in which fusions actually are complete but are followed rapidly by a new segregation of different components of the fused structures into separate compartments (Chapter 3).

2.4.3. Transfer of Contents: By Transport or by Transformation?

The last possibilities relate to broader concerns. In the earlier days of work on heterophagy, it was simply assumed that when endocytotic structures transferred their contents to lysosomes, the transfer was the result of direct fusions that resulted in a digestive chamber of hybrid (lysosomes plus endocytotic body) origin. This chamber, it was presumed, then evolved directly into a residual body that might later be defecated, reused as in Section 2.1.3.3, or stored. In favorable cell types, key elements of this story can be verified by viewing living material. Phagosomes, for instance, can be seen to fuse with lysosomes. But for pinocytotic processes such as receptor-mediated endocytosis, the grounding of assumptions about fusions and maturations is less firm. For example, though fusions among new endocytotic bodies and between such bodies and preexisting intracellular structures can be seen in living cells, frequently it is difficult to be sure of the nature of the participants [e.g., whether they do or do not contain acid hydrolases (see Section 2.4.4.1)].

A fair number of investigators therefore remain wary about assuming that endosomes actually fuse as such with preexisting secondary lysosomes or that early heterophagic structures physically mature into later ones. For example, upon finding that endocytosed materials in a given cell type are reliably detected in predictable time sequences within structures of different buoyant density, many cell fractionators speak of the "transfer of content" from low-density endocytotic structures to higher-density secondary lysosomes. This convention arose simply as an operational description of the findings but some investigators now speak consistently as though they believe the compartments to be stable and the content to pass from one long-lived structure to another (see Section 2.4.4.1). Could it be that endosomes and secondary lysosomes actually remain separate from one another and communicate by intermediary structures, such as vesicles that carry the contents of one structure to another, perhaps even shuttling repeatedly back and forth (Fig. 2.21)?

Such models echo prevailing views about how much of the transport is accomplished between other types of compartments—from the endoplasmic reticulum to the Golgi apparatus for example, or from one sac of the Golgi apparatus to another. And they can readily account for the maintenance of differences in membrane composition between communicating compartments like the plasma membrane, endosomes, and mature lysosomes (Sections 3.2.3 and 3.3). Even so, for many microscopists the proposed vesicular intermediaries have seemed superfluous. There is little clear positive evidence for stable endosomes through which endocytosed materials pass on their way to other structures. In contrast, one can observe movements and fusions of endosomes in living cells, and electron micrographs readily reveal all the intermediate structures expected for an assumed sequence in which endosomes with a vacuolar and tubular morphology convert directly into MVBs, acquire cytochemically demonstrable acid hydrolases, and by accumulating residues of digestion, mature into residual bodies. Still, confirmed "seeing-is-believing" types like myself are made especially hesitant by the strong evidence that movement of membranes to and from the surfaces of lysosomes and endosomes is far more extensive than we had originally suspected from our microscopic images (Section 3.3). If endosomes do transform into secondary lysosomes, as present opinion still favors on the whole, this "maturation" almost certainly occurs after portions of the endosome surface have separated off as vesicles and tubules that participate in processes not involving direct interactions with lysosomes (Section 3.3.4). Among these processes are membrane cycles through which the same membrane molecules repeatedly enter endosomes from the plasma membrane and then return to the cell surface, seemingly without moving on to the lysosomes. In this functional sense at least, membrane molecules of the endosomes do manifest features of prolonged persistence rather than of unidirectional flow to the lysosomes. It could also well be that some of the membranes budding from maturing endosomes persist as cytoplasmic vesicles that contribute to forming new endosomes. If this is so, then the distinctions between transformation of endosomes into lysosomes and communication by vesicular carriers may hinge upon the relative volumes of the tubules and vesicles that leave the endosomes versus the volumes of the endosomal "remainder" that acquires hydrolases. In other words, it may devolve upon semanticists to decide which description better fits the process.

At the time when new buds form from yeast zygotes (Section 7.5.2.2), the vacuoles contributed by the two parental cells can exchange an endogenously produced fluorescent material, believed to be a polymer derived from adenine precursors, without fusions. The vehicles of this exchange are presumed to be small vesicles, although this has not yet been demonstrated directly.

2.4.4. Timing

The timing of heterophagic fusions and of lysosomal digestion of heterophagocytosed materials varies considerably among cell types. With mammalian phagocytes, fusions of phagosomes with lysosomes are observed by light microscopy of living cells within one or a few minutes of the onset of phagocytosis. In *Paramecium* at room temperature, about 8 min elapses between separation of digestive vacuoles from the cell surface, and the fusion of acid hydrolase-containing bodies with the vacuoles. In the protozoan *Entamoeba histolytica,* it takes many hours before lysosomes accumulate appreciable levels of endocytosed materials; the cells behave as if their endocytotic apparatus has only minimal interest in the lysosomes (Section 3.3.4). And some cells store endocytosed molecules in more-or-less intact form for weeks, or months, before exposing them to the full fury of the degradative system (Section 4.2).

2.4.4.1. More Trouble with Endosomes: Early Acquisition of Hydrolases; Presumed Sequences of Transformation; MVBs Again

For receptor-mediated endocytosis and related forms of pinocytosis, questions of the timing of the processes leading to the entry of endocytosed material into degradative compartments are caught up in the uncertainties about mechanisms considered in Section 2.4.3. There is wide agreement that the materials appear sequentially in: (1) "early endosomes" (tubular and vacuolar structures isolable in cell fractions of relatively low density); (2) "late endosomes" (these often include MVBs; their density varies); (3) "definitive" secondary lysosomes [hydrolase-rich bodies of distinctly higher density than most endosomes (Fig. 1.5) and of appearance ranging between MVBs and residual bodies].

Only for the last category of structures is there general agreement about the presence of lysosomal enzymes: the hydrolases are easily demonstrable, biochemically, cytochemically, and immunocytochemically. Cell fractions enriched in early or late endosomes often do contain some hydrolase activity, but at levels low enough as to make it very difficult to decide whether this is really endosomal or reflects contamination of the fractions. Cytochemistry and immunocytochemistry frequently detect hydrolases in some of the MVBs or other endosome-like bodies in a given cell, especially in bodies of the "late" variety. But there are as yet no clear criteria for deciding whether the reactive bodies are best thought of as endosomes with some hydrolases (see below) or instead, as lysosomes newly charged with their acid hydrolases by fusion with preexisting lysosomes. Analogous problems, arising from the fact that the structures being studied are changing rapidly, afflict cell fractionation. Fraction density is often used as a key criterion for classifying organelles as "endosomal" or "lysosomal." But there is a lack of solid information from which to decide whether alterations in the buoyant density of structures in the heterophagic pathway reflect changes in the structures themselves, fusions among structures of different densities, or passage of the tracers being used from bodies with one range of densities into bodies of another range. How, for example, is one to predict the densities expected for hybrid bodies formed by the fusion of dense lysosomes with less dense endosomes, without knowing much more than we do about the state of the contents of the hybrids and about their osmotic and ion-transport properties?

For a while, most investigators took it for granted that fusions of endosomes with primary or secondary lysosomes are rare in the first few minutes after endocytotic internalization. The magnitude of this "delay" was estimated from findings that 5–10 min or more elapses before significant amounts of the content of endosomes of macrophages, perfused liver, cultured fibroblasts, and other mammalian cells, begin to appear in definitive lysosomes identified biochemically or cytochemically. As much as 30 min or more may pass before the bulk of a pulse of endocytosed material is present in compartments in which degradation to low-molecular-weight products is extensive. (The spread in these times is conceived of as a sort of probabilistic distribution mirroring the likelihoods of encounters between incoming endocytotic bodies and preexisting lysosomes.) The existence of an appreciable delay before endosomes containing newly endocytosed molecules interact with lysosomes seemed confirmed by the fact that exposure of cells to suitable inhibitory conditions, such as lowered temperature, can slow or minimize entry of the endocytosed materials into those compartments that are unambiguously identifiable as lysosomes, even when the materials are readily detectable in early, intracellular endocytotic structures (Section 2.4.7).

The fly in this ointment is that certain hydrolytic activities do seem to be present in the compartments that endocytosed materials enter within a minute or two after their uptake. Both small synthetic protease substrates, and natural substrates, such as EGF, undergo proteolytic cleavages within a very few minutes of the initiation of their uptake by mammalian cultured cells. Endocytosed phosphatase substrates also are split rapidly by these cells. Macrophages similarly initiate degradation of certain endocytosed proteins very rapidly. Hepatocytes begin to hydrolyze some molecules of endocytosed asialoglycoproteins with no appreciable lag after uptake, though for most of the molecules taken up at a given time, 10–20 min passes before degradation is detectable.

At least some of the enzymes responsible for these activities have acid pH optima, though others, like ones that can fragment proteoglycans, may be active near neutrality. For macrophages, an enzyme that closely resembles lysosomal cathepsin D is detectable in cell fractions enriched in the structures in which endocytosed proteins appear within 1–2 min after their uptake. But other hydrolytic activities characteristic of lysosomes, notably the glycosidases, are very difficult to find in "early" endosomes; many are so dilute, in endosome-rich cell fractions, that present analytical approaches cannot distinguish their levels from background.

The proteolytic capacities of endosomes considerably blur the distinction between what were supposedly prelysosomal compartments and lysosomes. Some important potential biological implications will arise in Section 4.2.4.2 and closely related issues will be discussed again in several other contexts (e.g., Sections 3.1.3 and 7.3.2.2). For now the following points are central.

1. It is likely that the delivery of hydrolases and the evolution of endosomes are spread out in time so that different endosomes containing tracers taken up over a given interval will come to be heterogeneous in enzyme content and other properties. The implications of this for the matters at hand have not yet been adequately explored.

2. Purification of endosomes by seeking fractions with minimal hydrolase content—as some of the methods are designed to do—needs reexamination.

3. Very likely, even if they do contain hydrolases, early endosomes cannot engage in extensive or promiscuous degradation: In several of the key cases studied, only a few peptide bonds are hydrolyzed during the initial minutes after endocytosis. Furthermore, viruses and other biologically active materials that are rapidly inactivated by lysosomes survive a sojourn within endosomes (Section 3.1.3.3). Perhaps the pH or other conditions in the endosomes do not favor full activity of some of the enzymes present (Section 3.1.3.1), or perhaps the enzymatic spectrum present in endosomes truly is much more limited than that in lysosomes.

4. The source(s) of the enzymes in early endosomes are not known. Several candidates are obvious on *a priori* grounds: lysosomes; other endosomes; endocytotic vesicles transporting endocytosed hydrolases bound to M6P receptors (Sections 6.4.1.4 and 7.2.3); vesicles and tubules carrying membrane shuttling

back and forth between endosomes and other compartments (Section 3.3). If it is primary lysosomes (Sections 2.1.3.1, 2.1.3.2, and 7.3.2) that provide the enzymes and if endosomes do contain only a few species of active hydrolases, one might postulate the existence of a specialized primary lysosomal subpopulation bearing only a few types of lysosomal enzymes. Sections 7.1.1., 7.4.3, and 7.5.1 will outline information suggesting the possibility that different enzymes could be delivered to nascent lysosomes at different times; hydrolase-containing endosomes might represent an early stage in this sequence. It might instead be argued, as above, that the full lysosomal complement of enzymes is delivered at the same time but that for one reason or another, many types of hydrolases entering the endosomes are inactive, at least for a while (see Section 7.4.1.1). Perhaps more likely is that the charge of hydrolases received by an endosome when it fuses with a primary or secondary lysosome includes most or all of the enzymes characteristic of lysosomes (see Chapter 7) but that somehow, while most of these hydrolase molecules soon come to locate in more mature structures, molecules of certain of the enzymes persist in the endosome population for longer times than others. (For example, departure might be delayed because the enzymes are bound to membranes and thus participate in some phase of the membrane recycling to be discussed in Chapter 3. The macrophage's "endosomal cathepsin D" seems to differ from its lysosomal counterpart in being tightly attached to membranes even though the endosomal enzyme appears to be a precursor of the lysosomal cathepsin; see Section 7.4.)

5. Few students of lysosomes are ready yet for wholesale abandonment of conventional wisdom, which holds that distinguishing "definitive" lysosomes from "prelysosomal" endosomes is still conceptually worthwhile, even if the two categories eventually turn out to designate overlapping stages in a continuum. Operationally, endosomes are usually discriminated from definitive lysosomes by their lower degradative activity, their higher pHs (especially in early endosomes; Section 3.1.3.4), their morphology (see below), and their lower density. Low temperature (Section 2.4.7.1) and agents that raise intra-compartmental pHs (Section 3.3.4.2) are used to retard or inhibit the transitions by which endocytosed ligands pass from exposure to an "endosomal environment" to exposure to a "lysosomal environment."

For many investigators, the MVBs represent the "dividing line" or "transitional state" between endosomes and lysosomes—they are endosome-related but at least some of them clearly do acquire a range of lysosomal hydrolases and initiate extensive digestion. In hepatocytes, cultured fibroblasts, and other cell types, the "early" vacuolar and tubular endosomes are found largely near the cell periphery while forms of the MVB type are common deeper in the cells, appearing, for example, in the vicinity of the Golgi apparatus or in the juxtanuclear cytoplasm of cultured cells. Many of the MVBs containing tracer molecules relatively deep in the cell still lack acid hydrolase activities demonstrable with conventional techniques for acid phosphatase or other enzymes. These bodies are thought to represent maturing endosomes just before they become lysosomes. Others of the MVBs have a spectrum of lysosomal hydrolases demonstrable by enzyme cytochemistry and by immunocytochemistry [M6P receptors are present in some, as well (Section 7.3.2.2)]; presumably these structures have recently fused with lysosomes or hydrolase-transport vehicles of one type or another (see Section 7.3.2.2) and now will transform, probably rapidly (Section 5.1.5.2), into residual bodies (see Fig. 2.3).

As expected for this sequence of events, when hydrolases such as acid phosphatase or aryl sulfatase are demonstrable cytochemically in MVBs participating in heterophagy, the enzymes are visualized in the space between the internal vesicles (Fig. 2.15). This is the same space as is occupied by endocytosed materials.

Summarizing the last two sections: Much of what is known about intracellular transfers of the hydrolases and other contents of lysosomes, about the control of lysosome numbers, and about the effects of various inhibitory conditions is still tentatively rationalized by the majority of observers in the following terms: Endosomes containing very recently endocytosed materials exhibit, at most, limited hydrolytic activities. These bodies mature into structures, often of the multivesicular type, that transform into active degradative compartments partly as a result of fusions with preexisting hydrolase-containing bodies (see Section 7.3.2.2). The degradative compartments—definitive secondary lysosomes—in turn mature into residual bodies.

2.4.4.2. Fusion Sequences: Changes in Fusion Capacities?

The sequence of fusions and membrane departures undergone by *Paramecium* digestive vacuoles before they fuse with lysosomes (Fig. 2.5) significantly transforms the vacuolar membrane in microscopic appearance and immunocytochemical reactivity (Section 3.2.3.3); perhaps this transformation is needed to prepare the vacuoles for encounters with the lysosomes.

Similarly, on the assumption that there actually is a delay of a few minutes before endosomes and related bodies begin to fuse with preexisting lysosomes, some researchers posit functional transformations of early endosomes affecting their inherent capacities to fuse with other bodies. In concert with their morphological evolution, the endosomes seemingly undergo a progressive decrease in internal pH (Sections 3.1.3 and 3.1.5.4) and there are major changes in the association of ligands with receptors (Section 3.3.4.2). Are such changes obligatory as preludes to fusion with older lysosomes? Are they, that is, accompanied by timed alterations in the actual **competence** of different structures to fuse with one another—endocytotic structures with endocytotic structures, and then endosomes with lysosomes?

Experiments to evaluate the fate of endocytotic tracers taken up at two closely spaced time points have yielded conflicting data on the extent to which older endocytotic bodies fuse with newer bodies or the degree to which very new endocytotic structures fuse directly with old secondary lysosomes. Endocytotic structures formed at a given time do often seem to mix their contents with structures formed a minute or two later. But several observers have concluded that at least a proportion of the molecules taken in at a given time point soon enter structures that are not readily accessible to endocytosed molecules taken up 5–10 min later (see also Section 3.3.4). In addition, it is claimed that observations on living cells show that when newly formed endosomes encounter bodies thought (from their size and behavior) to be mature secondary lysosomes, fusions fail to take place if the endosomes are too young. And new pinosomes have been reported not to fuse directly with new phagosomes in the same cell, despite the fact that pinosomes and phagosomes eventually contribute their content to the same lysosomes.

None of these observations, however, provide clear-cut evidence that the timing of heterophagic fusions depends principally on inherent capacities of the participants. Much of the timing of fusions could be a result of the geometry of the cell, and of patterns of intracellular motility. Early endocytotic structures clustered near the cell surface may be more likely to fuse with one another than with lysosomes in part because newly made endocytotic neighbors are present in relative profusion whereas there are not many lysosomes nearby. As endosomes move deeper into the cell, they may become less immediately accessible to newer endocytotic bodies while entering regions where lysosomes are more available. For instance, many endocytotically active cell types in multicellular organisms focus endocytosis at particular cell poles relatively far from most of their lysosomes. (The fact that small quantities of endocytotic tracers sometimes are detected in lysosomes within a very few minutes after uptake might reflect the presence of occasional lysosomes in appropriate locales. Even in nerve terminals that are far from the site of most of the cell's lysosomes, at least a few of the MVBs contain acid phosphatase.) The migrations of digestive vacuoles in *Paramecium* mentioned at the beginning of Section 2.4, bring the vacuoles into different cytoplasmic regions at different stages in their history; the other organelles that interact with the vacuoles show organized patterns of movement as well so it may be that coordinated movements are as important as inherent fusion capacities in governing the organelle fusions.

In rabbit neutrophils, alkaline phosphatase—an enzyme characteristic of the specific granules—is demonstrable cytochemically in new phagocytotic digestion vacuoles a few minutes before peroxidase and other azurophilic granule (lysosome) contents are detectable. The sequence of granule fusion—specifics before azurophils—implied by these last observations may be tied to the lapse of several minutes that occurs before the pH in new phagocytotic vacuoles falls into the optimal range for acid hydrolase activity (Section 3.1.3.1). That the two neutrophil granule types differ in membrane proteins (Section 3.2.2) obviously might be germane to regulation of the fusions. But there is no strong evidence that azurophilic granules are actually precluded from fusing with phagosomes until fusions with the specific granules have begun. The specific granules are two or three times more numerous and are smaller (hence more mobile?) than the azurophilic granules so perhaps they fuse first simply because they have more opportunities.

These kinds of explanations should not, of course, be taken to extremes. Section 2.4.6 will summarize the strong evidence from *in vitro* studies that endosomes do undergo changes in their intrinsic fusion capacities. And factors of abundance and mobility probably cannot account entirely for the differences in fusion behavior between the neutrophil granule types: When exposed to the proper concentrations of phorbol esters or to other stimuli (see Section 4.4.1.2), neutrophils can be induced to exocytose the contents of specific granules while retaining the lysosomal hydrolases and other constituents of their azurophilic granule. And specific granules undergo exocytosis at lower concentrations of Ca^{2+} than are needed for release of azurophilic granules. Such differential secretion argues that the controls governing fusions in neutrophils can distinguish between the different granule types.

2.4.5. Cytoskeletal Involvement

As emphasized in the beginning of Section 2.4, the patterns of organelle movement that precede heterophagic fusions are not random. Figure 2.22 and

FIG. 2.22. When cultured rat ovarian granulosa cells take up fluorescent ligands—derivatives of LDL or of Con A—and then are studied by fluorescence microscopy using time-lapse video intensification methods, the label is seen to accumulate in endocytotic structures ("endosomes") that move about in saltatory fashion for a few minutes, but then move in a seemingly steadier fashion to the cell center (the region near the nucleus where the Golgi apparatus and centrioles, among other organelles, are located). (Some of the other studies of

movement of endosomes toward the nucleus report that all phases are saltatory.) Meanwhile, "lysosomes" (identified as phase-dense bodies that accumulate the dye Acridine Orange; Section 3.1.2), previously having been distributed more or less randomly in the cytoplasm, migrate to the cell center; both before and after presentation of the ligands to the cell, the lysosomes move largely by saltation. At the cell center the lysosomes, many of which get there before most of the endosomes arrive, "await" the endosomes and fuse with them. (Characteristics of the movements are given in Table 2.1.)

Shown is a light (Nomarski) micrograph illustrating the distribution of lysosomes in the presence of taxol; this microtubule-stabilizing agent causes the lysosomes to congregate in regions near the nucleus (N), where they fall into linear arrays, along "fibrils" (arrows) that probably are bundles of microtubules. The lysosomes can migrate in either direction along these fibrils. Bar = 10 μm. From Herman, B., and Albertini, D. F. (1984) *J. Cell Biol.* **98:**565.

Table 2.1 present data from one influential investigation of the mechanism of this motion. Other studies suggest that details and observed rates of the motion vary among cell types and depend to an extent on experimental design and observational techniques. But most agree that, as viewed in living cells, the movements of secondary lysosomes and of recently formed endocytotic bodies include saltations (rapid, linear movements followed by pauses or slower movements) of the sort often reported for motion of structures along paths

Table 2.1. Characteristics of Organelle Movement in Cultured Granulosa Cells[a]

Organelle	Type of movement	Treatment[b]	Mean velocity (μm/sec ± SEM)	Mean distance (μm/sec ± SEM)
Lysosome	Saltatory	None	0.30 ± 0.03	8.7 ± 0.9
	Saltatory	Taxol	0.12 ± 0.02	4.4 ± 0.4
	Brownian	Nocodazole	0.09 ± 0.01	2.2 ± 0.2
Endosome	Saltatory	None	0.03 ± 0.01[d]	4.6 ± 0.7
	Centripetal	None	0.05 ± 0.004	36.6 ± 1.1
	—[c]	Taxol	—	—
	—	Nocodazole	—	—

[a] In the case of saltations, the distance moved and the velocity are means per saltations. From Herman, B., and Albertini, D. F. (1984) *J. Cell. Biol.* **98:**565.
[b] Taxol is a microtubule-stabilizing agent; nocodazole is a microtubule destabilizer.
[c] Movement inhibited completely.
[d] Studies on other cell types in which endosomes were found to move largely by saltation have estimated the rates of movement during each jump at about 1 μm/sec [see especially De Brabander *et al.* (1988) *Cell Motility* **9:**30].

delineated by microtubules. Agents that interfere with microtubules can markedly perturb the movements of heterophagic structures and certain of the agents that promote disassembly of the tubules—nocodazole, for example—inhibit delivery of endocytosed materials to definitive secondary lysosomes, at least sometimes. Such "antimicrotubule" agents also reduce the experimentally induced secretion of hydrolases from neutrophils, though this may be more a question of the agents disorganizing the geometry of the cell than a matter of their inhibiting movements. The elongate lysosomes of Fig. 2.20a owe their distribution in the cell to alignment along microtubules; fixing the cells with customary preservatives at conventional low temperature leads to fragmentation of the lysosomes perhaps in part because of suboptimal preservation of the microtubules. After injury or stress, elongate lysosomes also are found in axons of neurons, paralleling the longitudinally oriented microtubules (Fig. 2.20b). The large vacuoles in yeast cells fragment into smaller structures when the cells are exposed to nocodazole.

Cell biologists have been confident for some time that microtubules help **orient** movements of other structures. Recent studies on transport in axons and other cell regions have lent much strength to the idea that the tubules also **propel** structures; it now seems clear that, with the intervention of soluble components, microtubules are capable of long-distance translocation of membrane-delimited bodies in both directions along the length of the tubules. Endocytotic structures that move up axons were among the bodies for which such microtubule-motivated movements were first suspected.

One can, therefore, easily imagine that microtubules orient movements involved in heterophagy and otherwise participate in controlling the locations of various participants; in so doing, the microtubules could contribute, for example, to promoting fusions among lysosomes and endosomes and related organelles. As already indicated, students of many cell types have concluded that lysosomes tend to aggregate near the Golgi region or, more broadly, in the juxtanuclear region, and that endocytotic bodies migrate preferentially to this region (in the experiments from which Fig. 2.22 and Table 2.1 were taken, endosomes and lysosomes both seemed to migrate to the Golgi region after the endocytosed ligands were presented). These migrations—which presumably enhance the probability of fusions among endosomes and with lysosomes—might be generated by the sets of microtubules that radiate from the Golgi region into the cytoplasm. The microtubules focused on the Golgi region probably gain their orientation from the centrosomal materials, which act as microtubule organizing centers: in many cell types of animals the centrioles, which form part of the centrosomes, are located near the Golgi apparatus (see Fig. 2.2). But in cultured NRK cells, bodies thought to be endosomes, and perhaps lysosomes as well, can move in both directions along the microtubules radiating from the Golgi apparatus region. Thus, explanations must still be found for the seeming clustering of secondary lysosomes near the apparatus, for the apparent net transport of endocytotic bodies toward the clusters, and for findings that the lysosomes disperse into more random distributions when microtubules are disrupted experimentally. Dispersal is also seen, normally, during cell division when the distribution of the cell's microtubules changes as the mitotic spindle is assembled. One line of speculation is that relatively large bodies, like lysosomes, can contact several microtubules simultaneously where the microtubules are closely spaced, as in the Golgi region. This might impede the migration of bodies away from the Golgi zone, in contrast to the freer migration in the periphery of the radiating arrays of microtubules, where the tubules are too widely spaced to permit multiple contacts with a given lysosome or endosome.

Agents that inhibit the production of ATP reduce or abolish the movements of endosomes and the other heterophagic structures but, as compared with agents acting on microtubules, cytochalasins and antibodies against intermediate filaments have much less effect on movements of heterophagic structures. It is, however, premature to say that microtubules alone are responsible for all aspects of the movement of lysosomes or endocytotic bodies. As Section 2.3.2 summarized, microfilaments are involved in early phases of at least some forms of endocytosis. And there have been sporadic reports of associations of microfilaments, or elements of the still-controversial "microtrabecular" lattice, with lysosomes or with endocytotic structures moving in the cytoplasm. Claims also have been made that the membranes bounding lysosomes bind actin and in a few of the experiments in which perturbation of microtubules has been used to alter movements of lysosomes or endosomes, changes in the distribution of intermediate filaments accompany the alterations in the microtubules. At a minimum, these findings are reminders that the several subsystems of the cytoskeleton interact with one another.

Section 2.3.2.3 alluded to the actin-rich filament system that closely surrounds newly forming phagosomes. Some of the membrane-delimited granules or vesicles that fuse with newly forming phagosomes seem to adhere initially to the filament meshwork. But for phagosomes and lysosomes to approach one another close enough to fuse, the meshwork must be substantially modified. In fact, when neutrophils degranulate during frustrated attempts to phagocytose particles attached to coverslips, or after adhering to surfaces coated with immune complexes, the cytoplasmic side of the plasma membrane becomes locally denuded of its filamentous covering at the cell surfaces adhering to particles or opsonized substrate. The denuding spreads from focal points of initial disaggregation or loosening of the meshwork, and it is followed rapidly by fusion of granules with the membrane.

2.4.6. Cell-Free Fusion

Section 2.3.2.8 mentioned that induction of fusions of coated vesicles in cell-free systems requires that the clathrin coats be dismantled. In these studies, Ca^{2+} was required for the uncoated vesicles (isolated from brain) to fuse with isolated kidney lysosomes.

Fusions of phagocytotic vacuoles in crude homogenates of *Acanthamoeba* have been monitored by mixing homogenates from cells that had phagocytosed recognizably different particles (yeast, erythrocytes, or latex beads) and counting the numbers of vacuoles that subsequently show mixed content. ATP and cyclic AMP were found to increase the numbers of fusions observed. Vacuoles with mixed content also were more frequently found if the homogenates were allowed to age for a few hours before being mixed, as if some sort of "maturation" facilitates fusion.

Cell-free fusions in mixtures of endosomes or related endocytotic structures from cultured mammalian cells have been demonstrated by using electron microscope or immunological tracers or by demonstrating that suitable components taken up initially into different compartments come to interact with one another *in vitro*, enzymatically, or immunologically. ATP and "cytosolic" (soluble) factors are needed for the fusions. Low temperature inhibits the fusion; there are preliminary signs of a threshold near 20°C, like that seen in intact cells. In some of the experiments, requirements for added Ca^{2+} have not been evident; nor has it been essential to maintain the low intracompartmental pH normally characteristic of endosomes (Chapter 3; see Section 2.4.7.2). In at least one case, however, Ca^{2+} was found to induce endosomal fusions in the absence of ATP, although when ATP was present, Ca^{2+} had little stimulatory effect. Studies on fusions among the endocytotic structures in which tracers are detected at different times after uptake (presumably corresponding to the "early" and "late" structures of Section 2.4.4.1) suggest that endosomes do undergo changes in inherent fusion proclivities as they evolve. In the cell-free preparations, early endocytotic structures can fuse with one another and with slightly "later" structures, but they fuse poorly, if at all, with endocytotic bodies a few minutes older, or with mature lysosomes.

2.4.7. Perturbing Fusion

2.4.7.1. Low Temperature

With cultured mammalian cells, the entry of endocytosed materials into compartments that degrade them extensively, largely ceases when the temperature is lowered to 20°C or a bit below; the materials accumulate within compartments that look like endosomes microscopically and behave like endosomes in cell fractionation. This inhibition of "transfer to lysosomes" is reversed when the temperature is returned to normal. With *Paramecium*, delivery of cytochemically demonstrable hydrolases to newly formed phagocytotic digestive vacuoles is drastically slowed when the temperature falls below 20–22°C.

These effects have widely been regarded as due chiefly to the inhibition of fusions of recently formed endocytotic bodies with lysosomes or related hydrolase-transport vehicles. Correspondingly, lowering the temperature to 20°C or below has often been employed as a convenient experimental means for impeding such fusions with relatively little perturbation of other processes of heterophagy.* But though low temperature clearly does have dramatic effects on the evolution of heterophagic structures, Section 2.4.4.1 indicated that the delivery of hydrolases to endosomes now seems more complex than once was thought and thus there is less confidence that the effects of low temperature are adequately understood. Some findings suggest, for example, that very early endosomes in cultured mammalian cells still can exhibit proteolytic activity at temperatures below 20°C. It may be that the inhibitions of fusions sometimes are only partial—a slowing of fusion, rather than its complete cessation. And a number of investigators are exploring possibilities that temperature-induced

---

*In cultured mammalian cells held at 20°C or a bit less, digestion of material already in the lysosomes is not prevented, pinocytosis continues at an appreciable rate, and the major steps of the "prelysosomal" evolution and migration of endosomes are still observed. Acidification of lysosomes and endosomes may, however, be slowed or impaired. In addition, some of the recycling of membrane from endosomes to the cell surface may be impeded, which would slow the normal evolution of the endosomal membrane (Chapter 3), perhaps thereby affecting endosomal fusions or other functions.

In macrophages, the larger classes of endocytotic vacuoles largely cease their saltatory motion at temperatures below 25°C. Movement of smaller bodies, which some studies report to be less obviously saltatory than that of vacuoles, is less affected by the low temperature.

reductions in degradation of endocytosed materials are partly a matter of changes in the cycling of membranes from the endosome surface (Chapter 3), or of a modified environment within degradative structures [e.g., failure to establish or maintain a low enough pH (see footnote on p. 88)]. It also has been found that low temperature affects Golgi functions, which could influence the delivery of newly made hydrolases (Section 7.2.2.2). Still, it is not yet time to go overboard in doubting that low temperature actually does retard heterophagic fusions significantly—the mixing of contents of different secondary lysosomes, for example, does seem to be profoundly inhibited in mammalian cells at 15–20°C.

2.4.7.2. Lectins; Polyanions; pH Changes

The presence of certain materials within lysosomes or endocytotic structures, sometimes changes the apparent fusion behavior of these structures. These effects are particularly interesting because they could reflect membrane changes that originate in the interior of the affected compartments, where the substances are present, but become expressed on the external surface, where potential fusion partners confront one another. Speculative interpretations have focused on proposed conformational changes in transmembrane proteins and on possible shifts in membrane potential, realignments of membrane lipids, modifications of membrane fluidity, or displacement of Ca^{2+} from binding sites in the membrane.

The experimental literature is, lamentably, too full of contradictions and conflicts for ready testing of such hypotheses. For instance, when macrophages are exposed to the lectin Con A (see footnote on p. 63), uptake of presumed fluid phase markers is enhanced, but the Con A-containing pinosomes fail to fuse with lysosomes, as evaluated both by acid hydrolase cytochemistry and by preloading the secondary lysosomes with recognizable markers. This inhibition of fusion is relieved by exposure of the cells to mannose, a sugar that presumably binds to Con A and in so doing frees the cells' macromolecules from the lectin; thus, Con A's inhibition of fusion is attributed to binding of the lectin to glycoconjugates at the cell surface and the persistence of this binding in endocytotic structures. But in contrast to its inhibitory effect on fusions involving pinosomes, Con A has little effect on lysosome fusions with macrophage phagosomes. Additionally complicating the picture, Con A can alter the form or fate of endocytotic structures, so that they give rise to extensive interconnected networks. And in cytochalasin B-treated neutrophils, rather than inhibiting fusions, it helps induce the selective exocytosis of specific granules.

For a time, polyanions and agents that raise intracompartmental pH were used to affect heterophagic fusions, but the extensive literature on such effects too has come to be increasingly confusing. When macrophages, for example, are grown in the presence of polyanions that enter their lysosomes endocytotically—dextran sulfates, the polysulfonate drug **suramin,** or sulfated glycolipids from bacterial cell walls—the passage of phagocytosed materials into lysosomes seemingly is inhibited. (The intracellular processing of fluid phase pinocytotic tracers sometimes is only minimally affected.) In such preparations, the transfer of fluorescent vital dyes, or of electron-dense tracers, from secondary lysosomes to newly forming phagosomes appears to be drastically reduced, leading investigators to assert that the polyanions directly inhibit lysosomal fusions. Certain of the bases known to raise lysosomal pH, such as chloroquine and ammonium ions (Section 3.1.2.1), were thought also to alter rates of fusion, judging from their impact on transfers of lysosome-borne tracers to endocytotic compartments—either enhancements or inhibitions of the transfers could be produced, depending on which amine was used and at what concentration. Guesses about the mechanisms of such effects centered, for example, on the likely impact of pH changes on the behavior of endosomal membranes (see Chapter 3). Recently, however, strong doubt has arisen as to whether polyanions, or changes in the pH within lysosomes or endocytotic bodies actually influence fusion processes *per se.* Several of the agents and conditions found to decrease the rates or extent of transfer of dyes or comparable tracers among compartments have now been shown also to retard or abolish saltatory motions and other movements of the participating organelles; thus, lysosomes and endocytotic bodies might fail to fuse simply because they fail to encounter one another. In addition, the considerations raised in Section 2.4.2 have eroded confidence in the methodology used to estimate fusion rates, especially by light microscopy.

Among the most intriguing agents affecting heterophagic lysosomal fusion, are the microorganisms that dwell within persistent vacuoles in professional phagocytes or other cell types. Certain of these microorganisms suppress the propensities of their dwellings to fuse with the lysosomes of the host cells. Present ignorance about how they do this will be summarized in Section 4.4.3.3.

Acknowledgments

This chapter owes much to recent communications from and discussions with D. Albertini (microtubules), R. D. Allen (protozoa, especially *Paramecium*), D. Bainton (neutrophils), B. Bowers (*Acanthamoeba*), W. J. Brown (M6P receptor), P. D'Arcy-Hart (lysosome movement and fusion and perturbations thereof), S. Diment (proteases in endosomes), M. G. Farquhar (neutrophils; M6P receptor), A. Fok (protozoa, especially *Paramecium*), H. Geuze (immunocytochemistry), M. Goren (lysosomal fusions), K. Haussman (protozoa), A. Helenius (endocytotic mechanisms; viral uptake), K. Howell (endosomes *in vitro*), J. Kaplan (endocytotic mechanisms), F. Maxfield (endosomes), I. Mellman (endosomes), I. Pastan (endocytotic mechanisms), S. Silverstein (phagocytosis; macrophages), J. Slot (immunocytochemistry), P. Stahl (receptor-mediated endocytosis; hydrolases in early endosomes), B. Storrie (early endosomes; behavior of endocytotic tracers), J. Swanson (endocytotic mechanisms), M. Willingham (endocytotic mechanisms).

Contacts, in earlier years, with Z. Cohn, R. Steinman, M. Muller, and members of their research groups were valuable in informing me about endocytosis and lysosomes in macrophages, fibroblasts, and protozoa. P. Oates and O. Touster were generous in clarifying information about fusions in *Acanthamoeba.*

In addition to providing specific information, as acknowledged above, B. Storrie, J. Swanson, and F. Maxfield read much of the chapter and helped considerably with their comments.

Further Reading

Ashwell, G., and Harford, J. (1982) Carbohydrate specific receptors of the liver, *Annu. Rev. Biochem.* **51**:531–554.

Baggiolini, M., Horisberger, U., Gennaro, R., and Dewald, B. (1985) Identification of three types of granules in neutrophils of ruminants, *Lab. Invest.* **52**:151–158.

Bainton, D. F., Nichols, B. A., and Farquhar, M. G. (1976) Primary lysosomes in blood leukocytes, in *Lysosomes in Biology and Pathology* (J. T. Dingle and R. T. Dean, eds.), Vol. 5, North-Holland, Amsterdam, pp. 3–32.

Bowers, B., and Olszewski, T. E. (1983) *Acanthameba* discriminates internally between digestible and indigestible particles, *J. Cell Biol.* **97**:317–322.

Boyles, J., and Bainton, D. F. (1981) Changes in plasma membrane associated filaments during endocytosis and exocytosis in polymorphonuclear leukocytes, *Cell* **24**:905–914.

Braell, W. A. (1987) Fusion between endocytic vesicles in a cell free system, *Proc. Natl. Acad. Sci. USA* **84**:1137–1141.

Bretscher, M. S. (1987) How animal cells move, *Sci. Am.* **257**(6):72–90.

Buckmaster, M. J., Ferris, A. L., and Storrie, B. (1988) Effects of pH, detergent and salt on aggregation of CHO cell lysosomal enzymes, *Bioch J.* **249**:921–930.

Chapman-Andresen, C. (1977) Endocytosis in freshwater amebae, *Physiol. Rev.* **57**:371–384.

D'Arcy Hart, P., Young. M. R., Jordan, M. M., Perkins, W. J., and Giesow, M. J. (1983) Chemical inhibitors of phagosome–lysosome fusion in cultured macrophages also inhibit saltatory lysosomal movement, *J. Exp. Med.* **158**:477–492.

Davy, J., Hurtley, S. M., and Warren, G. (1985) Reconstitution of an endocytic fusion event in a cell free system, *Cell* **43**:643–652.

De Brabander, M., Nuydens, R., Geerts, H., and Hopkins, C. R. (1988) Dynamic behavior of the transferrin receptor followed in living epidermoid carcinoma (A431) cells with nanovid microscopy, *Cell Motility* **9**:30–47.

Diaz, R., Mayorga, L., and Stahl, P. (1988) In vitro fusion of endosomes, following receptor mediated endocytosis, *J. Biol. Chem.* **263**:6093–6100.

Diment, S., Leech, M. S., and Stahl, P. (1988) Cathepsin D is membrane associated in macrophage endosomes, *J. Biol. Chem.* (in press).

Dunn, W. A., Hubbard, A. L., and Aronson, N. N. (1980) Low temperature selectively inhibits fusion between pinocytic vesicles and lysosomes during heterophagy of ^{125}I-asialofetuin by the perfused rat liver, *J. Biol. Chem.* **255**:5971–5978.

Dunn, W. A., Connolly, T. P., and Hubbard, A. L. (1986) Receptor mediated endocytosis of epidermal growth factor by rat hepatocytes: Receptor pathway, *J. Cell Biol.* **102**:24–36.

Fenchel, T. (1980) Suspension feeding in ciliated protozoa: Structure and function of feeding organelles, *Arch. Protistenkd.* **123**:239–260.

Ferris, A. L., Brown, J. C., Park, M. D., and Storrie, B. (1987) Chinese hamster ovary lysosomes rapidly exchange content, *J. Cell Biol.* **105**:2703–2712. (See also *Proc. Nat. Acad. Sci. USA* **85**:3860–3864, 1988.)

Fok, A. K., and Paeste, R. M. (1982) Lysosomal enzymes of *Paramecium caudatum* and *Paramecium tetraurelia*, *Exp. Cell Res.* **139**:159–169.

Froese, A., and Paraskevas, F. (eds.) (1983) *Structure and Function of Fc Receptors*, Dekker, New York.

Goldstein, J. L., Brown, M. S., Anderson, R. G. W., Russell, D. W., and Schneider, W. J. (1985) Receptor mediated endocytosis, *Annu. Rev. Cell Biol.* **1**:1–40.

Goren, M., Vatterm, A. E., and Fiscus, J. (1987) Polyanionic agents as inhibitors of phagosome–lysosome fusion in cultured macrophages: Evolution of an alternative interpretation, *J. Leukocyte Biol.* **41**:112–121. (See also *J. Cell Biol.* **104**:1749–1756, 1987.)

Gruenberg, J. E., and Howell, K. E. (1986) Reconstitution of vesicle fusions occurring in endocytosis with a cell free system, *EMBO J.* **5**:3091–3101. (See also *Proc. Natl. Acad. Sci. USA* **84**:5728–5762, 1987.)

Halberg, D. R., Wager, R. F., Farrell, D. C., Hildreth, J., Quesenberry, M. S., Loeb, J. A., Holland, E. C., and Drickamer, K. (1987) Major and minor forms of the rat liver asialoglycoprotein receptor are independent galactose binding proteins, *J. Biol. Chem.* **212**:9828–9839.

Haussmann, K., and Patterson, D. J. (1982) Pseudopod formation and membrane production during prey capture by a heliozoon, *Cell Motility* **2**:9–24. (See also *Protoplasma* **115**:43–51, 1983.)

Herman, B., and Albertini, D. F. (1984) A time-lapse video image intensification analysis of cytoplasmic organelle movement during endosome translocation, *J. Cell Biol.* **98**:565–576.

Iacopetta, B. J., Rothenberger, S., and Kuhn, L. C. (1988) A role for the cytoplasmic domain in transferrin receptor sorting and coated pit formation during endocytosis, *Cell* **54**: 485–489.

Kaplan, J. (1985) Patterns in receptor behavior and function, in *Mechanisms of Receptor Regulation* (G. Poste and S. T. Crooke, eds.), Plenum Press, New York, pp. 13–36.

Lazarovits, J., and Roth, M. (1988) A single amino acid change in the cytoplasmic domain allows the influenza virus hemagglutinin to be endocytosed through coated pits, *Cell* **53**:743–752.

Leung, J. O., Holland, E. C., and Drickamer, K. (1985) Characterization of the gene encoding the major rat liver asialoglycoprotein receptor, *J. Biol. Chem.* **260**:12523–12527.

Marsh, M., Schmid, S., Kern, H., Harms, E., Male, P., Mellman, I., and Helenius, A. (1987) Rapid analytical and preparative isolation of functional endosomes by free-flow electrophoresis, *J. Cell Biol.* **104**:875–886.

Matteoni, R., and Kreis, T. E. (1987) Translocations and clustering of lysosomes depends on microtubules, *J. Cell Biol.* **105**:1253–1265.

Nichols, B. A. (1982) Uptake and digestion of horseradish peroxidase in rabbit alveolar macrophages, *Lab. Invest.* **47**:235–241.

Nilsson, J. R., and van Deurs, D. (1983) Coated pits and pinocytosis in Tetrahymena, *J. Cell Sci.* **63**:209–222.

Oates, P. J., and Touster, O. (1980) In vitro fusion of Acanthamoeba phagolysosomes. III. Evidence that cyclic nucleotides and vacuole subpopulations respectively control the rate and extent of vacuole fusion in Acanthamoeba homogenates, *J. Cell Biol.* **85**:804–810.

Pearse, B. M. F. (1987) Clathrin and coated vesicles, *EMBO J.* **6**:2507–2512.

Peck, R. K. (1985) Feeding behavior in the ciliate, *Pseudomicrothorax dubius* is a series of morphologically distinct events, *J. Protozool.* **32**:492–501.

Prusch, R. D. (1981) Bulk solute extrusion as a mechanism conferring solute uptake specificity by pinocytosis in *Amoeba proteus, Science* **213**:668–670.

Roederer, M., Bowser, R., and Murphy, R. F. (1987) Kinetics and temperature dependence of exposure of endocytosed material to proteolytic enzymes and low pH: Evidence for a maturation model for the formation of lysosomes, *J. Cell. Physiol.* **131**:202–209.

Rothman, J. E., and Schmid, S. L. (1986) Enzymatic recycling of clathrin from coated vesicles, *Cell* **46**:5–9.

Salisbury, J. L., Condeelis, J. S., and Satir, P. (1986) Role of coated vesicles, microfilaments and calmodulin in receptor mediated endocytosis by cultured B lymphoblastoid cells, *J. Cell Biol.* **87**:132–141.

Salzman, N. M., and Maxfield, F. R. (1988) Intracellular fusion of sequentially formed endocytic compartments, *J. Cell Biol.* **106**:1083–1092.

Severs, N. J. (1988) Caveolae:Static inpocketings of the plasma membrane, dynamic vesicles or plain artefact, *J. Cell Sci.* **90**: 341–347.

Silverstein, S. C., Steinman, R. M., and Cohn, Z. A. (1977) Endocytosis, *Annu. Rev. Biochem.* **46**:669–722.

Stockem, W. (1976) Endocytosis, in *Mammalian Cell Membranes* (G. A. Jamieson and D. M. Robinson, eds.), Butterworths, London, pp. 151–192.

Takemura, R., Stenberg, P. E., Bainton, D. F., and Werb, Z. (1986) Rapid redistribution of clathrin onto macrophage plasma membrane in response to Fc receptor–ligand interaction during frustrated phagocytosis, *J. Cell Biol.* **102**:55–69.

Trends in Biochemical Sciences, Vol. 10, No. 11, 1985, has a useful (if slightly outdated) centerfold (pp. 438–439) on molecular interactions in receptor cycling and endocytosis via clathrin-coated vesicles.

Ullrich, A., Bell, J. R., Chen, E. Y., Herrera, R., Petruzelli, L. D., Dull, T. J., Gray, A., Loussens, L., Liao, Y.-C., Tsubokawa, M., Mason. A., Seeburg, P. H., Granfeld, L., Rosen, O. M., and Ramachandran, J. (1985) Human insulin receptor and its relationship to the tyrosine kinase family of oncogenes, *Nature* **313**:756–761.

Wiley, H. S., Van Nostrand, W., McKinley, D. N., and Cunningham, D. D. (1985) Intracellular processing of epidermal growth factor and its effects on ligand–receptor interactions, *J. Biol. Chem.* **260**:5290–5295. (See also *Curr. Top. Membr. Transp.* **24**:369–412, 1985.)

Wright, S. D., and Silverstein, S. C. (1986) Overview: The function of receptors in phagocytosis, in *Handbook of Experimental Immunology* (4th ed.) (D. M. Weir and L. A. Hertzenberg, eds.), Blackwell, Boston, pp. 41.1–41.14.

3

Acidification; Membrane Properties; Permeability and Transport

3.1. Acidification

Given the pH optima of the acid hydrolases, one would expect the intra-lysosomal pH to be well below 6. This expectation was actually "verified in advance," long before the enzymatic equipment of lysosomes had been ana-lyzed. Metchnikoff pioneered by using litmus to demonstrate an acid pH in phagocytotic digestion vacuoles. From studies with indicator dyes bound to phagocytosable particles such as yeast, later investigators of protozoan and mammalian phagocytes estimated the pHs within the vacuoles at 3–5 (some higher figures were reported and some estimates were as low as 1.5). pH values in this range would sustain near-optimal activities of most lysosomal hydro-lases. Acid conditions also could have other important effects, by altering the state of aggregation of lysosomal contents (Section 2.1.4), by helping to kill the prey of protozoa and the microbes taken up by defensive phagocytes, and by promoting denaturation or other conformational changes in macromolecules that might facilitate attack by lysosomal enzymes.

3.1.1. pH Measurements *in Situ*

Visual determination of the color of indicator dyes is subject to intrinsic limitations of accuracy. In addition, any given dye is useful only over a limited range of pH, complicating quantitative study when pH changes are extensive. Uncertainties also have arisen about the degree to which the dyes are affected by adsorption to particles or to lysosomal contents (the ionic environment in the immediate vicinity of a surface can be strongly influenced by the properties of the surface). Indicator dye procedures have, therefore, been supplanted by methods that overcome some of these difficulties. With all the available meth-ods, however, it is difficult to ensure that the behavior of pH probes concen-trated in the intralysosomal environment, or bound to materials that undergo profound changes during digestion, is comparable to the behavior of the same probes in the model systems that are used for design and calibration of the methods. For example, the reactive oxygen metabolites that accumulate in phagolysosomes of professional phagocytes (Section 4.4.2.2) may alter some of the molecules used to estimate pH. Thus, although the procedures in current use are providing valuable information about directions of pH change and about large-scale differences, the precision with which they yield absolute values for intracompartmental pH needs to be evaluated further.

3.1.1.1. Fluorescence; FITC

The fluorescence of a number of the compounds widely used as tags for microscopy is responsive to pH and this responsiveness has been the basis for the most influential of the newer methods for estimating pHs in lysosomes and related compartments. There are simple techniques for conjugating molecules like fluorescein isothiocyanate (FITC; Section 2.2.2.3) to endocytosable macromolecules; FITC–dextran conjugates have been used most often, partly because they are not degraded within lysosomes, but conjugates with proteins that are ligands for receptor-mediated endocytosis have also been employed (see Fig. 3.1). The conjugates readily gain access to heterophagic apparatus of living cells by the usual endocytotic mechanisms. Video image enhancement and microspectrophotometric methods make it possible to follow individual fluorescent structures for long periods and to estimate their pH as they move around in a living cell. Complementary data on large numbers of cells and on isolated organelles can be obtained by adapting flow cytometric methodologies such as those used to discriminate among cells in fluorescence-activated cell sorting; this has been done, for example, to analyze the behavior of two different fluorescent molecules present simultaneously in cultured cells.

Care must be taken in purifying and employing the conjugates, to minimize the presence of fluorescent, low-molecular-weight molecules not attached to the macromolecules, such as free fluorescein or carboxyfluorescein. Sometimes, low-molecular-weight fluorescent contaminants seem able to cross membranes (see Section 4.7.1), which can greatly complicate interpretations especially with cells that take up only small amounts of the larger conjugates.

Representative applications of the fluorescence approach are illustrated in Fig. 3.1, Fig. 3.2, and Fig. 3.4. In Fig. 3.1, an FITC–protein conjugate (see Section 4.5.1.1) is used to show the basis for estimating pH by comparing the ratio of the intensity of emission of fluorescein *excited* by illumination of 490–495 nm with the intensity at 450 nm. (**Emission** wavelengths of 510–530 nm are customarily used for the measurements with fluorescein.) The 495/450 ratio of fluorescein or its conjugates declines tenfold as the pH drops from 8 to 4; the ratio therefore can provide a sensitive measure of pH change. The accuracy of the technique is, however, limited by the problem that emission at 450 nm is relatively weak so that autofluorescence and other effects can severely impede measurement. One of several ways around this problem is to compare the emission of compartments of interest at a pH-sensitive fluorescein excitation wavelength, such as 495 nm, in normal, intact cells with the emission of the same structures, at the same wavelength, after the cells are subjected to treatments that alter the pH of the compartments to a known value. The treatments used aim at collapsing the pH gradients across cellular membranes by increasing the permeability, thereby raising intracellular pHs to levels near that of the medium. This has been tried with low concentrations of detergents or relatively high concentrations of ionophores that transport monovalent cations, such as monensin (which is particularly effective at promoting exchanges of Na^+ and H^+) or nigericin (promotes exchanges of H^+ and K^+).

With *in situ* approaches using endocytosed fluorescent conjugates, the pH in secondary lysosomes of macrophages and cultured fibroblasts has been estimated at 4.5–5.5.

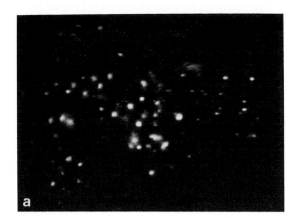

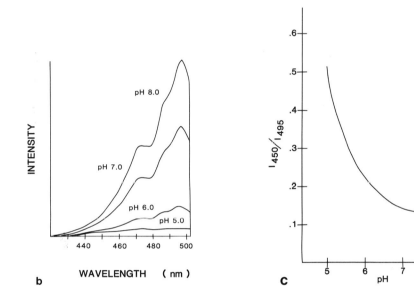

FIG. 3.1. (a) Fluorescence micrograph of a cultured fibroblast that had endocytosed FITC conjugates of α_2-macroglobulin and therefore shows fluorescence in its heterophagic apparatus. $\times$ 1500 (approx.).

(b) Fluorescence spectra of FITC–α_2-macroglobulin at different pHs. (The wavelengths indicated are excitation wavelengths; emission intensity was monitored at 520 nm.)

(c) pH dependency of fluorescence intensities of FITC–α_2-macroglobulin expressed as the ratios of emissions at excitation wavelengths of 450 and 495 nm.

From Maxfield, F. (1985) In *Endocytosis* (I. Pastan and M. Willingham eds.), Plenum Press, New York.

The conjugates have also been used to obtain information about acidification of lysosomes isolated by cell fractionation (Fig. 3.2).

3.1.2. "Weak Bases"; Vital Dyes; Chloroquine

Lysosomes and related structures are among the easiest intracellular bodies to stain in living cells with vital dyes of relatively low molecular weight. Some of the dyes, like Trypan Blue, enter the lysosomes by endocytosis in

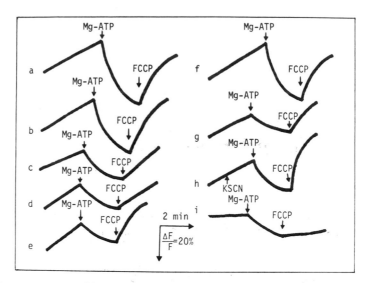

FIG. 3.2. Some properties of lysosomal acidification. Lysosomes, preloaded with FITC–dextran by injection of a rat, were isolated from liver and changes in their internal pH were monitored by changes in fluorescence, using excitation at the pH-sensitive wavelength of 495 nm (see text and Fig. 3.6).

Addition of ATP (with Mg^{2+}) to the medium produces a decrease in fluorescence corresponding to a decline in pH. That this is counteracted by the subsequent addition of the protonophore, FCCP, demonstrates that protons (H^+) had, in fact, accumulated after ATP was added. **a** shows the "standard" response when the lysosomes are suspended in KCl. In **f**, NaCl has been substituted for KCl: there is little difference, demonstrating the relative insensitivity of the acidification system to the identity of the predominant cations. The remaining curves illustrate the responsiveness of the system to anions: In place of KCl, the medium contained KBr (**b**), K phosphate (**c**), KF (**d**), K_2SO_4 (**e, h**), Na_2SO_4 (**g**), or sucrose (**i**).

From Ohkuma, S., Moriyama, Y., and Takano, T. (1982) *Proc. Natl. Acad. Sci. USA* **79**:2758.

company with proteins or other components to which the dyes become bound in the incubation media. But dyes like the aminoacridines (e.g., Acridine Orange) or Neutral Red stain the lysosomes faster, and more intensely than would be expected were endocytosis the primary route of access. One of several drugs that show similar behavior is the antimalarial agent **chloroquine,** whose rapid accumulation in lysosomes is readily visualized because it is fluorescent. Chloroquine, aminoacridines, and Neutral Red (Fig. 3.3) are regarded as "weak bases" because in solution they exhibit equilibria among species associated with H^+ ions and species lacking this association—the equilibrium shifts with pH so that H^+-associated (protonated) forms predominate at low pH.

Accumulation of weak bases in lysosomes may be due, in some degree, to their electrostatic binding to the polyanions that often seem to be present in lysosomes, but the principal requirement for accumulation of the bases is a low intralysosomal pH. At near-neutral pHs, as in the cytoplasm outside the lysosomes, the equilibria among the different forms of a given base are such that a significant number of the molecules are uncharged. These uncharged species traverse membranes with reasonable ease, either because they are lipophilic enough to move by passive diffusion or, perhaps sometimes, because

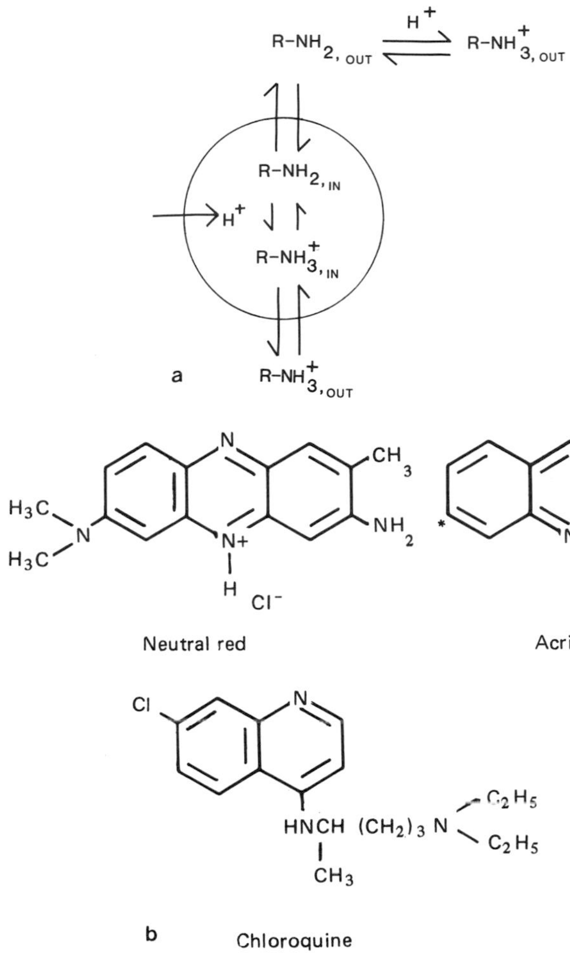

FIG. 3.3. (a) The distribution of a weak base between an acidified compartment and its surroundings depends on the rates of passage of the charged forms ($R\text{-}NH_3^+$ in our example) and the uncharged forms ($R\text{-}NH_2$) through the membrane and on the pH-dependent equilibria among the forms, within the compartment and outside. On the assumptions that an equilibrium distribution can be reached, that the permeability of the membrane to the charged form is negligible, and that there are no complications such as the binding of the bases to materials in the compartment (e.g., to polyanions), the distribution of the base will be given by:

$$[R\text{-}NH_2]_{in} = [R\text{-}NH_2]_{out} \qquad \frac{[R\text{-}NH_2]_{in}[H^+]_{in}}{[R\text{-}NH_3^+]_{in}} = \frac{[R\text{-}NH_2]_{out}[H^+]_{out}}{[R\text{-}NH_3^+]_{out}}$$

So

$$\frac{[R\text{-}NH_3^+]_{in}}{[R\text{-}NH_3^+]_{out}} = \frac{[H^+]_{in}}{[H^+]_{out}}$$

If the pH in the compartment is much lower than the pKa of the base so that $[R\text{-}NH_2]_{in}$ is much less than $[R\text{-}NH_3^+]_{in}$, the total concentration of base in the compartment will provide a reasonable estimate of the concentration of $R\text{-}NH_3^+$ there; such estimates are used to measure intracompartmental pHs. After Maxfield (see Fig. 3.1).

(b) Neutral Red and dyes derived from acridine [e.g., Acridine Orange, which has $-N\,(CH_3)_2$ groups attached at the two points indicated by asterisks] are among the weak bases most used as vital stains for acidified compartments. Chloroquine is a weak-base antimalarial agent often used to manipulate the pH in lysosomes and related compartments. (Formulas based on the Merck Index and on Pearse, A.G.E. (1980) *Histochemistry* (4th ed.), Churchill-Livingstone, London.)

they can exploit channels, or carriers and other facilitated diffusion devices in the membranes. When they move into low-pH compartments, the molecules tend to acquire H^+ ions and thereby become charged. These charged species penetrate membranes poorly and hence "escape" from acidified compartments much more slowly than their uncharged counterparts tend to enter. The net effect is the buildup of much higher concentrations of these molecules within low-pH compartments than the concentrations outside; the ratios of concentrations depend on the differences in pH between inside and outside* (Fig. 3.3).

3.1.2.1. Acidotropic Molecules Can Be Used to Measure pH and to Change It

The dyes and amines under consideration, and other small, weakly basic substances, are prominent among the molecules referred to as **acidotropic** because there is evidence that their accumulation is due to low intracompartmental pH. (Operationally, this is demonstrated by showing that their accumulation is prevented or reduced by circumstances that raise the internal pHs of the compartments.) The term **lysosomotropic,** originally coined to emphasize the predominance of lysosomes as depots for certain types of exogenous material, was used for a while with much the same set of connotations as "acidotropic." Now that it is clear that lysosomes are not the only important intracellular compartments with low pH, the two terms are being appropriately separated and refocused.

When cultured cells are exposed to media containing chloroquine at concentrations of 10–100 μM or to ammonium salts (usually NH_4Cl) at concentrations on the order of 10 mM, their lysosomes accumulate very high concentrations of the basic molecules. Chloroquine is concentrated particularly effectively, due in part to its being a dibasic ("diprotic") molecule; from fluorescence measurements *in situ,* intralysosomal levels have been estimated to reach roughly 100 mM. Concentrations this high should raise the intralysosomal pH unless the lysosomes have implausible passive buffering capaci-

---

*To understand this, it may be easiest to think first of the extreme situation that would prevail if H^+-associated forms of the weak bases could not traverse membranes at all and if association with H^+ took place only under acid conditions. Molecules of a given base placed in a neutral medium surrounding an acidified compartment would continue to diffuse into the acidified compartment until the concentration of base molecules *not associated with H^+* is the same within the compartment as it is outside; at this point, passage out will balance passage in. But this transmembrane equilibrium will not be reached until the total concentration of the base molecules inside is higher than the total concentration outside. This is true because many of the molecules that enter the acidified compartment will convert to the *H^+-associated* form and hence in essence will have been removed from the population that contributes to the equilibrium of transmembrane movement. The relative abundance of H^+-associated and unassociated species inside the acidified compartment, and therefore the ratio of total concentration inside to that outside, will depend on the pH and on the equilibrium constants governing the association and dissociation of H^+ from the molecules, as in Fig. 3.3.

Certain weak bases—benzylamine and tributylamine for example—are thought to be able to penetrate membranes in both protonated and nonprotonated forms, which would permit them to "carry" protons, almost like ionophores, from more acidified compartments to less acidified ones. This probably explains why these bases seem to raise intracompartmental pHs without accumulating extensively in the compartments, as indicated by the relative lack of the osmotic swelling mentioned in Section 3.1.2.1.

ties or are unexpectedly efficient at actively acidifying their interiors. Indeed, intralysosomal pHs in fibroblasts or macrophages treated with chloroquine or ammonium ions often rise to as high as 6 or 6.5. The cells frequently appear "vacuolated" as a consequence—the lysosomes and some other low-pH compartments swell considerably owing to osmotic influx of water (see Section 3.4.1.3). But the cells survive for at least a number of hours, and the effects are reversible upon removal of the agents; therefore, treatment with ammonium ions, chloroquine, or other weak bases has been extensively used to interfere with lysosome functions (see below; much less work has been done with the bases that produce little swelling; see footnote on p. 98). In addition, the effects of chloroquine on endocytosis or heterophagy in the malaria-causing *Plasmodium* parasites that dwell within erythrocytes (Section 4.4.3) may explain the antimalarial effects of the drug. Competing hypotheses attribute chloroquine's effects to inhibition of the degradation of macromolecules or to starvation of the parasites for iron (Section 4.1.2). However, it is not yet established that the usual therapeutic doses of chloroquine are high enough to alter intralysosomal pH drastically in the plasmodia in infected hosts, at least through the simple mechanisms of "acidotropy" considered thus far. Some investigators are convinced that the plasmodia concentrate chloroquine by special transport devices, whereas others believe that chloroquine's ability to inhibit certain lysosomal proteases or lipases directly may be of greater import than its impact on pH.*

Once it was understood that dyes like Acridine Orange and Neutral Red often stain compartments by acidotropy, these dyes came to be used to identify acidified intracellular structures in the light microscope. The dyes, and other weak bases, are also used to measure the extent of acidification within membrane-delimited bodies in cell fractions and *in situ*. An early experiment with this approach demonstrated that isolated lysosomes can maintain a low internal pH: it was shown that lysosomes obtained from liver ("tritosomes"; Section 1.3.1) concentrate radioactive methylamine. Figure 3.6 exemplifies quan-

---

*More generally, though weak-base effects on cell metabolism often are attributed to changes in lysosomal (or endosomal) pH, this attribution needs critical, case-by-case justification. The bases alter cytosolic pH, with potentially profound impact on metabolism, and they may also affect the nucleus. Certain of the bases (e.g., methylamine) seem to perturb protein synthesis and DNA synthesis. Others, like chloroquine, can affect lysosomal proteases, lipases, or other enzymes more or less directly. Many of the weak bases alter the osmotic distribution of water within the cell (Section 3.1.2.1), change the distribution of ions additional to H^+, affect membrane cycling, or have other influences. It is, therefore, often difficult to tease out the functional consequences that are properly ascribed to altered intracompartmental pH *per se*.

The ionophores, like monensin, sometimes used to influence pHs in lysosomes and endosomes, also have multiple effects on cells. Monensin, which mediates exchanges of H^+ and Na^+, was long used to interfere selectively with Golgi functions because microscopists observed that it induced dramatic swelling and dispersal of Golgi sacs and cell physiologists discovered that it inhibits secretion. It is only relatively recently that monensin's impact on pH has been widely recognized and it is still not certain how this impact relates to the ionophore's influence on Golgi functions.

Though in this book the weak bases will be of interest principally as experimental tools, speculations have been advanced that the ability of lysosomes to "soak up" components like ammonia could help control the natural levels of such components in extracellular spaces and in intracellular media with resulting influences on physiology and on metabolic reactions such as ammonia-dependent carbamoylations. (Parallel proposals argue that the lysosomal proton pump helps regulate pHs in the extralysosomal cytoplasm and that the distributions of other ions, such as Cl^- or HCO_3^-, are also strongly affected by lysosomal activities.)

titative work on cell fractions with fluorescent dyes that show quenching of fluorescence in acidified compartments.

Weak bases now are being employed by electron microscopists to identify low-pH compartments at the ultrastructural level and to begin to get some idea of their relative pHs. **Primaquine** (an amine) and **DAMP** (3-(2,4-dinitroanilino) 3'-amino-N-methyldipropylamine, a weak-base derivative of dinitrophenol) accumulate in low-pH compartments as do other weak bases and, because these molecules contain available amino groups, they can be immobilized by the conventional aldehyde electron microscope fixatives (these fixatives cross-link amino-containing molecules). The immobilized bases then can be detected immunocytochemically; efforts are now being made to quantify the immunocytochemical reactions so as to estimate the relative amounts of the bases that accumulate at different places in a cell or under different circumstances.

3.1.3. Timing of Acidification; Catching the Flu

3.1.3.1. Phagosomes and Endosomes

When rabbit neutrophils phagocytose yeast stained with pH-indicator dyes, a few minutes elapses before the dyes in newly formed phagocytotic vacuoles indicate a fall in pH. In macrophages the pH may actually rise to 7.5 or more in the first minute or two of a phagosome's intracellular life; then the vacuoles become acidified (Fig. 3.4). Periods at neutral pH or higher could enable enzymes with relatively high pH optima to initiate the attack on microorganisms engulfed by the professional phagocytes (see Section 4.4.2.1).

Interesting as such "delays" in acidification are, it is also striking that in several cell types the compartments containing newly endocytosed materials become acidified at times before there are indications of lysosomal fusion with the compartments. In *Paramecium*, the pH of new food vacuoles drops to 4 or even less within a few minutes after their separation from the cell surface, well before acid hydrolases are demonstrable in the vacuoles. (The pH estimates for *Paramecium* compartments were arrived at both by dye methods and by observing the pH-induced loss of the cytochemically demonstrable enzyme activity of endocytosed HRP.) Analogous findings of acidification of seemingly prelysosomal phagosomes have been made on amebas, with fluorescence meth-

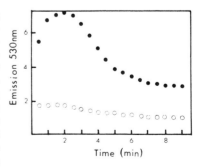

FIG. 3.4. Changes in fluorescence emission seen when yeast, coated with FITC–antibody conjugates, are phagocytosed by macrophages. The emissions at 493-nm excitation (●) and those at 450 nm (○) are presented separately to illustrate the much greater changes in the former than in the latter. The yeast were adsorbed to the macrophages at 4°C in a pH 6.5 medium and then, at time 0 in the illustration, the cells were warmed to 37°C to initiate phagocytosis. The pH of the intracellular compartments occupied by the yeast exhibits an initial transient rise (at 2 min the ratio of emissions at the two wavelengths corresponds to a pH of about 7.5) and then falls to 5.5–6. From Giesow, M. J., and Evans, W. H. (1984) *Exp. Cell Res.* **150**:36. (Copyright: Academic Press.)

ods. A still unexplained feature of heterophagy in *Paramecium* is that by the time lysosomes do fuse with phagosomes, the pH of these phagocytotic vacuoles is on the rise, so that during much of the period when digestion by acid hydrolases within a phagolysosome should be most intense, the pH inside the body is approaching neutrality or is already nearly there.

With endosomes, of course, it is difficult to know when they are truly "prelysosomal" (Section 2.4.4.1) or even just how to talk about their forming and maturing (Section 2.4.3). This aside, "early" endosomes of several types of cultured mammalian cells have been reported as having pHs of 6–6.5; endocytosed tracers suitable for estimating pH encounter such pHs within 1–5 min of their entry into the cell. A few minutes later, many of the ligands taken up by receptor-mediated endocytosis are found in compartments with notably lower pHs; commonly, pHs of 5–6 are encountered by tracers at 5–30 min after uptake. [Certain ligands, such as transferrin, deviate importantly from this pattern (Section 4.1.2.2).] The lower pH compartments are taken to include "late" endosomes and definitive, functional lysosomes.

Microscopic demonstrations of low "prelysosomal" pH *in situ* have been backed by initial reports that the structures in endosome-rich or MVB-rich cell fractions can maintain low internal pHs.

3.1.3.2. Coated Vesicles; Golgi Compartments

Does acidification begin immediately during endocytotic entry itself? Cell fractions highly enriched in coated vesicles contain structures that accumulate weak bases. And biochemical analyses of these fractions have revealed the presence of proton-pumping machinery very similar to that described below for lysosomes and endosomes. But which of the categories of coated vesicles would these findings apply to?

Data on this question are only beginning to come in. So far, there is little positive evidence that endocytotic coated vesicles maintain a low pH; the few efforts to demonstrate acidification in partly purified preparations of such vesicles have given negative results. Perhaps then it is some of the Golgi-derived vesicles that maintain a low pH. This would fit well with evidence, both biochemical and microscopic, for capacities of some of the membrane systems of the Golgi apparatus to maintain a moderately low pH (e.g., Section 6.1.5). But it is much too early to draw the firm conclusion that only the Golgi-derived coated systems are acidified. And even if this conclusion eventually proves to be correct, fundamental questions—such as whether the primary lysosomes of Section 2.1.3.1 carry acidification machinery or whether the Golgi apparatus is a source of proton pumps for other cell structures—will need to be addressed.

3.1.3.3. Membrane-Enveloped Viruses

Viruses infecting animal cells have confirmed that acidification of newly formed endocytotic compartments begins within a few minutes and have provided key testimony that acidification can precede the accession of the compartments to a state in which endocytosed materials are efficiently degraded. Membrane-enveloped viruses such as influenza virus and Semliki Forest virus have been studied most intensively. Each of these virus types has evolved so as to bind to particular sets of cell surface molecules ("receptors"; Section 1.5.1.3) that mediate uptake of the virus, via coated pits, into endosomes. Endocytosis, however, does not constitute effective viral entry into the cell because it leaves the viruses enclosed within both their own envelopes and a cellular membrane. To proliferate, the viral genome and some enzymes or other components from the viral contents, must traverse these membranes and pass into the cytosol. They must do so without prolonged exposure to active lysosomal enzymes, as the hydrolases can be fatal to the viruses.

The viruses are capable of fusing directly with the plasma membrane, which would produce the needed transmembane transfer of the genome. However, fusions with the plasma membrane occur only if the pH of the growth medium is lowered to 5–6, suggesting that such direct penetration is not a principal route of viral entry under ordinary circumstances, at least in the tissue culture systems usually studied. Instead, the "escape" of the genome occurs after endocytosis (Fig. 3.5). That the escape nonetheless depends on acidification is shown by the inhibition of viral proliferation observed in cells exposed to the weak bases or ionophores that raise endosomal and lysosomal pH. Escape is detectable within 5–7 min after endocytosis.

pHs in the 5–6.5 range observed in endocytotic structures are known to change conformations in certain of the viral surface macromolecules involved in entry into cells including the hemagglutinin of influenza virus and proteins of the "spikes" of Semliki Forest virus. These changes reveal hydrophobic domains that are conjectured to mediate interaction of the viral surface with the endosome membrane, initiating fusions that open paths to the cytosol.

Are the structures involved in penetration of membrane-enveloped viruses "prelysosomal" in the sense that the compartments lack extensive degradative capacity? The rapidity of penetration suggests they are. Moreover, viruses fail to proliferate when they are experimentally directed to functioning lysosomes. This can be done by coating them with antibodies, which seem to block the membrane interactions needed for fusion and escape. [If the viruses under discussion actually escaped chiefly after they reach lysosomes as some other viruses seem to (Section 4.4.3.1), one might even expect the degradation of the antibodies by lysosomal proteases to restore fusion capacities and allow the viruses out, rather than to promote viral destruction.]

3.1.3.4. Endosomal Changes

Mutant viruses have been isolated that differ from one another in the pH requirements for penetrating membranes, probably because the mutations in-

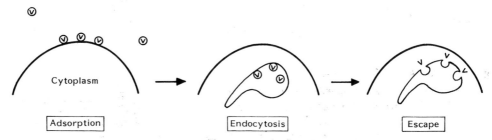

FIG. 3.5. Schematic diagram of the entry of membrane-enveloped viruses (V) into cells. The viruses first adsorb to specific molecules on the plasma membrane and are endocytosed. The viral "cores" escape into the cytoplasm when the viral membrane fuses with the membranes delimiting the endocytotic compartments. After work by A. Helenius and colleagues. (See also p. 205.)

fluence the conformational effects mentioned above. Mutants requiring unusually low pH take longer than others to escape. This correlation is regarded as evidence that as time passes, the viruses are exposed to an increasingly acid environment. Most observers assume that this means that the pH of virus-containing structures drops progressively as the structures evolve, rather than that the viruses move from structures with a higher pH, to separate lower-pH structures.

More generally, endosomes, whether involved in viral penetration or not, are widely believed to undergo a decline in pH from above 6 to near 5, as they pass from the "early" endosome state to the "late" state (Section 2.3.1.4 and 2.4.4.1). (Details vary from cell type to cell type. And even the endosomes in a given cell seem to show appreciable heterogeneity in the timing of pH changes among different endosomes, as indicated by the kinetics of escape of viruses and by microscopic estimates of pHs in individual bodies.) Estimates of the average time it takes for endosomes and related structures in mammalian cell cultures to reach the lower pH values usually fall in the range of 10–30 min or a bit more (Section 3.1.3.1). By this time many of the endosomes have traveled a considerable distance from the cell surface. Often they have assumed a multivesicular morphology.

It would be very helpful to know how their enzyme content correlates with their internal pH: The relatively high pHs in early endosomes are not optimal for many of the lysosomal enzymes, but they would permit appreciable action by certain of the hydrolases. The lower pHs attained by bodies regarded as late endosomes overlap with the pHs taken as typical of functioning secondary lysosomes. Following the logic of Section 2.4.4.1, reigning views are that highly acidified "late" endosomes are still not "true" lysosomes. This conclusion stems partly from observations that substantial acidification of compartments containing recently endocytosed tracers is seen before the time when many of these compartments acquire acid phosphatase activity demonstrable with conventional cytochemical techniques. These findings are paralleled by cell fractionation data indicating a paucity of activity for most acid hydrolases in fractions thought to correspond to acidified late endosomes. The strength of the conclusions to be drawn is reduced, however, by the heterogeneity in timing of acidification of endocytotic structures in a cell (and by the likely heterogeneity as well, in timing of hydrolase entry). There have been reports that acidification of endosomes is appreciable even at temperatures below 20°C, which are thought to inhibit fusions with lysosomes. (Temperatures near 10°C can virtually abolish the falls in pH.) But, as pointed out in Section 2.4.7.1, the effects of low temperature on fusion are themselves somewhat ambiguous. And in certain cultured cell types, temperatures below 20°C retard or reduce the later stages of endosomal acidification.

3.1.4. Plants; Fungi; Yeast

The pH in the large, central vacuoles of plant cells was thought to be low (Fig. 3.6) from observations such as the one that the colors exhibited by anthocyanin pigments stored in the vacuoles of flowers of petunias and other plants are those expected for an acid environment. Plant cell vacuoles and the vacuoles of fungi are commonly reported to have pHs of 4–6.5. This is the range obtained with diverse types of measurement: microelectrode penetration of large vacuoles; direct determination of the pH in the "sap" isolated from vac-

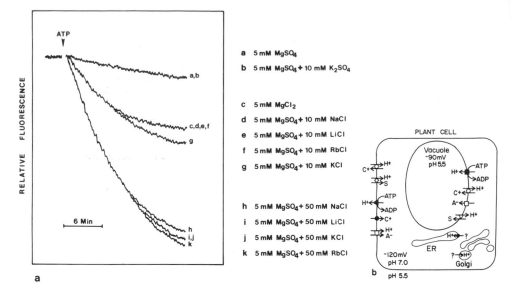

FIG. 3.6. (a) ATP-dependent acidification of sealed vesicles prepared from corn root tonoplasts monitored by measuring the fluorescence of quinacrine. (As with other fluorescent amines, the fluorescence of quinacrine is "quenched"—i.e., the fluorescence declines—as the amine accumulates within low-pH compartments; this effect can be calibrated to provide a measure of pH in the compartments.) The experiment also illustrates (see Fig. 3.2) the sustenance of acidification by the presence of chloride ion in the medium, and the relative lack of effect of cations (except that Mg^{2+} is needed for optimal activity of the ATPase). Relative fluorescence is $\Delta F/F$ where F = initial fluorescence intensity. The quenching has not been fully explained as yet but may relate to self-quenching phenomena that become prominent as concentrations of the fluorescent molecules rise, or to binding of the molecules to membranes or other surfaces. From Bennett, A. B., and Spanswick, R. M. (1983) *J. Membr. Biol.* **71**:95. (Copyright: Springer-Verlag, New York.)

(b) Summary of widespread types of major pumps and transport processes (symports and antiports) thought to influence or to utilize the distribution of protons in cells of higher plants. Along with the vacuole, the extracellular medium in the spaces enclosed by the cell wall can be at a markedly lower pH than the extravacuolar cytoplasm. The potential difference of the vacuole with respect to the extracellular medium is shown as less than that of cytosol in order to emphasize the electrogenic effect of the proton pump; it is not meant to imply that the vacuole maintains a positive, large, steady-state potential difference with respect to cytosol. A⁻, anions; C⁺, cations; S, uncharged solutes such as sugars. From Sze, H. (1985) *Annu. Rev. Plant Physiol.* **36**:175. (Copyright: Annual Reviews Inc.)

uoles; NMR spectroscopy of pH-sensitive molecules that are naturally concentrated in the vacuoles of some cells [e.g., basic amino acids in the vacuoles of *Neurospora* (Section 3.4.2)]. Estimated values have been as low as 3 or less, for vacuoles of lemon, which store citric acid (Section 3.4.2.1). Values of 5.8 to as high as 6.5 have been reported for vacuoles of yeast, partly from studies with endocytosed fluorescent conjugates [an approach not usable for most plant cells and not without its serious problems for yeast, given the low rates of endocytosis in these cells (Section 4.7)]; supposedly the modest acidification of the yeast vacuole correlates with the relatively high pH optima of yeast vacuolar enzymes.

3.1.5. Mechanisms of Acidification

Among the first circumstantial evidence that the low pH in lysosomes depends on active transport was the realization, from work with metabolic inhibitors, that metabolic energy is required for cells to maintain accumulations of Acridine Orange and other weak bases in their lysosomes. Subsequently, reports appeared that addition of ATP to isolated lysosomes enhances rates of proteolysis within the organelles. On the other hand, in the methylamine experiments mentioned in Section 3.1.2.1, tritosomes were found to maintain their internal pH as much as 1 unit below that of the medium even in the absence of an energy source.

3.1.5.1. "Donnan Effects"; First Studies of H^+ Permeability

This last finding contributed to a decade-long flirtation with the possibility that the low pH in lysosomes is maintained chiefly by means other than energy-dependent transport of ions. Transmembrane differences in concentration of inorganic ions, including H^+, are known sometimes to occur when a membrane permeable to these ions separates compartments that differ in concentration of nondiffusible or impermeant charged species such as macromolecules. The gradients in inorganic ions are often referred to as "Donnan gradients." They do not depend on active transport but rather they owe their origin to passive or facilitated diffusion across the membrane, plus the physicochemical "constraint" that a compartment containing a solution or suspension of charged materials must contain equal numbers of negative charges and positive charges. H^+ might accumulate in lysosomes because the lysosomal membrane is permeable to H^+ but not to the large anions, such as sulfated glycosaminoglycans or sialylated glycoproteins, that sometimes abound in the lysosome interior.

Initial studies on tritosomes, with classical osmotic and tracer methods (Section 3.4.1.1), seemed to show that the lysosomal membrane is, in fact, moderately permeable to H^+. These studies, however, also indicated a substantial permeability to K^+. Because K^+ concentrations normally are high in the cytoplasm, if lysosomes are fairly permeable to this ion one would expect Donnan gradients to be dominated by K^+ rather than H^+, meaning that the Donnan effects would produce only very small pH differences. Next it was found that the high lysosomal permeabilities to monovalent cations—H^+ and K^+ included—were due partly to peculiarities of tritosomes and partly to the use of temperatures near 0°C for preparing particles and studying their permeabilities (Section 3.4.1.1). These considerations cast strong doubt on the importance of Donnan effects for lysosomal properties.

Newer studies have swung the pendulum back a bit toward convincing some researchers that lysosomes do have significant—even if low—permeabilities to monovalent cations like H^+, Na^+, and K^+ at physiological temperatures. Thus, for example, isolated mammalian lysosomes, suspended in media lacking energy sources, do seem to exchange internal H^+ for external K^+, though this exchange is at a leisurely pace: The lysosomes maintain their low interior pH for many minutes when placed in solutions containing high concentrations of K^+; and even when the antibiotic **valinomycin** is added, the lysosomal pH rises more slowly than might be expected if the membrane were highly permeable to H^+. (Valinomycin molecules can insert in membranes and serve as effective carriers for K^+. The antibiotic therefore should selectively speed the influx of K^+ into the lysosomes, which would tend to raise the internal concentration of positive charges and drive out H^+ as rapidly as the permeability of the membrane to H^+ permits.)

Expectations and interpretations in such experiments, however, depend heavily on many factors that are difficult to assess: the purity and homogeneity of the lysosome preparations studied; the state and concentrations of the participating ions within the compartments being studied; the osmotic behavior of the compartments (e.g., the osmotic consequences of Donnan effects can be severe); movements of anions that can accompany the cations; the behavior of electrical potentials across the membrane (Section 3.1.5.3). We are ignorant about so many such matters that sharp disagreements persist about how to interpret the experiments in the last paragraph. More generally, no one pretends that the current picture of lysosomal permeability is at all adequate (see Section 3.4). Reliable information even about the normal ionic composition of the lysosome interior is minimal. Analytical electron microscopic findings, based on X-ray emission analysis, suggest that lysosomes of mammalian cells resemble the rest of the cytoplasm in being low in Na^+ and Ca^{2+} and high in K^+. In contrast, vacuoles of certain plant cells are known to maintain much higher internal concentrations of Ca^{2+} and other ions than are present in the cells' surroundings (Section 3.4.2).

3.1.5.2. Pumping Protons

Overall, few investigators now believe that "passive" mechanisms like Donnan effects are dominant in governing the normal acidification of lysosomes and attention to such effects is therefore shifting to situations where lysosomes are abnormally stuffed with polyanions like LDL particles or glycosaminoglycans. Under ordinary circumstances, the magnitude of plausible Donnan effects is not sufficient to account for the very low pHs in lysosomes. And control of intralysosomal pH by passive mechanisms would leave unexplained the impact of inhibitors of energy metabolism on lysosomal acidity mentioned at the outset of Section 3.1.5. In addition, when cells accumulate high concentrations of weak bases in their lysosomes, after the initial very rapid uptake, which discharges much of the pH gradient across the lysosome membrane, there ensues a prolonged, slower, continued accumulation of the bases. This second phase is difficult to explain without postulating active "efforts" by lysosomes to reacidify their interior by accumulating H^+, an effort that is "thwarted" by a continued counterbalancing influx of base.

From time to time it is speculated that the lysosomes acidify their interior by liberating acidic digestion products as they degrade macromolecules. But like the Donnan effects, this process, if it contributes at all to lysosomal acidification, is unlikely to be quantitatively significant under most conditions. Instead, lysosomal acidification is now believed, almost universally, to result predominantly from the operations of an active "proton pump"—an ATPase, located in the membrane bounding the lysosomes, that uses energy to translocate H^+ ions (protons). Fundamentally similar proton pumps have been identified and partially analyzed in the membranes of lysosomes of kidney, liver, and other animal tissues, in the bounding membranes (tonoplasts) of plant cell vacuoles, and in the vacuolar membranes of yeast and fungi. The known properties of the pumps (plus a few judicious assumptions to fill in the gaps) can account convincingly for the major phenomena of acidification. Figures 3.2 and 3.6 present representative experiments and information.

A loose and still-shifting consensus has been achieved about the outlines of the pumps' properties. The ATPase that drives transport is activated by Mg^{2+} (see Figs. 3.2 and 3.6) and has a pH optimum at 7.5–8, consistent with the expectation that ATP comes into play from the cytoplasmic side of the membrane rather than from the side facing the acidified lysosomal interior. The ATPase and the associated proton transport have features distinguishing them from other major cellular ATPases and proton transport systems. For instance, in contrast to the well-known Na^+/K^+ ATPase responsible for ion transport across the plasma membrane, the lysosomal system is not inhibited by ouabain. Nor is it affected by vanadate, an ion that inhibits a number of ATPases whose functioning involves transfers of phosphates from ATP to polypeptide components of the enzyme system. The lysosomal system differs from the mitochondrial oxidative phosphorylation sequence (the "F_0F_1" system, which can act as an ATPase in appropriate assays) in that the lysosomal system is not sensitive to oligomycin or azide, is more sensitive to alkylating agents like N-ethylmaleimide (NEM), and is less sensitive to artificial proton ionophores [e.g., dicyclohexylcarbodiimide (DCCD) or carbonylcyanide p-fluoromethoxyhydrazone (FCCP)].

As yet, no inhibitors are known to be uniquely specific for the lysosomal ATPase or proton-translocating mechanism. The sensitivities of the lysosomal system to NEM and to NBD-Cl (chloronitrobenzo-oxadiazole, an adenine analogue) are often used to help identify the system and to discriminate it from others. The pump, especially in some plant vacuoles, is more readily inhibited by nitrate than are others of the cell's ATPases; nitrate may extract essential components from the membranes or it may act on anion-sensitive sites (Section 3.1.5.3). Currently, there is a stirring of interest in the possibility that two antibiotics—**dauromycin,** which inhibits acidification of coated vesicles, and **bafilmycin,** which inhibits proton transport into plant vacuoles—will prove to be general inhibitors of the proton pumps of heterophagic compartments.

The most highly purified preparations of the proton pumping machinery prepared thus far, are relatively large complexes, including several polypeptides; the smallest total molecular sizes claimed are about 250 kD and most complexes range up to 500 kD or more. The complexes include integral membrane proteins and initial successes have been reported in reconstituting them into artificial membranes. The complexes from several sources are similar in including a polypeptide with a molecular size of 10–20 kD and two species of 60–75 kD; other polypeptides seem to be present as well. The 10–20 kD component binds DCCD, leading to the suspicion that it participates directly in moving protons. In yeast, photolabeling with ATP derivatives and studies on the effects of NBD-Cl suggest that a subunit of 70–90 kD participates centrally in the catalytic site responsible for the ATPase activity; similar proposals have been made for polypeptides of approximately 75 kD from other organisms.

Models currently discussed propose that most of the mass of the pump is located on the "cytoplasmic" side of the membrane, i.e., facing the cytosol. A pump complex may involve six to 12 or more polypeptide chains: several copies of the 10–20 kD component are inserted in the membrane where they provide the path for passage of protons; some of the larger polypeptides are also inserted in the membrane, probably as transmembrane components; additional polypeptides, probably including those involved in binding or hydrolyzing ATP, are not directly inserted in the membrane but are instead associated with cytoplasmic domains of the chains in the membrane.

Along with the ATPase, plant vacuoles possess a system able to transport protons into the vacuoles using energy coming from the hydrolysis of pyrophosphate.

3.1.5.3. Electrical Problems

Isolated liver cell lysosomes and vesicles fragmented from plant vacuoles can exhibit potential differences from the medium as great as 90 mV or more (the vacuole interior is negative) when suspended in suitable low-salt media (e.g., in sucrose) in the absence of an energy source.* These potential differences, like most transmembrane electrical potentials in biological systems, are expressions of the gradients in concentrations of diffusible ions across the membrane bounding the lysosome, and of the differences in permeability of the lysosomal membrane to different ions. When ATP is added to the cell fractions, the potential differences alter sharply: the lysosomal interior becomes relatively more positive so that potential differences decline to -50 to -20 mV in the preparations above; and isolated plant vacuoles, kept in media without permeant anions (see below), accumulate probes of membrane potential to the extents expected for positive changes of several tens of millivolts. These experiments involve rather artificial conditions and do not, for example, imply that the lysosomal interior *in situ* is at a negative potential with respect to the cytosol (which generally contains a high concentration of K^+ and Cl^-). Neither the sign nor the size of the potential difference *in situ* is reliably known presently. But the findings on the direction of the ATP-induced changes, in these experiments and others using various media and materials, have convinced most investigators that the lysosomal proton pump is **electrogenic:** The pump produces a net transport of positive charge to the interior of the lysosome.

The lysosomal proton pump is relatively insensitive to the concentration of other cations in the medium (Figs. 3.2 and 3.6). Taken together with the electrogenic capacity of the pump, this probably signifies that the mechanism by which protons are actually translocated does not directly couple this transport to the movement of other species of ions. In this respect the pump mechanism differs from systems like the plasma membrane's Na^+/K^+ system, which exchanges Na^+ for K^+, or the Na^+/H^+ exchange systems, recently recognized as plasma membrane transport devices important for the regulation of cytoplasmic pH. (Note that like the proton pump, some of these coupled cation transport systems can be electrogenic if the number of ions they transport in one direction is different from the number they transport in the other; for instance, the Na^+/K^+-ATPase of the plasma membranes of certain cell types moves three Na^+ out of the cell for each two K^+ ions it moves in.)

*Electrical polarization of the membrane bounding small closed compartments is usually measured by determining the extent to which the compartments accumulate certain cationic tracers, such as radioactive triphenylphosphonium ions (TPP+) and the fluorescent carbocyanine dye, diS-C3-(5), or anions (e.g., SCN− and certain negatively charged optical probes such as oxonals). These molecules are small enough or lipophilic enough to traverse membranes. The cations are mostly used to study compartments that maintain negative potentials (relative to the surrounding medium) and the anions for compartments with positive interior potentials. As with the weak bases used to estimate pH, differences in experimental detail can markedly affect results (e.g., for cell fractions, the magnitude and sometimes even the "sign"—positive or negative—of the potential depend on the composition of the suspension medium), and studies can be additionally complicated by the binding of probe molecules to the membranes or to the contents of compartments. See also footnote on the facing page.

Were the full electrogenic capacity of the lysosomal proton pump to be expressed in the cell, the lysosome interior would presumably be at a markedly more positive potential relative to the surrounding cytosol than it is in actuality; this would operate against the effective acidification of the lysosome interior.* How the cell "solves" this problem is still being debated. Likely, the influx of H^+ into lysosomes is accompanied by the entry of anions, especially Cl^-, moving through routes separate from the proton pump but driven, it is assumed, by the electrogenic effects of proton pumping. Figures 3.2 and 3.6 show that some other anions are not as effective as $Cl-$ in promoting lysosomal acidification *in vitro*, probably reflecting, in part, differential permeabilities of the lysosomes to the different ions (Section 3.4.1.1). Presumed anion-channel-blocking agents like the stilbene derivatives, DIDS and SITS, inhibit acidification of lysosomes, probably by blocking passage of anions through the lysosomal membrane (though the possibility is not yet fully ruled out that such agents also directly inhibit the proton pump). In plant cells, Cl^- has been found to stimulate the vacuolar ATPase directly, complicating analysis of how this ion contributes to acidification.

Along with the entry of Cl^-, movements of other anions such as phosphate have occasionally been claimed to "help" reduce the electrical potential produced by the proton pump. More widely accepted is the possibility that part of the electrogenic effect is countered by movement of Na^+ (or K^+) out of the compartments, by still unknown routes (Section 3.1.5.4).

3.1.5.4. Endocytotic Vesicles and Endosomes; Other Acidified Compartments

Though I have treated the proton pumps of animal cell lysosomes and the pumps of the vacuoles of plant cells and fungi as fundamentally similar, it

---

*It may be useful, for the arguments that follow, to conceive of the situation that would prevail were H^+ the only pertinent cation and were the lysosomal membrane completely impermeable to anions. When the energy is provided, the proton pump would begin to move H^+ into the lysosomes, initiating a difference in H^+ concentration (a ΔpH) across the membrane. But the effect also would be to change the electrical potential of the interior of the lysosomes in a positive direction and this would make further net import of positively charged ions such as H^+ increasingly unfavorable in energetic terms. Relatively few H^+ ions might have to be moved for the positive potential change to become a significant impediment to the development of a ΔpH large enough to fulfill the lysosome's needs.

In reality, H^+ is not the only ion involved and the membrane is not completely impermeable to ions. Nonetheless, the establishment of gradients in permeant ions between the interior of lysosomes and the cytoplasm outside tends to create a potential difference ($\Delta\Psi$) across the membrane as well as a difference in the concentrations (C) of ions dissolved in the interior and exterior media. (For H^+, the difference—CH_{in}^+ versus CH_{out}^+—expressed in logarithms yields the ΔpH.) [For passively moving permeant cations, the relations between the electrical potentials and the concentration gradients at electrochemical equilibrium are expressed by the equation $\Delta\mu = RT \ln(CIon_{in}/CIon_{out}) + zF\Delta\Psi = 0$, where $\Delta\mu$ is the difference in "electrochemical" potential across the membrane. (Strictly speaking, the "activity" of the ion should be used in the equation instead of the concentration.)]

Under the conditions used to generate active proton pumping, as in recent experiments with isolated lysosomes incubated with ATP, one usually expects changes in the pH difference and the electrical potential to vary reciprocally as prevailing conditions alter. Thus, in contrast to the situation above, if for every H^+ imported by the proton pump, a negative charge (e.g., a monovalent anion) is simultaneously spirited into the lysosomes, or some other positively charged ion is expelled, the proton pump would generate no change in electrical potential, expressing itself entirely in the pH gradient.

would be surprising were no interesting differences in detail to turn up. Plant vacuoles, for instance, probably are more versatile in membrane transport than are lysosomes of animal cells, because the vacuoles play more diverse roles—in some plants they concentrate organic bases or acids, and they probably rely on their proton gradients to drive the entry or exit of other ions and molecules (Section 3.4.2.1).

As Section 3.1.5.2 pointed out, the systems that establish a low pH in endosomes of animal cells show properties fundamentally similar to those of the lysosomal systems. And the lysosomal and endosomal proton pumps are at least generically similar to ATP-responsive acidification mechanisms operating in coated vesicles (Section 3.1.3.2), Golgi compartments (Sections 3.1.3.2 and 6.1.5.1), and secretory structures such as the chromaffin granules of the adrenal medulla (Section 6.1.5.1). But do all these membrane systems and coated vesicles share precisely the same machinery for active transport of H^+? The final answer will not be in until purer and more homogeneous cell fractions are available for relevant physiological and molecular studies. For instance, lysosomes, in partially purified cell fractions, are able to use GTP as a substitute for ATP in driving acidification whereas fractions enriched in endosomes use GTP much less effectively. But until this difference is verified with highly purified material it would be premature to conclude that the ATPases of the two types of structures differ.

One type of evidence that heterophagic structures vary in the machinery for maintaining low pH is that different mutant lines of CHO cells, 3T3 cells, and other mammalian cultures can be obtained with different patterns of defects in acidification. The best known of these cell lines fail to produce normal low pHs in endocytotic structures and Golgi compartments but show less severe effects on acidification of lysosomes (Table 3.1 and Fig. 3.7). Other lines are differentially affected in the "later" phases of acidification: the pHs encountered by the usual tracers are normal at early times, but not at the later times when the tracers should be in lysosomes, or very late endosomes. It cannot be taken for granted that such differences reflect variation in the proton pump itself from one compartment to another. Perhaps lysosomes, endosomes, Golgi compartments, and coated vesicles overlap closely in the set of proteins

TABLE 3.1. Endocytosis in Wild-Type CHO Cells (WTB), Mutants (DTG), and Revertants (Rev)[a]

	Inhibition: EC_{50}[b] (ng/ml)				
	WTB	DTG 1-5-4	DTG 1-5-4-122	Rev 123	Rev 211
Diphtheria toxin	40	1000	700	50	30
Modeccin	2	>3000	>3000	1	1
Pseudomonas toxin	350	3000	3500	240	560
Ricin	140	3	2	60	90
Man 6-P uptake (cpm/μg protein)	110	0	0	80	100

[a]The mutants were selected on the basis of defects in endocytosis and for resistance to certain toxins that enter the cell via acidified compartments (Section 4.4.3.2). The table demonstrates that the mutants resist three toxins (diphtheria toxin, modeccin, and *Pseudomonas* toxin) but are hypersensitive to another (ricin). The mutants also fail to endocytose lysosomal enzymes via the M6P receptor (Section 6.4.1.4). From Robbins, A. R., Oliver, C., Bateman, J. L., Krag, S. S., Galloway, C. J., and Mellman, I. (1984) *J. Cell Biol.* **99**:1296.
[b]EC_{50} is the dose required to inhibit protein synthesis to 50% of that measured in parallel samples of untreated cells.

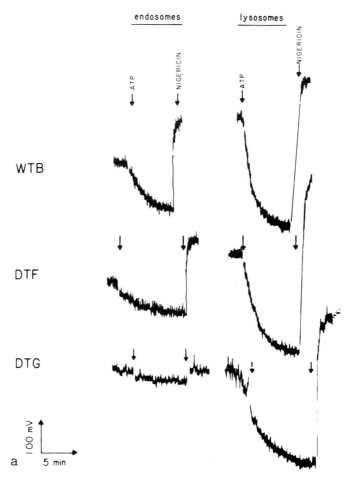

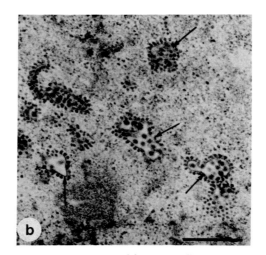

FIG. 3.7. CHO cell mutants selected as in Table 3.1 also show defects in endosome acidification and in Golgi functions.

(a) Acidification of Percoll-isolated endosomes and lysosomes in wild-type (WTB) and mutant (DTG and DTF) cells. pH was monitored by the decrease in fluorescence of FITC–dextran within the organelles. Note that on adding ATP (left arrow) the pH in endosomes and lysosomes declines in all preparations except the endosomes from the DTG cells; lysosomes from the DTG cells do show acidification. [At the right arrow the ionophore nigericin was added to collapse ion gradients (Section 3.1.1.1), thereby confirming that the ATP-driven fluorescence changes are dependent on these gradients.] The scale indicator at the lower left, calibrated in millivolts, scales changes in fluorescence intensity measured by spectrofluorometry.)

(b) Late stages in processing of Sindbis viral proteins are delayed or prevented in DTG mutants, and release of virus from the cells is decreased. Viral nucleocapsids accumulate in association with sacs of the Golgi apparatus. × 40,000.

From Robbins, A. R., Oliver, C., Bateman, J. L., Krag, S. S., Galloway, C. J., and Mellman, I. (1984) *J. Cell Biol.* **99:**1296.

directly responsible for proton pumping but vary in the membrane environment of these proteins, or in factors such as the ionic composition within the compartments. Or perhaps there are differences in other proteins—such as anion channels—with which the pump components associate or interact.

A matter especially requiring exploration is that the different compartments maintain quite different pHs (see Section 3.1.3). The Golgi compartments seem least acid, their reported pHs being well above 6. Early endosomes generally have higher pH than do the definitive lysosomes of the same cells. From the considerations raised in the last section, if the same species of proton pumps operated in both types of structure, endosomal pH would tend to be higher than lysosomal pH if endosomes had a lower permeability than lysosomes to crucial anions such as Cl^- or if they had a higher permeability to H^+. Scattered data support possibilities along both of these lines.

It has also been pointed out that the plasma membrane regions from which endocytotic vesicles and vacuoles arise are often lined with polyanions and contain ion channels or transport systems. If small numbers of these polyanions, channels, and transport systems were incorporated into endosomes, they could profoundly influence the internal ionic compositions and transmembrane electrical potentials; endosomes are small enough that very few ions need be moved or bound for the effects on concentration to be significant. One group believes that early endosomes contain operating, electrogenic, plasma-membrane-derived Na^+/K^+ pumps and that indirect effects—perhaps involving the endosome's membrane potential—of these pumps are largely responsible for the higher pH of early endosomes as compared to structures later in the heterophagic pathway. The pumps presumably are lost, or inactivated as the endosomes evolve (see Section 3.2). Possibly pertinent to this proposal is that one of the mutant cell lines that acidifies its late components very slowly (see above) is also resistant to **ouabain,** an inhibitor of the Na^+/K^+ ATPase.

3.1.5.5. Sources of Proton Pumps

How do lysosomes and prelysosomal compartments acquire their acidification capacities (and how do they keep them in view of the extensive movements of membranes to and from their surfaces to be described in Section 3.2)? In *Paramecium*, small vacuoles, which stain intensely with Neutral Red and Acridine Orange, fuse with new phagosomes at about the time that the phagosomal pH drops (Fig. 2.5). Do these vesicles ("acidosomes") produce acidification by contributing acid to the phagosome interior, or proton pumps to the membrane? Granules isolated from neutrophils seem able to acidify their interiors, as judged from their abilities to accumulate visible bases such as aminoacridines or DAMP. Several reports claim this property for the primary lysosomal (azurophilic) granules but another assigns especially low pHs to a set of **tertiary** granules, supposedly separable both from the azurophilic granules and from the specific granules. Do any of these granule types produce the acidification of phagosomes upon fusion?*

Three lines of speculation have been put forth to explain the acquisition of acidification capacities by endosomes and other "prelysosomal" structures in

---

Dictyostelium amebae contain numerous vacuoles that stain with Acridine Orange and that fuse with new phagosomes before the phagosomes acquire cytochemically demonstrable hydrolases. The vacuoles are also accessible to pinocytosed materials. For various other cell types, acidified vesicles or vacuoles, whose functions are as yet unknown, have been reported and could eventually be implicated in acidifying other bodies.

mammalian cells. One suggestion is that vesicles from the Golgi apparatus provide these capacities by fusing with the endocytotic structures; candidate vehicles for this include acidified coated vesicles (Section 3.1.3.2). A second proposal is that compartments derived from preexisting endosomes acidify new endocytotic bodies when the older and the newer bodies fuse (Section 3.3). This class of proposal is particularly attractive for those who posit that endosomes or their derivatives are in some sense long-lived compartments, which retain a separate existence while acquiring newly endocytosed contents and passing them on to the lysosomes (Section 2.4.3). [One possibility is that the proton pumps occupy specialized patchlike subdomains in the surfaces of lysosomes and endosomes (Section 3.2.3.3) and that these patches of membranes can bud off as specialized vesicles that later can fuse anew with endosomes or lysosomes.] A third idea ascribes major importance to hypothetical proton pumps resident in the plasma membrane or contributed to the plasma membrane by fusions of intracellular bodies. These pumps would move directly into endocytotic bodies as the bodies form. The known resident plasma membrane pumps that transport H^+ are different from the lysosomal pump, though in at least one case antigenic overlaps have been reported (Section 6.1.1; see Fig. 3.10). On the other hand, in turtle bladders and the collecting ducts of the kidney, proton pumps of the same genus as the lysosomal system do seem to cycle back and forth between the plasma membrane and the cell interior; the pumps are "stored" in the cytoplasm within the membrane of vesicles, enter the cell surface by exocytosis, and then can be withdrawn back into cytoplasmic vesicles by endocytosis. (These pumps secrete H^+ to the urine; their exocytotic passage to the cell surface is Ca^{2+} dependent and is responsive to circulating levels of CO_2. The membrane in which they are carried has a distinctive ultrastructure owing to the abundance of short, roughly cylindrical or "studlike" structures, 10 nm in diameter, inserted in it.)

The reported sequential changes of pH in endocytotic structures as they evolve before becoming definitive lysosomes (Section 3.1.3.4) may find explanations in progressive changes in the endosomal membrane: As Section 3.3.4 will describe, it is known that the endosomal membrane does lose and gain proteins selectively as endosomes mature so that one can envisage a remodeling of the membrane's makeup with effects on the distribution of ions across the endosomal membrane (see the discussion of the Na^+/K^+-ATPase at the end of Section 3.1.5.4). Conversely, Section 2.4.7 mentioned speculation—based partly on dubious grounds—that changes in the pH or membrane potential of heterophagic structures might govern fusions of endocytotic bodies and lysosomes, perhaps by influencing conformations or other properties of membrane molecules.

The transient initial **rises** in pH observed for phagosomes of mammalian phagocytes need explanation. Conceivably, they are due to consumption of protons during the "respiratory burst" or other metabolic activity that accompanies phagocytosis in these cells (Section 4.4.2.3) rather than stemming directly from behavior by the acidification machinery.

3.2. Membranes of Lysosomes and of Endocytotic Structures; Tonoplasts of Vacuoles

The membranes bounding lysosomes do not look particularly remarkable and their overall composition is not known to be extraordinary. They are

114

lipoprotein structures, 7–10 nm thick in conventional microscopic preparations; they have the customary trilaminar appearance in thin sections; and they contain a goodly population of the intramembrane particles (IMPs) visualized by freeze-fracture microscopy.

3.2.1. Survival

One perplexing question is how the membrane at the lysosomal surface evades the degradative fate that befalls membranes internalized within lysosomes. An obvious *a priori* possibility is that the membrane is resistant to degradation by virtue of its possessing certain proteins (enzyme inhibitors?) or because the unusual lipids thought to be present (Section 3.2.2.1) may not readily be attacked by lipases. This postulated resistance is, however, difficult to reconcile with observations that membranes prepared from isolated lysosomes are eventually degraded when they are "fed" to other lysosomes via endocytosis.

A popular alternative hypothesis about the maintenance of the surface integrity of lysosomes is that the bounding membrane is "protected" from the dangers nearby, by a layer of material interposed between the membrane and the digestive space inside the lysosome. Residual bodies and other secondary lysosomes generally show a clear-looking "halo" immediately below the limiting membrane (Fig. 3.8); this zone reacts cytochemically as though it is rich in carbohydrates. Though the provenance and nature of the halo material are unknown, it is located where a barrier would be useful and is usually thought

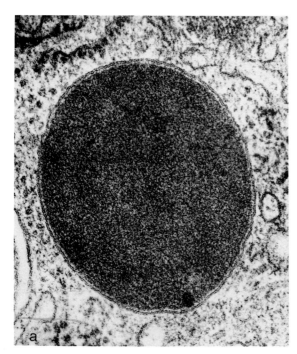 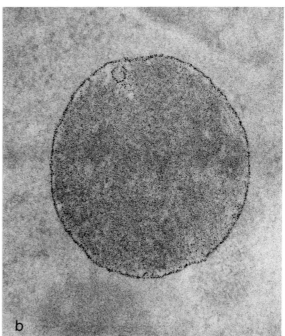

FIG. 3.8. Lysosomes from rat kidney. In **a** the zone of low electron density ("halo") separating the bounding membrane from the lysosomal contents is seen as it appears in conventionally prepared material. In **b** this zone has been stained by a method (periodic acid–thiocarbohydrazide– silver proteinate) thought to demonstrate polymeric carbohydrates or glycoconjugates. From Neiss, W. F. (1984) *Histochemistry* **80**:603. × 100,000. (Copyright: Springer-Verlag, Berlin.)

to be constructed of domains protruding from integral glycoproteins and glycolipids of the lysosomal membrane. Perhaps the unusually high content of acidic oligosaccharides found on certain of the membrane proteins from lysosomes (Section 3.2.2.2) somehow protects the membrane; consistent with this are findings that several lysosomal membrane proteins exhibit reduced life spans in cells treated with tunicamycin to inhibit glycosylation (Section 7.4.3.7).

Another proposal is that the membrane actually is continually being degraded, but gradually enough that the membrane can be preserved by its "dynamics," i.e., by the continual replacement of components in the course of lysosomal functioning.

3.2.2. Biochemical Features

Progress in analyzing the membranes that delimit lysosomes has been very slow and the available data are of uncertain reliability. It is not even understood why the lysosomes show such marked attraction to the anode in free-flow electrophoresis of cell fractions. Biochemical analysis has been hindered by the fact that functioning secondary lysosomes often contain an abundance of internal membranes. These enter the lysosomes as components of autophagocytosed or heterophagocytosed materials or arise from invaginations of the surface of the lysosome itself or from infoldings of the membranes bounding prelysosomal structures (see Section 5.2.1 and Fig. 2.14). Therefore, even with relatively pure cell fractions, bulk analyses of membranes prepared by disrupting lysosome-rich fractions can give a misleading portrait of the membrane at the lysosomal surface.

The primary lysosomes of neutrophils, and phagolysosomes loaded with indigestible, inert particles such as polystyrene latex beads, are the most common sources of lysosomal membranes for which problems of "contamination" with internal membranes are minimal. Vacuoles of some plant cells and yeast and fungi, also may provide relatively straightforward information because they show few membranes other than the **tonoplast** (the name given to the membrane that bounds plant vacuoles and often used for yeast and fungi as well).

3.2.2.1. Lipids

Reported ratios, by weight, of lipid to proteins range from less than 0.5 for membranes from tritosome preparations to 1.4 or more for yeast vacuoles. Ratios for phagolysosomes and plant vacuoles fall between these extremes. The extreme values are, in fact, suspect. On the one hand, it is difficult to remove all proteins adsorbed artifactually to membranes obtained in cell fractions (see Section 3.2.2.2). On the other, preparations of tonoplasts and other lysosomal membranes may sometimes be slightly contaminated with lipid-rich structures.

Phospholipids are prominent in all preparations. They account for most of the weight of the lipids in animal cell lysosomal membranes. The predominant phospholipids in these membranes are the usual glycerolipids—phosphatidylcholine, phosphatidylserine, phosphatidylethanolamine, and phosphatidylinositol—and their relatives; varying proportions of these species are found in different material. In animal tissues, such as liver, sphingomyelin is present in lysosomes, as it is in the plasma membrane. Glycosphingolipids have been found in moderate amounts in granule-rich fractions from neutrophils, though whether this means that these lipids are present in the lysosomes is not known. Glycolipids abound in tonoplasts, as they do in other membranes of plant cells.

Tonoplast preparations of yeast reportedly have quite high concentrations of "neutral" lipids among which ergosterol is prominent. Tonoplasts of certain plants also contain some sterols. Estimates for several animal cell preparations, including neutrophil granules and tritosomes, give molar ratios of sterols to phospholipids ranging up to 0.3–0.4 or more. (Here too the higher values may need reexamination.)

Lysosomal membranes of liver cells reportedly differ from many other cellular membranes in containing high concentrations of dolichols and particularly in possessing an unusual phospholipid, **bis-monoacylglyceryl phosphate** (lyso-bis-phosphatidic acid). The latter lipid is thought to be produced in the lysosomes from the mitochondrial lipid, cardiolipin. This exemplifies the possibility that lipid molecules delivered to the lysosomes as components of structures to be degraded, and intermediates in the digestion of these molecules, become extensively refashioned into components that eventually insert in the lysosome's delimiting membrane. Perhaps along with their degradative capacities, some of the lysosomal hydrolases can also catalyze low levels of synthetic or exchange reactions that contribute to this refashioning (see also Section 6.1.2).

Membranes isolated from certain plant cell vacuoles—but not those from yeast—show appreciable concentrations of phosphatidic acids, whereas preparations of yeast tonoplasts are surprisingly rich in lysolecithin. The lysophospholipids and phosphatidic acids could conceivably arise *in vivo*, or during cell fractionation, through action of phospholipases of the A and D class. Unfortunately there is little secure information about the lipase content of vacuoles, although some preparations seem relatively rich in phospholipase D (Table 1.2).

3.2.2.2. Proteins

Because adsorption—natural or artifactual—of proteins from the lysosome interior to the lysosomal membranes can be very extensive, and because criteria for discriminating adsorbed proteins from proteins belonging to the membrane are often difficult to apply, special care must be taken in preparing and washing membranes intended for analyses of protein content. From experience with membranes of secretory granules and other structures, rinsing with carbonate buffers is one of the purification steps now often employed. Where feasible, biochemical data must be corroborated by other methods demonstrating that suspected membrane proteins are, in fact, localized to membranes. Immunocytochemical approaches are most widely used for this (see Figs. 3.10, 7.10, and Section 3.2.3.3) though they too are subject to artifacts that can exaggerate the extent to which proteins in a structure appear to be associated with membranes.

Membranes isolated from secondary lysosomes of liver and other animal cells, and tonoplasts of plants, yeast, and fungi, reveal a variety of proteins when analyzed by conventional techniques such as detergent solubilization and polyacrylamide gel electrophoresis (PAGE). A given preparation may show two dozen readily detectable polypeptide chains, of molecular sizes ranging from less than 20 kD to more than 200 kD; often additional polypeptides are present in smaller or trace amounts. Acronyms such as LIMP or LAMP [lysosomal integral (associated) membrane protein] and others used below are being coined to designate the proteins, but terminology is still unsettled.

Both animal and plant membranes contain glycoproteins, some fairly large (150 kD or more). Liver lysosomal membranes have a number of such proteins. Most of the glycoproteins are presumed to be integral membrane proteins and to have their oligosaccharide chains ("glycans") facing the interior of the organelles rather than facing the cytoplasm outside. (The sidedness of the membranes has been studied only superficially, which is particularly regrettable in that the cytoplasmic face of the lysosomal membrane is where one would hope to encounter protein segments that might help account for recognition or fusion phenomena.) The glycoproteins of liver lysosomal membranes from several animals include species with impressive numbers of N-linked oligosaccharide chains (15–20 along a single polypeptide chain, contributing half or more of the total molecular weight). Among these chains are complex, sialic acid-terminated structures, like the oligosaccharides of secretory and plasma membrane glycoproteins (Chapter 7). That O-linked oligosaccharides might also be present has been put forth as a possible explanation for the resistance of certain of the chains to hydrolysis by endoglycosidase F.

A 120-kD lysosomal membrane glycoprotein, called LGP-120 and identified first in rat liver, is thought, from immunocytochemistry (see Fig. 7.10), to be widespread among mammalian cell types and to typify a type of lysosomal protein common to many organisms. LGP-120 probably extends most of its mass on the side of the membrane facing the lysosome interior and this part of the polypeptide chain contains a fair number of cysteine residues, which could be important in establishing the domain's three-dimensional structure. Only a short "cytoplasmic" tail is present, and judging from sequence data, there is only one hydrophobic membrane-spanning region. The 18 N-linked oligosaccharides on this protein are on the part of the chain facing the lysosome interior (lumen); they are clustered in two groups, separated by a domain relatively rich in prolines and serines. A comparable chick protein (Section 3.2.3.3), LEP-100, has 414 amino acids, 11 of which (COOH-terminal) probably are the cytoplasmic tail; the extensive luminal domain has 8 regularly spaced cysteine residues, 17 sites of N-glycosylation, and a region, near its middle, that is rich in proline, serine, and threonine.

Membranes isolated from the azurophilic granules of neutrophils by freeze-thawing and washing show only a few prominent protein species along with a number of minor ones; one recent report on rabbit cells suggests that two proteins account for over 60% of the total mass of proteins and that glycoproteins are not quantitatively prominent. As judged from PAGE, the specific granules of the same cells have a distinctly different membrane protein composition, in which five proteins account for 70% of the mass (Fig. 3.9). The azurophilic granules of human neutrophils are believed, from immunocytochemistry studies (Section 3.2.3.3), to lack the principal membrane glycoproteins characteristic of secondary lysosomes in other cell types and present in MVBs of neutrophils.

Findings that components of the membranes of rabbit azurophilic and specific granules are represented also at the neutrophil plasma membrane were quite puzzling at first, and often were laid to fraction impurities. It now seems that neutrophils and other mammalian phagocytes may normally secrete some

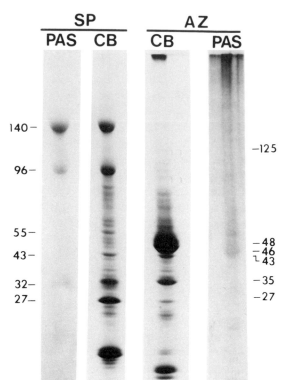

FIG 3.9. Gel electropherograms stained with Coomassie Blue (CB) or the periodic acid–Schiff reaction for carbohydrates (PAS) showing the membrane proteins extractable from the specific (SP) and azurophilic (AZ) granules of rabbit neutrophils. The molecular weight marks indicate proteins thought to be authentic membrane constituents in the pair of gels adjacent to the marks. Note the marked differences between the two types of granules. From Brown, W. J., Shannon, W. A., and Snell, W. J. (1983) *J. Cell Biol.* **96**:1030.

of their granule contents (Section 4.4.1.2); exocytotic secretion would incorporate the membranes bounding granules into the plasma membrane, at least transiently.

3.2.2.3. Enzymes and Transport Proteins?

The proton-pumping system described in Section 3.1.5.2 is one of the very small set of lysosomal membrane proteins whose functions are even partly understood. Another is the acetyl-transferase system that seems to be needed for degrading some glycosaminoglycans (Section 6.4.1.1). For neither of these systems is there a clear picture of the pertinent protein organization. Nor is much known of the proteins that take part in the passage of digestive products out of lysosomes (Section 3.4.1.4) or of the presumed ion channels in the lysosomal membrane (Section 3.1.5.3). Tonoplasts are known to have many transport activities (Fig. 3.6) but the responsible membrane molecules have not been identified. Phagocytotic vacuoles of mammalian phagocytes acquire a membrane-associated oxidase system (Section 4.4.2.3) whose microbicidal roles are very important. Other oxidation–reduction systems have occasionally been reported for tonoplasts or animal cell lysosomal membranes but their existence, locales, and roles are in dispute.

Because they are difficult to solubilize by treatments that do not dissolve membranes, several lysosomal hydrolases are thought to exist in special associations with the lysosomal membrane. (As before, ruling out binding of the enzymes to the lysosomal **content** can be tricky.) Even acid phosphatases have been claimed to occur in such associations, but a less controversial example is the principal glucocerebrosidase of animal cells (Section 7.3.3.5). In yeast, α-mannosidase seems to be associated with tonoplasts, although from evaluations in cell fractions, this association has been reported to be tighter in stationary cells than in rapidly growing cells. None of these enzymes have been shown definitively to be an integral membrane component; for most, a more superficial relationship to the membrane is likely. In certain cases the membrane-associated enzymes are components of multiprotein complexes that are able to bind to membrane surfaces (Section 6.4.1.3). For lipases or oligosaccharide-degrading enzymes, associations with membranous structures could result from binding of the enzymes to their substrates.

3.2.3. Membranes of Endocytotic Structures

What does the extensive traffic from the plasma membrane to endocytotic structures and lysosomes imply for the composition of the membranes of heterophagic bodies? Because the traffic is based on the movements of membranes—the separation of vesicles and vacuoles from the cell surface, fusions of membrane-delimited compartments with one another, and so forth—a first guess might be that the membranes bounding lysosomes, those bounding endocytotic structures, and the plasma membrane would have similar compositions. Initial direct comparisons seemed to confirm this expectation, especially for phagocytotic structures. But doubts soon arose. For example, although enzyme activities characteristic of the plasma membrane often are represented in hepatic tritosome preparations, some of the activities are so feeble as to preclude deciding whether the activity derives from contaminants, from the lysosome surface, or from the lysosome interior. Activities resembling 5'-nucleotidase, an enzyme present at the cell surface, do seem to be detectable in some preparations of tritosomes and other lysosomes, although care must be taken not to confuse the nucleotidase, which is optimally active at neutral or mildly alkaline pH, with acid phosphatases, some of which retain activity at pHs near 7.

When yeast tonoplast preparations were found to have a lower sterol content and other differences in lipid composition from the cell surface membranes, some observers attributed the differences to inadequacies in fractionation (e.g., to difficulties in purifying plasma membranes) whereas others invoked the supposed nonoccurrence of endocytosis and related membrane cycling in yeast. These explanations are being reexamined in light, for instance, of current beliefs that appreciable endocytotic communication takes place between the yeast cell surface and the vacuole (Section 4.7.1).

In general, it now appears that even though there may be some components common to the plasma membrane, to newly formed endocytotic structures, and to the membranes bounding secondary lysosomes, each of these three sets of membranes also shows significant differences from the others, differences that can be quite extensive. Concepts based on the data summarized in the next few sections generally incorporate assumptions that endocytotic structures start out with a delimiting membrane resembling the region of the plasma membrane from which they derive; this region sometimes resembles the average plasma membrane of the cell, but often has a more specialized composition. Once in the cell, changes in the membrane are rapid.

3.2.3.1. Phagosomes and Phagolysosomes

Gel electrophoresis of membrane proteins obtained by solubilizing macrophage phagosomes and phagolysosomes containing indigestible particles, demonstrates that the vacuole membranes start out substantially similar to the plasma membrane in polypeptide composition. This similarity can persist for the first hours of the vacuole's existence. Plasma membrane activities such as the 5'-nucleotidase are present in the phagosomes and phagolysosomes. And when macrophages phagocytose latex beads, after the cells had been pretreated with exogenous peroxidases and radioactive iodine so as to label proteins at the cell surface, the beads come to reside in intracellular compartments whose membranes show patterns of iodinated proteins closely resembling those of the plasma membrane.

Given the large proportion of its cell surface that a macrophage can invest in phagosomes (Section 3.3.2), these findings are not surprising. But even for actively phagocytosing macrophages there are signs of specialization of the membranes involved in heterophagy. For instance, migration of plasma membrane receptors to the regions directly involved in internalization (Sections 2.3.2.1, 2.3.2.6, and 2.3.2.7) can locally enrich the receptor content of these regions and correspondingly deplete other regions of the surface. As a result of such phenomena, intensive bouts of phagocytosis of IgG-opsonized, indigestible particles can leave the macrophage surface impoverished in Fc receptors until new receptors can be recruited from intracellular sources (see Section 3.3). Selective "sparing" of specific plasma membrane components during phagocytosis has also sometimes been claimed, as if phagosome membranes tended to exclude these components. But very few cell types and very few components have been studied in such terms, and the prejudice persists that ameboid cells like macrophages and amebas remodel regions of their plasma membranes only to a quite limited degree prior to using the membranes for phagocytosis. In apparent contrast, the plasma membrane bounding the specialized regions where food vacuoles form in ciliates differs immunocytochemically and in microscopic appearance from the membrane at adjacent zones of the cell surface.

Little is known in detail about the impact of lysosomal fusion on the composition of the membranes bounding phagocytotic vacuoles or about membrane changes during the vacuoles' later evolution into residual bodies. The end product is presumably a membrane of the lysosomal type (see below).

3.2.3.2. Endosomes and Related Structures

Adsorptive pinocytosis involves significant selectivity as to the membrane components that become internalized. This is seen most obviously at coated pits initiating receptor-mediated endocytosis; as Chapter 2 pointed out, these pits are highly enriched in endocytotic receptors. That coated pits and the vesicles to which they give rise, also can be impoverished for specific cell surface components, has been concluded, for example, from immunocytochemical demonstrations that the pits on cultured mammalian cells have much lower levels of **Thy1** or other cell surface antigens, than do adjacent regions of the cell surface. Speculation as to how this comes about includes the suggestion that proteins anchored to membrane lipids, as Thy1 seems to be, behave differently from other proteins, and the suggestion that the selective clustering of receptors in coated pits creates such high local concentrations of the receptor proteins as to crowd out other membrane proteins.

In absorptive cells of the kidney, coated pits contain little transpeptidase, an enzyme demonstrable at the surfaces of the microvilli adjacent to the sites of coated vesicle formation; evidently this enzyme is "anchored" by the plasma membrane or cell coat so as to be unable to diffuse into adjacent membrane domains. Interestingly, one cell surface antigen detectable immunocytochemically in coated pits of kidney cells, is not evident in endosomes of the same cell. (See also Section 3.2.3.3.)

Endocytotic coated vesicles were once thought to have little cholesterol, unlike the plasma membrane, but the evidence for this has been called into question. The conclusion was based principally on freeze-fracture electron microscopy and it has turned out that the probes used to demonstrate sterols microscopically, are artifactually impeded from access to the membrane by the presence of clathrin.

Some initial reports on endosome-enriched cell fractions (Section 2.3.1.6) from cultured cells or liver stressed similarities to the plasma membrane in overall lipid and protein composition and in PAGE patterns of proteins. Differences in detail, such as the differential presence or prominence of a few specific proteins, were noted but seemingly were of limited extent. For macrophages, on which plasma membrane molecules were prelabeled by using exogenous galactosyl transferase to attach radioactive galactoses, intracellular bodies arising in early phases of pinocytosis were found to resemble the plasma membrane in the pattern of radioactive glycoproteins. Other studies, however, revealed greater differences between endosome-rich fractions and plasma membranes. These discrepancies among studies are not surprising given the problems in purifying endosomes (Section 2.3.1.6) and the likelihood that the endosomes of a given cell are heterogeneous. Nowadays, most emphasis is being placed on the differences between the endosomes and the plasma membranes of given cell types revealed, for example, in two-dimensional electrophoretic analysis of the proteins in carefully prepared cell fractions. Beginnings have also been made in demonstrating differences between "early" and "late" endosomes; the earliest strong data along these lines concerned several classes of endocytotic receptors, which are much more abundant in early endocytotic structures than in later ones (Section 3.3).

Cytochemical studies on fibroblasts suggest that 5'-nucleotidase is internalized during fluid phase endocytosis and passes through most of the endocytotic compartments, including endosome-like bodies. An issue that badly needs addressing is the extent to which this, or other enzymes or transport systems of the plasma membrane, continues to function within endocytotic structures, and the timing of such function if it occurs. As already pointed out, the operation of ion transport systems from the plasma membrane could influence lysosomal or endosomal pH (Sections 3.1.5.4 and 3.1.5.5). Later, we will consider the transport of low-molecular-weight digestion products out of lysosomes, a process in which one might imagine participation by plasma membrane-derived systems (though as yet there is little evidence for such participation). Plasma membranes also possess hydrolytic activities, additional to the nucleotidase, which could contribute to the observed, limited degradative abilities of early endosomes: cell surface aminopeptidases and other enzymes capable of cleaving peptides or proteins are widely distributed among cell types (Section 6.2.1) and there may also be cell surface glycosidases.

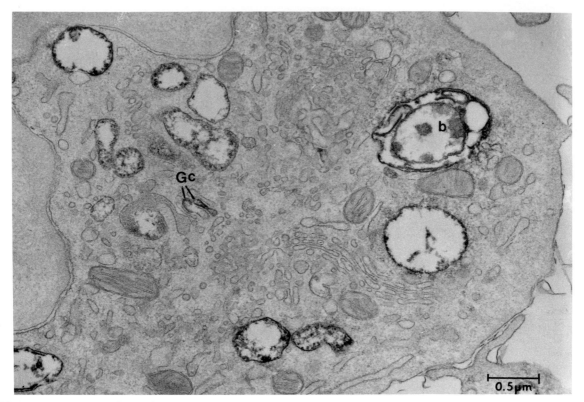

FIG. 3.10. Immunocytochemical preparation (HRP technique; Section 7.3.2.3) showing staining of vacuoles in a rat macrophage by antibodies against rat liver lysosomal membrane proteins. One of the vacuoles contains a recently endocytosed bacterium (b). Two sacs of the Golgi apparatus (Gc) also react. In hepatocytes, the same antibodies bind to the plasma membrane as well as to lysosomes. The principal lysosomal antigen recognized by the antibodies is a protein of about 100 kD and the antibodies also cross-react with a purified H^+/K^+-ATPase from pig gastric mucosa. From Reggio, H., Bainton, D., Harms, E., Coudrier, E., and Louvard, D. (1984) *J. Cell Biol.* **99**:1511.

3.2.3.3. Membrane Antigens; Immunocytochemistry

As in other realms of cell biology, antibodies directed against membrane constituents are expected to be of special value in examining the membranes of lysosomes, endosomes, and related structures. One goal is to seek antigens unique to lysosomal or endosomal membranes. The corresponding antibodies would be useful for immunocytochemistry and biochemical analyses and for perturbing function.*

Among the antibodies that have been raised against rat liver lysosomes, several do seem to bind selectively to membrane-associated antigens, to cross-react with lysosomes of kidney and other cells, and to stain lysosomes almost exclusively. With some of these "antilysosomal" antibodies, however, careful study reveals staining at sites additional to the lysosomes, such as endosomes and the plasma membrane bordering bile canaliculi (see Fig. 3.10). [The extent

*In immunocytochemical studies of endocytotic and lysosomal structures, thought must be given to the possibility of partial degradation, "masking," or other effects that could render antigens immunocytochemically undetectable even when the antigens are present. These possibilities are difficult to assess in detail. However, the antigens considered in this section mostly seem to be proteins with relatively long life spans. For this reason, their digestion by lysosomal or endosomal enzymes is viewed as contributing in only a minor way to the observed distributions.

of such staining varies with the immunocytochemical procedure used (see Section 7.3.2.3) and the cell type studied.] Comparable findings have been made with antibodies that were raised initially against coated vesicles from chicken liver but have been found to bind to the heavily glycosylated, LEP-100 lysosomal membrane proteins described in Section 3.2.2.2: these antibodies stain lysosomes of cultured chick fibroblasts heavily, but under ordinary circumstances, at least 10–20% of their binding is at nonlysosomal sites—endosomes, plasma membrane, and Golgi region. How many of the "lysosomal membrane antigens" identified in different material have a similarly dispersed distribution is not clear, because the sensitivities of immunocytochemical approaches are rarely understood well enough to predict whether small amounts of antigens would be detectable.

One straightforward explanation for the presence of "lysosomal" antigens at nonlysosomal locales is that components of the lysosomal membrane can move to other compartments when, say, liver lysosomes are defecated at bile canaliculi. There are other possibilities, such as that lysosomes, plasma membranes, and endosomes each receive the same, or related, antigens from a common source and thereafter do not exchange them with other compartments, at least under ordinary circumstances. But, for example, when cells are treated with chloroquine (see Section 3.3.4.2), the chick antigen mentioned in the last paragraph undergoes rapid and drastic redistributions to its nonlysosomal locations, which are most readily explained by changes in the patterns of membrane exchanges among compartments. (The distribution of this protein at the cell surface parallels that of clathrin, so perhaps coated membranes mediate its transport.) And, as will be seen in Section 3.3, the notion that membrane antigens concentrated in lysosomes move among the several compartments that participate in heterophagy accords well with other types of evidence.

Along with antigens confined principally to the lysosomes, there seem to be ones that are present in early endocytotic compartments but are largely excluded from definitive secondary lysosomes. In *Paramecium*, some of the membrane antigens acquired by a food vacuole upon its formation or in its initial prelysosomal fusions remain with the vacuole for much of its subsequent life, but other antigens become much less evident as the vacuole evolves. During this evolution, the pattern of intramembrane particles in the vacuole membrane also changes, suggesting alterations in the content or organization of proteins in the membrane. Analogous results have been obtained with mammalian systems in which immunologically detectable receptors and other proteins observed in the plasma membrane and in endosomes or comparable compartments, are not found at appreciable levels in the lysosomes to which the endosomes contribute. Most strikingly, the M6P receptors, believed to carry recently made lysosomal enzymes to functioning lysosomes, are much more readily demonstrated in structures with the morphology of endosomes (e.g., in MVBs) than they are in "late" secondary lysosomes such as residual bodies (Section 7.3.2.2).

One antibody, raised against a 100-kD protein of rat liver lysosomal membranes, cross-reacts with the gastric mucosal H^+/K^+-ATPase involved in acidification of the contents of the gut. This antibody stains lysosomes in mammalian liver cells, macrophages, and other cell types (Fig. 3.10) and also shows some staining of endosomes, coated vesicles, and Golgi-associated compartments, as well as of plasma membrane. (In certain cell types, notably osteoclasts, this last staining is intense; see Section 6.1.1.) The best current guess is that the corresponding antigen is a component of the H^+ pumping system discussed earlier. The finding that the antigen is immunologically related to a component of the gastric system, even though the gastric proton pump is markedly different from the lysosomal one in properties such as vanadate sensitivity, suggests that the cell uses overlapping or evolutionarily related components to assemble sets of pumps with differing properties.

In cultured porcine kidney cells, monoclonal antibodies, raised against kidney proton pumps, stain lysosomes and several of the other compartments we have discussed. In the light microscope, the staining often gives the impression of being patchy, as if local regions of the surface of the lysosomes or endosomes were highly enriched in the antigen while adjacent regions of the same membrane were markedly poorer. If borne out by electron microscopy, this might mean that the membrane zone bearing proton pumps constitutes a long-lived, differentiated domain (see e.g., Section 3.1.5.5).

3.3. Membrane Recycling in the Heterophagic Pathway

That some membrane antigens are chiefly confined to definitive secondary lysosomes whereas others are found principally in endosomes has strongly swayed investigators toward conceptualizing receptor-mediated endocytosis in terms of a distinct endosomal stage prior to the lysosomal stages (see Section 2.4.4). Indeed, one line of argument suggests that in the long run it may be better, or at least easier, to define endosomes, lysosomes, and other bodies by the presence of such antigens than in the more traditional terms of morphology and enzyme activity. Decisions about this viewpoint must, however, await better analysis of the antigens, especially as regards their lysosomal or endosomal functions.

Demonstrations of certain antigens in both lysosomes and endosomes as well as at the cell surface, and findings that antigen distribution among these compartments can change with circumstances that alter heterophagic traffic, have also spurred intense interest in the cycling of membranes to and from endosomes and lysosomes. Specifically: Lysosomes, endosomes, and related bodies are now believed to participate in an extensive multidirectional cycling of membranes, also involving the cell surface, Golgi compartments, secretory bodies, and some other structures. Defecation of secondary lysosomes and fusion of old lysosomes with nascent digestive structures are the aspects of this circulation to which I have alluded most so far, but they are far from the only processes involved, or necessarily the most extensive ones. One challenge is to understand how compartments that seem to fuse with one another or send membrane-delimited transfer vehicles to one another can nevertheless differ significantly in the properties of their bounding membranes. Another is to determine the relations between movements of the **membranes** and movements of the **contents** of the compartments.

3.3.1. Rudiments Evident from Studies on Phagocytes

The changes in volumes and surface areas undergone by phagosomes and phagolysosomes during their life histories are tied to membrane cycles. After sealing off from the cell surface, phagocytotic vacuoles can grow transiently, increasing their volumes and surface areas by fusions with lysosomes or other bodies. Presumably as a result of such fusions, the close fit between the phagosome membrane and the engulfed particles in ameboid cells (Section 2.3.2.1) often is loosened temporarily. Subsequently, vacuole size diminishes, which can be partly accounted for in many protozoa by the departure of membrane from the vacuole surface in the form of tubular and vesicular structures that bud off into the cytoplasm (Fig. 2.5). Cytoplasmic structures arising in this way recycle into later rounds of heterophagy, as is exemplified by the reuse of the acid phosphatase-containing bodies that bud from late stages of protozoan food vacuoles and fuse with new food vacuoles (Section 2.1.3.4).

Macrophage lysosomes clad in membranes fitting closely over enclosed, indigestible latex spheres can subsequently receive considerable charges of pinocytosed HRP without long-term loosening of the membrane fit. Apparently, the lysosome is able to reduce its surface to compensate for continued fusions of membranes delivering endocytosed material.

3.3.2. Quantitative Considerations

Phagocytes can engulf striking quantities of food or other particles, and in so doing invest immense expanses of membrane in forming phagocytotic vacuoles. Single protozoan food vacuoles like the one in Fig. 2.7 attain a surface area of 8000 μm^2 or more. In other organisms, dozens or hundreds of smaller vacuoles are generated in a short time: *Paramecium* can form a vacuole, with a surface area of 800 μm^2, every minute for many minutes. In the ciliate *Pseudomicrothorax dubius*, which feeds on filamentous cyanobacteria (Section 2.2.1.1), lengths of algae up to 15 μm can be ingested per second; this requires use of as much as 200 $\mu m^2/sec$ of membrane for food vacuoles. But production of vacuoles even at these rates does not result in a correspondingly rapid overall shrinkage of the cell surface or diminution of specialized "feeding areas" of the cell surface.

Not only ciliates, but also free-living amebas like *Acanthamoeba* or *Dictyostelium* can use membranes for endocytosis at prodigious rates.

Rates of pinocytotic internalization of membrane by mammalian cell types have been estimated by determining the pace of uptake of supposed fluid phase tracers (this is done to estimate the total volumes of fluid being internalized) and measuring the areas and volumes of the intracellular compartments through which the tracers move. Some of the calculations need reexamination: In a few cases, tracers thought to be taken up solely in the fluid phase have turned out to bind to plasma membranes (Section 2.2.2.4), which would lead to overestimation of the rates of endocytotic use of plasma membranes. On the other hand, some of the earlier studies probably underestimated the rates of pinocytosis by failing to take into account the very rapid reflux of a proportion of the tracer now known to occur (Sections 3.3.3.1 and 3.3.4).

Despite these problems, it is firmly established that mammalian cells can sustain rates of pinocytosis requiring the cells to replace very large areas of cell surface membrane rapidly (Tables 3.2 and 3.3). Fibroblasts seemingly can replace an area of membrane equivalent to their entire cell surface every hour or two. Macrophages, when active in phagocytosis or pinocytosis, may have to replace plasma membrane areas equivalent to their cell surface area two or more times an hour. (Estimates for actively endocytosing *Acanthamoeba* fall in the same range.) Mouse macrophages can pinocytose at rates that incorporate fluids equivalent to at least 25% of the cell's volume per hour. As estimated

TABLE 3.2. Quantitative Data on Pinocytotic Internalization[a]

Cell type	Cell dimensions				Internalization			
	V (μm^3)	S (μm^2)	ϕ (μm)	ξ	PM/hr	v (μm^3/min)	s (μm^2/min)	ϕ' (μm)
Mouse peritoneal macrophages	395	825	9.1*	3.2*	1.9	1.2	25.6	0.28*
Mouse macrophage cell line, J774	1662	3925	14.7*	5.8	2.8	4.5	180	>0.2
Guinea pig alveolar macrophages	1750	2250*	15*	3.2[b]	5.3*	8.6	198*	0.26[b]
Mouse macrophage cell line, P388D$_1$	900	1300	12	2.9	2.8	2.5	61.1	0.25
Mouse fibroblasts, L cells	1765	2100	15*	3*	0.5	0.9	18.3	0.30*
Human fetal lung fibroblasts, IMR-90	2500	2670*	16.8*	3	2.5*	5.3	107*	0.3[b]
Chinese hamster ovary cells	—	—	—	—	1*	1.1	—	—
Acanthamoeba castellanii	2540	2388	17	2.6	9*	33	—	0.6*
Dictyostelium discoideum	4200	6300	20	5	1.5	10	160	0.38
Mast cells	—	—	—	—	0.1	—	—	—
Polymorphonuclear leukocytes	395[b]	—	—	—	—	2.9	—	—
Chinese hamster ovary cells	—	—	—	—	—	1.3 + 5.6	—	—
Lymphoid cells	—	—	—	—	—	0.6×10^{-3}	—	—

[a]Internalization presented in terms of membrane areas and volumes. V and S refer to the overall cell volume and surface area; ϕ, the cell diameter; ξ, the factor by which the cell surface area exceeds that for a sphere of similar dimensions; PM/hr, the rate of internalization in terms of the multiple of the cell's plasma membrane area taken in per hour; v and s, the volumes and membrane surface areas taken in by a cell per minute; ϕ', the volume-weighted average diameter of the initial endocytotic vehicles in which tracers are seen. Data recalculated and estimated from the literature by L. Thilo (*Biochim. Biophys. Acta* **822**:243, 1985).
[b]Assumed value based on similar data in table, used for further calculations (-).

with HRP, under conditions where this tracer is not bound extensively to the cell surface, 25–50 μm^2 of membrane—3–5% of the macrophage's cell surface—is internalized per minute. Eventually, the tracer accumulates in a population of lysosomes and related structures occupying 2–3% of the cell's volume, and having a surface area equal to 15–20% of the cell's surface area. The lysosomal compartment starts receiving tracer about 5 min after internalization commences. By 45 min to 1 hr the compartment is "saturated" in the sense that the volume and area of structures labeled with HRP show no further increases, as if the tracer has reached all the members of a population of intracellular structures that is in a steady state.

3.3.2.1. The Cell Surface Is Replenished with "Old" Membrane Materials

One might postulate that the lysosomes continually degrade the membranes that reach them through endocytosis but that the cells sustain their surface area by replacing the degraded membrane with newly assembled membrane. Indeed, the interior vesicles within certain MVBs do include membrane components that come from the cell surface and are on their way to degradation (Section 5.2.1).

But phagosomes and phagolysosomes rarely show the large populations of interior vesicles that would be expected if shrinkage of phagocytotic vacuoles depended substantially on the vacuole's internalizing their own surface membranes. Microscopic observations, particularly on protozoa, suggest that the loss of membranous tubules and vesicles in the opposite direction—into the cytoplasm—is much more extensive. Furthermore, endocytosis has often been found to continue actively for many hours under conditions that inhibit synthesis of new membrane macromolecules. And even where MVBs are central participants in endocytosis, rates of degradation and resynthesis of membrane macromolecules usually are much lower than would be expected if membranes taken into the cell by endocytosis were simply degraded and new membranes made to replace them.

None of this is to say that membranes involved in endocytosis are immortal. When, for example, macrophages take up large immune complexes or opsonized particles, they degrade the Fc receptors responsible for uptake. This degradation is much faster than the breakdown of Fc receptors by "resting" cells; the "half lives" (Section 4.5.2.2; see also footnote on p. 218) of the Fc receptors decline from resting values of 10 hr, to 2 hr (see Section 5.2). The receptor responsible for adsorptive endocytosis of EGF also seems to be degraded fairly rapidly when EGF is being internalized (Section 5.2.1.2).

But even in rapidly endocytosing cells, lysosomal membrane proteins and many of the relevant plasma membrane proteins are synthesized and degraded at the relatively relaxed pace characteristic of the steady-state turnover of many other cell proteins (half lives of a half-day to a few days; Chapter 5, Sections 7.1 and 7.4.3.7). And many of the receptors responsible for receptor-mediated endocytosis differ from the EGF or Fc receptors in that their degradation rates are little affected by the presence or absence of the corresponding ligands (Section 5.2). Under some conditions of endocytosis, it is possible to deplete the cell surface of these receptors, but restoration often takes place within a minute or two. Transient depletion and rapid restoration is frequently observed, for example, when endocytosis is restarted after presentation of ligands during blockade of uptake by low temperature.

The explanation for the rapid replacement is that, during endocytosis, the cell draws upon "pools" of preexisting membrane materials to replenish its cell surface components. The existence of large intracellular pools was suspected, for example, from the ability of cells to undergo very rapid changes in shape that require major increases in cell surface area as, for example, when cultured macrophages are induced to flatten by treatment with phorbol esters. From studies of such shape changes it is calculated that ameboid protozoa and macrophages have internal sources sufficient at least to double or triple their surface area without new synthesis. Estimated pool sizes for specific proteins—endocytotic receptors and the like—are similarly large (see Section 3.3.4.1).

Under unusual circumstances, the pools can eventually be drawn down. Depletion of intracellular membrane stores is, for example, the explanation most frequently advanced for findings that macrophages induced to take up very large amounts of indigestible particles in a short time, may virtually cease phagocytosis for a period after accumulating a large load; the cells seemingly must wait until they have synthesized new membrane components to replace those tied up in intracellular vacuoles. Ordinarily, however, cells are able to compensate for endocytotic withdrawal of cell surface materials for quite prolonged periods. As will be seen shortly, this is not chiefly because they have set aside supplies of membrane soon after its assembly, to await use when needed (though such storage may sometimes occur). Much of the replacement relates to the repeated recycling of membrane components to and from the cell surface; the intracellular "pools" detected in the experiments above often are largely of membrane molecules that undergo such cycling and "happened" to be in the cell interior at the time the experiment was done.

Most schemes propose that cells resupply their surfaces by bulk movements and fusions of assembled membranes rather than by the recruitment of individual protein and lipid molecules. In ciliated protozoa, replacement is thought to be carried out from swarms of cytoplasmic vesicles that accumulate near the oral regions and can insert in the cell surface by exocytotic-like fusion. In amebas and mammalian cells as well, there are plenty of small vesicles and tubules in the places appropriate for these structures to be responsible for replenishing the cell surface, though there is not much direct evidence yet as to which of these vesicles and tubules actually do the job (see Sections 3.3.4.4 and 3.3.4.5).

3.3.3. Recycling from Lysosomes

In *Paramecium* the cytoplasmic vesicles that accumulate at the oral region include a population with a distinctive discoidal shape (Fig. 3.11). At least some of the members of this population represent membrane "retrieved" from old digestive vacuoles—more precisely, from membrane reinternalized at the "cytoproct" subsequent to the defecation of residual bodies. Evidently the residual bodies' membrane added to the cell surface by exocytosis is gathered back into the cytoplasm by endocytosis-like processes that eventuate in the formation of discoidal vesicles. (We will see in Section 5.3.3 that such coupling of exocytotic insertion of membrane in the cell surface with endocytotic retrieval is common.) The vesicles then migrate, along routes defined by microtubules, from the cytoproct to the cytopharynx where food vacuoles arise (Fig. 3.11).

These observations were among the early indications that one reason why

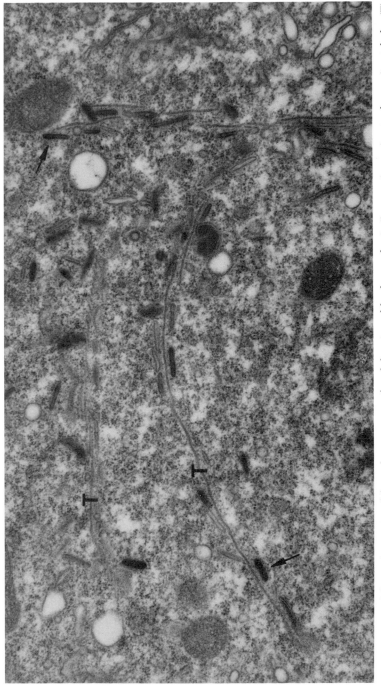

FIG. 3.11. Discoidal vesicles retrieved from the cytoproct after defecation of food vacuoles in *Paramecium* migrate along microtubules (T) back to the cytopharynx for reuse. The presence of electron-dense reaction product for HRP in many of the vesicles (arrows indicate two) testifies to their endocytotic origins. (See also Fig. 2.5.) From Fok, A. K., and Allen, R. D. (1980) *J. Cell Sci.* **45:**131. (Copyright: The Company of Biologists, Cambridge.)

cells can replenish their surfaces with old membranes for extended periods is because the pools available for such replacement can themselves be recharged with recycling membranes. For many cell types, however, it is unlikely that this recirculation depends heavily on lysosomal defecation: defecation by actively endocytosing cells is often minimal, as with macrophages that retain indigestible materials for days or weeks. Even for protozoa, the residual bodies that undergo defecation are usually quite a bit smaller than the endocytotic structures from which they evolve so that membrane retrieval after defecation could provide only a fraction of the membrane needed to maintain the cell's surface area during sustained feeding.

3.3.3.1. Nondefecatory Recycling

Lysosomes almost certainly can recycle membranes to the cell surface without undergoing defecation. In one experiment pointing in this direction, macrophages were allowed to take up latex beads with lactoperoxidase covalently attached, for periods long enough to permit most of the beads to accumulate in lysosomes. The cells were next exposed at 4°C to peroxide and ^{125}I; under these conditions, the lactoperoxidase in the lysosomes catalyzes the attachment of radioactive iodine to proteins in the lysosomal membranes. When the cells were then warmed to 37°C, much of the iodine-labeled membrane protein redistributed from the vacuole to the cell surface but the latex beads remained within the lysosomes. In other investigations, *Acanthamoeba* amebas were allowed to take up fluorescein covalently linked to inert beads by bonds that can be hydrolyzed by lysosomal enzymes. The fluorescein was extruded from the cells much more rapidly than were the beads. On the assumption that the fluorescein cannot cross membranes even after being liberated from its bondage to the beads, the escape of the tracer from the lysosomes to the cell exterior was presumed to depend on membrane-bounded vehicles that do not simultaneously transport the beads. Similar conclusions have been drawn from experiments with endocytosing macrophages or CHO cells, in which small, highly soluble, and readily diffusible indigestible materials, such as sucrose or Lucifer Yellow, were observed to efflux from the cells much more rapidly after they were endocytosed into lysosomes than did inert beads or HRP taken into the same compartments. And cells digesting proteins in their lysosomes sometimes regurgitate small amounts of partially digested molecules, apparently without releasing other lysosomal contents. (For a number of the experiments summarized in this paragraph, the presumed inability of the tracers to cross membranes under the conditions prevailing still needs rigorous testing; see Section 3.4.1.2.)

The vehicles believed to mediate the lysosome-to-cell-surface links in these various phenomena have not been identified definitively. The task of making this identification is vastly complicated by the fact that lysosomes are not the only compartments from which membrane recycles to the surface (Sections 3.3.4 and 5.3.3). However, when tracers like HRP or cationic ferritin (Section 2.2.2.3) are allowed to accumulate in lysosomes and then are followed during subsequent intervals, some of the internalized tracer returns to the cell surface and small tracer-containing vesicles and tubules are seen in the distributions appropriate for structures mediating this return. Also possibly germane, though the involvement of lysosomes is not clear, are findings on *Amoeba proteus*: After forming fluid-filled intracellular vacuoles containing the cationic dye Alcian Blue, mostly bound to polyanions of the vacuole membrane, this organism can subsequently return much of the Alcian Blue to the cell surface by forming transient channels connecting the vacuoles to the exterior of the cell.

But if vesicles or transient channels do account for the return of membrane from the lysosomes to the cell surface, how is it arranged that such transfer is selective, so that only certain lysosomal contents efflux rapidly from the cell? Is it plausible to think of mechanisms of vesicle formation in which diffusion limits the entry of materials into a nascent vesicle so that when (hypothetical) very small vesicles bud from lysosomes at rapid rates, only the most diffusible of the lysosome's contents have time to enter the vesicles in appreciable amounts? Or are some materials prevented from departing by being bound within the lysosomes (Section 2.1.4.1 and 2.4.2.2) or being unable to filter through the "halo" zone (Section 3.2.1)?

In the experiments outlined above, in which macrophages ceased endocytosis after taking up large loads of inert beads, the presence of the indigestible beads within the lysosomes may have limited the shrinkage of the lysosomes and in so doing, affected the recruitment of membrane for recycling. Such recruitment shows hints of interesting controls. Thus, the work illustrated in Fig. 2.9 has been followed up by further studies in which cultured macrophages were made to diminish the volume of their lysosomes very rapidly, by first exposing the cells to sucrose, which increases lysosomal volume (Section 3.4.1.3), and then exposing the cells to invertase, which reverses the increase. As the lysosomes shrank, Lucifer Yellow, previously incorporated into the lysosomes by pinocytosis, continued to efflux from the cells, but the rates of efflux were essentially unchanged from those prevailing before the rapid shrinkage of the lysosomes' volume. The dye is not expelled wholesale, as would be expected, for example, if the cell got rid of its sudden excess of lysosomal membrane by sending the membrane to the cell surface via transport vesicles (or by defecation). Interpretation of this sort of finding is difficult as we know so little of the pertinent regulatory mechanisms. But one possibility is that most of the lysosomal membrane is somehow precluded from circulating to the cell surface and must be disposed of in other ways (e.g., by degradation within the lysosome itself; see Chapter 5). Circulation to the surface might be restricted to specialized membranes whose amounts do not change much in the experiment just described. More generally, if lysosomal membranes do in fact have antigens not shared much with the plasma membrane or with endosomes (Section 3.2.3.3), these antigens must either be ones that somehow are selectively restricted from participating in cycling out of the lysosomal surface or ones that return to the lysosomes very efficiently and rapidly when they do cycle away.

3.3.4. Recycling from Endosomes; Receptor Recycling

Detailed kinetic studies of the uptake of fluid phase tracers, such as sucrose or Lucifer Yellow, by macrophages, fibroblasts, or slime mold (*Dictyostelium*) amebas, show that the markers enter sequentially into at least two different sets of compartments. The second set, which probably corresponds to definitive lysosomes and late endosomes, takes at least 30–60 min to reach a steady-state level of "filling" with the tracers; the tracers efflux from these structures to the cell surface at a correspondingly "slow" rate. The earlier set of structures, presumed to correspond to endocytotic vesicles and early endo-somes, fills to steady state within 5–10 min or even less. Many molecules of the tracers in the first compartment do move to the second, but 25–50% (some-times more) of the tracer molecules pass, instead, back to the cell exterior, within a few minutes after they are taken up.

An extreme case of such behavior is encountered in *Entamoeba histolytica* in which soluble tracers, such as dextrans or HRP, enter lysosomes in apprecia-ble amounts only after hours of exposure of the cells to the tracers (see Section 2.4.4). The cells do take up the tracers at reasonably rapid rates, but most of the initial accumulation is in a set of compartments (believed to be "prelysosomal" and nonacidified) that readily return the tracers to the extracellular space.

The conclusion from such studies is that recycling to the cell surface from "mature" lysosomes may be quantitatively less important than is recycling from endosomes and related structures.

3.3.4.1. Recycling of Endocytotic Receptors

Along with the return of fluid phase solutes from "early" endocytotic structures to the cell exterior, there is also recycling of membrane-associated components. When cells are allowed to initiate receptor-mediated uptake and then, a few minutes later, are subjected to treatments that dissociate ligands from their receptors, 10–50% of the ligand molecules that had most recently been internalized can be released from the cells. (Dissociating conditions gen-erally used in these experiments include exposure to competing ligands or to EGTA, for Ca^{2+}-sensitive receptors.) As the interval before the dissociating treatments is increased, the proportion of endocytosed ligand that can be re-leased declines. Evidently in the first minutes after endocytotic compartments seal off from the surface, a significant proportion of the ligand–receptor com-plexes in these compartments return to the surface. (It cannot readily be deter-mined whether all of the complexes remain intact throughout this rapid cycle or whether some dissociate within the cell, transiently or permanently; see below.) Other experiments showed that when proteins at the cell surface are labeled by covalent attachment of radioactive galactose residues, the label is internalized, endocytotically, with biphasic or multiphasic kinetics roughly comparable to the kinetics of internalization of fluid phase tracers. This is taken to indicate that, of the membrane internalized during endocytosis, some returns very rapidly to the surface, whereas some stays in the cell for a longer time.

At the core of the evidence for recycling of membranes to the cell surface from endosomes and related structures are demonstrations that various of the receptors responsible for receptor-mediated endocytosis cycle repeatedly be-tween the cell surface and the cell interior even while delivering ligands, most of which are degraded within the cell. The data were obtained chiefly with tissue culture or organ culture systems encouraged to endocytose at very high rates, but few doubt that the situation for cells under normal conditions *in situ* is similar, at least qualitatively.

The cell types studied most often and therefore treated as "typical," gener-ally have tens of thousands to hundreds of thousands of copies per cell of given species of receptor, such as the fibroblast's LDL receptor or the hepatocyte's asialoglycoprotein receptor (Table 3.3). Hundreds of thousands or even mil-lions of molecules of a given ligand can be taken up per hour. At any given

TABLE 3.3. Distributions and Cycling Behavior of Endocytotic Receptors[a]

Receptor for	Cell type	K (10^{-9} M)	R_{out}/cell ($\times 10^{-3}$)	R_{in}/R_{out}	τ (min)	Round trip (min)
Transferrin	Human hepatoma cell line, HepG2	5	50	2	4	16
	Mouse teratocarcinoma stem cells	6.7	5.7	3	6	24*
	HeLa cells	—	180	3	7	28
	Human leukemic cell line, K562	~1	160	3	6.7	27*
Asialoglycoprotein	Human hepatoma cell line, HepG2	7	200	0.14[b]	2.3 + 8.7[c]	16
	Isolated rat hepatocytes	5.6	450	4.9[d]–1.2[e]	6.3	3.7[d]–14[e]
Chemotactic peptide	Rabbit peritoneal polymorphonuclear leukocytes	20	(33–10)[f,*]	(0.9–7.3)[f,*]	6.7	(13–56)[f,*]
Mannose residues	Rat alveolar macrophages	10	≈ 100	2*	3*	9*
Low-density lipoprotein	Cultured human skin fibroblasts	1	7.5–15	1.7–2.3	≈ 6	12
α-Macroglobulin	Rabbit alveolar macrophages	~1	≈ 120	≥ 2	6	≥ 18*
	Cultured fibroblasts, NRK-2T cells	40–100[g]	600–800	—	≈ 2	20

[a] R_{out}, receptor number on cell surface; R_{in}, receptor number on intracellular membranes; τ, lifetime on cell surface; round trip estimated by product of $(R_{in}/R_{out} + 1) \cdot \tau$. Data recalculated and estimated from the literature by L. Thilo (*Biochim. Biophys. Acta* **822**:243, 1985).
[b] In the absence of ligand.
[c] Binding of ligand takes 8.7 min.
[d] Freshly isolated cells.
[e] Isolated cells in culture at 37°C.
[f] Depending on ligand concentration: $(1–20) \times 10^{-8}$ M, respectively.
[g] Low-affinity receptor type; high-affinity type: $(K_d < 0.2 \times 10^{-9}$ M), $R_{out} \approx 10^4$/cell.
* Estimated by the author, based on data given in the literature.

time, some of the receptor molecules are present at the cell surface, where they can bind ligands directly from the extracellular space. Other receptors of the same type are in intracellular locations; extracellular ligands ordinarily cannot reach them "directly" (i.e., without endocytosis).* Under steady-state conditions, the intracellular populations generally are as large as the cell surface populations, or larger—sometimes by severalfold.

From the relatively long life spans of many endocytotic receptors and for example, from observations that oocytes of *Xenopus* continue to endocytose the yolk protein, vitellogenin (Section 4.2.1.1), even after 12–24 hr of exposure to inhibitors of protein synthesis, it is inferred that the intracellular populations are not comprised primarily of recently made receptors on their way to the cell surface. Rather, in large measure, the receptors inside the cell represent ones internalized during endocytosis and will sooner or later return to the cell surface to be put to work again. This last conclusion is strikingly supported by bookkeeping: the known rates of receptor-mediated endocytosis require repetitive and rapid reuse of the receptors. Mammalian fibroblasts, for example, can take up LDL at rates requiring that each of their LDL receptors be used, on the average, 100–150 times during its life span of roughly 30 hr. The asialoglycoprotein receptors of cultured hepatocytes may each be reused more than 250 times. The intensity of reutilization reportedly can entail that an average LDL or asialoglycoprotein receptor undergo an entire endocytotic cycle—from coated pit to cell interior and back to coated pit (Figs. 3.12 and 4.1)—every 10–20 min, sometimes spending as little as 1–5 min of each cycle awaiting ligand in a coated pit (Table 3.3).

3.3.4.2. Intracellular Dissociation of Ligands and Receptors; Inhibitory Effects of Bases

The cycling of receptors between the cell interior and the cell surface has been manipulated experimentally in several ways (Fig. 3.13). Dramatic in-

---

*Estimates of receptor number at the cell surface usually are based on measuring the capacity of cells to bind corresponding ligands under conditions where the ligands cannot pass into the cell interior. For intracellular receptor populations, determinations are made of the ligand binding capacities of cell fractions, or of detergent-permeabilized or -solubilized cells. Sometimes, antibodies to the receptors are used in place of ligands.

These methods are prone to technical problems (e.g., with intracellular receptors it is difficult to be sure that all the receptors have been made accessible or that the means used to provide access have not also altered receptor affinities for ligands). But even when the numbers are not suspect, there are nuances that will eventually have to be considered in perfecting the concepts to be discussed in this section. For example, though we will deal with all the receptor molecules with similar ligand specificities as though they were interchangeable, this may well be oversimplified. Thus, EGF receptors with different binding constants for EGF can coexist in the same cell, and for other receptors, subpopulations with similar binding specificities but different behaviors in endocytosis seem to be present in the plasma membrane. The explanations for such situations are not firm, but at least for the human asialoglycoprotein receptor, one factor could be that there are at least two active genes coding for distinguishable forms of the receptor polypeptide (see Section 2.3.2.9); perhaps others of the receptors also exist as families of related proteins differing in intimate details. In addition, a given receptor molecule might change its properties as it moves among different cellular locales (see footnote on p. 136), or at different stages in its life history.

The M6P receptors that bind lysosomal enzymes have intracellular roles as well as roles in endocytosis (Section 7.3). How these roles are related and whether the receptors are in identical forms in both activities are not known.

FIG. 3.12. Summary of the likely involvement of endosomes and MVBs in sorting and cycling during receptor-mediated endocytosis. Ligands (L) bind to receptors (R) at coated pits and are internalized, soon entering the tubules and vacuoles of the endosomal system. In this system, many types of ligands lose their association with receptors owing chiefly to the low pH, but certain ligands [notably transferrin (Tf)] remain bound. From the endosomes, various receptors return to the cell surface (step 5) while the endosomes transform into MVBs. Now, lysosomes (ly) fuse with the MVBs, which consequently evolve into residual bodies; certain receptors (e.g., those for EGF), having been internalized within the MVBs, are degraded. In reticulocytes (see Fig. 5.14), MVBs exocytose their contents, apparently without having fused with lysosomes (step 7b). From Harding, C., Aach, M., and Stahl, P. (1985) *Eur. J. Cell Biol.* **36**:230.

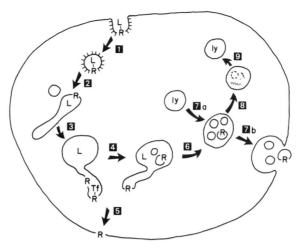

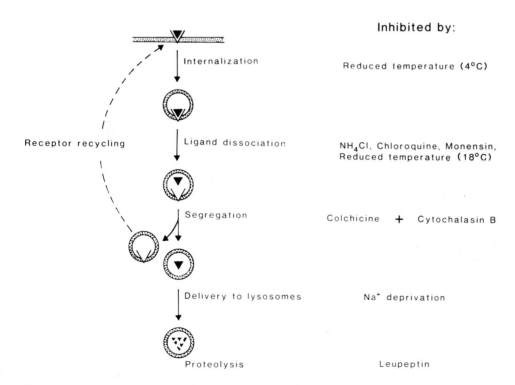

FIG. 3.13. Conditions and agents that inhibit different steps in the endocytotic cycle mediated by the asialoglycoprotein receptor in cultured rat hepatocytes. From Wolkoff, A. W., Klausner, R. D., Ashwell, G., and Harford, J. D. (1984) *J. Cell Biol.* **98**:375.

creases in receptor number at the cell surface can be provoked in some cells, within a few minutes after stimulation: For example, when exposed to phenylarsine oxide, alveolar (lung) macrophages rapidly insert into their plasma membranes many molecules of receptors selective for the protease inhibitor α_2-macroglobulin (Section 4.5.1.1). A customary means to trap cycling receptors within cells and correspondingly to deplete the surface receptor populations, is by exposure to ionophores such as monensin, or to the "weak base" amines discussed in Section 3.1.2. These exposures also inhibit degradation of ligands carried by the receptors.

Findings with bases and ionophores were historically important in the evolution of views as to how receptors recycling repeatedly to and from the cell surface can be used to take up ligands that are retained by the cell. The most obvious guess as to how the bases and ionophores act is, of course, that they raise the pH of endosomes and lysosomes. In following up this idea, it was shown that many of the receptors we have considered, including the asialoglycoprotein receptor, the LDL receptor, and the α_2-macroglobulin receptor, bind their ligands much more avidly at neutral pH than they do at the acidic pHs characteristic of endosomes or lysosomes. Correspondingly, ligands visible in the microscope are seen to be intimately associated with the membranes of very early endocytotic structures like coated pits and vesicles, but to lose this association when they enter endosomes—here the molecules or particles are more randomly distributed with most appearing to be free in the compartment's interior. It is this dissociation that permits the receptors and ligands subsequently to follow separate paths—the receptors returning to the cell surface and the ligands being disposed of in the digestive interior of lysosomes. In cells inhibited with weak bases, ligands and receptors remain bound together for prolonged periods. Often they stay in compartments with the microscopic appearance of somewhat swollen endosomes, an effect described as the prevention (or slowing) of passage of ligands from endosomes to definitive lysosomes.

Plausible as notions of acidification-controlled dissociation are, bear in mind that the concentrations of ligands in endosomes can be quite high so that even relatively low-affinity binding would be expected to result in appreciable occupancy of receptors. In addition, though the pH at the cell surface probably is higher than that in endosomes, the extent of the differences needs detailed evaluation; the concentrations of H^+ in the immediate vicinity of the cell surface could be strongly influenced by the charged groups of the plasma membrane and cell coat. It may be that factors additional to acidification help foster ligand–receptor dissociation in endocytotic compartments. Postulates have been put forth that changes in membrane potential or in concentrations of ions other than H^+ might be important (e.g., for those receptors that are Ca^{2+} sensitive, like the LDL and asialoglycoprotein receptors). Or enzymatic alterations, such as the limited proteolysis observed with EGF very soon after its uptake (Section 2.4.4.1), might modify ligands so as to discourage reassociation with receptors.*

---

*For receptors that undergo cycles of phosphorylation and dephosphorylation or comparable modifications, it will be important to determine whether the modifications are directly pertinent to the processes we are considering in this section and also how they relate, in timing, mechanism, or other respects, to the acidifications of endocytotic structures. Of immediate interest here are observations with cell fractions or other in vitro preparations suggesting that certain of the receptors are subject to substantial pH-dependent conformational changes. These changes have been detected in vitro as altered accessibility of parts of the molecules to proteases or to reagents

Receptors capable of recycling efficiently to the cell surface can be inhibited from so doing, and even "directed" to lysosomal compartments in which they are degraded, by multivalent ligands, antibodies, or other experimental agents that cross-link them in the plane of the membrane. The principal Fc receptor of macrophages, for example, is "unorthodox" in that it is not very sensitive to pH so that receptor–ligand binding persists at endosomal and lysosomal pH. When this receptor binds antibody-coated particles or comparable multivalent immune complexes such that the complexes cross-link the receptors, both the receptors and their ligands are digested within lysosomes (Section 5.2). But when the Fc receptors form univalent complexes, as with appropriate fragments of immunoglobulins, the receptors and the ligands cycle together into endocytotic compartments and then return to the cell surface, apparently unaltered. The differences in receptor behavior in these two situations are thought to be due largely to the fact that when multiple Fc receptors are bound in the same immune complex, the receptors are severely restricted in their mobility within the membrane; for this reason, or because the complexes are too large, the receptors might not be able to segregate into the vehicles presumed to carry membrane back to the surface. Influences of other types of cross-linking agents on receptor recycling are similarly laid to impacts on receptor mobility in the plane of membranes or to related influences on the capacity of membrane regions to undergo the shape changes needed to bud off as vesicles or tubules. By extension, comparable perturbations are hypothesized to explain the failure of pH-sensitive receptors to recycle when the pH is kept high enough to prevent dissociation of ligands from receptors, and to account for effects of low temperature on receptor recycling.

The serum protein **transferrin** normally cycles into endocytotic compartments and back to the cell surface, in company with the corresponding receptor (Section 4.1.2.3). But when transferrin is adsorbed to the surfaces of colloidal gold particles, so that multivalent ligand–receptor complexes are formed, endocytosed transferrin–gold is delivered largely to lysosomes.

Asialoglycoprotein receptors cross-linked by exposing cells to antireceptor antibodies become subject to partial intracellular degradation, but this may not be mediated by "typical lysosomes," for the degradation, though inhibitable by raising the pH, is not as sensitive to low temperature or to inhibitors like leupeptin as would be expected for lysosomal processes. (See Section 5.2.1.2; Section 2.4.4.1 mentioned that some molecules of endocytosed asialoglycopro-

---

that attach covalent labels to proteins. Similar conformational alterations in receptors as they pass into different environments during their functioning in cells may not only change the receptors' affinities for ligands but also may help "protect" the receptors from denaturation or enzymatic attack while in the cell.

Because approaches like those summarized above indicate that the asialoglycoprotein receptors in cell fractions enriched in intracellular compartments differ from those at the cell surface in the details of orientation of the receptor in the membrane, speculation has arisen that the ligand binding site undergoes cyclical changes in conformation and affinity. At one point it was even suggested that a portion of the receptor molecule moves to face the cytoplasmic side of the membrane in intracellular compartments and then shifts back to face the extracellular space when the receptor returns to the cell surface. Another imaginative proposition is that cells modulate the availability of asialoglycoprotein receptors to exogenous ligands by changing the glycosylation patterns of the endogenous glycoproteins that neighbor the receptors in membranes; to reduce receptor availability, the glycoproteins might be made in forms that bind to the receptors.

teins may be subjected to proteolysis very much faster than others, suggesting that at least two degradation routes may exist.)

These cases must, of course, be studied further, to test whether receptor mobility is in fact at the center of the phenomena. Note also that despite their historical importance, the inhibitory effects of bases or ionophores on receptor recycling and on the behavior of ligands have yet to be explained satisfactorily in detail (see the discussion of the impact of the bases on lysosomal fusions in Section 2.4.7.2). The effects themselves vary considerably in different cell types, for different ligands and perturbing agents and for different concentrations of the same agent. Often the inhibitions are only partial, or are manifested relatively slowly. Most importantly, the assumption that the impact of weak bases and ionophores on recycling of receptors to the cell surface is mediated mainly through inhibition of ligand dissociation from receptors has been questioned: with the LDL receptor and others, doubts have arisen because depletion of detectable receptors, interpreted as due to inhibition of receptor recycling to the cell surface, is seen even when ligands are absent. (The recycling of these receptors, like their internalization, is presumed normally to be constitutive, in the sense defined in Sections 2.2.2.6 and 2.3.2.7.) As the membrane of a given endocytotic compartment generally contains several different types of receptors, one can "save the hypothesis" by arguing that the behavior of receptors whose ligands are absent is being controlled by the effects of other, adjacent, receptors whose ligands are present. There are, however, also cases in which preventing the normal dissociation of ligands from receptors by means not expected to alter endosomal pH (e.g., by linking the two covalently) results in the return of receptors to the surface with their ligands still bound.

One tentative conclusion: Adequate explanations for the effects of bases and ionophores on receptor recycling will depend on better understanding of how these agents affect overall ionic and osmotic balances in endocytotic structures and how these balances bear on the membrane phenomena and other processes of endosomal maturation (see also Section 5.3.3.1).

3.3.4.4. Tubules and Vacuoles; CURL

I have repeatedly mentioned that small vesicles and tubules are regularly seen near endosomes and lysosomes, and that continuities between the vesicles or tubules and the larger bodies are often encountered. In earlier days such continuities were almost always interpreted as demonstrating fusions of the tubules or vesicles with the larger bodies, or the existence of channels conveying materials to the lysosomes or endosomes. But now it seems likely that at least some of the configurations represent the genesis of small recycling vehicles—the budding of tubules and vesicles, which then move away.

One attractive proposal ascribes particular importance to tubules that commonly protrude from the surfaces of early endosomes and from MVBs (Sections 2.3.1.4 and 2.3.1.5; Figs. 2.11 and 2.15). Immunocytochemical and other microscopic findings suggest that the tubules protruding from endosomes are rich in receptors and that this is the case from relatively early in the endosome's evolution. Perhaps the tubules bud off before a degradative environment is established within the maturing bodies, carrying the receptors away from exposure to danger.

Tubules of narrow bore have notably higher surface-to-volume ratios than do more spherical vacuoles or vesicles. So, the separation of such tubules from more distended, vacuolar bodies would be favorable for transporting membranes away from the bodies while leaving most of the luminal contents behind. As ligands within endosomes dissociate from their complexes with receptors within endosomes, the ligands will tend to accumulate chiefly in the distended, vacuolar portions of endosomes because these portions contain most of the volume of the structures. The receptors will abound in the tubular extensions because these extensions contain much of the membrane area. Most models assume that a given tubule can carry several different receptors (there is only a little firm evidence on this point yet) and some suggest that the receptors are randomly dispersed in the plane of the endosomal membrane, distributing simply according to the relative surface areas of the tubular and vacuolar regions. But different receptors do show different distributions—those for EGF, for example, often are interiorized within maturing endosomes (Section 5.2.1.2) rather than remaining in the bounding membranes—so that investigators have toyed also with proposals that receptors concentrate in the tubules, in part through nonrandom, selective mechanisms.)

The term CURL (**C**ompartment for **U**ncoupling of **R**eceptors and **L**igands) has been proposed for the endosomal bodies combining vacuolar dilations with tubular extensions. As with other sobriquets for lysosome-related structures, this one has yet to be universally accepted. Nor is it proper to assume that all tubules bearing endocytosed materials or endocytotic receptors pertain to recycling phenomena. Noted already is that endocytosis may itself occur via tubular structures or generate small tubular intermediate structures directly (though many of the instances in which this was reported need reexamination, as most of the studies were done before the extent of recycling was appreciated).

3.3.4.5. More Questions; Golgi Involvement?

How do the routes and mechanisms responsible for the recycling of membranes, ligands, and soluble molecules to the cell surface in the first minute or two after internalization differ from the routes and mechanisms involved at later times? The migrations and progressive acidification of compartments carrying endocytosed materials probably are among the relevant factors but what are the others?

How does the cell manage to produce vacuoles with tubular extensions (and later also invaginate regions of the same structure to give rise to the internal vesicles of MVBs)? Could part of the explanation reside in differential distributions of transport systems or channels in the plane of the membrane? Resultant, transient local differences in osmotic balances might be such that some regions of the compartment balloon out or that other regions lose volume.

What paths are followed by recycling vehicles in returning to the cell surface? In epithelial cells and others in which particular types of receptors function almost exclusively at one or another cell pole, how is recycling organized so as to maintain this polarity (see Section 4.3.2.2)? When recycling receptors are reinserted in the cell surface, do they remain clustered in the patch of plasma membrane produced by the fusion of the recycling vehicle with the cell surface, or are they free to diffuse away? The answer to this last question is not known, but probably varies with receptor type. If, as they function, some receptors become linked to one another by covalent bonds or strong noncovalent interactions, the resulting cluster might not diffuse apart as it cycles. On the other hand, those receptors whose congregation in coated pits requires driving by ligands might be much less constrained. Even the receptors that recycle constitutively probably must move laterally to at least some extent, in order to find coated pits after returning from the interior; this, at least, is the usual explanation for the coexistence of receptors in pits with seemingly identical receptors spread out in adjacent regions of the plasma membrane.

For lysosomes, recycling of membrane to the cell surface is only part of the story; there also is the intracellular "recycling" wherein older lysosomes or derivatives thereof fuse with newer structures. Are there comparable multiple paths for endosome-derived structures such that some return rapidly to the surface whereas others have a longer residence in the cell? Do some, for example, fuse with new endocytotic bodies rather than with the plasma membrane? (This last possibility has already been raised both in considering the question as to whether endosomes are in some sense long-lived structures rather than transient states in the transformation of endocytotic bodies into mature lysosomes, and in connection with suggestions that derivatives of endosomes help acidify newer endocytotic structures.)

A related unknown that will vex us repeatedly in the remainder of the book is the possible involvement of the Golgi apparatus in endocytotic recycling. It may be more than fortuitous that the tubules and vesicles deemed likely to participate in recycling from endosomes and heterophagic lysosomes often mingle with and sometimes closely resemble Golgi elements (Sections 7.2.2.1 and 7.2.2.2). The Golgi apparatus itself participates in bidirectional exchanges of membrane with the cell surface (Section 5.3.2), and there seem to be Golgi or Golgi-derived compartments in which acidification and/or degradation are functionally important (Section 6.1.5.1). Sections 5.3.3.2, 5.3.3.3, 7.2, and 7.4 will explore the relations of the Golgi apparatus to lysosomes and endosomes in more detail. Points from those discussions worth noting here are that some of the receptor molecules recycling from endosomes seem to move into compartments with enzymatic capacities typical of the Golgi apparatus and that this passage sometimes is enhanced by exposure of the cells to weak bases.

Few studies have focused on the routes of membrane **lipids** during endocytosis. Fluorescent derivatives and analogues of gangliosides and phospholipids have been inserted in the plasma membranes of cultured cells and their fate followed by light microscopy: Sphingomyelin molecules recycle to the cell surface after accumulating transiently in aggregates in the cytoplasm—the aggregates probably represent membranes that collect near the Golgi apparatus or the "centrosomes" from which microtubules radiate (see Section 2.4.5). The labels do not accumulate much in lysosomes. Microscopically visualizable conjugates of cholera toxin, which bind to GM1 gangliosides in the cell surface, are endocytosed (sometimes via noncoated vesicles; Section 2.3.2.11) and move into several compartments including MVBs and, in some cell types, Golgi-associated systems like those considered in Section 7.2.2.

3.4. Escape from the Lysosomes

3.4.1. What Sorts of Materials Penetrate the Membranes Bounding Lysosomes of Animal Cells and How Do They Do It?

Like most other biological membranes, the lysosome's delimiting membrane is a "semipermeable" structure that permits water to cross much more rapidly than most other substances. This property has been made use of in a variety of experiments on the permeability of lysosomes. For example, suspensions of lysosomes exposed to millimolar concentrations of methyl esters of amino acids, swell and can burst if maintained under conditions that encourage the hydrolysis of these esters by lysosomal enzymes. Though occasionally it is suggested that the bursting results from production of membrane-disruptive agents such as methanol, the accepted explanation is that the esters, being weak base-like, penetrate the lysosomal membrane in un-ionized forms and then are hydrolyzed to liberate amino acids. The amino acids are much less able to penetrate the membrane than the esters, water enters osmotically as a consequence of the imbalance between ester influx and amino acid efflux, and it is this entry that eventually bursts the lysosomes. The implication—that the passive permeability of the lysosomal membrane is not exceptionally high, even to small molecules like amino acids that are expected to be liberated by intralysosomal digestion—is reassuring. It would, for example, be very difficult to understand how the organelles could maintain their low internal pHs by the mechanisms described in Section 3.1.5 were the delimiting membranes very porous.

Nevertheless, under ordinary circumstances in the cell or in cell fractions, the degradation products produced in the lysosomes do escape rapidly enough that lysbsomes do not swell osmotically to a marked degree even when they are degrading macromolecules and thereby liberating large number of smaller molecules. Instead, as digestion proceeds in the cell, the lysosomes normally shrink, at least eventually. And, for instance, when cultured cells take up proteins labeled with radioactive iodine, iodinated amino acids appear in the medium within 10 min or so of the onset of endocytosis. (Iodinated amino acids cannot be reused for protein synthesis and, once free in the cytoplasm outside the lysosomes, they tend to be released from the cells relatively rapidly.) On the assumption that vesicle-mediated efflux of solutes from lysosomes makes a negligible contribution, such "leakage," and similar release from lysosomes in cell fractions, presumably indicates that molecules of the sort and size of monoiodotyrosine (M_r 307) can traverse the lysosomal membrane reasonably well. (See also the comments on thyroid hormones in Section 4.2.4.1.)

With amebas or *Tetrahymena*, labeled sugars like radioactive glucose appear to have more ready access to the cytoplasm when the cells are actively endocytosing than is the case in resting cells. An intriguing interpretation is that the sugars can escape from compartments they enter after endocytosis—perhaps from the lysosomes—more readily than they can cross the plasma membrane. This notion needs buttressing by studies determining whether the plasma membrane itself alters in permeability to sugars or its transport of sugars during endocytosis. Even so, the speculation raises interesting questions about the similarities and differences in permeability and transport properties of the membranes involved at successive stages of heterophagy. Do any of the transport systems of the plasma membrane reach the lysosomal surface in a functional state? The weight of present evidence is negative, but the data are scanty. Protozoan food vacuoles undergo stages at which their surfaces exhibit extensive folds and protrusions. Does this provide increased surface area for egress of digestion products? And does release of the products continue from the small vesicles and tubules that bud from the food vacuoles? (E.g., do these small structures have surface-to-volume ratios more favorable for solute escape than do the larger vacuoles from which they originate?)

3.4.1.1. Permeation into Lysosomes of Animal Cells

Systematic evaluations of the abilities of molecules and inorganic ions to traverse the lysosomal membrane began with "osmotic disruption" studies. These were done on tritosomes isolated from rat liver, and on less highly purified lysosomal preparations from liver of animals not exposed to Triton, or from kidney, or from protozoa such as *Tetrahymena*. The usual approach was to determine the extent to which lysosomal enzymes lose their association with sedimentable particles, or their latency (Section 1.2), when suspended in isotonic or hypertonic concentrations of solutes of interest. If latency is lost in a given solution, and the solute has no known disruptive effects on membranes, it is concluded that the solute can penetrate into the lysosomes, raising the total intralysosomal concentration of osmotically active materials and thus producing influxes of water that eventually burst the organelles. (Impermeant solutes used alone, or along with permeant ones, "protect" against such osmotic disruption.) Reliance on changes in the sedimentability or accessibility of enzymes that are specifically lysosomal permits differential evaluation of lysosomal responses even in cell fractions containing other particles.

The data obtained in such studies must be evaluated with due attention to the abnormal environments to which the lysosomes are exposed and in the case of tritosomes, to the possible influences of their detergent load. Nonetheless, the osmotic disruption investigations and a few additional early studies using radioactive tracers or other methodology led to a rough consensus that provided the framework for present views about the passive permeability of lysosomes: The lysosomal membrane is believed to permit ready passage of small lipid-soluble molecules. Hydrophilic molecules with molecular weights up to about 200 also can enter lysosomes, though this movement is impeded if the molecule bears charged groups. Marked restrictions are evident above molecular weights of 200, and above the molecular weight range of 200–400, only distinctly hydrophobic molecules can penetrate. Monosaccharides enter lysosomes more readily than do hexitols, though *Tetrahymena* lysosomes show easier entry of hexitols than do mammalian preparations. (This last fact may relate to the presence of hexitols in the "diet" of the protozoan.) Disaccharides, oligosaccharides, and charged saccharide derivatives, like glucuronic acid, cannot cross the lysosomal membrane (whence, the utility of sucrose media for isolating lysosomes). Nucleosides (molecular weights of about 250) probably can cross the membrane, whereas nucleotides cannot. The data on amino acids are ambiguous. With *Tetrahymena* homogenates, most amino acids seem able to enter lysosomes; most dipeptides cannot. With mammalian preparations, results differ among different laboratories but in some hands at least, entry of the larger amino acids seems quite restricted though even a few small dipeptides such as glycyl-glycine do seem able to penetrate.

Components of lipids, such as fatty acids, have been studied much less than amino acids or sugars in part because, at the high concentrations utilized for osmotic disruption studies, many lipid components tend to insert into membranes and to alter them severely.

Monovalent inorganic ions, like Na^+ or K^+, offer little osmotic "protection" to isolated lysosomes at 0°C but in osmotic disruption tests at 25 or 37°C, the ions seem to enter lysosomes much more slowly than at 0°C. Like the supposed effects of low temperatures on the fusions of endosomes and lysosomes (Section 2.4.7.1), the effects on permeability to monovalent cations may reflect changes in the structure or fluidity of the lysosomal membrane at temperatures near 20°C. Divalent cations such as Ca^{2+} and Mg^{2+} enter very slowly, if at all, either at 0°C or at higher temperatures. For a time, some investigators assumed that none of the inorganic cations could permeate passively across the lysosomal membrane at significant rates under normal conditions. But this view has moderated somewhat in light of the effects of cations on intralysosomal pH and on lysosomal membrane potentials. Section 3.1.5.1 pointed out that lysosomes may in fact be modestly permeable to H^+, K^+, and Na^+ judging, for instance, from interactions of the pH gradient across the membrane with gradients in other ions.*

That anions like Cl^- can enter lysosomes is inferred from effects of the ions on membrane potentials (Section 3.1.5.3), and from osmotic experiments (see also Figs. 3.2 and 3.6). I^- penetrates into lysosomes faster than Cl^-. Larger and more complex biological anions such as phosphate enter markedly more slowly, often at rates very near the limits of reliability of the detection methods. Sulfate enters somewhat faster than phosphate; bicarbonate and especially acetate, an ion that may be needed for intralysosomal acetylations (Section 6.4.1.1), enter distinctly more rapidly than phosphate.

3.4.1.2. Hints of Carriers or Other Selective Transport Devices

Only a few experiments on isolated lysosomes of animal cells have attempted systematically to sort out the possible mechanisms for entry across the membrane, other than the proton pump. As earlier implied, familiar transport systems of the plasma membrane, such as the ones that couple transmembrane movements of inorganic ions to movements of other ions or of organic molecules, are not prominent in carefully prepared lysosome-rich cell fractions. But hints that carrier mechanisms or similar transport devices may be involved in passage of some molecules into isolated lysosomes have been obtained in osmotic disruption experiments in which rat liver lysosomes were exposed to supposedly isotonic concentrations (250 mM) of monosaccharides. Different sugars appeared to compete with one another for entry into the lysosomes and there were marked differences in apparent rates of entry among sugars of similar size and overall properties, and between D and L sugars: Of the hexoses,

---

*No firm agreement has been reached about the relative permeabilities of the lysosomes to different monovalent cations. (Remember that the cations differ not only in size, but also in other properties affecting their permeability behavior, such as the strength of the associations of the ions with water molecules.) The issue has been approached experimentally by, for example, evaluating the relative effects of ionophores on lysosomal osmotic properties, electrical potential, or pH (using ionophores that selectively increase the penetration of specific cations).

At the outset, some findings in these and other studies seemed to indicate that the passive lysosomal permeability to H^+ is notably less than that to the other common cations, but more recent reports have challenged this conclusion. Uncertainty persists partly because it is technically difficult to evaluate permeabilities precisely, in mixtures of structures that differ in permeabilities, pH, osmotic fragility, and membrane potential. For example, the heterogeneous content of most lysosome-rich cell fractions limits the utility of traditional methods that use radioactive ions to evaluate ion entry into compartments: one cannot be sure about which structures in the mixture dominate the patterns of ion behavior observed. Additional cloudiness arises from the presence of facilitated or active transport systems like the lysosomal proton pump. And charged macromolecules or particles in the lysosomal entrails provide sites with which different ions might associate with different affinities.

mannose and galactose (which are principal products of glycoconjugate degradation) enter more readily than glucose; glucose enters more readily than fructose or 2-deoxyglucose. Entry of the sugars is inhibited, to an extent, by classical transport inhibitors such as phlorizin or cytochalasin B.

These experiments on sugars involved quite high concentrations of solutes and of inhibitors and were done on crude preparations of lysosomes in which substantial disruption is evident even without the osmotic stress. But findings on the behavior of radioactive sugars, in protocols involving tritosomes and relatively low (20 mM) concentrations of sugar, seem to confirm several of the crucial inferences and the broad conclusion—that the "permeation" of some products of digestion may well involve specific transport systems rather than "simple" pores or channels—fits well with the findings to be discussed in subsequent sections.

When ATP is added to the medium surrounding cells of the J774 cultured mouse macrophage line or to mouse peritoneal macrophages, the cells increase their permeability to molecules with molecular weights up to 800. Dyes like Lucifer Yellow, which under usual conditions are thought to enter cells exclusively by endocytosis, can therefore pass directly from the extracellular space into the cytoplasm and nucleoplasm. The permeability change reverses once the ATP is removed and within hours thereafter much of the dye is expelled from the cell, while the remaining dye molecules become segregated largely to endosomes or lysosomes. Lucifer Yellow is a small anion, and there is increasingly convincing evidence that, in these experiments, it is moving into the intracellular compartments via transmembrane "transporters" for organic anions. In a variety of cell types, channels or other "transporters" for anions are present in the plasma membrane. In the ATP-treated cells it may be that they are present as well in intracellular lysosomal or prelysosomal compartments, enabling these compartments to accumulate the dye from the "cytosol." The alternative explanation—that the Lucifer Yellow in these experiments gets to the lysosomes and endosomes from small endocytotic vesicles forming at the same time as the dye penetrates the plasma membrane—seems less likely from microscopy and from the fact that expulsion of the dye from the cell and its accumulation in the endosomes and lysosomes are inhibited by probenecid, an agent that blocks anion transport in epithelia. Whether cells not treated with ATP also have significant intracellular populations of the anion transporters is yet to be determined. Perhaps, however, the systems for anion movement detected in this work are among the devices that contribute to the anion balances across lysosomal or endosomal membranes alluded to in Section 3.1.5.3. We will see also, in Section 3.4.2, that movements of organic anions into the vacuoles of some plant cells are extensive and physiologically important.

A voltage-gated cation channel has recently been identified in yeast vacuoles, and plant tonoplasts have been shown to possess transmembrane channels responsive to concentrations of cytoplasmic ions such as Ca^{2+} and to changes in potential.

To account for their ease of entry, it has been argued that certain acidotropic or lysosomotropic agents, chloroquine among them (Section 3.1.2.1), enter the cell by exploiting carriers or other transport systems "intended" for other materials. If so, do these agents enter lysosomes, or endosomes, in like manner? And if they do, are some of their effects due to their competing for carriers needed to transport other molecules?

The experiments in the last sections mostly involve penetration in the "wrong" direction—from outside the lysosome to inside. Lysosomal transport mechanisms might well have strongly vectorial aspects, biasing them toward movements from inside to out and, perhaps, using the transmembrane pH and potential differences or other ionic gradients to affect the affinities of carriers or even to drive transport.

Highly influential evaluations of the normal movement of digestion products from within the lysosome have relied on observations of the osmotic behavior of lysosomes in living cells, similar to the one illustrated in Fig. 2.9. The lysosomes there are thought to owe their "swollen" appearance to the fact that sucrose (a disaccharide) has entered them by endocytosis and can neither be digested nor escape; the sucrose in the vacuoles generates osmotic effects that prevent the vacuoles from undergoing their normal shrinkage. When the lysosomes are supplied, endocytotically, with invertase, this enzyme hydrolyzes the sucrose into the monosaccharides, glucose and fructose. These sugars can exit from the lysosomes, accompanied by osmotically motivated water whose departure "encourages" the lysosomes to diminish their volumes.

A series of such observations with sugars and sugar polymers led to the conclusion that the lysosomal membrane does not permit the passage of disaccharides, oligosaccharides, and other carbohydrate polymers, but that the membrane can be traversed by the usual pentose and hexose monosaccharides and by many smaller carbohydrates and carbohydrate derivatives, especially ones that lack ionizing groups. Similar experiments to evaluate the movement of amino acids and peptides utilized small peptides of L-amino acids as digestible probes, and peptides of D-amino acids as indigestible probes. The L forms produced little or no "swelling": as they are released by hydrolysis, most amino acids seem able to escape from lysosomes. The results with peptides of D-amino acids depended on size; small dipeptides evidently could escape but larger peptides such as tripeptides of alanine, or highly charged dipeptides, such as ones of glutamic acid, could not. [Perhaps, in vivo, some small dipeptides exit from the lysosomes to be hydrolyzed eventually by the ill-understood peptidases thought to exist in the extralysosomal cytoplasm (e.g., Section 5.6).]

Lysosomes isolated from fibroblasts shortly after the cells have endocytosed glycosaminoglycans (GAGs) lose digestion products from the GAGs—sulfate and hexosamine derivatives—to the incubation medium at faster rates than can be accounted for by leakage from lysosomes damaged during isolation.

Overall the results summarized in this section conform to expectations from the known digestive capacities of lysosomes.

3.4.1.4. Amino Acid Transport

The escape rates of amino acids from isolated lysosomes have been monitored by using methyl esters of the amino acids to load the lysosomes under conditions (e.g., low ester concentrations) that do not produce bursting (see the beginning of Section 3.4) and then measuring the efflux of the amino acids liberated by lysosomal hydrolysis. Isolated rat liver lysosomes exposed to leucine ester subsequently release leucine with a half-time of 2–3 min at 35°C. Other amino acids exit from human leukocyte lysosomes with half-times of 15–

30 minutes (but see the next section, and footnote on p. 147). That these movements of amino acids may be via specific carrier systems is suggested by evidence of "countertransport" phenomena similar to phenomena seen with carrier-mediated transport across the plasma membrane. For instance, lysosomes preloaded with micromolar concentrations of radioactive lysine increase their rate of loss of this labeled amino acid to the incubation medium when millimolar concentrations of unlabeled lysine or other basic amino acids are added to the medium, From the specificities of such effects in studies on mammalian liver, fibroblasts, and cultured thyroid cells, lysosomes are thought to contain a carrier for basic amino acids, a separate one for tyrosine and some other aromatic and neutral amino acids, and perhaps a carrier for proline. Each of these presumed carriers differs in details of requirements and behavior from seemingly analogous plasma membrane systems.

The carrier for basic amino acids may be particularly important for lysosomal function, as these amino acids are weak bases and thus might show tendencies to remain accumulated in the acidified, intralysosomal environment (see also Section 3.4.2.1).

Is any of this transport energy dependent? A few studies have found that transport of the basic amino acids, and perhaps of other amino acids (see below), can be diminished by weak bases and enhanced by ATP, suggesting that the proton gradient could be important for transport. But this work is still in its infancy, as are most investigations of the natures, specificities, and mechanisms of the amino acid carriers.

3.4.1.5. Defects in Transport; Cystinosis

Degradation of proteins containing disulfide links between cysteine residues should liberate cystine (M_r 240). That cells sometimes encounter special problems in dealing with cystines present in their lysosomes is manifested in the human disease, **cystinosis.** In this disorder the lysosomes of many cell types accumulate stores of cystine that can reach concentrations 100-fold greater than normal. (Intralysosomal cystine concentrations have been estimated at 20–50 mM or more in cystinotic cells, and precipitation of the amino acid as intralysosomal "crystals" is sometimes seen.) As far as is known, most of the cystine comes from ordinary degradation of endogenous and exogenous proteins: The hydrolytic capacities of the lysosomes of cystinotic individuals are normal, amino acid transport by the plasma membrane is unaffected, the cells metabolize all the amino acids other than cystine properly, and there is no obvious defect in lysosomal acidification.

Cystinosis is currently being blamed on defects in the abilities of lysosomes to transport cystine to the cytoplasm, most likely stemming from genetically based problems in a lysosomal cystine carrier system. Cultured normal cells loaded with cystine or cystine analogues, by growing them in cystine-methyl diesters or in "mixed disulfides" of cysteine with glutathione or penicillamine, clear their lysosomes of the load within an hour or two upon placement of the cells in cystine-free medium. Cells from cystinotic material take very much longer to accomplish such clearance. Lysosomes isolated from cystine-loaded normal human cells, or isolated lysosomes loaded *in vitro* by placing them in methyl esters of cystine or cystine analogues, release cystine to the medium with a half-time of less than 30 min; these rates are not much greater than the rates of efflux of other amino acids from the same preparations.

Corresponding lysosomal preparations from cystinotic sources clear their cystine loads severalfold more slowly, but they handle other amino acids normally. Normal lysosomes, when heavily loaded with cystine compounds, seem able to exchange these loads for radioactive cystine from outside, apparently by countertransport; this observation is taken as among the strongest evidence for the operation of a cystine carrier.* No such countertransport is seen with cystinotic material.

For a time, the idea that cystinosis flows from a defect in a cystine carrier was rivaled by suggestions that cystine normally is reduced to cysteine in endosomes or in lysosomes; the defect in cystinosis might affect either this reduction or the ability of cysteine to traverse the lysosomal membrane. Cysteine, however, is small enough (M_r 121) to escape readily from lysosomes, and cystinotic cells seem to deal with it normally. Furthermore, though knowledge about the reduction of disulfide bonds by cells is inadequate, there is no good evidence that **extensive** reduction normally precedes lysosomal hydrolysis or that disulfides ordinarily are reduced in lysosomes.† The prevailing view is that reduction of cystine is carried out chiefly by a glutathione-dependent system in the cytoplasm outside the lysosomes.

Depletion of cystine stores is promoted by exposing cystine-loaded cells or lysosomes to cysteamine, a small (M_r 77), sulfhydryl-containing weakly basic compound that normally is present in cells at low concentrations. Cysteamine seems able to enter lysosomes by penetrating the membrane, as well as by endocytosis. In the lysosomes it can engage in "exchange" reactions with

---

*Differences in observations made with different preparations have led to disputes about the nature and the specificities of the presumed cystine carrier, and about the extent to which cystine transport is responsive to ATP or to the proton gradient across the lysosomal membrane. For instance, efflux of cystine from isolated lysosomes of normal lymphocytes or leukocytes is stimulated, to a degree, by ATP and is inhibited by protonophores like CCCP, which collapses the proton gradient. But granule preparations isolated from leukocytes (principally neutrophils) can transport cystine without added ATP, and effects of weak bases on efflux of cystine from these granules are seen only with relatively high concentrations.

These findings may richly reward follow-up because they might be due to important special features of the neutrophil granules, such as a postulated ability to maintain a proton gradient for an unusually long time without needing ATP. The hows and whys of neutrophil accumulation of cystine are not entirely clear, however. For example, the cells studied in the work on cystinosis and cystine uptake were obtained from the human peripheral circulation, but presumably, in ordinary circumstances they would have accumulated most of their lysosomal cystine by digesting proteins acquired after the cells left the circulation. And it is not definitively known which type of neutrophil granule is most affected in cystinosis. When the neutrophils from normal subjects are loaded with cystine under culture conditions, about 25–30% of the cell's granules are visibly affected by the loading. Peroxidase-containing structures in the neutrophil are among those that "swell" upon such loading, but it is not self-evident whether these structures are altered primary lysosomes or, rather, forms of secondary lysosomes.

Under some experimental conditions, isolated lysosomes of normal fibroblasts seem to have more difficulties in dealing with "mixed disulfide" cystine analogues than do lysosomes from other cell types.

†There may be more to the story than this. Investigators of the biological fate of insulin, a protein hormone that is endocytosed by many cell types, have long debated the possibility that the disulfides essential for insulin's activity and integrity are reduced by a plasma membrane enzyme or by an intracellular system encountered very soon after endocytotic internalization. Moreover, drugs linked by disulfide bonds to endocytosable carriers (e.g., to polylysine; Section 4.8.2), behave as though at least some of the drug molecules are liberated from these linkages before the complexes reach mature lysosomes; one of several speculative explanations for this behavior posits reduction by glutathione taking place in nonlysosomal compartments. See also the discussions of toxins in Section 4.4.3.2.

cystine; the products are one molecule of cysteine and one of a "mixed disulfide" of cysteine and cysteamine. The cysteine exits from the lysosomes. At first it was believed that the remaining mixed disulfide reacted with a second molecule of cysteamine, liberating another cysteine, which exits from the lysosomes, and producing cystamine (a pair of cysteamines linked by a disulfide), which may also be able to escape because of its relatively small size (M_r 152). This reaction may take place to an extent, but more likely most of the cysteine–cysteamine compound exits the lysosome via a membrane-associated carrier and then is reduced in the cytoplasm. The carrier responsible for this may be one that normally transports basic amino acids, as a countertransport-like efflux of lysine from isolated lysosomes can be stimulated by suspending the lysosomes in solutions containing the mixed disulfide.

The ability of cysteamine to help clear cystine stores from the lysosomes is being exploited in experimental therapies for cystinosis.

3.4.1.6. Salla Disease; Vitamin B_{12}

Two other human disorders are likely to be due to defects in lysosomal transport. In Salla disease and related conditions, fibroblasts and other cells accumulate sialic acid molecules (chiefly N-acetyl neuraminic acid) in their lysosomes; the cells seem to hydrolyze glycoproteins and other relevant macromolecules normally, but fail to transport the sialic acids released by this hydrolysis to the cytoplasm.

Fibroblasts from a patient with a vitamin B_{12} deficiency disorder have been found to accumulate the vitamin in lysosomes rather than freeing it from the lysosomes to the cytoplasm, as is seen in normal cells (Section 4.1.2.1), Perhaps this too reflects a transport defect.

Other mutants useful for analysis of the lysosomal or endosomal membrane probably will turn up in human material and elsewhere. Section 3.1.5.4 mentioned mutant cell lines defective in the acidification of some of the compartments involved in heterophagy, presumably owing to alterations affecting membrane components that might include ion channels or carriers. There has also been speculation that certain of the storage diseases (Section 6.4) in which lipids accumulate might involve perturbed movement of cholesterol or other lipids out of lysosomes.

3.4.2. Transport across the Tonoplasts in Plant Cells, Yeast, and Fungi

Vacuoles of plant cells, yeast, and fungi engage in degradation of macromolecules (Sections 5.1.6.1 and 5.4.2) but efflux of vacuolar digestion products has been little studied. On the other hand, transport of small molecules and ions **into** the vacuole is a much examined phenomenon, crucial to the biology of plant cells, yeast, and fungi.

The familiar **turgor pressure** in plants is an osmotically based pressure, owing its origin to the accumulation of solutes in the cytoplasm at concentrations much higher than those in the extracellular fluids: Cells of algae and higher plants contain K^+, Na^+, and Cl^- and other inorganic ions, at concentrations as much as 50- to 100-fold or more greater than the concentrations in fresh water or in water that has percolated into soil. The fine details of the distribution of the ions within the cytoplasm remain to be worked out, but the vacuoles of a broad variety of plants are known to be substantially hypertonic, in ion concentrations, with respect to the extracellular environment. The plasma membrane and the tonoplast both participate in accumulating the ions by mechanisms that include energy-dependent transport.

The osmotic effects consequent upon the high intracellular solute concentrations, press the plant cell against its wall. This pressure provides mechanical support for mature plants. In addition, during growth, when cell walls are still plastic, the pressure generates force for morphogenetic shape changes such as the marked elongations that accompany cell differentiation in shoots, roots, and other organs and tissues. Growing plant cells can enlarge their volume enormously without investing heavily in new macromolecular components; most of the increase is accounted for by osmotic expansion of the vacuoles. [As the vacuole expands its volume, the tonoplast obviously must increase its area; see Sections 5.4.2 and 7.4.3.7. And specific morphological features must be created and maintained, such as the presence of the channels ("strands") of cytoplasm that traverse the large vacuoles of some plant cells and apparently facilitate communication among the cytoplasmic regions bordering different portions of the vacuole.]

Many other cellular phenomena—the protrusion of daughter buds in dividing yeast for example—have been laid to vacuole-generated osmotic force. Often the schemes are speculative but changes in turgor pressure do seem to be responsible for reversible, physiological modulations of cell shape such as in the opening and closing of stomata (the structures at the surfaces of leaves through which gases move to and from the interior).

3.4.2.1. Examples of Substances Transported across the Tonoplast

Plant species differ from one another in the relative importance of different solutes for vacuole osmotic pressure, but the principal osmotically relevant solutes in most plants include ions, sugars and sugar derivatives, amino acids, and organic acids. The usual plant growth factors such as auxins, and other regulatory substances, including phytochrome, help control relevant behavior of the cells. Several of these agents are known to regulate transport across the plasma membrane or to affect the metabolic mobilization of organic solutes, thereby controlling the amounts of solutes available for movement into the vacuole. It is less certain which of the regulatory agents, if any, act directly on the vacuole but in growing plants, auxins and other agents do influence the plasticity of the cell wall, and in this way at least, contribute to the ability of vacuolar pressure to elongate the cells.

In yeast and fungi (and reputedly in some plants), vacuoles build up concentrations of amino acids 5- to 20-fold or more greater than those in the surrounding cytoplasm. Along with contributing to the osmotic state of the vacuole, these amino acids represent pools, potentially available for metabolism during periods of stress or starvation. (Animal cells might build up such lysosomal stores as well, but this has not been much studied.) Basic amino acids such as arginine are particularly concentrated in vacuolar pools. The side chains of these amino acids involve nitrogen-containing groups so that the selective enrichment of the vacuolar pools in basic amino acids could help in storage of nitrogen as well as of carbon. The stored amino acids are drawn upon, for example, when the fungus *Neurospora crassa* is depleted of available glutamine as happens under conditions of nitrogen starvation.

Amino acids are transported across the tonoplast by H^+/amino acid antiport systems dependent upon the proton gradient, and therefore, upon metabolic energy (Fig. 3.6). In the yeast *Saccharomyces cerevisiae*, several such transport systems have been detected, each responsible for a specific group of amino acids, and overall, enabling the entry of a dozen or so different amino acids, including the basic ones. In *N. crassa*, the basic amino acids in the vacuole bind, by ionic interactions, to the polyphosphates, which also are stored in the vacuole (Section 5.4.1); such complexing reduces the osmotic impact of the amino acid stores and so may enable the vacuoles to contain higher concentrations of amino acids than would otherwise be manageable.

Divalent cations evidently also enter vacuoles by transport coupled to the H^+ gradient. Ca^{2+}, which can reach concentrations of 10mM or more in the vacuoles, probably by a Ca^{2+}/H^+ antiport, is sometimes stored in complexes with phosphates or oxalate; in certain plants, the oxalate complexes form intravacuolar crystals. Movements of ions with possible regulatory roles, like Ca^{2+} or H^+, into or out of vacuoles could markedly affect metabolic events that are influenced by these ions in the extravacuolar cytoplasm.

Acidic amino acids are largely excluded from the vacuole, perhaps reflecting the fact that their entry into the acidified vacuole interior is less energetically favored than is the entry of basic compounds. Growth factors and similar agents are also rarely present as such in appreciable concentrations within the vacuole, though occasionally derivatives of certain of the agents are found there. Could it be that one evolutionary "reason" why several key plant regulatory agents are acids is that the accumulation of weak acids in vacuoles is "discouraged," energetically? This could be advantageous for avoiding the excessive "waste" of the agents that might ensue were they to penetrate readily into the large vacuolar volumes, where they cannot accomplish their biological tasks. (Some investigators, however, have advanced still-controversial proposals that ascribe major roles to vacuoles in handling various growth factors, including participation in synthesis of agents like ethylene.)

On the other hand, organic "acids" such as oxalate, malate, and citrate ions are commonly found in plant cell vacuoles. It might be thought that these compounds serve simply as anions that "accompany" the H^+ transported by the tonoplast's proton pump (Section 3.1.5.3). But in some plants the movements of the organic anions are physiologically regulated in ways that suggest a more complex transport. Thus, in **crassulacean acid metabolism,** a metabolic pathway whereby malate is formed by fixation of CO_2 and later decarboxylated to supply photosynthesis, malate moves cyclically, passing into the vacuole in the dark and out into the surrounding cytoplasm in the daytime. The mechanisms governing the direction of net movement of solutes across the tonoplast in cycles like this one, or in the storage and mobilization of amino acids are not known in detail (see below).

The passage of malate or amino acids out of the vacuole is down concentration gradients. Conceivably, such efflux could be used to foster influx of other components, if suitable antiport systems exist.

Concentration of sucrose into vacuoles is especially evident in sugar-storing tissues of sugarcane, sugar beets, and many fruits. Not only is such passage important for accumulating sucrose stores in cells but it also helps govern the osmotic movements of water that are important in motivating or sustaining the transport of sucrose and other materials in the vascular systems. Most investigators now believe that a tonoplast-associated H^+/sucrose antiport system takes part in accumulating sucrose in vacuoles, but there is also evidence for a "group translocation" system, in tonoplasts, that uses UDP-glucose from the cytosol to synthesize sucrose and deposit it within the vacuoles.

Plant cell vacuoles are large enough and accessible enough that quite detailed studies of the tonoplast in terms of its channels, transport systems, ion gradients, and potentials should be possible with powerful methods such as patch clamping. Pilot work along these lines has already led, for example, to the speculative hypotheses that, in crassulacean acid metabolism, malate moves into and out of the vacuoles through calcium-regulated channels. The directions of movement may depend upon the extent of the proton gradient, which, in turn, could be governed by light-responsive changes in the availability of ATP.

3.5. Storage of Toxic Materials, Poorly Digestible Residues, Wastes, and the Like; Lipofuscin

In these sections I will briefly consider some materials that normally do not escape readily from lysosomes and I will return to the subject of the fate of lysosomes (see Section 2.1.3).

3.5.1. Plant Cells

The vacuole of plant cells is used as a depot for storage of proteins and of polymeric carbohydrates that can be mobilized when called for by physiological triggers (Sections 5.4.1 and 5.4.2). In addition, the vacuole is utilized to sequester a variety of the small molecules, metabolic by-products, that often are called "secondary metabolites." Some of these stored materials are inert; others are potentially toxic to the cells.

Phenolic compounds, such as anthocyanin pigments and tannins, accumulate in the vacuoles of many plants. The immense vacuolar systems of the giant laticifers found in some of the rubber-producing plants are rich in polyterpenes.* Alkaloids, with membrane-disruptive capacities, are found in the vacuoles of a number of plants, perhaps using selective carriers to get there. In some plants the alkaloids complex with tannins; this is thought to render the alkaloids innocuous and to help retain them in the vacuoles. Other secondary metabolites are found in vacuoles in the form of covalent conjugates, often as glycosides and especially as glucosides; this is the case with phenols in many plants. Some of these conjugates can be hydrolyzed by vacuolar hydrolases, freeing the products to return to the cytoplasm, whereas others seem to represent long-term sequestrational forms for biologically active materials.

Certain of the secondary metabolites and other materials stored in vacuoles serve in defense against pathogens or predators. For example, the mustard oil and glycosides of cyanide-related compounds present in vacuoles of certain plants are altered when mixed with nonvacuolar components so as to release unpleasant or toxic products. Mixing takes place during chewing of the plants by animals and the products are deterrents to ingestion.

*In the rubber tree *Hevea*, however, the vacuoles ("lutoids") are small and the polyterpenes accumulate in the cytoplasm largely outside the vacuoles. *Hevea* vacuoles are readily isolated from the plant's latex, and therefore have been used extensively for analysis of vacuole properties in higher plants.

Overall, sequestration in the vacuole may have evolved as a device important for the many plant cells whose cell walls would impede the effective extracellular disposal of macromolecules or insoluble products and of molecules that are dangerous to the cell or are otherwise biologically active. Even when such materials can permeate through the walls to an extent, the organization of multicellular plants severely limits the possibilities for excretion from deep-lying tissues. Yeast and fungi, however, commonly secrete peptides and proteins to growth media, suggesting that their cell walls pose fewer problems for excretion than do the walls of higher plants. Thus, materials like polyphosphates present in yeast or fungal vacuoles are presumed to be there as mobilizable stores rather than for disposal. Molecules, such as allantoin, often thought of chiefly as a "waste" product in animal cells, have been reported in yeast vacuoles, but the extent and significance of the presence of such materials remain obscure.

3.5.2. Animal Cells

Most protozoa would seem to have little need for intracellular storage compartments in which to sequester digestive residues or toxic materials; they can defecate lysosomal contents rapidly and are not surrounded by walls that would impede diffusion of defecated materials away from the cell. But even among the protozoa there are cases in which lysosomal residual bodies do accumulate in quite large numbers. For example, *Tokophyra* is one of a few protozoa that generate offspring by the budding of smaller cells from a large "parent"; in time, the parent *Tokophyra* cell exhibits numerous residual bodies, presumed to be of lysosomal origin. The parent cell eventually dies.

With multicellular organisms, there is considerable variation in net accumulation of lysosomal residues among cell types and circumstances. For instance, cultured cells that are rapidly growing and dividing show far fewer "mature" residual bodies than do the same cells under conditions of slow growth and division. Such observations most likely reflect differences in the intensity of lysosomal reuse, in the production of new lysosomes, and in lysosome defecation (see Section 2.1.3). That defecation of lysosomal contents by hepatic or renal cells into the bile or urine takes place at appreciable rates, is inferred, for example, from findings that rat liver depletes its lysosomes of exogenous indigestible materials (e.g., Triton WR-1339; Section 1.3.1) over the course of a number of days (but see also Section 7.5.1.3). The liver normally releases lysosomal hydrolases to the bile at rates that can reach 5% of the hepatic content per day, suggesting that each day, on the order of 5% of the lysosomal volume is defecated by hepatocytes at the surfaces where these cells abut the bile canaliculi.* But there are limits to the abilities of cells of multicellular organisms to clear their lysosomes by defecation. This is shown by the existence of the storage diseases in which neurons, muscle cells, phagocytes, and many other cell types exhibit huge accumulations of materials in their lysosomes as a consequence of defects in the lysosomal enzymatic complement (Section 6.4).

---

*The release of hydrolases at these rates from the liver has, however, not been rigorously shown to depend on defecation as opposed to other secretory processes (Chapter 6). This is the case as well for the kidney, which releases hydrolases to the urine at rates up to 10–20% of the cells' content per day.

In a few situations, the lysosomal contents of one cell wind up in the lysosomes of another. Neutrophils containing phagocytotic residues are disposed of, as intact cells or after disintegrating, through phagocytosis by macrophages (Section 4.4.1). In the testis, Sertoli cells phagocytose the lysosome-containing cytoplasmic droplets shed by maturing sperm.

3.5.2.1. Fats

Of the materials that normally enter lysosomes of animal cells, lipids and metals are most often seen to persist in appreciable quantity. Lipids are believed to be digested more slowly than other molecules, partly because of special problems enzymes face in attacking the membranes and other organized structures in which most lipids are encountered (e.g., Section 6.4.1.3). Frequently, cells confronted with quantities of lipid-containing materials to digest (e.g., the cells that degrade myelin sheaths after injury to nerves) accumulate visible droplets of lipid in their residual bodies as digestion proceeds (Fig. 2.3). Subsequently, similar droplets are seen in the cytoplasm outside the lysosomes; the customary interpretation is that these droplets are extruded or pinched off from the residual bodies or that residual bodies transform entirely into lipid droplets. However, bear in mind that when only modest amounts of lipids such as cholesterol are entering their lysosomes (Section 4.1.1.1), cells seem able to move the lipids out into the cytoplasm as individual molecules; these molecules can later be modified and assembled into extralysosomal droplets. Eventually, lipid droplets that arise after heavy bouts of lysosomal digestion disappear from view, through utilization in metabolism.

3.5.2.2. Metals; Ferritin and Hemosiderin

Metal-rich secondary lysosomes are encountered, normally and in several diseases. Iron-containing **hemosiderin** granules ("siderosomes") are prominent in the populations of macrophages that degrade aging and abnormal erythrocytes (Section 4.6.1) and in erythroblasts, especially in anemias and some other disorders. Lysosomes of normal hepatocytes and of other cell types commonly show accumulations of electron-dense grains similar in appearance—at least superficially—to **ferritin,** an iron storage protein found in the extralysosomal cytoplasm of many cell types (Fig. 5.6; see also Fig. 1.10b). Copper-containing lysosomes have been reported in the liver of normal toads, and are especially evident in the hepatocytes of humans suffering from Wilson's disease.

The iron compounds in lysosomes of "erythrophagocytes" such as the macrophages mentioned in the last paragraph, derive from the heme groups of hemoglobin in the phagocytosed erythrocytes. But few of the other details of the mechanisms by which metals accumulate in lysosomes, or later depart, have been determined, in part because there are so many ways in which metal compounds could enter lysosomes during autophagy and heterophagy. Heme groups, for instance, are common in many of the cellular membranes known to be taken into autophagic lysosomes (see Chapter 5) and ferritin is abundant enough in some cell types that considerable uptake by autophagic lysosomes is likely. Copper is present in several relatively common proteins of mitochondria, which are subject to autophagic digestion, and copper is also carried in blood proteins like **ceruloplasmin,** which the liver's lysosomes acquire by endocytosis. Perhaps autophagic lysosomes also pick up the cytoplasmic proteins known as **metallothioneins,** which form stable complexes with copper, zinc, and other metals and are reputed to be major agents in controlling metal balances in the cell.

When metal ions or metal-containing compounds are introduced experimentally in excess or in unusual locations, a portion of the introduced load often winds up in the lysosomes (Fig. 5.6). For instance, when iron compounds are infused into the blood, hepatic lysosomes show increases in their iron content within a few tens of minutes. The mechanisms of this loading are not understood in detail but there are molecules in the circulation, such as the protein **hemopexin** that can scavenge metal-containing groups such as hemes, and might deliver them to endocytotic compartments (Section 4.5.1).

The presence of ferritin-like grains inside lysosomes has often been thought to imply that lysosomes store iron within their own ferritin molecules, which they might acquire through autophagy, or somehow assemble *de novo*; native ferritin is relatively resistant to hydrolysis by lysosomal enzymes, and some ferritin has been detected in lysosome-rich cell fractions and by immunocytochemistry of lysosomes *in situ*. But the characteristic microscopic appearance of ferritin as a small, electron-dense grain, is due to the iron-rich core of the ferritin molecule. The protein "hull" that surrounds this core in native ferritin is not readily seen, and it is not known how many of the grains visible in lysosomes have such a hull. In fact, although hemosiderin is still very poorly characterized, hemosiderin deposits in cells seem to arise largely through degradation of ferritin's protein moiety, leaving more or less intact cores in the deposits.

The sequestration of metals in lysosomes may help avoid toxic effects that metals or metal-containing groups like hemes can engender when free in the wrong places. But the lysosomes seem also to be involved in conservation and recycling pathways. For example, iron from the erythrocytes destroyed by macrophages, eventually is sent to the bone marrow for reuse in the synthesis of new hemogloblin molecules. The cycling of iron from lysosomes of hepatocytes to other sites probably accounts for the fact that, in patients with hemochromatosis, lysosomal loads of hepatic iron are reduced by bleeding, which imposes increased demands on iron stores to make new hemoglobin.

Yeast seem able to store iron in their vacuoles and to mobilize it from these stores for uses elsewhere in the cell.

Metals and metal-containing compounds "escape" from lysosomes to some extent through defecation into the bile or other extracellular fluids. Identification of other "escape" routes and analysis of the relations of lysosomal metal stores to pools in other parts of the cell, could have practical importance for rational therapies to deal with metal overloads (metals such as plutonium remain dangerous even within lysosomes). Chelating agents have been used in several such attempts and penicillamine is known to increase the secretion of stored copper as soluble complexes, but therapeutic successes have been very limited.

3.5.2.3. Lipofuscin

Lipofuscin, ceroid, hemofuscin, and several other "pigments" were first noticed by cytologists and pathologists as granules with distinctive color, cytochemistry and fluorescence properties, and with characteristic distributions in pathological and normal organisms. Several such "pigments" have turned out to represent lysosomes full of indigestible residues. Where to draw the boundaries among the different types of accumulations are partly matters of convenience; conventions as to what a given type of pigment should be called have varied among different research groups and at different times. The best known category is **lipofuscin** (Fig. 3.14), in which the granules have a golden-brown color and characteristic fluorescence and have been found to contain acid hydrolases both by cytochemistry and by biochemical analysis of cell fractions.

Lipofuscin accumulates strikingly in such slowly dividing or nonmitotic cells as hepatocytes, muscle cells, neurons, or steroid-secreting cells, especially in aging animals, where the granules can occupy 20% or more of the cellular volume. This association with age has led to lipofuscin being called "aging pigment" or "wear and tear" pigment. Strictly speaking, such terms are misnomers; In organs like the adrenal gland, lipofuscin-like granules normally abound at all ages and granules strongly resembling lipofuscin also accumulate in several tissues of young animals under various abnormal conditions (e.g., in the brain when lysosomal inhibitors such as chloroquine or leupeptin are introduced).

Lipofuscin is difficult to characterize in biochemical detail because it is rich in hard-to-handle insoluble polymeric materials. Some of the materials of lipofuscin and related pigments can be generated from unsaturated lipids and from proteins by peroxidatic or autooxidatic reactions *in vitro*, or by nonenzymatic oxidations of melanin. Hence, such reactions are widely believed to participate in the genesis of the pigments *in vivo*. Lipofuscin granules contain iron and other metals, so it may be that the materials in the granules are subjected to nonenzymatic peroxidations, catalyzed perhaps by heme-like groups.

The resemblance of lipofuscin to the residual bodies of many cell types (compare Figs. 1.3, 2.3, 3.14, and 7.7) has led to suggestions that the granules arise through the slow accumulation of indigestible residues within lysosomes.

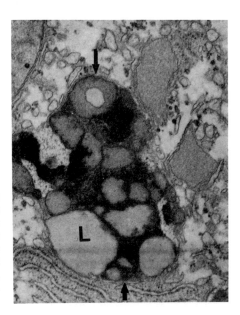

FIG. 3.14. A lipofuscin granule from a human hepatocyte; the arrows indicate the top and bottom of the structure. The body contains globules with the microscopic appearance of lipid (L); at suitable magnification, the areas of high electron density are seen to include small grains and amorphous material. From Novikoff, A. B., Novikoff, P. M., Quintana, N., and Davis, C. (1973) *J. Histochem. Cytochem.* **21:**1010. × 30,000 (approx.).

The concept, in essence, is of the gradual accretion of such residues in the lysosomal populations of long-lived cells that reuse their secondary lysosomes repeatedly over prolonged times. Thus, the fact that neurons and glial cells of the central nervous system are prone to lipofuscin accumulation is explained by the presumed tendency of these cells to retain their lysosomes rather than defecate them (Section 2.1.3.4). (But why then do hepatocytes, which seemingly can defecate, accumulate lipofuscin?)

Centophenoxane and a few other agents reduce the stores of lipofuscin in neurons; how they do so is not known.

3.5.2.4. Telolysosomes?

Cathepsin activities in lipofuscin granules are reasonably high but overall levels of lysosomal hydrolase activities are reputed to be low by comparison with other types of lysosomes. One view is that lipofuscin granules are "debilitated" or aged, perhaps to the extent that many of the granules are virtually incapable of further useful participation in digestion. The term **telolysosome** has been proposed for types of lysosomes that are supposedly "demobilized" from digestive activity.

Telolysosomes might be sought in cells engorged with abnormally enlarged lysosomes (see Section 6.4.2). In such cells, many "old" lysosomes could be largely precluded from receiving new supplies of substrates and of hydrolases by being jammed in the center of lysosomal aggregates whose more peripheral members fuse with most of the available transport vehicles. But notions that cells contain lysosomal bodies that are "resting," or inherently unable to function or mechanically prevented from so doing, have not really been tested. Most lysosomes that have been studied in detail do seem accessible to new substrates in that they can be labeled with endocytotic tracers, and for usual cells and circumstances, the working assumption has been that all the lysosomes are potential participants both in heterophagy and in autophagy. Few detailed observations contradict this assumption. When cell fractions from endocytosing cells are examined for the distribution of the materials they have recently taken up, small populations of acid hydrolase-rich structures sometimes are encountered that lack these materials. These structures could include telolysosomes, and occasionally have been tentatively proferred as such, but for the most part they are compartments of relatively low buoyant density that are more likely to represent primary lysosomes or other derivatives of the systems through which newly made lysosomal hydrolases are packaged (Chapter 7). Macrophages of mouse spleen reportedly possess numbers of large secondary lysosomes that are poorly accessible to endocytosed tracers under conditions where the tracers readily enter smaller bodies suspected to be lysosomes. If these observations are correct, perhaps the controls that normally govern the sizes of individual lysosomes (Section 2.4.1) operate so as to favor interaction of new endocytotic structures with the smaller lysosomes or perhaps the larger ones are less mobile than the others and therefore are less likely to fuse with incoming endocytotic bodies.

On the whole, as with most other questions about the extent of heterogeneity of the lysosomes in a given cell (Section 7.5.1), the evidence bearing on the possible widespread existence of telolysosomal subpopulations is not yet convincing one way or the other. Thus, though most observers are skeptical about telolysosomes, many have suspended judgment until critical studies are done on the behavior of suitable tracers and of hydrolases in the lysosomal population of a suitable cell type over long periods. Further investigation of the long-term storage of endocytosed materials by oocytes and insect cells detailed in the next chapter, could be particularly illuminating.

Acknowledgments

Drs. Q. Al-Awqati, M. Chrispeels, S. Goldfischer, F. Maxfield, and R. Murphy read sections of this chapter and made quite useful suggestions. I am grateful also to Drs. R. G. W. Anderson and L. Taiz for their direct help and for organizing the informative 1987 conference on Acidic Intracellular Compartments in Plant and Animal Cells sponsored by the American Society for Cell Biology.

In addition, I am indebted, for specific information and discussions, to B. Bowers (*Acanthamoeba*), M. G. Farquhar (endocytosis in kidney), A. Fok and R. Allen (*Paramecium*), M. Forgac (acidification), W. A. Gahl (cystinosis and related matters), M. Geisow (acidification), H. Ginsburg (chloroquine), S. Gluck (proton pumps), S. Goldfischer (metal storage; lipofuscin), K. Howell (endosomes *in vitro*), A. Hubbard (endosomes in hepatocytes), J. Kaplan (receptor properties), J. Lippincott-Jones and D. Fambrough (lysosomal antigens in chicken cells), P. Matile (plant cells and yeast), F. Maxfield (acidification), I. Mellman, A. Helenius, R. Fuchs, and their colleagues at Yale (acidification and other endocytotic mechanisms; macrophage receptor cycling; lysosomal antigens), R. Murphy (acidification; early endosomes), S. Ohkuma and B. Poole (acidification), A. Robbins (acidification mutants), S. Silverstein, J. Swanson, and their colleagues at Columbia (macrophage lysosomes and endocytosis), R. Spanswick (plant cells), T. Steck (*Dictyostelium* acidification), D. Stone (acidification), H. Sze (plant vacuoles), L. Thilo (quantitative analysis of endocytosis), J. Thoene (cystinosis and other aspects of transmembrane transport), S. Wiley (EGF).

Many of the other people mentioned at the end of Chapter 2 deserve acknowledgment here as well. As with Chapter 2, contact over the years with the groups of R. Steinman, Z. Cohn, and their colleagues at the Rockefeller University has proved especially helpful (here, particularly for their work on quantitative aspects of endocytosis and membrane cycling, and on macrophages).

Further Reading

Al-Awqati, Q. (1986) Proton-translocating ATPases, *Annu. Rev. Phys. Chem.* **2**:179–199.

Aley, S. B., Cohn, Z. A., and Scott, W. A. (1984) Endocytosis in *Entameba histolytica*: Evidence for a unique, non-acidified compartment, *J. Exp. Med.* **160**:724–737.

Allen, R. D. (1984) Paramecium phagosome membrane: From oral region to cytoproct and back, *J. Protozool.* **31**:1–6.

Anderson, R. G. W., and Taiz, L. (organizers) (1987) *Acidic Intracellular Compartments in Plant and Animal Cells*. Abstracts published by American Society for Cell Biology, Bethesda, MD.

Anderson, R. G. W., and Orci, L. (1988) A view of acidic intracellular compartments, *J. Cell Biol.* **106**:539–543.

Arai, H., Terres, G., Pink, S., and Forgac, M. (1988) Topography and subunit stoichiometry of the coated vesicle proton pump, *J. Biol. Chem.* **263**:8796–8802.

Bernar, J., Tietze, F., Kohn, L. D., Bernardini, I., Harper, G. S., Grollman, E. F., and Gahl, W. A. (1986) Characteristics of a lysosomal membrane transport system for tyrosine and other neutral amino acids in rat thyroid cells, *J. Biol. Chem.* **261**:17107–17112. (See also *Biochem. J.* **228**:545–550, 1985.)

Besterman, J. M. (1985) Endocytosis–exocytosis coupling, in *Developments in Cell Biology 1: Secretory Processes* (R. T. Dean and P. Stahl, eds.), Butterworths, London, pp. 58–74.

Biochemical Society Transactions **12**:899–915 (1984) *The Lysosome and Its Membrane*. (Collection of articles by several authors.)

Bleistein, J., Heidrich, H. G., and Debuch, H. (1980) The phospholipids of liver lysosomes from untreated rats, *Hoppe-Seylers Z. Physiol. Chem.* **361**:595–597.

Bowman, B. J., and Bowman, E. J. (1986) H$^+$-ATPases from mitochondria, plasma membranes and vacuoles of fungal cells, *J. Membr. Biol.* **94**:83–97.

Brown, W. J., Shannon, W. A., and Snell, W. J. (1983) Specific and azurophilic granules from rabbit polymorphonuclear leukocytes. I. Isolation and characterization of membrane and content subfractions. II. Cell surface localization of granule membrane and content proteins before and after degranulation, *J. Cell Biol.* **96**:1030–1039, 1040–1046.

Courtoy, P. J., Quintart, J., and Baudhuin, P. (1984) Shift of equilibrium density induced by 3,3'-diaminobenzidine cytochemistry: A new procedure for the analysis and purification of peroxidase-containing organelles, *J. Cell Biol.* **98**:870–876.

Cramer, C. L., and Davis, R. H. (1984) Polyphosphate cation interaction in the amino acid containing vacuole of *Neurospora crassa*, *J. Biol. Chem.* **259**:5152–5157.

Crichton, R. R., and Chaloteaux-Wanters, M. (1987) Iron storage and transport, *Eur. J. Biochem.* **164**:485–506.

Dean, R. T., Jessup, W., and Roberts, C. R. (1984) Effects of exogenous amines on mammalian cells with particular reference to membrane flow, *Biochem. J*, **217**:27–40.

Draye, J. P., Quintart, J., Courtoy, P. J., and Bauduin, P. (1987) Relations between plasma membrane and lysosomal membrane, *Eur. J. Biochem.* **170**:395–405.

Evered, D., and Collins, G. M. (eds.) (1982) *Membrane Recycling*, Pitman Press, London.

Fok, A. K., Ueno, M., and Allen, R. D. (1986) Differentiation of Paramecium phagosome membrane and stages using monoclonal antibodies, *Eur. J. Cell Biol.* **40**:1–8.

Forgac, M., and Cantley, L. (1984) Characterization of the ATP-dependent proton pump of clathrin-coated vesicles, *J. Biol. Chem.* **259**:8101–8105.

Geisow, M. J. (1984) Fluorescein conjugates as indications of subcellular pH: A critical approach, *Exp. Cell Res.* **150**:29–35.

Geuze, H. J., Slot, J. W., and Strous, G. J. A. M. (1983) Intracellular site of asialoglycoprotein receptor–ligand uncoupling: Double label immunoelectronmicroscopy during receptor mediated endocytosis, *Cell* **32**:277–287.

Ginsburg, H., and Geary, T. G. (1987) Current concepts and new ideas on the mechanisms of action of quinoline-containing antimalarials, *Biochem. Pharmacol.* **36**:1567–1576.

Haylett, T., and Thilo, L. (1986) Limited and selective transfer of plasma membrane glycoprotein to membrane of secondary lysosomes, *J. Cell Biol.* **103**:1249–1256. (See also *Biochim. Biophys. Acta* **822**:243–266, 1985.)

Hedrich, R., and Neher, E. (1987) Cytoplasmic calcium regulates voltage-dependent ion channels in plant vacuoles, *Nature* **329**:833–835.

Hohman, T. C., and Bowers, B. (1984) Hydrolase secretion is a consequence of membrane recycling, *J. Cell Biol.* **98**:246–252. (See also *Eur. J. Cell Biol.* **39**:475–480, 1985.)

Hopkins, C. R. (1986) Membrane boundaries involved in the uptake and intracellular processing of cell surface receptors, *Trends Biochem. Sci.* **11**:473–477.

Jonas, A. J. (1986) Studies of lysosomal sialic acid metabolism: Retention of sialic acid by Salla disease lysosomes, *Biochem. Biophys. Res. Commun.* **137**:175–181.

Kielian, M., Marsh, M., and Helenius, A. (1986) Kinetics of endosome acidification detectedly mutant and wildtype Semliki Forest virus, *EMBO J.* **5**:3103–3109.

Krogstad, D. J., and Schesinger, P. H. (1987) Acid vesicle function: Intracellular pathogens and the action of chloroquine against *Plasmodium falciparum*, *N. Engl. J. Med.* **317**:542–549.

Lennarz, M. R., Leman, T. E., and Stahl, P. D. (1987) Isolation and characterization of a mannose specific endocytic receptor from rabbit alveolar macrophages, *Biochem. J.* **245**:705–711.

Lewis, V., Green, S. A., Marsh, M., Vihko, P., Helenius, A., and Mellman, I. (1985) Glycoproteins of the lysosomal membrane, *J. Cell Biol.* **100**:1839–1847.

Lippincott-Schwartz, J., and Fambrough, D. M. (1987) Cycling of the integral membrane protein LEP100 between plasma membrane and lysosomes: Kinetic and morphological analysis, *Cell* **49**:669–677. (See also *J. Cell Biol.* **106**:61–67, 1988.)

Lloyd, J. B., and Foster, S. (1986) The lysosome membrane, *Trends Biochem. Sci.* **11**:365–368.

McNeil, P. L., Tanasugarn, L., Meigs, J. B., and Taylor, D. C. (1983) Acidification of phagosomes is initiated before lysosomal activity is detected, *J. Cell Biol.* **97**:692–702. (See also *J. Cell Biol.* **94**:143–149, 1982.)

MacRobbie, E. A. C. (1979) Vacuoles: The framework, in *Plant Organelles* (E. Reid, ed.), Wiley, New York, pp. 61–68.

Maguire, G. A., Docherty, K., and Hales, C. N. (1983) Sugar transport in lysosomes, *Biochem. J.* **212**:211–218.

Marin, B. P. (ed.) (1985) *Biochemistry and Function of Vacuolar Adenosine-Triphosphatase in Fungi and Plants*, Springer-Verlag, Berlin.

Marin, B. (ed.) (1987) *Plant Vacuoles: Their Importance in Solute Compartmentation and Their Applications in Plant Biotechnology*, Plenum Press, New York.

Maxfield, F. R. (1985) Calcium and pH in cytoplasmic organelles, *Trends Biochem. Sci.* **10**:443–444. (See also *J. Cell Biochem.* **26**:231–246, 1984.)

Mellman, I., Pluttner, H., and Ukkoness, P. (1984) Internalization and rapid recycling of macrophage Fc receptors tagged with monovalent antireceptor antibodies: Possible role of a prelysosomal compartment, *J. Cell Biol.* **98**:1163–1169.

Mellman, I., Fuchs, R., and Helenius, A. (1986) Acidification of the endocytic and exocytic pathways, *Annu. Rev. Biochem.* **55**:663–700.

Mueller, S. C., and Hubbard, A. L. (1986) Receptor mediated endocytosis of asialoglycoproteins by rat hepatocytes: Receptor positive and receptor negative endosomes, *J. Cell Biol.* **102**:932–942.

Ohkuma, A. (1987) The lysosomal proton pump and its effect on protein breakdown, in *Lysosomes: Their Role in Protein Breakdown* (H. Glaumann and F. J. Ballard, eds.), Academic Press, New York, pp. 115–148.

Pagano, R. E., and Sleight, R. G. (1985) Defining lipid transport pathways in animal cells, *Science* **229**:1051–1057.

Pastan, I., and Willingham, M. (eds.) (1985) *Endocytosis*, Plenum Press, New York.

Pisoni, R., Thoene, J. G., and Christensen, H. N. (1985) Detection and characterization of carrier mediated cationic amino acid transport of cystinotic and normal human fibroblasts, *J. Biol. Chem.* **260**:4791–4798. (See also *J. Biol. Chem.* **262**:15011, 1987.)

Raguzzi, F., Lesyisse, E., and Crichton, M. (1988) Iron storage in *Saccharomyces cerevisiae*, *FEBS Letters* **231**:253–258.

Rausch, T., Butcher, D., and Taiz, L. (1987) Active glucose transport and proton pumping in tonoplast membranes of *Zea mays* coleoptiles are inhibited by anti-H^+-ATPase antibodies, *Plant Physiol.* **85**:996–999.

Reeves, J. P. (1984) The mechanism of lysosomal acidification, in *Lysosomes in Biology and Pathology* (J. T. Dingle, R. T. Dean, and W. Sly, eds.), Elsevier, Amsterdam, pp. 175–199.

Reggio, H., Bainton, D. F., Harms, E., Coudrier, E., and Louvard, D. (1984) Antibodies against lysosomal membranes reveal a 100,000 molecular weight protein that cross reacts with purified H^+-ATPase from gastric mucosa, *J. Cell Biol.* **99**:1511–1526.

Reijngoud, D.-J., and Tager, J. M. (1977) The permeability properties of the lysosomal membrane, *Biochim. Biophys. Acta* **472**:419–449. (See also *Biochim. Biophys. Acta* **508**:15–26, 1978.)

Renlund, M., Tietze, F., and Gahl, W. A. (1986) Defective sialic acid egress from isolated fibroblast lysosomes of patients with Sallas disease, *Science* **232**:759–762.

Robbins, A. R., Oliver, C., Bateman, J. L., Krag, S. S., Galloway, C. J., and Mellman, I. (1984) A single mutation in Chinese hamster ovary cells impairs both Golgi and endosomal functions, *J. Cell Biol.* **99**:1296–1308.

Rodman, J. S., Seidman, L., and Farquhar, M. G. (1986). The membrane composition of coated pits, microvilli, endosomes and lysosomes is distinctive in the rat kidney proximal tubule cells, *J. Cell Biol.* **102**:77–87.

Rome, L. H., Hill, D. F., Bame, K. J., and Crain, L. J. (1983) Utilization of exogenously added acetyl coenzyme A by intact isolated lysosomes, *J. Biol. Chem.* **258**:3006–3011. (See also *Biochem. J.* **235**:707–713, 1986).

Rosenblatt, D. S., Hosack, A., Matiaszuk, N. V., Cooper, B. A., and LaFramboise, R. (1985) Defect in vitamin B12 release from lysosomes: Newly described inborn error of vitamin B12 metabolism, *Science* **228**:1319–1321.

Rudnick, G. (1986) ATP-driven H^+ pumping into intracellular organelles, *Annu. Rev. Physiol.* **48**:403–413.

Sato, T., Ohsumi, Y., and Ansaku, Y. (1984) Substrate specificities of active transport systems for amino acids in vacuolar membrane vesicles of *Saccharomyces cerevisiae*, *J. Biol. Chem.* **259**:11505–11508.

Schmid, S. L., Fuchs, R., Male, P., and Mellman, I. (1988) Two distinct subpopulations of endosomes involved in membrane recycling and transport to lysosomes, *Cell* **52**:73–83.

Schneider, D. L. (1981) ATP-dependent acidification of intact and disrupted lysosomes, *J. Biol. Chem.* **256**:3858–3864. (See also *J. Biol. Chem.* **261**:1077–1082, 1986.)

Schneider, J. A., Jonas, A. J., Smith, M. L., and Greene, A. A. (1984) Lysosomal transport of cystine and other small molecules. *Biochem. Soc. Trans.* **12**:908–910. (See also *J. Biol. Chem.* **262**:1244–1253, 1987.)

Steinberg, T. H., Newman, A. S., Swanson, J. A., and Silverstein, S. C. (1987) Macrophages possess probenecid inhibitable organic anion transporters that remove fluorescent dyes from the cytoplasmic matrix, *J. Cell Biol.* **105**:2695–2702.

Steinman, R. M., Mellman, I. S., Muller, W. A., and Cohn, Z. A. (1983) Endocytosis and the recycling of plasma membrane, *J. Cell Biol.* **96**:1–27.

Stoorvogel, N., Geuze, H. J., and Strous, G. J. (1987) Sorting of endocytosed transferrin and asialoglycoprotein occurs immediately after internalization in HepG2 cells, *J. Cell Biol.* **104**:1261–1268.

Swanson, J. A., Yirinec, B., Burke, E., Bushnell, A., and Silverstein, S. C. (1986) Effects of alterations in the size of the vacuolar compartment on pinocytosis in J 774.2 macrophages, *J. Cell. Physiol.* **128**:195–201.

Sze, H. (1985) H^+-translocating ATPases: Advances using membrane vesicles, *Annu. Rev. Plant Physiol.* **36**:175–208.

Uchida, E., Ohsumi, Y., and Anraku, Y. (1988) Characterization and function of catalytic subunit of H^+ translocating adenosine triphosphatase from vacuolar membrane of *Saccharomyces cerevisiae*, *J. Biol. Chem.* **263**:45–51.

Van Dyke, R. W., Hornick, C. A., Belcher, J., Scharschmidt, B. F., and Havel, R. J. (1985) Identification and characterization of ATP dependent proton transport by rat liver multivesicular bodies, *J. Biol. Chem.* **260**:11021–11026.

Weintraub, L. R., Edwards, C. Q., and Krikker, M. (eds.) (1988) Hemachromatosis, *Ann. N.Y. Acad. Sci.*, **526**.

Wileman, T., Harding, C., and Stahl, P, (1985) Receptor mediated endocytosis, *Biochem. J.* **232**:1–14.

Xie, X.-S., and Stone, D. K. (1986) Isolation and reconstitution of the clathrin coated vesicle proton-translocating complex, *J. Biol. Chem.* **261**:2492–2495. (See also *J. Biol. Chem.* **262**:14790, 1987.)

Yusko, M. A., and Gluck, S. (1987) Production and characterization of a monoclonal antibody to vacuolar H^+ ATPase of renal epithelium, *J. Biol. Chem.* **262**:15770–15779.

4

Uses and Abuses of Endocytotic and Heterophagic Pathways

Endocytosis and the degradation of endocytosed material in lysosomes are used extensively by unicellular organisms and by some "lower" multicellular forms for handling food and other necessities of life (Chapter 2). Even the ion-induced pinocytosis in amebas might represent an adaptation to fresh-water habitats, contributing to rapid uptake of inorganic ions when the protozoan stumbles upon a rich source. In "higher" multicellular animals, endocytosis and lysosomes are used for feeding in some special circumstances: intestinal cells of suckling rats, for example, contain very large lysosomes in which they digest components of milk delivered from the intestinal lumen by endocytotic vesicles or tubules. But in general, digestion of the food of vertebrates and most invertebrates takes place in specialized extracellular locales, especially in the lumen of the digestive system. Although acid hydrolases are involved in this digestion, the battery of enzymes is quite different from that typifying the lysosomes.

Why then do almost all cell types of multicellular animals show capacities for endocytosis and the consequent ability to degrade extracellular macromolecules in lysosomes? The answer is that a variety of uses have evolved for endocytotic and heterophagic pathways, in which lysosomal digestive capacities and/or the transport possibilities of endocytosis are employed for diverse cellular processes. This chapter provides examples of these uses. I will also discuss the exploitation of endocytotic and heterophagic pathways by pathogens, toxins, and experimenters, and briefly examine endocytosis and heterophagy in plant cells and yeast.

4.1. Nutrition

Among the first lysosomes to be described were the large vacuoles (diameters of 1–5 μm or more) in mammalian kidney tubule cells (Fig. 1.7). These vacuoles rapidly acquire proteins endocytosed from the lumen of the nephron. Demonstrations of this uptake by W. Straus and others were of unusual historical importance because they were among the first modern cytochemical uses of HRP, helping to introduce this workhorse tracer to a wide audience. Endocytosis of proteins by the kidney tubule cells, and their degradation in lysosomes, apparently is a conservation device enabling the body to retain and reuse amino acids that otherwise would be lost in the urine. This was among the earliest recognized clear sign that lysosomes are used in the economy of "nutrients" in "higher organisms." Many other indications have followed.

The blood contains diverse macromolecular and multimolecular vehicles serving in transport to the tissues. Among these are abundant serum proteins like albumin, a protein often thought of chiefly as contributing to the osmotic properties of blood, but which also complexes with and thus carries several types of smaller molecules—fatty acids, bilirubin, and probably some steroid and thyroid hormones. Some other materials carried in the circulation, notably cholesterol and other lipids, are poorly soluble in aqueous media; inclusion in lipoprotein particles with hydrophilic surfaces permits them to be transported in suspension.

Bear in mind that some blood-borne materials can be dangerous unless complexed with carriers. Thus, for example, heme compounds free in the blood or tissue fluids might produce a variety of toxic effects but when such compounds are liberated, as by pathological hemolysis, they are scavenged by binding to proteins such as **hemopexin** and **haptoglobin** (Section 4.5.1.1) which subsequently deliver the hemes to the liver for reuse or excretion.

Many of the transport vehicles of the blood interact with specific cell surface receptors, permitting efficient, selective delivery to the proper tissues and cells. Delivery need not always be endocytotic. For instance, adipose tissues use cell surface-associated or extracellular lipases in their capillaries to release fatty acids from the triglycerides in circulating lipoprotein particles.* For the materials carried by albumin, proposals about delivery envisage release of the carried components at the cell surfaces to which albumin binds, and subsequent passage of the materials across the plasma membrane. But endocytosis is found to be responsible for the acquisition by cells, of several types of blood-borne molecules important for cellular nutrition. A number of receptor-dependent endocytotic mechanisms for such uptake are exemplified in the next few sections (see Fig. 3.12).

4.1.1. Lipoproteins

Along with carrying lipids from one site to another, the lipoprotein particles in the vertebrate circulation may transport fat-soluble molecules, including vitamins or their precursors. The major sources of circulating lipoprotein particles are the intestine (which produces such particles from components of recently digested lipids) and the liver (which synthesizes and secretes lipoproteins and also takes them up, as part of its function in regulating the composition of the blood). Each of the classes of blood-borne lipoproteins is characterized by a particular spectrum of lipids and each has, as well, specific proteins that control its interactions with cellular receptors and other regulatory or enzymatic systems. The particles evolve in the course of their life history

---

*As Section 2.3.2.11 mentioned, adipocytes in some physiological states have abundant populations of small, noncoated vesicles. These were initially thought to be endocytotic and some probably are. But the importance of endocytosis for adipocyte function has yet to be clarified (see also Section 4.5.2.1). Some of the smooth vesicles may retain long-term continuities with the cell surface (Section 4.3.3). In addition, the storage and depletion of fat can require the cells to undergo large-scale changes in size and shape to which the vesicles might contribute by adding membrane to or removing it from the cell surface. Some investigators suppose that the vesicles provide the principal means by which adipocytes endocytose insulin (Sections 4.5.2.1 and 5.2.1.2).

through processes that include exchanges among different types of particles, changes in the complement of proteins present, modification by enzymes such as the lipase-mediated removal of triglycerides already mentioned, and exchanges of lipids with the plasma membranes of cells.

4.1.1.1. LDL

LDLs are particularly rich in cholesterol and have received special attention both because of their abundance and because elevated levels of LDLs are associated with atherosclerosis and related circulatory disorders. LDL particles arise from the very-low-density lipoprotein particles (VLDLs) secreted by the liver. VLDLs contain both triglycerides and cholesterol esters. Once in the circulation, they are believed to evolve first into intermediate-density lipoproteins (IDLs) and then into LDLs, through a sequence of processes that includes loss of their triglycerides to adipose tissues, muscle, and other tissues. In humans, the LDL particles that ultimately form are about 20 nm in diameter and contain a core of about 1500 cholesterol molecules esterified to fatty acids. The particle's surface is a hydrophilic monolayer comprised of a few hundred molecules of phospholipid and nonesterified cholesterol; associated with this layer is a single large (M_r 400,000) protein molecule known as apoprotein B-100.

LDL receptors (Sections 2.3.2.7 and 2.3.2.9; Figs. 1.10 and 2.19) have been detected on many cell types. Some of these cells have special roles in storage, metabolism, or secretion of sterols or steroids—cells of liver or adrenal cortex, for example. Others, including fibroblasts, use cholesterol largely for more mundane housekeeping, such as the genesis of membranes. The receptors of fibroblasts have been studied most intensively. Each of these receptors binds a single LDL particle, recognizing the particles by their characteristic apoprotein. The particles are internalized and delivered to the lysosomes, while the receptors recycle to the cell surface (Fig. 4.1; see Section 3.3.4). In the lysosomes, the particles are disrupted and the cholesterol is deesterified by hydrolysis; the cholesterol then passes into the cytoplasm where it is either utilized metabolically, or reesterified for intracellular storage by intervention of the enzyme **acyl-CoA cholesterol-acyl-transferase** (ACAT).

The LDL receptor also can mediate uptake of certain other lipoproteins, such as IDLs, which it recognizes principally through the apoprotein E molecules present, in company with the apoprotein B-100 molecule, in the surface of each IDL particle.

Accumulation of LDL-carried cholesterol by cultured fibroblasts has the effect of reducing rates of endogenous sterol synthesis in the cells. This is based on diminution of the levels of the enzyme **hydroxymethyl-glutaryl-CoA reductase** (HMG-CoA reductase) through phenomena currently thought to include both enhanced degradation of the enzyme, and reduced transcription of the relevant mRNA. Cholesterol uptake by the cells also can lead to reduced transcription of mRNAs coding for the LDL receptor, which eventuates in reduced rates of LDL uptake. Apparently these are mechanisms to prevent the cells from being overloaded with excessive stores of cholesterol.

4.1.1.2. HDL; Chylomicra; Lipophorins

High-density lipoproteins (HDL; another type of cholesterol-carrying particle) are endocytosed through the operation of a receptor system different from

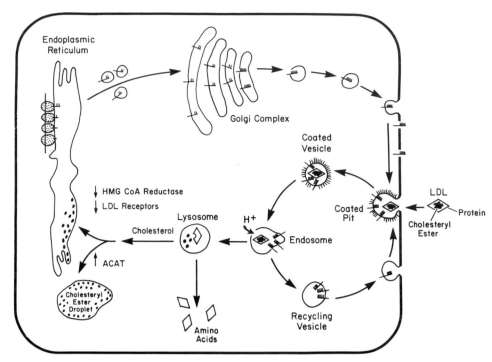

FIG. 4.1. Summary of the cycling and functioning of the LDL receptor in human fibroblasts. The receptor molecules are symbolized by the small toothbrush-like structures. They are produced in the ER and move to the cell surface via the Golgi apparatus, where their glycosylation is completed. They carry LDLs into the cell by endocytosis and then recycle to the cell surface while LDLs undergo degradation in lysosomes. The cholesterol from the LDLs is freed by the degradation to enter the cytoplasm where, among other effects, it helps regulate the levels of several enzymes and of LDL receptors (see text). From Brown, M. S., and Goldstein, J. (1985) *Curr. Top. Cell Reg.* **26**:3. (Copyright: Academic Press.)

the one used for LDL. Cells of luteinized ovarian follicles and certain other cells that produce steroid hormones can acquire considerable amount of cholesterol from the HDLs. HDLs are of particular interest for the understanding of overall sterol balances because in addition to their participation in cholesterol **delivery,** circulating HDL particles seem able to **acquire** cholesterol from cells. This acquisition is most often thought of as occurring at the cell surface although recent speculation, based in part on findings that the HDL receptor in macrophages can cycle repeatedly between the cell interior and the surface, has suggested that the particles actually pick up the cholesterol intracellularly as they cycle into endocytotic compartments and then back out to the circulation. There is, however, not much evidence yet that HDLs actually enter most cell types, or, that if they do, that they can come back out.

Dietary cholesterol is packaged by the intestine within the lipoproteins known as **chlyomicra,** the principal lipid-rich particles released by the intestine to the bloodstream. Normally these particles are transformed into circulating "remnant" particles, by hydrolysis of their triglycerides, and then are endocytosed, chiefly by hepatocytes.

Lipid transfer via the hemolymph in insects utilizes "lipophorin" proteins, which complex with and deliver glycerolipids and cholesterol. In insects such as the locust, the proteins have life spans of several days, whereas the

165

USES AND ABUSES
OF ENDOCYTOTIC
AND
HETEROPHAGIC
PATHWAYS

times required to clear newly introduced lipids from the hemolymph are much shorter. Evidently then, the "lipophorins" are reusable carriers (Section 4.1.2) that are not obligatorily degraded as they deliver lipids to cells. It is not known whether endocytosis is essential for their functioning in delivery to somatic tissues, although there have been reports that certain lipophorins may be taken up by growing oocytes (section 4.2.1) that seem not to return these proteins to the circulation for reuse.

4.1.1.3. Atherosclerosis

The elevated levels of circulating LDL cholesterol that correlate with susceptibility to atherosclerosis can result from a variety of factors, including diet and heredity. Some of the dramatic hereditary predispositions to elevated cholesterol levels in humans are due to mutations that reduce the abundance or efficiency of LDL receptors (Fig. 2.19). Among the known mutations, some affect the synthesis of LDL receptor, some slow transport of receptor molecules to the cell surface, some reduce the capacities of receptors to bind lipoproteins, and some abolish the clustering of receptors into coated pits (Section 2.3.2.7). All of these mutations reduce the rates at which cholesterol-rich lipoproteins are cleared from the circulation.

Much less is certain about the details of how elevated cholesterol levels in the circulation engender the deposits in arterial walls that are fundamental to the pathology of atherosclerosis. Cholesterol is so extensively deposited in some of the cells at atherosclerotic lesions that the cells frequently assume a "foamy" appearance. Macrophages, smooth muscle cells, and fibroblasts all can acquire the excess cholesterol, and though macrophages are now thought to be the major "foam cell," there is still disagreement as to the relative importance of the different cell types in the eventual pathology. The lesions involve adhesion of platelets, entry of monocytes from the blood, and changes in the endothelial lining of the artery, including transient or longer-lasting compromising of the integrity of the lining. Alterations also occur in the patterns of smooth-muscle and other cells' proliferation. Some of the events at the lesions resemble phenomena seen at inflammatory sites (Section 4.4.1) including, for example, the release of biologically active molecules from cells such as platelets and phagocytes.

Current conceptions of atherogenesis suggest that deposition of cholesterol in the artery wall depends on the combination of increased levels of circulating lipids with chronic or episodic increases in "leakage" of circulating lipoproteins to interior layers of the artery wall. Why the cells in the wall store so much cholesterol is not clear, however. Macrophages, for example, normally are not particularly avid in their uptake of LDL particles. They do avidly endocytose experimentally modified LDL particles such as acetylated or peroxidated forms or ones linked to glycosaminoglycans, using receptors other than that for LDL but it is not yet known whether the modified LDLs resemble natural materials. There must also be further evaluation of hints that the pathways for endocytosis of the modified lipoproteins are less tightly regulated than are the LDL pathways.

Rational preventative and therapeutic measures are being designed or interpreted on the basis of the increased understanding of the involvement of LDL and LDL receptors. For instance, reduction of circulating cholesterol upon repeated administration of resins that bind bile acids in the intestine probably follows, in part, from increases in the levels of LDL receptors at cell surfaces in the liver. These increases reflect compensatory responses by the liver to the increased loss, via resin-containing feces, of cholesterol-derived bile acids. (Normally the bile acids secreted by the liver are largely retrieved by the intestine and eventually find their way back to the liver.) Regimens in which LDL receptor levels are increased through administration of inhibitors of HMG-CoA reductase also seem promising.

Other human diseases thought to be based on defects in lipoprotein metabolism include Tangier disease, in which HDLs may be handled abnormally.

4.1.2. Transport of Vitamin B_{12} and of Iron

4.1.2.1. Transcobalamin

Vitamin B_{12} (cyanocobalamin and its relatives) is carried in the circulation by proteins known as transcobalamins, which are recognized by specific cell surface receptors. Transcobalamin II is regarded as the principal transporter of vitamin B_{12} to many tissues. Vitamin-bearing molecules of this protein are cleared from the circulation with a half-time similar to that of the vitamin molecules they carry, suggesting delivery by endocytotic uptake followed by degradation of the protein within the cell. Thus, as with LDLs, it appears that lysosomes effect release of a transported component from its carrier, by enzymatically attacking the carrier. One difference from LDLs is that the sacrificed carrier is much more massive (M_r 38,000) than the material being delivered (M_r of the vitamin is about 1350).

Section 3.4.1.6 mentioned a human disorder in which vitamin B_{12} accumulates, perhaps because of a defect in the system by which the vitamin is transported out of the lysosomes.

4.1.2.2. Transferrin

Much of the iron transferred via the mammalian circulation is carried by an abundant glycoprotein, **transferrin.** Receptors with strong affinity for transferrin are found on many mammalian cell types, often in numbers exceeding 10^5 per cell. Predictably, erythroblasts and closely related cells are particularly rich in such receptors.

Each transferrin molecule can carry two atoms of iron, in the form of ferric ions. Altered temperature and pH have been used (Section 3.3.4.3) to establish that delivery of these irons to the cytoplasm depends upon endocytosis of transferrin and its entry into acidified compartments (Fig. 4.2). However, the delivery does not require degradation of the protein. Within a few minutes after uptake is initiated, iron has been released within the cells and transferrin molecules that had been internalized by tissue culture cells begin to emerge in the extracellular medium. They emerge in the form of **apotransferrin**: without the irons, but otherwise intact and ready to be reused. Eventually most of the endocytosed molecules return to the outside either in this form or, for a proportion of the molecules that varies with cell type and situation, still containing one or two irons. These findings correlate well with *in vivo* observations that the average life span of circulating transferrin molecules is many hours whereas a pulse of radioactively labeled iron bound to transferrin is cleared from the circulation in an hour or two. In other words, iron is delivered at rates that imply multiple reuse of the transferrin molecules.

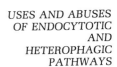

167

USES AND ABUSES
OF ENDOCYTOTIC
AND
HETEROPHAGIC
PATHWAYS

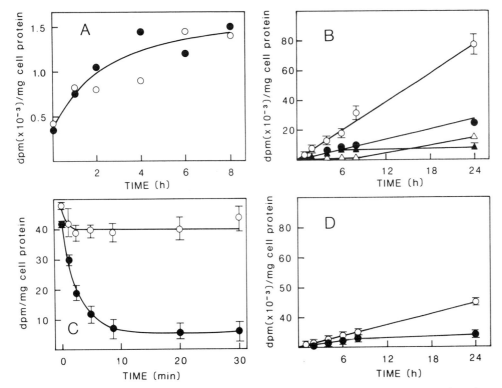

FIG. 4.2. (A) Adsorption of transferrin, detected as cell-associated radioactivity, to cultured rat fibroblasts exposed to transferrin at 4°C. The polypeptide chain of the transferrin had been labeled with ³H (●) and the molecules were also labeled with ⁵⁹Fe (○) to approximately the same specific activity as the ³H. **(B)** In cells exposed to transferrin at 37°C, retention of iron from the transferrin endocytosed by the cells (○) rises considerably more rapidly than the sum of the retention and degradation of the protein (●) (▲, the retained ³H label; △, the ³H label released to the medium as degradation products). In **C**, cultures were exposed to labeled transferrin at 37°C for 1 min and then washed and incubated in the absence of added transferrin. Iron stays with the cells to a much greater extent than does the polypeptide, which is released rapidly in a form still reactive with antitransferrin antibodies. **(D)** Methylamine inhibits accumulation of radioactive iron by the cells at 37°C (compare with B). From Octave, J. N., Schneider, Y.-J., Trouet, A., and Crichton, R. R. (1983) *Trends Biochem. Sci.* **8:**217. (Copyright: Elsevier, Cambridge.)

In vitro studies show that iron dissociates from transferrin as the pH is lowered to pH 5–6; the ferric ions seem to be bound to transferrin at sites from which they can be displaced by protons. Binding of anions—bicarbonate or carbonate—at nearby sites in the protein molecule accompanies binding of the irons. But whereas the irons are released at low pH, the association of transferrin with its receptors has been found to remain tight in acid media, as well as at neutral pH. At neutral pH, however, the affinity of the transferrin receptor for apotransferrin is notably weaker than it is at lower pH. From these observations, and from experiments like those in Fig. 4.2, it is proposed that transferrin's intracellular sojourn includes a stay spent in acidified compartments where the irons dissociate but the protein moiety of transferrin remains associated with the transferrin receptor (see Fig. 3.12). The apotransferrins are then carried back to the cell surface along with the receptors, to be released from the cell upon encountering the near-neutral pH of the extracellular milieu. Meanwhile, the irons somehow find their way out of the acidified compartments into the cytoplasm by crossing the compartmental membrane.

Some investigators believe that there are cells and circumstances in which transferrin in the circulation, or in culture media, delivers much of its iron to cells without undergoing endocytosis, but this view has only a few strong adherents at present.

4.1.2.3. Recycling of Transferrin and Its Receptor

In cultured erythroid precursor cells the majority of transferrin receptors and transferrin molecules make the complete cycle from the cell surface to the interior and back to the surface in 10–15 min or less, providing the very large iron delivery capacity needed to service the assembly of hemoglobin. In lymphocytes or HeLa cells, the average cycle time is roughly twice as long; with these cells at steady state, the ratio of intracellular transferrin receptors to cell surface receptors is about 3 : 1.

Endocytotic uptake of transferrin is via the same coated vesicles that take up other ligands. In fact, under artificial conditions in which transferrin molecules come to be taken up by binding to the "asialoglycoprotein" receptor of hepatocytes, the transferrins can later reemerge at the cell surface in association with transferrin receptors. Apparently, the two receptor types occupy the same acidified intracellular compartments, so that transferrins released from the asialoglycoprotein receptor (whose affinities for protein ligands, recall, diminish at low pH) can be picked up quickly by the transferrin receptors. Under normal conditions, intracellular transferrins and transferrin receptors are readily detectable in early endosomes and related structures by the usual microscopic approaches. Far fewer are seen in definitive lysosomes. [As mentioned earlier, however, transferrin receptors can be "driven" into lysosomes, when for example they bind large multivalent ligands like transferrins adsorbed to gold particles (several transferrins attach to each gold particle) or when anti-transferrin receptor antibodies cross-link the receptors.]

Along with their endosomal localization, intracellular transferrins are also seen in small membrane-delimited sacs, tubules, and (noncoated) vesicles, some of which are aligned along microtubules. These structures are thought to include the elements that mediate return of the transferrins and transferrin receptors to the cell surface (Section 3.3.4.4): The time course with which they acquire labels is appropriate for such a role. Present proposals are that the sacs, tubules, and vesicles mostly bud directly from the endosomes, carrying transferrin receptors in their membranes, along with receptor-associated transferrin molecules (chiefly in apo form). Some of these structures are thought to move rapidly to the plasma membrane and fuse with it. Interestingly, however, as is true of other ligands and receptors, from times very soon after their uptake, some endocytosed transferrin molecules and their receptors are detected in compartments closely associated with the Golgi apparatus. What this means is hotly disputed (see Sections 3.3.4.5 and 7.2.2). The best current guesses are that only a small proportion of transferrin molecules departing from the endosomes reach compartments that are functionally part of the Golgi apparatus, and that the Golgi apparatus is not an obligatory way station on the route back to the plasma membrane.

Section 5.2.1.3 will describe another variation of the transferrin cycle—the release of transferrins and their receptors from cells by defecation of the contents of MVBs.

The vehicles used for exit of transferrin receptors and their ligands from

169

USES AND ABUSES
OF ENDOCYTOTIC
AND
HETEROPHAGIC
PATHWAYS

the cell are presumed to be shared by other endocytotic receptors on their way back out of the cell. This is quite difficult to prove definitely because of problems in being sure that a given small tubule or vesicle seen in the microscope was actually on its way to the plasma membrane when the cell was killed to · prepare it for microscopic demonstration of receptor locations.

Canine kidney (MDCK) cells, cultured under conditions in which they establish a highly polarized "epithelial" organization, cycle transferrin selectively at the basal surfaces of the epithelium. By unknown targeting or sorting devices, the cells maintain much higher population densities (receptors per unit membrane area) of transferrin receptor at their basal poles than at other regions of the cell surface, despite the continued cycling of receptors to and from the plasma membrane.

Judging from changes in the fluorescence properties of fluorescein–transferrin conjugates in CHO cells and from similar data on other experimental systems, by 2–10 min after their uptake, transferrins begin to pass into compartments with pHs between 6 and 7. This contrasts with the behavior of ligands destined for degradation, which at this time are beginning to be exposed to much lower pHs. From studies on purified transferrin and similar *in vitro* assays, pHs much above 6 would not be expected to be optimally effective in promoting the dissociation of the irons, or the optimal retention of binding of apotransferrin to the transferrin receptor. Most likely by the time they reach the compartments with such a pH, the transferrins have already undergone a brief exposure to lower pHs, in an earlier endocytotic compartment they shared with other ligands. The higher pH compartments would be vehicles of the exit pathway. pHs of 5.5 or below have been reported for the compartments that accumulate transferrin in erythroid cells.

4.2. Storage and Processing

Here we will deal with two overlapping variations on themes of the uses of endocytosis and heterophagy: cases in which proteins and other materials destined for eventual degradation are stored for prolonged periods before being broken down, and cases in which products of heterophagic degradation serve as biological signals rather than in nutrition.

4.2.1. Intracellular Storage of Yolk

Oocytes of many species, vertebrate and invertebrate, contain prominent yolk bodies in which are stored quantities of proteins, often along with lipids, for use after fertilization. Frequently, these bodies are the most abundant structures in the cell; in the amphibian *Xenopus laevis*, yolk components account for well over 80% of the cytoplasmic protein. Study of the deposition of yolk and of its mobilization by the embryo has, however, been slowed by the considerable variation in these processes among different species of animals and by technical limitations. Yolky eggs and early embryos are difficult to preserve well for morphological or cytochemical studies, and components of yolk, such as the phosphoproteins, can produce artifactual deposits of cytochemical reagents, misleadingly mimicking enzymatic reaction products. In cell fractionation studies, there often are problems in obtaining fractions of adequate purity, and the adsorption of materials from other cytoplasmic sources onto yolk bodies during or after homogenization also is a frequent complication.

4.2.1.1. Endocytosis of Yolk Components

Despite these difficulties, it is firmly established that oocytes of most familiar animals acquire the bulk of their yolk components through endocytosis of products secreted by other cells.

Yolk proteins known as **vitellogenins** have been studied most carefully. Proteins of this class (the ones from *Xenopus* are described in more detail in Section 4.2.1.3) are released in large amounts to the circulation—by the liver in female nonmammalian vertebrates such as frogs or birds, by the fat body in insects, and by comparable tissues in other animals. In vertebrates and in many invertebrates, the vitellogenins typically are phosphorylated glycoproteins that circulate in company with modest amounts of lipid. Their secretion is regulated by estrogens in vertebrates and by ecdysteroids and juvenile hormone in insects. The vitellogenin genes of different vertebrate species are closely related—and some of the relations may extend into the invertebrate kingdom as well. Nonetheless, vitellogenins of different species of animals can differ substantially in details. In some species a small family of related proteins is produced rather than a single type.

In molluscs, a ferritin-like molecule is among the principal endocytosed components stored in the egg. Along with vitellogenins, chickens store immunoglobulins (IgGs) and oocytes of other animals store diverse other proteins.

Oocytes can take up massive amounts of circulating proteins and lipoproteins. In chickens the total uptake rates approach 1 g/day per oocyte. Often the oocytes endocytose vitellogenin much more rapidly than they do most other proteins (Table 4.1) and the uptake mechanism shows the saturability and other features expected for endocytosis dependent on specific receptors. Receptors with affinity for vitellogenin abound at oocyte plasma membranes and in intracellular pools that cycle receptors to the cell surface: There are 10^{13}–10^{14} vitellogenin receptors per chicken oocyte and 10^{10}–10^{11} per *Xenopus* oocyte; the numbers per unit area are not astonishingly high by comparison with other endocytotic receptors in other cell types, but the oocytes are immense and have correspondingly enormous surface areas. Uptake is initiated at the coated pits that line much of the oocyte surface (Fig. 4.3); it has been claimed that the coated pits and vesicles present at a given time have a surface area amounting to 20–50% or more of the cell surface of a chick oocyte. The subsequent steps of endocytosis lead initially to the formation of vitellogenin-containing intracellular vesicles, vacuoles, and tubules by steps that are not startlingly different from the conventional endocytosis of other molecules by other cell types.

TABLE 4.1. Relative Uptake of Various Proteins by *Xenopus laevis* Oocytes *in Vitro*[a]

Protein	M_r	Uptake	
		ng/mm² per hr	fmoles/mm² per hr
Ferritin	465,000	0.7	2
Hemoglobin	64,000	0.3	4
Serum albumin	68,000	0.3	5
Vitellogenin	460,000	49.2	107

[a]Source as in Fig. 4.3. Data are amounts of protein taken up per mm² of cell surface.

171

USES AND ABUSES
OF ENDOCYTOTIC
AND
HETEROPHAGIC
PATHWAYS

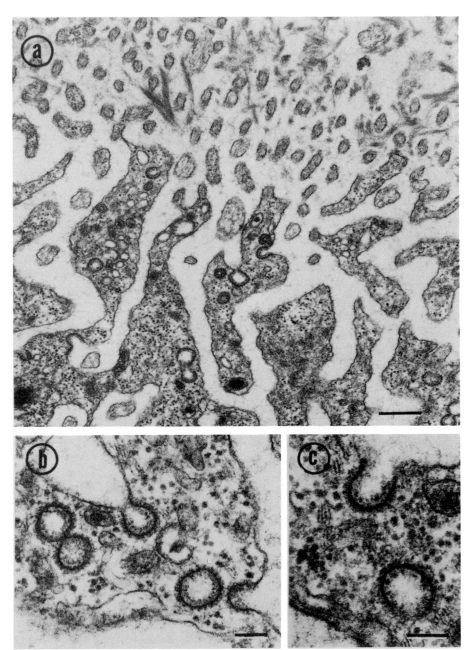

FIG. 4.3. The surface of a *Xenopus* oocyte during vitellogenesis is very highly infolded **(a)**. Numerous coated pits are present along the surface **(b)** and bound to their extracellular surfaces are ''particles'' with the diameter (ca. 10 nm) expected for vitellogenin molecules **(c)**. Bars = 500 nm (a), 100 nm (b, c).

From Wallace, R. A. (1985) In *Developmental Biology: A Comprehensive Synthesis*, Vol. 1 (L. W. Browder, ed.), Plenum Press, New York, p. 127. (From work by Wallace, Jared, Opresko, Wiley, and Selman. See *J. Cell Biol.* **69**:345, 1976, and *Ciba Found. Symp.* **98**:228, 1983.)

Some of the other yolk proteins reach the oocytes through the circulation. In chickens, the yolk lipids acquired from the blood by oocytes are carried by blood-borne lipoproteins, especially ones of the VLDL class, which are secreted by the liver and taken up by receptor-mediated endocytosis. But other routes of supply also operate; for instance, in insects, follicle cells present near the oocytes contribute materials to the eggs' stores.

Of interest in terms of the evolution of endocytotic yolk-storage mechanisms are reports that the reproductive cells of some "lower" forms, such as sponges, phagocytose other cells or portions thereof, to provide stored material for development. Symbiotic bacteria may be acquired by the reproductive cells in like manner (Section 4.4.3).

4.2.1.2. Storage of Endocytosed Yolk Proteins

Eventually, yolk proteins are deposited in highly concentrated stores within membrane-delimited, mature yolk bodies. In many animals these bodies have the form of granules that are poorly soluble in aqueous solutions but in other species they are fluid-filled globules. Quite frequently, they contain prominent crystal-like aggregates of protein. The bodies in a given cell can vary in diameter from a few micrometers to tens of micrometers. It can take some time before recently endocytosed yolk proteins are detectable in the larger of these bodies; for example, in *Xenopus*, accumulation of newly endocytosed vitellogenin in mature yolk bodies ("platelets") is spread out over several hours. Much of the interval between endocytosis and deposition in mature yolk bodies, is thought to be spent in "intermediate" ("transitional") yolk bodies that either transform into or fuse with the mature bodies.

Intermediate bodies are identified in cell fractionation studies of *Xenopus* and some other species, by virtue of their lower density than the mature bodies (Fig. 4.4). Morphologically, some of the suspected intermediate compartments call to mind presumed "prelysosomal" and early lysosomal structures of other cell types: MVBs are involved in yolk storage in many species of animals and, in certain oocytes, such as those of mosquitoes and *Drosophila*, the intermediate compartments include tubules, and vacuoles with tubular extensions.

In silk moths and *Xenopus*, the functioning of the intermediate compartments seems to be sensitive to ionophores expected to affect their pH. With monensin, for example, vitellogenin remains in intermediate structures of *Xenopus* oocytes rather than accumulating in more mature yolk bodies, and vitellogenin receptors are inhibited from recycling to the plasma membrane. However, studies of this sort are still at too early a stage to provide answers to fundamental issues, such as the routes and mechanisms by which vitellogenin receptors recycle to the cell surface, or the mechanisms by which the very high concentrations of protein in the mature yolk bodies are produced. Vitellogenin and its derivatives (see below) tend to be poorly soluble at low pH, which could be an important factor in their storage if the compartments through which they pass do prove to be acidified. But measuring the pH of yolk bodies, and making sense of the measurements, are likely to be problematic; the yolk bodies of amphibians, for example, contain large crystalline aggregates of phosphoproteins, which can interfere with the usual dye procedures.

One finding that is perplexing, in light of the prevailing models for receptor functioning and cycling, is that the affinity of chicken oocyte membranes for vitellogenin peaks at pH 6, the apparent pH of the yolk itself. In *Xenopus* the

173

USES AND ABUSES
OF ENDOCYTOTIC
AND
HETEROPHAGIC
PATHWAYS

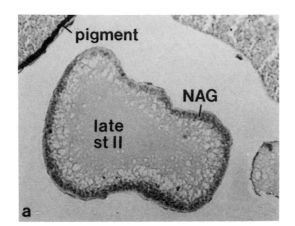

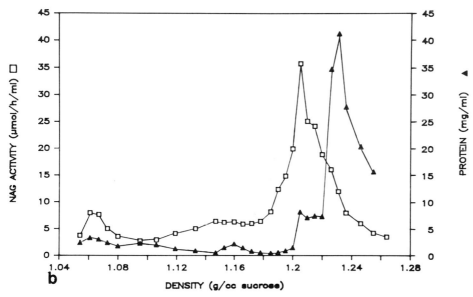

FIG. 4.4. (a) Light micrograph showing a section of a *Xenopus* oocyte in which N-acetylgluco-saminidase (NAG) activity has been demonstrated cytochemically with an "azo-dye" procedure (see Fig. 1.6). Reaction product (NAG) is seen in bodies at the periphery of young oocytes (here: "late stage II"). At later stages, evaluation of the hydrolase activity is more difficult because pigment granules accumulate at the periphery of the oocyte. × 140.

(b) Sucrose density gradient preparation, in which the principal peak of NAG activity (□) corresponds in position to the "light" yolk platelets (▲, density of 1.21); the light platelets probably are an intermediate stage in the maturation of yolk into the "heavier" platelets, which contribute the larger peak of protein (density near 1.23) but have little NAG activity.

From Wall, D. A., and Meleka, I. (1985) *J. Cell Biol.* **101**:1651.

affinity of the vitellogenin receptor for its ligand is more obviously dependent on the presence of Ca^{2+} than on pH; there are reports, in fact, that the protein fails to dissociate at the usual endosomal pHs. Contrastingly, *Drosophila* oocytes take up vitellogenin very poorly when the pH of the medium is low, or when amines expected to alter the pH of intracellular compartments are present.

Over the course of the many hours (or days or even weeks) of oocyte growth, fusions among maturing yolk structures and continued delivery of yolk from intermediate bodies to preexisting older ones produce increasingly large yolk platelets or, in some fish eggs, a coalesced yolk mass. But in large, yolky eggs such as those of amphibians, "old" yolk platelets may eventually become displaced to regions deep within the forming yolk and this may make them much less readily accessible to recently endocytosed proteins than are "younger" platelets located nearer the cell surface.

4.2.1.3. Are Endosomes and Lysosomes Involved? Processing of the Proteins

How lysosomes fit into yolk storage is frustratingly mysterious. Lysosomes are demonstrable in oocytes of many species, cytochemically and by cell fractionation. In most cases, however, it is not certain whether they contribute hydrolases to the forming yolk bodies. Cytochemical studies suggest that yolk in oocytes of some insects, including *Drosophila,* contains maternally derived acid phosphatase. Contrastingly, cell fractions highly enriched in mature yolk platelets from *Xenopus* oocytes contain levels of acid hydrolase so low as to suggest that most platelets lack the enzyme. However, both cell fractionation and cytochemistry indicate that some of the intermediate yolk bodies and related immature yolk platelets in *Xenopus* contain acid N-acetyl-glucosaminidase and β-glucuronidase (Fig. 4.4), though other acid hydrolases generally found in lysosomes, such as proteases and acid phosphatases, are barely detectable, if at all.* In fish like the trout, MVBs involved in accumulating yolk protein are among the structures exhibiting cytochemically demonstrable acid phosphatase, though this activity sometimes seems to become undetectable relatively early in the maturation of the bodies.

The enzymatic capacities, pH, and other features pertinent to the potential degradative capacities of yolk storage organelles are of particular interest because in many animals vitellogenin undergoes proteolytic processing before it is finally stored. (Some of the chicks' VLDL proteins also may be hydrolyzed during lipid storage.) In *Xenopus* the vitellogenin polypeptides produced by the liver and present in the bloodstream, have molecular weights in the neighborhood of 200,000. The vitellogenin derivatives present in the mature yolk platelet are **lipovitellin** molecules [composed of large (M_r roughly 120,000) and small (30,000) peptides] and heavily phosphorylated **phosvitin** molecules (about 33,000). In certain invertebrates, proteolytic alterations of vitellogenin occur at stages before the uptake of the protein by oocytes, but in many cases the processing follows endocytosis.

The conversion of vitellogenin to lipovitellin and phosvitin in *Xenopus* oocytes is believed to take place in intermediate yolk bodies (perhaps in MVBs) because, for example, exposure of the oocytes to the protease inhibitor pepstatin, or to monensin, prevents the conversion and prevents the accumulation of recently endocytosed labeled vitellogenin in mature yolk platelets. The uncleaved vitellogenin molecules remain instead in endosome-like structures, which apparently are disabled both from developing further and from fusing

---

*The presence of proteases and other hydrolases in supposedly "prelysosomal" structures (Section 2.4.4.1) and the possibilities that the lysosomes of a given cell are heterogeneous in enzymatic capacities (Section 7.5.1.2) obviously could be germane here. It also has been reported that when cDNAs for human cathepsin D are introduced into *Xenopus* oocytes, the oocytes synthesize the corresponding hydrolase molecules and package some of them into bodies with the sedimentation behavior of immature yolk platelets.

175

USES AND ABUSES
OF ENDOCYTOTIC
AND
HETEROPHAGIC
PATHWAYS

with older platelets; these disabilities have been speculatively laid to a posited failure of intact vitellogenins to dissociate from their receptors.

Surprisingly, when HRP or radioactive bovine serum (BSA) or DNA is endocytosed by *Xenopus* oocytes, the corresponding labels are mostly lost from the cells within hours. Although full proof is lacking even that these losses involve intracellular hydrolysis, as far as is known, the difference in behavior of endocytosed HRP, BSA, or DNA from the behavior of vitellogenin is not based on entry into different intracellular compartments. Rather, the findings may mean that macromolecules endocytosed by the oocyte pass through compartments capable of proteolysis and other digestive activities but that vitellogenin somehow is spared from cleavages beyond the initial fragmentation into its stored derivatives. Perhaps as lipovitellin and phosvitin form, they aggregate into insoluble deposits not readily attacked by the enzymes. Or perhaps once vitellogenin is cleaved, the cleavage products move rapidly into the mature platelets where active hydrolases are too scarce to do much damage.

4.2.2. Digestion of Yolk

Yolk proteins eventually are degraded, but this often takes place days or even months after their storage, and at relatively late stages in development (e.g., in *Xenopus* embryos, yolk is digested largely after the onset of the heartbeat).* The phenomena of mobilization are diverse.

In marine invertebrates and amphibians, cleavages divide the egg into discrete cells from the start of development. Yolk is widely distributed among these cells, though it may be especially concentrated in some locales, such as in the cells of the vegetal pole of an amphibian egg. In birds and fish, the early embryo often is dwarfed by the yolk mass attached to it. In the most-studied insect eggs, early development involves nuclear division with no cytoplasmic cleavage. It is only after several rounds of nuclear division that the peripheral cytoplasm of the egg is parceled out among uninucleate cells separated from one another by plasma membranes. Meanwhile, the yolk forms a nucleus-poor mass in the center of the egg; this mass remains syncytial for at least a time, though in some insects it reportedly eventually cleaves into very large cells. In *Drosophila* the peripheral cells retain thin cytoplasmic continuities with the central yolk mass at least until gastrulation; during the period where continuities persist, there is appreciable digestion of yolk.

Cells containing yolk seem able to digest it in place and to utilize the products for their own metabolism. But in birds, amphibians, and insects, much of the yolk present in the early embryo comes eventually to lie in the developing digestive system, or in spaces continuous with the digestive system, as in the case of the yolk mass of developing birds. In these locations, yolk seemingly is digested extracellularly, with products of digestion being distributed to other tissues via the circulation.

4.2.2.1. Endocytosis?

Some embryos or early fetuses may use endocytosis to acquire yolk or other nutrients. Thus, developing forms of lower invertebrates are reputed to

---

*Occasional disconcerting experiments suggest that yolk, or at least its normal components, may be less important for development than would appear at first glance. For instance, eggs of molluscs appear to develop normally even after most of their yolk had been removed by centrifugation.

phagocytose "vitelline" cells in order to mobilize nutrients stored in these cells. The endoderm of birds and fish has been reported to endocytose yolk components from the large yolk masses, though this needs critical study. Endocytosis of nutrient-rich fluids from the extracellular compartments surrounding the developing egg has been claimed to supplement digestion of yolk in molluscs, among other invertebrates. Analogous endocytotic roles have been proposed for the visceral yolk sac of mammals, whose cells can be quite active in endocytosis and might therefore aid in nutrition of the embryo during the stages before placental connections are operative. (Recall that mammalian embryos lack substantial intracellular nutrient stores.) The yolk sac of avian eggs transports various yolk components, including stored IgGs, to the circulation, most likely using endocytotic mechanisms to do so (see Section 4.3.2).

In *Drosophila* and other insects, the yolk mass shows islands of nucleated cytoplasm often called **vitellophages.** Occasionally, these have been taken to be specialized cells that phagocytose yolk and digest it. But the notion that vitellophages actually are separate cells, rather than regions of the yolk syncytium, is not well supported and their participation in endocytosis seems dubious at present. Certain insects are, however, among the invertebrates that do phagocytose entire oocytes when starved, apparently thereby drawing upon the nutrients stored in the yolk. Follicle cells are among the chief agents of such resorption. (Follicle cells in fish also phagocytose unlaid eggs in situations of follicular atresia.)

4.2.2.2. Lysosomes?

Evidence that lysosomes are involved in digestion of yolk during development is considerable, though still fragmentary and circumstantial. When, for example, eggs of the mollusc *Barnea* are stratified by centrifugation so that their yolk is concentrated in a discrete layer, this layer shows readily demonstrable acid phosphatase and esterases, from the time of fertilization on. In many other species as well, vertebrate and invertebrate, cytochemically demonstrable acid phosphatase—and more rarely other acid hydrolases—reportedly are present in yolk bodies during their degradation.

As many children given their first microscope or biology experiment kit find out, brine shrimp (*Artemia*) survive hard times by arresting embryos at the gastrula stage and encysting them. When development of these arrested embryos is renewed, considerable degradation of yolk takes place. Increases in the levels of acid hydrolases in the embryo accompany this digestion. In other species too, including *Drosophila*, acid hydrolases are particularly abundant in yolk-rich regions during stages of yolk degradation.

For a few cases, there is preliminary evidence that weak bases inhibit degradation of yolk. However, neutral proteases and enzymes with relatively high pH optima frequently also are active during periods of yolk degradation and often the cytochemical reactions demonstrating the presence of acid hydrolases in yolk during its digestion are marginal, or otherwise unreliable. So it cannot be taken for granted that lysosomal enzymes are the principal digestive agents.

4.2.2.3. A Few Guesses about Controls of Yolk Storage and Digestion

If the ability of large, yolky oocytes, like those of *Xenopus*, to store proteins in undegraded form does reflect the relative lack of hydrolases in mature yolk

177

*USES AND ABUSES
OF ENDOCYTOTIC
AND
HETEROPHAGIC
PATHWAYS*

bodies, this impoverished state could result from the incapacity of lysosomes to fuse with the yolk bodies or from specific controls of lysosome movement. Alternatively, lysosomes may simply be too scarce in the oocyte to provide appreciable levels of hydrolases to most of the yolk bodies. Or perhaps lysosomes are precluded from reaching many bodies deep within masses of yolk, by the abundance and packing of the bodies or by other such "nonspecific" or "geometric" factors. Might it also be that maturing platelets selectively retain aggregated yolk proteins but cycle some of their surface membrane and their soluble contents, including hydrolytic enzymes, back to other cellular compartments or out to the cell surface?

For organisms in which hydrolases seem to be permanently packaged with the yolk during oogenesis or to be present in the intermediate bodies, the failure of the enzymes to degrade yolk components has occasionally been explained by the effects of putative inhibitors: Protease inhibitors have been sought—with provisional success—among the yolk components, though what they do is not yet certain. Another view is that the numerous phosphate groups or other features of the yolk proteins protect them from the enzymes. This might complement the possible protective effects of aggregation.

During development, digestion could be activated by disinhibition of the hydrolases (enzymes with this role are being sought) or by alterations of the ionic environment and pH so as to favor activities of the enzymes. Changes in the state of the yolk itself might render it more readily digestible. (Does, for example, the removal of phosphates by acid phosphatase facilitate attack by other enzymes?) And increased lysosome production, changes in cell geometry, or modifications in the cytoskeleton could permit more extensive fusions of lysosomes with the yolk stores.

In the silkworm *Bombyx mori*, a protease that specifically cleaves one of the main yolk proteins into several fragments is newly synthesized at the time when this protein undergoes digestion; the mRNA corresponding to the protease increases markedly in amount during this period, reflecting, apparently, a genetically programmed need to attack this particular yolk component. The protease is not known to be lysosomal but perhaps the products of its action become available to the lysosomes.

4.2.3. Other Storage Phenomena in Insects

Insects store several sorts of material within intracellular compartments in tissues like the fat body (Fig. 4.5). In the larvae of the lepidopteran *Calpodes* for example, the fat body periodically forms large vacuoles rich in tyrosine; during tanning of the cuticle these vacuoles are discharged exocytotically, presumably releasing the tyrosines known to be needed for tanning reactions. The origins of the vacuoles are incompletely understood, but at least some of their membranes arise from endocytotic-like infoldings of the plasma membrane.

Many insects, including cockroaches, produce membrane-delimited bodies rich in uric acid. For lepidopterans, acid phosphatase has been demonstrated, cytochemically, in such bodies, which have been reported to form via endocytosis and pass through intermediate, MVB stages.

Endocytotic delivery to storage structures is quite evident for the **protein granules.** These granules are particularly abundant in the fat bodies of several types of insect at the time of metamorphosis, when the fat body stores endocytosed storage proteins and lipophorin (Section 4.1.1.2) acquired from the hemolymph, and later uses the stores for its metabolism. In *Calpodes* the pro-

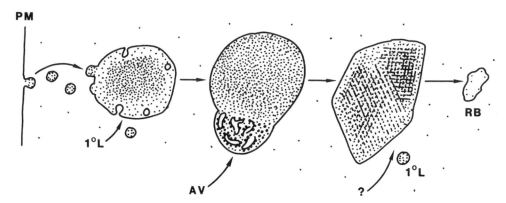

FIG. 4.5. In the fat body of insects like the butterfly *Calpodes ethlius*, protein storage granules form by incorporation of materials endocytosed from the hemolymph into MVBs. During certain developmental stages (Sections 5.1.4 and 5.1.6, Fig. 5.7), the MVBs transform into storage structures with crystal-like contents that later, through degradation, evolve into residual bodies. Lysosomal hydrolases may enter the MVBs and storage bodies by fusions with small lysosomes (1°L indicates the possibility that primary lysosomes may participate); during certain periods, fusions with autophagic vacuoles seem to occur. PM, plasma membrane, AV, autophagic vacuole; RB, residual body. From Locke, M. (1982) In *Insect Ultrastructure* (R. C. King and H. Akai, eds.), Plenum Press, New York, p. 151.

tein granules derive partly from large MVBs. [They may acquire additional materials by fusion with the autophagic vacuoles that abound in the fat body during certain developmental phases (Section 5.1.4).] Actually, MVBs accessible to endocytosed tracers are prominent in the fat body at almost all times but usually they degrade their contents rather than store them. The formation of protein granules at metamorphosis involves delaying the digestion of endocytosed proteins in the MVBs until pupal and adult stages of development. Acid phosphatase is demonstrable in the protein granules during the latter periods, so perhaps fusion with lysosomes brings about the degradation. But acid phosphatase also is sometimes seen in protein granules during stages of their formation, raising the same sorts of questions about the controls of degradation as were encountered in the considerations of yolk above.

4.2.4. Lysosomal or Endosomal Processing of Material Followed by Release of Biologically Active Products from the Cell

4.2.4.1. The Thyroid Gland

Cells of the thyroid epithelium synthesize the large glycoprotein **thyroglobulin** (M_r about 660,000 in rats), which the cells then store extracellularly by secreting the protein into the "colloid" that fills the space enclosed by each thyroid follicle (Fig. 4.6). As the protein is released to the colloid, oxidative reactions mediated in part by a peroxidase, iodinate tyrosine residues in the thyroglobulin and couple some of these iodotyrosines to form iodinated thyronine residues in the polypeptide chain. (Two tyrosines are linked to generate each thyronine.) The majority view is that most or all of this iodination and oxidative coupling takes place outside the cell, probably at or near the plasma membrane.

Iodinated thyronines, especially the triiodo (T3) and tetraiodo (T4; **thyrox-**

179

USES AND ABUSES
OF ENDOCYTOTIC
AND
HETEROPHAGIC
PATHWAYS

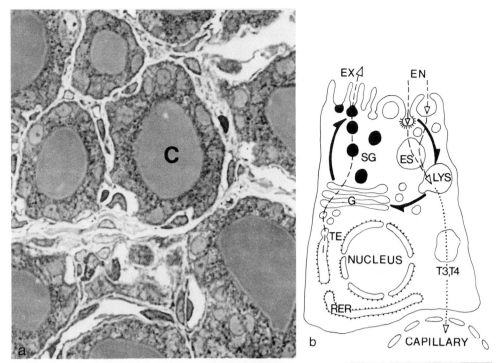

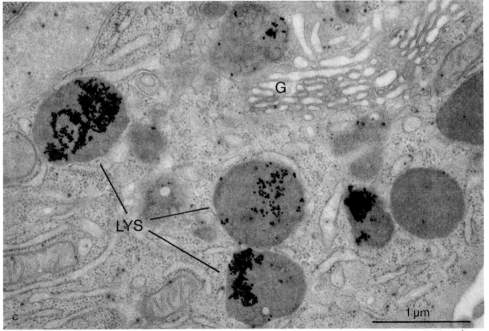

FIG. 4.6. (a) Light micrograph of a pig thyroid preparation showing the colloid (C) stored within the lumina of the follicles. The colloid is surrounded by a layer of cells like the cell diagrammed in (b). × 880.

(b) Diagram of likely routes involved in the production and secretion of the thyroid hormones, T3 and T4. EN, endocytosis; ES, endosomes; EX, exocytosis; G, Golgi apparatus; LYS, lysosomes; RER, rough endoplasmic reticulum; SG, secretion granules; TE, transitional elements of the endoplasmic reticulum. The cell pole at which thyroglobulin is exocytosed and later endocytosed faces the colloid.

(c) Accumulation of gold particles, from endocytosed thyroglobulin–gold complexes, in the lysosomes of an "inside-out" preparation of pig thyroid follicles.

From Herzog, V. (1983) *J. Cell Biol.* **97:**607.

ine) forms, are the major hormones released by the thyroid to the bloodstream. These hormones are "freed" from the polypeptide chain by proteolysis when the gland endocytoses colloid and degrades it in lysosomes. Then, T3 and T4 exit the cell at the opposite pole from the one at which thyroglobulin is secreted.

Endocytosis of colloid can occur both by "ordinary" coated-pit pinocytosis, and by macropinocytosis (Section 2.2.2.1), which, in the thyroid, takes place by the extension of folds or protrusions from the cell surface and the engulfment of a droplet of colloid in a manner reminiscent of phagocytosis. The diameters of the macropinocytotic droplets range up to a few micrometers. Ordinary pinocytosis is seen in the glands of many species when they are fixed for microscopy with no special prior treatment of the animals. Macropinocytosis is not very obvious under these circumstances, though it is observed to some extent in species such as rats. Macropinocytosis is much more evident in glands stimulated to increase their release of hormone by thyrotropin (TSH: thyroid-stimulating hormone). Conversely, treatment of animals with thyroxine, which reduces rates of thyroid hormone release, also reduces macropinocytosis.

Though isolated plasma membrane-enriched fractions from thyroid glands do show some affinity for thyroglobulin, the extent to which specific thyroglobulin receptors are needed for endocytosis is not known. Nor is it clear why, in certain species of animals, thyroglobulin has phosphorylated sugars in its oligosaccharide side chains, including M6P residues, which can be bound by the M6P receptors that handle lysosomal enzymes. (Another unexpected and unexplained recent finding is the presence in thyroid cells of a receptor whose high affinity for N-acetylglucosamines is favored by low pH.)

Even under conditions where thyroid glands have not been intentionally exposed to TSH, endocytosis at the cell surface bordering the colloid occurs at substantial rates. In rats, the area of plasma membrane internalized per hour from the borders of the colloid is estimated to be of the same order of magnitude as the total cell surface area at these borders. But endocytotic removal of membrane is balanced by exocytotic fusion of vesicles, including ones that carry thyroglobulin on its way to the colloidal stores (Fig. 4.7).*

In rats under routine laboratory conditions, recently synthesized and iodinated thyroglobulins diffuse, or are displaced, away from the cell surface into the interior of the mass of stored colloid within a few minutes after their secretion. This is detectable in autoradiographic studies of thyroid glands labeled by short exposures to radioactive amino acids or to radioactive iodine; eventually, the label becomes distributed more or less uniformly within the colloid. Dispersal is much slower after thyroxine treatment: iodine- or amino acid-labeled proteins remain relatively concentrated near the surfaces of the thyroid cells for at least several hours. This effect of thyroxine is associated with a net increase in the overall concentration of protein in the colloid. Under these circumstances, recently iodinated thyroglobulins, being close to the sur-

---

*Endocytosis by thyroid cells is understood as important not only for taking up thyroglobulin but also in retrieving membranes added to the surface during secretion of thyroglobulin; the retrieved membranes are recycled to the Golgi apparatus for reuse in secretion (see Section 5.3.2). Their higher surface-to-volume ratios would make the small coated structures that mediate ordinary pinocytosis better than the macropinocytotic structures, at retrieving membrane with minimal uptake of colloid; the macropinocytotic vesicles presumably specialize in engulfing large "gulps" of colloid when thyroid hormones are needed.

181

USES AND ABUSES
OF ENDOCYTOTIC
AND
HETEROPHAGIC
PATHWAYS

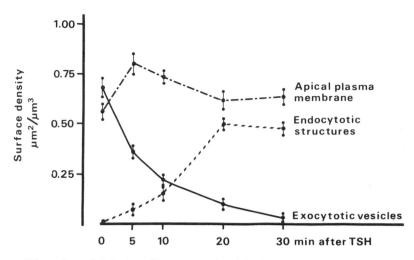

FIG. 4.7. When thyroid follicle cells are stimulated by hormones like TSH to release thyroid hormones, membrane in the apical cell region (bordering the colloid) is redistributed: Endocytotic structures (vesicles, macropinocytotic structures, and intracellular droplets of endocytosed colloid) increased markedly in abundance but the overall cell surface area of the region remains fairly constant, because membrane is also inserted exocytotically from cytoplasmic vesicles. Data are expressed in terms of the surface area of the indicated structures relative to the volume of the tissue. From Ericson, L. E. (1984) *Mol. Cell. Endocrinol.* **22**:1. (Data from Ericson and Engstrom.) (Copyright: Elsevier, Ireland.)

face of the cells, probably stand a markedly better chance of being endocytosed and processed into hormone than do molecules of older vintage. This sort of situation could help account for physiological and biochemical data indicating that the hormone molecules that turn up in the bloodstream appear to have been generated preferentially from the "younger" of the thyroglobulins (i.e., those that have acquired their iodines more recently than the average for the thyroglobulins stored in the gland).

On the other hand, macropinocytosis, which takes up molecules from deeper regions of the colloid than does ordinary pinocytosis, may permit the gland to avoid endocytosing exclusively those thyroglobulins that lie directly adjacent to the plasma membrane. This is important, it is argued, because newly released thyroglobulin molecules, which will be most numerous immediately adjacent to the cell, require some time to acquire their full complement of iodinated and coupled tyrosines and so are not as good sources of hormones as are older molecules.

Experiments with lysosomal extracts confirm the supposition that lysosomal hydrolysis is sufficient to liberate the iodothyronines from their peptide linkages. But how these hormones exit from the lysosomes and from the cell and how they come to associate with proteins in which company they seem to circulate are not certain. There is little evidence for extensive exocytosis-like release of the lysosomes' contents into the extracellular space bordering the capillaries. However, tyrosines and iodotyrosines do seem able to escape from lysosomes by traversing the lysosomal membrane (Sections 3.4.1 and 3.4.1.4). Though iodothyronines are larger than iodotyrosines, it may be that both types of molecule use the same transmembrane routes. Once outside the lysosomes, the hormones might escape through the plasma membrane, either passively or through intervention of a transport system yet to be identified. [The existence of a system by which thyroid hormones can cross membranes has often been postulated. Some models propose that thyroid hormones can move directly into the cytosol of target cells either by crossing the plasma membrane or by escaping from endosomes and lysosomes (see Section 4.5.2.1) and eventually enter the cells' mitochondria or nuclei.)

4.2.4.2. Endocytosis and Antigen Processing in the Immune System

That endocytosis and heterophagy are important for the functioning of the immune system is obvious in the case of the professional phagocytes (Sections 2.2.1.2 and 4.4.1). Lymphocytes also exhibit endocytotic and heterophagic activities important for immune responses, though these were not well appreciated until the last few years. For example, lymphocytes engage in ordinary-looking receptor-mediated endocytosis of many of the components taken in by other cell types, such as transferrin or insulin; uptake rates are minimal in resting cells but when lymphocytes are "activated" by immune agents or other stimuli, endocytosis is stimulated considerably. When antibodies, mitogens, or antigens bind in suitable multivalent forms to lymphocyte surfaces, the bound molecules and their corresponding receptors can undergo endocytosis after migrating along the plasma membrane to form scattered patches, or a "cap" at one cell pole (Section 2.3.2.6).

Most interest in endocytotic and heterophagic events in the immune system is now focused on the processing and presentation of antigens. In particularly instructive studies, antigens that had first been modified by actions of macrophages or lymphocytes of the B class were used to stimulate proliferation, in culture, of the class of lymphocytes known as "helper T cells." In experiments of this sort, both with soluble proteins and with particulate material such as heat-killed bacteria (Listeria), the macrophages or B cells are found first to bind and endocytose the antigens, and then to modify them through intracellular mechanisms that can be inhibited by chloroquine or other weak bases. Once such "processing" has occurred, the B cells or macrophages "present" the modified antigen to the T cells, i.e., the modified antigen returns to the surface of the B cell or macrophage and now takes part in interactions with the T cell surface. Effective presentation of antigens—after artificial processing as below—is possible even if the B cells or macrophages are killed by treatment with glutaraldehyde, the cross-linking agent generally used as a fixative for the electron microscope.

The result of these events is a stimulation of proliferation of helper T cells and the release of interleukins by these cells, phenomena which are central to a subsequent chain of events that leads, eventually, to the production of antibodies.

Antigen processing can sometimes be mimicked by exposing antigens to proteases, which generate peptide fragments. From this and other clues, "processing" within B cells or macrophages is thought to correspond to partial, intracellular, proteolysis. It is the resulting fragments of the antigen that are then returned to the cell surface for presentation to the T cells; the return, presumably, is via one or another of the membrane-bounded vehicles that cycle from intracellular compartments to the cell surface.

The information summarized above suggests that processing takes place in an acidified compartment, and it was initially assumed that ordinary lysosomes were responsible. This assumption is being reexamined, particularly in light of the observations that macrophages—one of the cell types most capable of effective processing—have proteolytic capabilities in their endosomes (Section 2.4.4.1). In addition, though the lysosome population of lymphocytes is not particularly abundant and is not unusual in appearance, a smattering of evidence indicates that these lysosomes, or endosome-like compartments in the lymphocytes, may have unusual features. For one thing, some cell fractions from rat lymphocytes have levels of cathepsin D-like protease activity that are unexpectedly high in comparison to the other hydrolases demonstrable in the fraction. For another, in endocytosis by lymphoblast cell lines, most of the protein molecules taken up seem to return relatively rapidly to the cell surface after spending a time in intracellular compartments that do not become markedly acidified. (This finding, however, contrasts with the apparent involvement of acidified compartments in the antigen processing described above, and with fluorescence data indicating directly that endocytotic compartments in lymphocytes do become acidified.)

Antigen presentation is effective only if the presenting cell has, in its plasma membrane, appropriate species of proteins specified by the "major histocompatibility (MHC) locus." The MHC locus is a set of genes that controls many cellular recognition phenomena by specifying a set of cell surface proteins. In cells such as fibroblasts, these proteins seem not to undergo extensive, rapid endocytotic recycling (at one point it was thought possible that they normally do not cycle at all, but this probably oversimplifies or overstates the actual situation). Lymphocytes, however, especially when they are in activated states like those involved in immune responses, internalize MHC proteins extensively into endocytotic compartments and then recycle them to the cell surface. Different types of lymphocytes cycle different classes of MHC proteins, corresponding to the sorts of antigens the cells handle and to the types of immune responses in which they are involved (e.g., some B cells cycle class II MHC molecules preferentially whereas some T cells cycle class I molecules). At least some of the compartments involved in the cycling of MHC proteins seem to be the same as compartments that accumulate endocytosed antigens. In fact, MHC proteins—either recycling ones, or newly made ones—are widely assumed to encounter processed antigens within intracellular compartments of endocytotic origin, with the two types of molecules then returning together to the cell surface where they collaborate in antigen presentation.

Macrophages may process endocytosed proteins additional to antigens by proteolyzing them within endosomes and then returning active fragments to the cell exterior. Parathyroid hormone, for example, can be converted to a highly active peptide fragment by cultured macrophages, which seem able then to return the fragment to surrounding media; perhaps the liver's Kupffer cells or other of the resident macrophage populations in the body carry out this conversion *in vivo*.

4.3. Transfer and Transport

The movement of yolk proteins from liver or follicle cells to oocytes (Section 4.2.1) exemplifies cell-to-cell transfer mediated by endocytosis. There are several other modes of use of endocytosis to transfer materials from one cell to

183

USES AND ABUSES
OF ENDOCYTOTIC
AND
HETEROPHAGIC
PATHWAYS

another or to move specific molecules through tissues. In these uses, as with yolk, degradation of the endocytosed materials does not take place, is delayed, or is of secondary importance.

4.3.1. Melanin

The melanocytes of skin or feathers produce membrane-delimited granules containing melanin pigments. The melanocytes transfer the granules to keratinocytes and similar cells that consequently come to contain deposits of the pigment though they cannot produce it. The melanocytes of the skin reside at the base of the multilayered epithelium, extending processes ("dendrites") toward the skin surface; melanin granules accumulate in the tips of the dendrites. Keratinocytes arise by cell division near the base of the epithelium and migrate, or are displaced, upward. As they move, the keratinocytes acquire pigment granules from the melanocytes' dendrites. That this transfer depends upon endocytosis is concluded from morphological studies and from observations that endocytotic tracers can enter the membrane-delimited compartments in which the keratinocytes accumulate melanin. Acid phosphatase activity is also seen, cytochemically, in some of the melanin-containing compartments of keratinocytes, leading to the tentative conclusion that these compartments are secondary lysosomes. Evidently the pigment can survive exposure to acid hydrolases for prolonged periods. [Acid phosphatase activity is demonstrable as well, in nascent pigment granules (**melanosomes**) of melanocytes as these bodies form near the Golgi apparatus (see Section 6.1.5 for discussion of acid hydrolases in Golgi-associated granules).]

How melanocytes pass melanin granules to keratinocytes or other recipients has been a controversial issue, in part because some of the morphological evidence initially adduced was obtained with suboptimal methods. Granules from dead and disrupted melanocytes can be endocytosed by other cells, but "drastic" transfer mechanisms based on melanocyte death probably are of negligible importance under normal conditions. The pigment inclusions in keratinocytes frequently appear as if comprising a composite structure in which several granules from the melanocytes have come to form a cluster. The individual granules within the cluster often seem no longer to be clad in their own membranes, as they were when produced, but the cluster as a whole is surrounded by a membrane. Some investigators have explained these observations by arguing that melanocytes release the pigment exocytotically. The granules, having lost their membranes during exocytosis, would then be picked up within endocytotic vacuoles by the keratinocytes, with the vacuole membranes persisting as the membranes seen surrounding the pigment in keratinocytes. The prevailing view, however, is that keratinocytes acquire melanin by phagocytosing granule-containing portions of melanocyte cytoplasm directly, particularly from the dendrites, and using lysosomal hydrolases to degrade the nonpigment components taken up with the melanin. To the extent that the individual granules in the keratinocyte clusters do lose their own membranes, this might simply come about through degradation within the phagolysosome.

Uptake of bits of cytoplasm from adjacent cells, or even apparent endocytosis of one part of a given cell by another part of the same cell (Fig. 4.8), has frequently been reported for nervous tissue under conditions in which the participating cells and cell regions show no signs of ensuing damage or death. It may be more than coincidence that melanocytes arise from cells of the embryo's neural crest and thus share common origins with several sorts of nervous tissue.

185

USES AND ABUSES
OF ENDOCYTOTIC
AND
HETEROPHAGIC
PATHWAYS

4.3.2. Transcytosis

In several tissues, vesicles are utilized to transport quantities of specific molecules across cells; the molecules are acquired by endocytosis at one cell surface, and released by exocytotic-like secretion at another cell surface. Such **transcytosis** has been described in greatest detail for selective passage of immunoglobulins across epithelia but may account for many other interesting phenomena, including, some believe, long-range transfers of materials in the nervous system.

4.3.2.1. Intestinal Transport of IgGs in Neonatal Mammals

For a period around the time of birth in mammals, antibodies (IgGs) from milk or colostrum cross the intestinal epithelium (mucosa) and pass into the circulation. The yolk sac of developing chicks probably mediates analogous transfer of antibodies from yolk to the circulation, and comparable processes may transfer antibodies [and other proteins? (Section 6.2.3)] across mammalian yolk sacs and placenta.

Intestinal epithelial cells of suckling rats bind IgGs from the intestinal lumen and internalize them via coated pits, endocytotic tubules, and small vacuoles—the usual cast of participants in receptor-mediated endocytosis. But studies with IgGs labeled with ferritin or HRP (Fig. 4.9) indicate that subse-

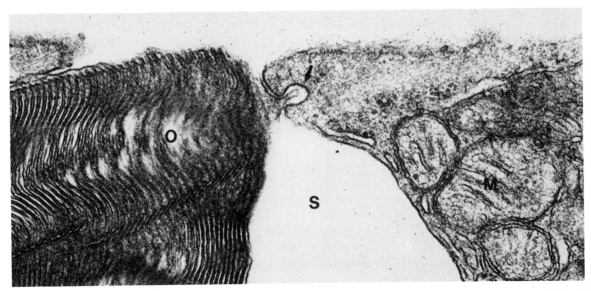

FIG. 4.8. Two portions of a single rod photoreceptor cell from a frog retina. In a three-dimensional reconstruction the "outer segment" at O would be seen to be attached, by a cilium, to the remainder of the cell whose edge occupies the right half of the picture. The cellular space at S would be seen to correspond to the "extracellular space" between the cell body and the outer segment (see Fig. 4.17). At the arrow, a vesicle, with coating faintly visible along its cytoplasmic surface, seems to be taking up part of the adjacent surface of the outer segment; if so, this would represent endocytosis of one part of a cell, by another part of the same cell. M indicates a mitochondrion. In other nervous tissue at particular points in development, adjacent cells often seem to endocytose portions of one another via configurations similar to the one shown here. Micrograph by M. L. Matheke in the author's laboratory.

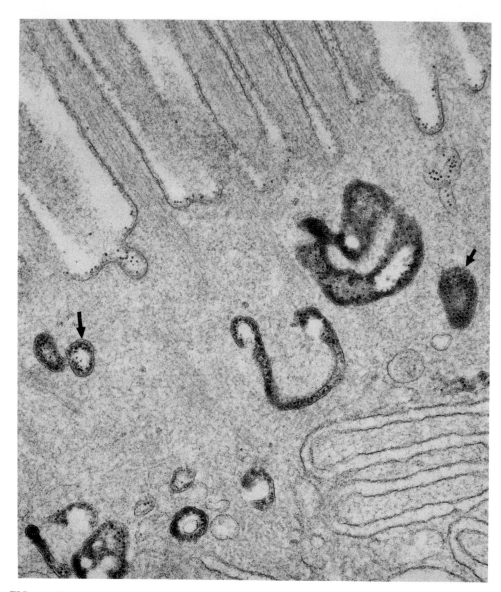

FIG. 4.9. From a preparation of the intestinal epithelium of a neonatal rat. The luminal surface of the epithelium (top) was exposed simultaneously to HRP and to ferritin conjugates of IgG. As seen here (e.g., at the arrows), both of these labels are present in many of the same endocytotic structures (the ferritin is seen as small electron-dense grains, and the HRP is detected by its diffuse, dark cytochemical reaction product). The labels do, however, eventually sort out into separate vehicles so that only the IgG conjugates are transported to the basal region of the cell; the HRP is delivered to lysosomes. From Abrahamson, D. R., and Rodewald, R. (1981) *J. Cell Biol.* **91**:270 × 80,000 (approx.).

quent to endocytosis, only a small proportion of the endocytosed molecules wind up in compartments containing cytochemically demonstrable hydrolases. Instead, most are transported, in vesicles, to lateral and contraluminal (basal) regions of the cells where they are released to the extracellular space by exocytotic-like fusions of the transport vesicles with the plasma membrane. In contrast, when HRP not conjugated to IgGs is taken up in the same endocytotic

187

*USES AND ABUSES
OF ENDOCYTOTIC
AND
HETEROPHAGIC
PATHWAYS*

vesicles as the IgGs, most of the tracer is delivered to lysosomes; little is trancytosed. The separation between the paths of IgGs and that of other endocytosed material like HRP probably occurs after delivery to endosome-like tubules and small vacuoles, and the divergence of pathways apparently depends upon whether the endocytosed material remains adherent to membranes. Most of the HRP does not. The IgGs do. So does cationic ferritin (see Sections 2.2.2.3 and 5.3.3.2), which therefore has been used to trace the transcytotic route in intestinal epithelial cells.

The capacity of the intestine of newborn rats to transport IgGs is attributable to properties of the epithelial cells' IgG receptors. When these special receptors are lost as the animal matures, transcytosis of IgG ceases. The receptors recognize the Fc portion of the IgG molecule, and bind IgGs avidly at pHs of 6–6.5, which correspond to the pH of the intestinal lumen. Binding affinity is much lower at pHs of 7–7.5, like those of the blood. Thus, the receptor can obtain IgGs from the intestinal lumen, retain them even if exposed to moderately acidified intracellular compartments, and then discharge the IgGs at basolateral and basal cell surfaces, where the intercellular fluids presumably have a pH like that of blood. The receptors are presumed to cycle ("shuttle") repeatedly between the opposite cell surfaces, but few details are clear about the vehicles in which they move.

4.3.2.2. Transcytosis of IgA and Other Materials in Adult Mammals

Considerable transepithelial transport of immunoglobulins take place in adult animals, but movement is in the opposite direction from that just described. By carriage from the basal regions of epithelia to the luminal surfaces, dimers, and higher multimers of IgA and polymers of IgM are secreted into milk, saliva, bile, and the lumen of the gut, where the immunoglobulins serve various protective functions. IgAs have been studied most (Fig. 4.10).

The IgAs are generated and initially secreted by plasma cells of the immune system. The epithelia that transport them acquire the proteins from the circulation and tissue fluids by endocytosis, using specific receptors, coated pits, and tubular and vacuolar structures to do so. For hepatocytes, electron microscopic and cell fractionation studies suggest that the endocytosed IgAs enter the same endosomes and related compartments as do other endocytosed materials but that subsequently the IgAs are segregated to a pathway that delivers them to the bile by exocytosis, while other materials from the same endosomes are being degraded in lysosomes. In rodent liver, peak secretion of transcytosed tracer molecules into the bile is seen relatively rapidly—within less than an hour after initiating exposure of the liver to the tracers.

The receptor that governs endocytosis of IgA is a transmembrane protein to which at least some of the IgAs eventually become linked via disulfide bonds. This covalent linkage probably ensures stable binding of the IgA to the membrane during transcytosis. Freeing of the IgAs into the bile or other extracellular fluids requires a proteolytic cleavage of the receptor; the immunoglobulin is released still attached to a fragment ("secretory component") of the IgA receptor. The proteolysis is presumed to take place after the IgAs have entered the compartments that transport them to the sites of exocytosis. In fact, because the cleaved form of secretory component is not readily found in intracellular compartments, the current tendency is to suppose that the proteolysis is carried out by peptidases encountered or activated only when the IgA has reached the cell

moves small amounts of many proteins to the bile even though most molecules of the same materials have other fates, such as degradation in lysosomes. This is found for EGF and other natural blood proteins, as well as for HRP and other foreign tracers. Such movement is most often explained as due to the "accidental" (nonselective) incorporation of small amounts of these materials in the wrong vehicles; "missorting" of a few molecules into transcytotic vehicles might take place at the endosomes or nascent lysosomes. Analogous phenomena have been thought to account for the movement of large molecules across tissues such as the kidney tubule epithelium.

In the regions of the intestinal epithelium known as Peyer's patches, "M cells" transport potential antigens to lymphoid tissue lying below the epithelium, probably by some form of transcytosis. Reovirus particles (Section 4.4.3.1) are suspected to exploit this route as a path for invasion of the organism. *Salmonella* bacteria may cross epithelia, at least in culture, by transcytosis in vacuoles.

One outcome of transcytosis may be a net, selective movement of plasma membrane molecules from one surface of a cell to another. Such a process could help establish or maintain differentiated regions of the plasma membrane.

4.3.3. Transport across Endothelia

Although the endothelial cells of blood capillaries (and small lymphatic vesicles) have large populations of vesicle-like structures, electron microscopy indicates that at any given moment, many—perhaps most—of these structures are continuous with the plasma membrane. As already mentioned, their membranes are "smooth" in that they lack clathrin coats; they have instead a distinctive "striated" ultrastructure visualizable by scanning electron microscopy. Those of the vesicles that are continuous with the plasma membrane, and thus are accessible to the extracellular medium, bind polycationic tracers from the medium much more poorly than do adjacent regions of the plasma membrane. But they do bind exogenous lectins of types with affinity for galactosyl and N-acetylglucosaminyl residues. Some of the lectin-binding materials in the vesicles probably are components of glycosaminoglycans.

Extending across the mouths of many of the open vesicles are thin, nonmembranous diaphragms of poorly understood nature.

The endothelial vesicles are viewed as membrane domains specialized for transport across the endothelium, i.e., for movement of materials between the blood and extracellular tissue fluids. For a time it was assumed that individual vesicles simply traversed the cytoplasm by budding off at one surface of the cell, and fusing with the plasma membrane at the opposite surface. Now it is thought likely that some of the vesicles may, rather, fuse temporarily with one another while retaining continuities with the cell surface; chains of two or more such interconnected vesicles could open transient, tubule-like channels across the endothelium. The relative importance of the two modes of vesicle functioning has yet to be worked out. But one or both of these mechanisms would seem to be the only means by which large molecules—albumin, for example—can cross the endothelia of those capillaries, such as the ones in muscle or brain, whose cells are tied together by tight junctions and whose walls lack fenestrae or the wider gaps found in modified capillaries such as the hepatic sinusoids.

The vesicles in question deliver little of their content to lysosomes or prelysosomal bodies. However, endothelial cells do form coated vesicles, MVBs, and other structures that participate in ordinary-looking endocytosis. The latter route seems more often taken by cationic tracers, whereas the transendothelial path may be favored for anionic molecules. (For brain, however, reports that cationic proteins cross the blood–brain barrier better than others, and findings that certain lectin–HRP conjugates that are transcytosed pass through the Golgi region on their way across the endothelium, have prompted speculation that transcytosis in brain capillaries differs from this picture.) From microscopic studies of the behavior of blood proteins linked to colloidal gold or other tracers, the smooth endothelial vesicles have also been postulated to contain receptors, enabling them to concentrate albumin and other serum proteins before transporting them. The operation, in this way, of receptors for the circulating proteins that carry small molecules—fatty acids, hormones, and so on—in the blood would increase the efficiency of transfer of such molecules to the media directly surrounding cells (see Section 4.1; but the hesitations about colloidal gold tracers noted in Sections 2.2.2.3 and 3.3.4.3 arise here as well).

It is still difficult to decide which materials are transported by endothelial vesicles under normal circumstances and to determine, directly, the extent of this transport when it occurs. In addition to albumin, claims have appeared that LDL and insulin are transferred extensively across capillaries by endothelial vesicles at various sites and that transferrin–iron complexes are delivered from the circulation to the bone marrow in this way. Some of the more provocative proposals relating to vesicle transport capacities arise from findings that endothelial cells isolated from liver exhibit active endocytotic-like uptake of several sorts of proteins. Though much of this uptake seems to be by coated pits and their derivatives and therefore differs from typical endothelial transcytosis, the endocytosis has been regarded by a few researchers as signifying extensive use of vesicular transport for movement of blood-borne materials to hepatocytes. Conventionally, hepatocytes are thought to exchange materials freely with the blood through the gaps readily seen in the sinusoid walls, but these researchers feel that the endothelial vesicles serve to select particular materials for special treatment. One possibility emerges from observations that when the copper-containing protein **ceruloplasmin** is taken up by isolated hepatic endothelial cells, some of the molecules are modified into forms that can now avidly be bound by the hepatocytes' asialoglycoprotein receptors. This has sparked speculation that the endothelial cells endocytose specific materials, modify them intracellularly, and then release them for subsequent high-affinity acquisition by the hepatocytes.

4.4. Defense

Most multicellular animals have endocytotically active cells that defend against invasive microorganisms and help sequester other potentially harmful materials. The phagocytes of vertebrates are best known, but analogous cells exist in invertebrates. For instance, hemocytes in insects and ameboid cells in the blood of *Limulus* and other arthropods are capable of phagocytosis. In addition, some defensive cells of invertebrates act by cellular adhesion and secretion, to generate a coagulum that immobilizes foreign organisms or other materials, calling to mind aspects of platelet functions in vertebrates. Invertebrates have hemagglutinins that probably function as opsonins, and their se-

191

USES AND ABUSES
OF ENDOCYTOTIC
AND
HETEROPHAGIC
PATHWAYS

cretory cells include types that seemingly play roles reminiscent of the functions of the B or T lymphocytes in vertebrates.

4.4.1. Mobilization and Activation of the Professional Phagocytes of Vertebrates

The activities of the two major vertebrate defensive phagocytotic cell groups, the mononuclear phagocytes (macrophages and their relatives; see footnote on p. 18) and the neutrophils and other granulocytic leukocytes, are coordinated and modulated by a system of cells and circulating factors to which the phagocytes respond by means of batteries of receptors. The phagocytes arise from dividing stem cells in the bone marrow and related sites but, as earlier mentioned, populations of phagocytes or their immediate progenitors are normally maintained both in the circulation and in many tissues.

In mice, which are widely used for studies of mammalian defenses, the circulation normally contains a steady-state population of monocytes at least several hundred thousand strong, which supplies a monocyte pool of similar size in the tissues. The latter pool is the immediate source of the functional macrophages that reside in splenic and lymphatic tissues; in the liver (the liver's Kupffer cells constitute the largest resident macrophage population in the body); in the lung (alveolar macrophages); in the adrenal cortex; and in connective tissues of many other sites. Monocytes that have recently arrived in tissues may sometimes undergo cell division but mature macrophages do not proliferate. The normal average half-times of residence of mouse macrophages in different tissues range from a few days to a few weeks, as estimated from thymidine labeling studies, from determinations that monocytes transit through the pool in the circulation with half-times of somewhat less than a day, and from data on the sizes of macrophage populations. Some macrophages almost certainly persist for longer periods; mouse macrophages can survive in tissue culture for months and macrophages that accumulate at sites of chronic inflammation (see below) can persist for at least that long. The eventual fate of macrophages under normal conditions is not fully understood: some may be disposed of by passage into routes of exit from the body, such as the gut or trachea; some may die in lymph nodes; and some may be phagocytosed by younger macrophages.

The neutrophil population in the human circulation is on the order of 10^{11}, with a half-time of residence in the blood of 6–12 hr. It was once assumed widely that neutrophils undergo a normal steady-state migration from the circulation to the tissues, where the cells survive as potential phagocytes for a day or two, but the evidence is equivocal for the existence of appreciable extravascular pools of neutrophils maintained in the absence of invasive microorganisms or comparable pathological stimuli. Mature granulocytes do not divide. The best known mechanism for their disposal is through phagocytosis by macrophages (see below); "aged" neutrophils normally may be removed from the circulation chiefly by spleen and liver macrophages.

In inflammatory responses,* locales of infection, injury, or intrusion of

---

* "Inflammation" is a cluster of responses that varies somewhat with the initiating stimuli. Though the influx of phagocytes is of primary interest in the context of this book, bear in mind that inflammatory sites commonly exhibit immune phenomena, release of many biologically active molecules from cells to extracellular fluids, changes in local vascular permeability promoting the accumulation of fluid, disruption of connective tissue matrices, relative anoxia and acidosis (pHs of 5.5 or below have sometimes been reported), and other alterations.

foreign objects exhibit markedly enhanced local accumulation of phagocytes from the bloodstream. Neutrophils are the first line of phagocytotic defense, especially in acute infections. They provide the most active, large population of phagocytes during the first few hours. Granulocyte types other than neutrophils contribute to inflammatory responses as well, though under most circumstances they are present in considerably smaller numbers than the neutrophils. Accumulation of eosinophils, however, is quite prominent in immune hypersensitivity responses and at some sites of persistent foreign bodies (e.g., in responses to parasites such as schistosomes and their eggs).

193

USES AND ABUSES
OF ENDOCYTOTIC
AND
HETEROPHAGIC
PATHWAYS

During the initial days of an inflammatory response, the neutrophil population is reinforced and then supplanted, by the macrophages. The macrophages phagocytose the remnants of "spent" granulocytes (many of which, having largely exhausted their granule supplies, may no longer be "useful") as well as taking up microorganisms, and sequestering persisting undigested materials, such as components of bacterial cell walls that are resistant to degradation. Macrophages generally constitute the major phagocytotic population at sites of prolonged infection or other chronic inflammations. They can give rise to specialized derivatives, such as multinucleate giant cells that help segregate materials too large to be readily surrounded by ordinary macrophages (see also Section 6.1.1). Indigestible materials—bacterial debris, mineral fibers (Section 6.3.3.1), and the like—can be found within macrophages, months or years after the materials enter the body. This persistence probably implies both that some macrophages have very prolonged life spans and that materials released upon death of a given macrophage are rapidly phagocytosed anew by surviving cells.

Invasive microorganisms that avoid being killed (Section 4.4.3) and indigestible materials often become included in clusters or nodules of cells and extracellular materials, known as **granulomas.** These are rich in macrophages and macrophage derivatives, lymphocytes, plasma cells, and, sometimes, granulocytes. Fibroblasts eventually appear, particularly near the periphery of the granuloma where they lay down extracellular materials that tend to isolate persistent foci of infection or help initiate healing of sites when inflammations are resolving. The fibroblasts also secrete proteases or other enzymes that condition the local environment (Sections 4.4.1.2 and 6.2.1).

4.4.1.1. Orchestration and Control

The vertebrate defensive system is tuned to respond to certain components of common invasive microorganisms, such as the lipopolysaccharides (once know as "endotoxins") of bacterial cell walls, and to materials released or generated by tissue damage. In addition to antibodies, the agents ("factors") responsible for initiating and coordinating the responses include small molecules, such as **histamine,** which is secreted by basophils and mast cells, **lymphokine** proteins, which are secreted by lymphocytes, and many other molecules secreted by platelets, lymphocytes and, other cell types, including the professional phagocytes (Section 4.4.1.2 and Table 6.2). Frequently, the **complement** system of blood serum plays a central role. This is a set of about 20 proteins that can be modified into biologically active forms by suitable activating agents. Immune complexes of IgGs with antigens initiate the so-called "classical" pathway for activating complement; components such as bacterial lipopolysaccharides set off an "alternate" pathway. These two pathways differ in details but in both proteolytically active derivatives of complement proteins are generated. These in turn split specific bonds in other complement proteins to produce molecules that can opsonize particles for phagocytosis, molecules that influence the migrations of phagocytes, and molecules that participate in membrane-active cytolytic complexes, which insert in cellular plasma membranes, opening holes that fatally compromise the permeability of target cells.

Chemoattractant factors such as the complement-derived protein **C5a,** a strong chemotactic agent for neutrophils, help encourage movement of phagocytes from the circulation to sites of infection or damage. Though mononuclear phagocytes are able to traverse the walls of intact, normal, small blood vessels, the passage of phagocytes to inflammatory sites is facilitated by the changes in the vascular walls that result from tissue damage and from the impact of agents such as histamine (which increases vascular permeability). Once neutrophils or macrophages have arrived at regions of tissue damage or bacterial influx, their departure may be inhibited by specific factors. Indeed, one set of proteins was once known as "migration inhibitory factors" because of their effects on motility of mononuclear phagocytes in culture. [These proteins are now thought to be lymphokines that control diverse aspects of macrophage differentiation (see below) and influence the retention and release of monocytes maturing in the bone marrow, as well as affecting the residency of functional phagocytes at peripheral sites of accumulation.]

One of the systems by which neutrophils move up chemoattractive gradients involves plasma membrane receptors that control aspects of the ionic permeability of the cell surface. This system enables the cells to respond to the presence of N-formyl peptides such as formyl-Met-Leu-Phe, thought to be or to mimic molecules arising from bacteria at sites of infection. Like the other mechanisms through which the phagocytes respond to lymphokines, complement-derived factors, and similar agents, this one needs to be better understood in specific detail. The phagocytes are known to have an extensive repertoire of mechanisms likely to be important in their responses, including just about all the devices currently popular in the debates as to how cellular movements and metabolism may be controlled: gates and channels that alter plasma membrane permeability and potentials; systems to modulate intracellular pH or the concentrations of Ca^{2+} and other ions; protein kinases; G-proteins and cyclic nucleotide systems; enzymes catalyzing turnover of inositol lipids and of related inositol phosphates; kinases (especially protein kinase C) responsive to phorbol esters; and others.

4.4.1.2. Signaling and Secretion by the Phagocytes

Vertebrate phagocytes release numerous biologically active molecules to their surroundings (see Table 6.2). For obvious reasons, release has been studied most under tissue culture conditions, but plausible surmises can be made about more natural conditions. For instance, lipids of macrophage plasma membranes are rich in the fatty acid **arachidonic acid.** When responding, in culture, to inflammatory stimuli such as yeast cell walls or complement derivatives, the macrophages release this fatty acid from their lipids through action of phospholipases, and use it, via lipoxygenase and cyclooxygenase metabolic pathways, to generate prostaglandins, thromboxanes, and leukotrienes. These products include chemotactic molecules likely to help control phagocyte influx from the circulation *in vivo,* and agents that alter vascular permeability and therefore could participate in evoking local edema.

195

USES AND ABUSES
OF ENDOCYTOTIC
AND
HETEROPHAGIC
PATHWAYS

As features of their antimicrobial activities, the professional phagocytes produce highly reactive "oxygen metabolites" related to peroxides, some of which are released to extracellular media (Section 4.4.2).

Release of macromolecular materials from the phagocytes occurs both by exocytosis from living cells, and upon cell death. Sections 6.2 and 6.3 will take up such release in more detail. For the present chapter, keep in mind that among the proteins living macrophages secrete (Table 6.2) are acid hydrolases, neutral proteases, lysozyme, lipoprotein lipase, several components of the complement system, an interferon, an interleukin that helps activate lymphocytes, factors that are mitogenic or chemotactic for cells such as fibroblasts, and factors that participate in regulating the production and mobilization of monocytes from the bone marrow. The proteases secreted by macrophages are capable, at least under *in vitro* conditions, of activating the complement system and also the thrombin and plasminogen systems involved in formation and dissolution of fibrin clots. Macrophage proteases thus could contribute to several cascades of events that impact on the course of inflammations. **Cachectin,** a macrophage secretory protein originally of interest for its capacities to lyse tumor cells, is now attracting considerable attention for its roles in the wasting of body tissues often associated with severe infections; one of cachectin's effects is inhibition of the lipases of adipocytes.

Among the proteins released by neutrophils, some reduce microbial viability or growth. At least one is "pyrogenic," which is of interest in view of the high local temperatures associated with inflammation and the wider effects of infection on body temperature. Macrophages also secrete pyrogens.

The proteases and other enzymes secreted by neutrophils and macrophages probably contribute to loosening or degrading connective tissues, including the basal laminae of blood vessels and other matrices through which the cells must penetrate to leave the bloodstream and move within tissues (see Chapter 6).

4.4.1.3. Activation of Macrophages

Macrophages sometimes are obtained for study simply by rinsing them out of the peritoneal cavity ("lavage"). More often, macrophage-rich exudates are elicited in the peritoneal cavity or elsewhere by insertion of inert glass or plastic surfaces such as the coverslips used for light microscopy, or by injection of more "active" materials such as thioglycollate broths or BCG (bacillus Calmette–Guérin, a strain of tuberculosis bacterium frequently employed in immunological adjuvants). The specific conditions used to prepare macrophages from the organism or the conditions to which macrophage-related cell lines are exposed in culture, markedly affect the properties of the cells obtained (Table 4.2). For instance, macrophages obtained from sites of inflammation or other challenge tend to be more active phagocytes than are the cells of normal resident populations. They possess more lysosomal hydrolases and exhibit considerably greater release of antimicrobial oxygen derivatives and other products. They exhibit different patterns of histocompatibility antigens and other surface molecules. And they show enhanced capacities to kill tumor cells, bacteria, and protozoan parasites (Section 4.4.2). In these respects and others, macrophage properties can be attuned to local circumstances, yielding somewhat different types of cells at different sites of normal residence, or for different sorts of pathological insult.

TABLE 4.2. Properties of Macrophages Activated or Elicited by Various Agents[a,b]

Properties	Resident	BCG or Corynebacterium parvum	Trypanosoma brucei		TPM	Endotoxin	MDP	Proteose peptone
			Day 9	Day 15				
Spreading	+	++++	++++	++++	++	++	++	++
Plasminogen activator	±	++++	++++	++++	+++	++	−	+
H_2O_2/O_2^-	±	++++	++++	+	++	±	++	−
Tumoricidal activity	−	++++	ND	ND	−	±	++	−
Enhanced antimicrobial activity	−	++++	ND	ND	−	±	++	−
5'-Nucleotidase	+++	+	ND	ND	++	+	ND	ND
Alkaline phosphodiesterase	+	++	ND	ND	+	++	ND	ND
Surface markers								
F4/80	+++	+	+	+++	++	+++	++	+++
Ia	−	+++	+++	+++	+	+	++	−
Mac-1	+++	+++	++	+++	+++	+++	+++	+++
FcR IgG2a	+	+++	+++	++	+++	+++	ND	+++
FcR IgG2b	++	+	+	++	+++	++	++	++
MFR	+++	+	+	+++	+++	+++	++	+++

[a]The cells were monitored for the extent and speed of spreading on solid surfaces, for levels of the indicated enzymes, for production of peroxide and superoxide (see Section 4.4.2.2), and for effectiveness against tumor cells and microorganisms. The surface markers were detected immunologically or by binding of appropriate ligands. Courtesy of S. Gordon. Based on Ezekowitz. R. A. B., and Gordon. S. (1984) Contemp. Top. Immunol. **13**:33.

[b]Abbreviations: BCG, bacillus Calmette–Guérin; FcR, Fc receptor; MDP, muramyl dipeptide; MFR, mannose receptor; TPM, thioglycollate peritoneal macrophages; ND, not determined.

Current classifications of the state of differentiation of macrophages generally refer to normal **resident** types, to **elicited** or **inflammatory** macrophages (as are obtained with "nonspecific" stimuli such as thioglycollate), and to **activated** types (which usually have been stimulated by the immune system). There are subcategories for each of these three classes. For the most part, the resident cells tend to be relatively quiescent in phagocytotic activities, the secretion of enzymes, and the production of antimicrobial agents. Elicited macrophages are more vigorous phagocytes and secretors than are the resident types. Activation is needed to bring out full antimicrobial capacities and other properties.

These modulations reflect the multistep nature of macrophage differentiation and the diversity of agents that influence this process. For example, activation of the capacities of cultured macrophages to kill microorganisms can be accomplished by first exposing the cells to lymphokines and then to bacterial cell walls or to lipopolysaccharides derived therefrom. The lymphokines "prime" the cells and the cell walls then "trigger" expression of several activated functions not seen in the primed cells. γ-Interferons are among the agents effective in such macrophage activation sequences.

Macrophages in different states respond differently to materials such as complement derivatives. For example, human monocytes or macrophages maintained under culture conditions, can use their C3 receptors (Section 2.2.1.2) to bind particles coated with protein **C3b,** which is generated in the complement reaction. But unlike the case with IgG opsonization, which itself suffices to promote particle uptake under normal circumstances, phagocytosis of C3b-coated particles bound to the cells does not occur unless additional conditions are met. With monocytes, phagocytosis ensues if the cells are exposed to the connective tissue matrix protein fibronectin, or to the phorbol esters (these esters induce secretion by many cells, including the exocytosis of the contents of neutrophil granules). One implication may be that mononuclear phagocytes accumulating at wounds are enabled to respond to C3b opsonization when injury results in the exposure or deposition of fibronectin. Resident-type peritoneal macrophages will ingest C3b-coated particles if the coating also includes IgGs, in the form of immune complexes that can interact with the Fc receptor. But ingestion will occur in the absence of IgG if the cells have been activated by treatments involving exposure to lymphokines. This effect probably is related to the fact that activation of macrophages greatly increases the mobility of receptors for C3b in the plane of the plasma membrane (see Section 2.3.2.1).

Neutrophils, upon exposure to formyl-Met-Leu-Phe and to fibronectin, also can phagocytose particles opsonized with C3 derivatives.

4.4.2. Microbicidal Systems

Though the mammalian phagocytes do take up many types of microorganisms and expose them to acid conditions and to lysosomal enzymes, and though, through their release of lytic enzymes, the phagocytes help the organism create an inhospitable extracellular environment, these mechanisms by themselves often are inadequate to kill the microbes or to stunt their growth and stem their proliferation. Fortunately, the phagocytes have antimicrobial enzyme systems that complement or supplement the lysosomal hydrolases.

197

USES AND ABUSES
OF ENDOCYTOTIC
AND
HETEROPHAGIC
PATHWAYS

4.4.2.1. Neutrophil Granules

As Section 2.1.2.2 outlined, the secondary lysosomes of neutrophils acquire a variety of enzymes and other proteins additional to those generally characteristic of lysosomes. Some of these proteins come from the specific granules, some from the azurophilic granules, and some may come from the other, less well-characterized granule types claimed to be present in neutrophils of certain animal species (see footnote on p. 27). The proteins equip neutrophil phagolysosomes and secretions better to deal with microorganisms and other materials than would be the case if the cells had to rely solely on their acid hydrolases.

In rabbits and other species, the specific granules and the azurophilic granules both contain lysozyme, which hydrolyzes important bonds in the peptidoglycan of bacterial cell walls. The specific granules also contain lactoferrin, which complexes with iron, and a protein that binds vitamin B_{12}; these proteins deny the microorganisms components needed for survival or proliferation and may also sequester potentially troublesome bacterial products. Neutral proteinases, including one with collagenolytic capacities, are present within the specific granules, and other such proteinases, including an elastase-like enzyme, probably are packaged in the azurophilic granules.

A set of cationic neutrophil proteins that are particularly prominent in bovine cells, has strong antibacterial effects when tested *in vitro*; the proteins may, for example, compromise the barrier normally posed by the outer membrane of gram-negative bacteria. Small arginine-rich proteins of this sort from the azurophilic granules of humans and rabbits have come to be called "defensins".

That specific granules in rabbit neutrophils begin to fuse with nascent phagolysosomes before azurophilic granules do, and that the pH during this initial period is high (Section 3.1.3), may facilitate attacks on phagocytosed bacteria by the specific granule components, many of which, like lysozyme and alkaline phosphatase, have relatively high pH optima. How general this pattern of fusion timing is and how important the neutral pH is for initial microbicidal attack have yet to be examined rigorously.

4.4.2.2. Reactive Oxygen Metabolites; Peroxidases

Neutrophils can rely upon anaerobic metabolism for most of their activities and therefore can function at sites where interrupted circulation or other factors reduce the oxygen supply. But, under culture conditions at least, oxygen is required for robust antimicrobial activities. Interest first focused on H_2O_2 as an oxygen derivative of probable antimicrobial importance, because the azurophilic granules and secondary lysosomes of neutrophils contain myeloperoxidase. In addition, the principal granules of eosinophils contain an eosinophil-specific peroxidase, and the lysosomes of monocytes also have a peroxidase (yet to be fully characterized). Remember, however, that many types of resident and activated macrophages lack demonstrable lysosomal peroxidases, although peroxidase activity, of uncertain significance, is shown, cytochemically, in their endoplasmic reticulum.

Neutrophils and macrophages—especially activated forms—generate H_2O_2 during phagocytosis, and in responses to opsonized microbial cell walls, arachidonic acid derivatives, chemotactic agents, or phorbol esters. Some of the peroxide is detectable, by chemical analysis, in culture media. Cytochemical methods, based on the use of cerium to precipitate peroxide, have also demonstrated H_2O_2 formation, both at the plasma membrane and within phagocytotic vacuoles. In tissue culture situations, peroxide itself can kill some strains of microorganisms, such as the protozoan parasite *Leishmania* but much stronger and more consistent microbicidal effects are produced by the phagocytes' peroxidases, which utilize halide ions along with the peroxide. The precise mechanisms of the halide effects are not known. The phagocytes' peroxidases can catalyze iodinations of the proteins in bacterial walls. But, whereas iodine is quite effective in tissue culture assays of neutrophil microbicidal activities and although eosinophils utilize bromine effectively, under ordinary circumstances *in vivo* Cl^- is by far the more prevalent halide. The phagocytes' peroxidases can generate "active" chlorine derivatives such as HOCl. Oxidative decarboxylations and chlorinations by these derivatives, or other actions of the peroxidases, might lead to debilitating alterations of proteins and other macromolecules (e.g., through formation of aldehydes and chloramines). The neutrophil peroxidase is active at pHs of around 5 and thus could act in direct synergy with the acid hydrolases within phagolysosomes.*

Peroxide is not the only reactive oxygen metabolite produced and released by the professional phagocytes, and it may not even be the most important one. In fact, the initial product of relevant oxygen metabolism in neutrophils and macrophages is superoxide (O_2^-). Superoxide gives rise to peroxide spontaneously, at low pH, or by enzymatic action of superoxide dismutase (see Fig. 4.11). Peroxide and superoxide can undergo additional reactions, catalyzed, for example, by metals such as iron, to yield hydroxyl radicals and other highly reactive oxygen radicals. Many of these oxygen derivatives are potentially toxic to microorganisms and several have been detected in culture media bathing suitably stimulated phagocytes.

Rough correlations have been noted between the abilities of cultured macrophages lacking lysosomal peroxidase to kill microorganisms and their production of various of the oxygen metabolites. Correspondingly, the macrophages' antimicrobial capacities are sometimes diminished when catalase (which decomposes H_2O_2), superoxide dismutase, or some other selective "scavenger" of one or another oxygen metabolite, is added experimentally or is present naturally in microorganisms that the phagocytes confront. Unfortunately, in many cases such correlations and inhibitory effects are not clean, and it is

---

*Different classes of defensive cells can utilize their antimicrobial machinery differently, even against the same microorganisms. For instance, eosinophils attack schistosomes at least in part by secreting peroxidase and antimicrobial cationic proteins, whereas neutrophils may actually fuse with the schistosomes.

In this connection, note also that the types of macrophages that produce peroxide but lack appreciable levels of lysosomal peroxidase, would appear unlikely, at first glance, to rely on halide-dependent peroxidase reactions as principal microbicidal devices. Macrophages have antimicrobial capacities that are quite effective against some microorganisms and that seem not to be dependent upon oxygen metabolites, but perhaps the peroxide the macrophages produce is a by-product arising when the cells generate the other reactive oxygen metabolites to be considered above. Or the peroxide might itself have antimicrobial effects or, on release from the cell, become available to cell-associated or extracellular peroxidases from other sources, like the enzymes of neutrophils or eosinophils or secretory lactoperoxidases such as those in salivary secretions and milk. Could it also be, however, that macrophages use peroxide in their secondary lysosomes to sustain activities of peroxidases acquired endocytotically (e.g., when the macrophages phagocytose dying granulocytes and thereby presumably pick up peroxidases as well as lactoferrin and other "defensive" proteins)?

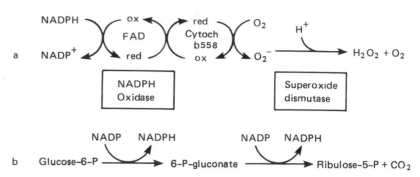

FIG. 4.11. (a) Neutrophils generate superoxide by an NADPH oxidase thought to involve a flavoprotein and a cytochrome of the *b* class; the participation of other electron transport molecules is in dispute. The superoxide can give rise to hydrogen peroxide by the action of superoxide dismutase, an enzyme found in many cell types. Among the chief sources of the NADPH is the "hexose monophosphate shunt" (also called the pentose phosphate pathway), relevant steps of which are schematized in **b**. (Glucose-6-P is glucose-6-phosphate.)

not yet universally agreed which of the oxygen metabolites are essential or how they act.

4.4.2.3. The Respiratory Burst and the NADPH Oxidase; Chronic Granulomatous Disorder

The importance of the reactive oxygen metabolites for coping with infection is underscored by the existence of human disorders resulting from genetic abnormalities in pertinent pathways. Affected individuals have diminished ability to resist infectious bacteria, fungi, and other organisms, which therefore proliferate within the professional phagocytes, and elsewhere in the organism.

Some of the disorders involve myeloperoxidase defects. The best known, however, are the **chronic granulomatous disorders** (CGD), which are characterized by drastically reduced abilities of the phagocytes to generate reactive oxygen derivatives.*

The metabolism through which reactive oxygen derivatives normally arise is signaled by a "respiratory burst"—oxygen consumption insensitive to cyanide and other mitochondrial inhibitors. The burst can be triggered in cultured phagocytes by appropriately opsonized particles or by agents such as phorbol esters; triggering is conveniently monitored by detecting reduction of tetrazolium dyes or through chemiluminescence due to oxygen metabolites. At least two metabolic sequences are crucial. The hexose monophosphate shunt (Fig. 4.11), probably operating in the cytosol, generates reduced coenzymes. These are then used to reduce oxygen via a membrane-associated oxidase sys-

---

*Different strains of microorganisms vary markedly in susceptibility to killing by CGD phagocytes and the differences sometimes correlate interestingly with features of peroxide-related metabolism in the microorganisms. For example, in culture, the phagocytes do relatively well against *Streptococcus* bacteria, which lack the catalase activity used by other microorganisms to decompose H_2O_2. The bacteria might either generate enough H_2O_2 themselves to "compensate" for the CGD deficit, or they might be unable to decompose residual levels of peroxide produced by the phagocytes.

Some microorganisms are thought to release superoxide, which might similarly prove useful to defensive systems under some circumstances.

tem. NADPH is the principal relevant product of the shunt, but over the years opinion has been divided as to whether this is used directly by the oxidase or whether there is an intermediate reduction of NAD yielding NADH for the oxidase. Present views strongly favor NADPH though it may still be that there is more than one pathway.

The NADPH oxidase of neutrophils has been tentatively analyzed as involving a chain in which NADPH is used to reduce a flavoprotein, and a b-type cytochrome is interposed between this protein and a terminal oxygen-consuming enzyme (Fig. 4.11). The possible participation of ubiquinone and other components is in dispute.

An active NADPH oxidase is detected in neutrophils, or in cell fractions therefrom, only if the cells had previously been stimulated as outlined in the last section. The active state resulting from a given round of stimulation persists for many minutes, but it does eventually decline, so that the production of the oxygen derivatives ceases. The cells then sometimes become refractory for a few hours to reevocation of the respiratory burst. That activated oxidase resides in the plasma membranes of stimulated phagocytes is concluded from the cytochemical evidence mentioned already, from the fact that there is considerable release of the oxygen metabolites to the extracellular space, and from the fact that internalization of particles is not required for the respiratory burst. Though phagocytotic vacuoles also seem to acquire an active oxidase, it is not known if this can be accomplished independently of the plasma membrane or simply results from the internalizing of activated oxidase with the plasma membrane used to generate the vacuoles.

What is needed to switch the oxidase on? Perhaps all that is essential are local movements of the components of the chain in the plane of the membrane, or conformational changes, leading to altered interactions of the components with one another, or with regulatory proteins present in the membrane or in the adjacent cytoplasm. Other kinds of models propose, however, that longer-range assembly steps are needed to activate the chain. For instance, cell fractionation observations suggest that the membranes bounding the neutrophils' specific granules contain cytochrome b of the type involved in the oxidase and that cells exposed to phorbol esters under conditions that activate production of the reactive oxygen derivatives, insert these molecules in the cell surface through exocytotic-like fusions. On the other hand, neutrophil fragments (cytoplasts) containing few granules can be stimulated to generate superoxide; this finding, plus some recent observations on activation in cell-free systems, could mean that translocation of components of the oxidase chain from the specific granules is not an absolute requirement, at least for short-term needs. Perhaps the translocation serves instead to replenish the supply of oxidase components at the cell surface.

All the usual regulatory agents have been charged with chief regulatory responsibility for the oxidase by one investigator or another. Most often, kinase-mediated phosphorylations, cyclic nucleotide systems and G-proteins, phosphoinositides, arachidonic acid derivatives, or Ca^{2+} are blamed.

In the X-linked form of CGD, the cytochrome b is absent or defective as a consequence of a defect in the corresponding structural gene. In autosomal forms of the disease, the defect may be in cytosolic proteins or occur early in the chain, perhaps at the level of the flavoprotein.

Because the reactive oxygen metabolites are relatively small molecules, it is likely that some cross vacuole membranes or plasma membranes, thereby leaking into the cytoplasm. Here the metabolites are assumed to be detoxified through the agency of peroxisomal catalase, of superoxide dismutase, and of glutathione-dependent pathways.

4.4.2.4. Chediak–Higashi Disease

In the human inherited disorder **Chediak–Higashi** disease, the lysosomes and other granules in many cell types, including the phagocytes, are abnormally enlarged and reduced in number. This is associated with a broad syndrome including distinctive pigmentation and reduced effectiveness of the phagocytes with resulting increased susceptibility to infection. Suspected contributory cellular aspects of the disease include dramatic reduction or inhibition of certain lysosomal proteases (cathepsin G and elastase activities are reduced in mature neutrophils, though they may be present in stem cells), altered fusion properties or permeability of organelles or membranes, and cytoskeletal abnormalities. But the underlying primary effect of the mutation is not known.

In mice as well as humans, mutations, recognized initially by their effects on pigmentation, have turned out also to alter lysosomes or the behavior of lysosomal enzymes. Of these, the "beige" mutation has effects closely resembling those of the human Chediak–Higashi mutation (Figs. 1.7d and 7.7).

4.4.3. Organisms and Toxins That Avoid Destruction

Bacteria and other microorganisms have evolved many means to avoid or overcome host cell defenses. These mechanisms need not be entirely successful for each individual invader; it is often more a question of altering overall balances between destruction, survival, and proliferation. The state of the phagocytes, such as whether macrophages are activated or not, frequently is crucial to the microorganism's chances.

Some microbes avoid triggering marked immunological responses or deflect them when they occur. For instance, the abundant protein A of *Staphylococcus* cell walls binds to the Fc portion of immunoglobulin molecules, thereby reducing the overall concentration of IgGs available to promote phagocytosis. Other pathogens shed quantities of antigens with similar effect.* Certain bacteria and viruses (of which the AIDS organism is best known) kill or disable lymphocytes. The ameboid protozoan pathogen, *Entamoeba histolytica*, turns the tables by secreting cytolytic agents that can kill macrophages and neutrophils, as well as other cells in the host organism. (These agents include a pore-forming protein comparable to ones utilized in the immune system. Presumably, they evolved originally to permit an ameba to kill its prey but now help it defend itself as well.) Several organisms either avoid evoking the production of reactive oxygen metabolites or minimize this production.

Viruses and microsporidia once were thought to evade destruction in the heterophagic apparatus by penetrating directly through the plasma membrane. However, as Section 3.1.3.3 indicated, under normal circumstances the most prominent supposed viral exemplars of this entry mode actually are taken up by endocytosis and escape, intracellularly, from endosomes; entry through the

---

*Complement-derived opsonins such as C3b (Section 4.4.1.3) may help the phagocytes remain effective in the face of such immobilization of IgGs or of the overloading of the cells' Fc receptors by excesses of immune complexes.

203

USES AND ABUSES
OF ENDOCYTOTIC
AND
HETEROPHAGIC
PATHWAYS

plasma membrane takes place only under unusual conditions, such as at low pH. (But see footnote on p. 205.)

Other microorganisms—those of principle concern in this part of the book—are endocytosed but through one device or another, survive and thrive within endocytotic or heterophagic structures (Table 4.3). This is the situation for various bacteria, for several sorts of obligate or facultative protozoan parasites, and for some intracellular symbionts. By living within a host cell, the organisms shield themselves from the immune system and other defense mechanisms. They may also have special access to nutrients. Migrations of host cells such as macrophages help disseminate the "guests" within the invaded organism.

Microorganisms that invade the endocytotic apparatus generally use host cell plasma membrane molecules as "receptors" to which they bind prior to entering the cells. Glycoproteins, or sometimes glycolipids, are the chief suspected receptors. The distribution of such receptors among different cell types is one of the factors influencing the host cell range of the organisms. Endocytotic capacities of potential targets are another. Quite often, the uptake of organisms into susceptible cells is unusual. For instance, when alveolar macrophages in culture phagocytose the *Legionella* bacteria responsible for Legionnaires' disease, the phagocyte's pseudopods coil around the bacterium rather than extending as a more or less symmetrical cup, like that seen with other bacteria or particles. The protozoan plasmodia, which cause malaria, enter the cytoplasm of red blood cells—normally endocytotically inactive—by establishing surface attachments to the host cells. These "junctions" move pro-

TABLE 4.3. Modes of Entry of Intracellular Parasites into Host Cells, and Their Subsequent Fate[a]

	Intracellular parasite											
	Bdellovibrios	Microsporidias	Plasmodias	Toxoplasma gondii	Trypanosoma cruzi (epimastigotes)	Trypanosoma cruzi (trypomastigotes)	Leishmania donovani[b]	Mycobacterium tuberculosis	Shigella flexneri	Rickettsiae	Coxiella burnetii	Chlamydiae
Parasite has specialized entry organelles	+	+	+	±[c]	0	0	0	0	0	0	0	0
Parasite makes hole in host cell envelope	+	+	0	0	0	0	0	0	0	0	0	0
Parasite enters nonprofessional phagocytes	−[d]	−	+	+	0	+	+	0	+	+	+	+
Parasite enters in host-derived vacuole	0	0	+	+	+	+	+	+	+	+	+	+
Parasite entry inhibited by cytochalasin B	−	−	+	+	+	+	+	−	+	+	+	0[e]
Parasite expends energy during entry	+	+	+	±	−	−	±	−	+	+	−	0
Parasite escapes vacuole after entry	−	−	0	0	0	+	0	0	+	+	0	0
Vacuole fuses with lysosomes	−	−	−	0	0	0	+	0	0	0	+	0

[a]See also Table 4.4. From Moulder, J. W. (1985) *Microbiol. Rev.* **49**:298. (Copyright: American Society for Microbiology.) +, yes; 0, no; other symbols, see footnotes.
[b]Both promastigotes and amastigotes.
[c]Conflicting reports.
[d]No record of test.
[e]Entry inhibited by cytochalasin D.

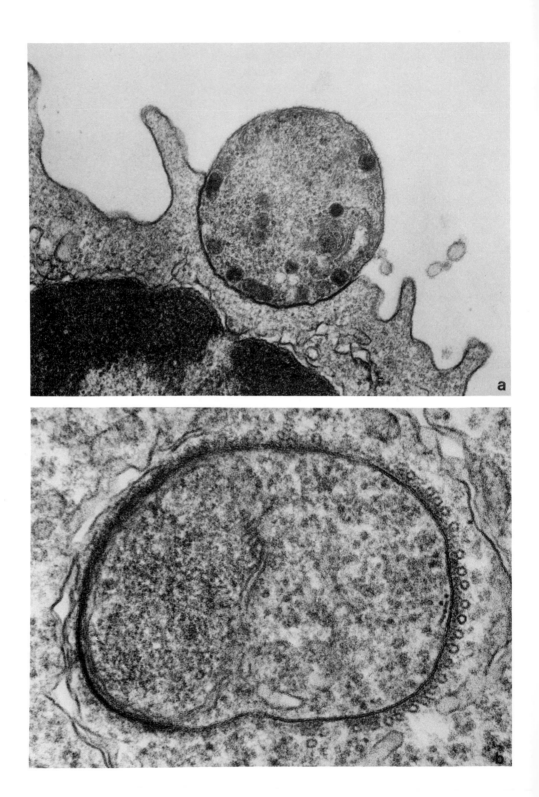

gressively along the surface of the parasite, leading to its engulfment. A comparable mechanism leading to invasion of lymphocytes is illustrated in Fig. 4.12; low temperature does not prevent this type of entry, suggesting that uptake is "passive," in the sense of not requiring phagocytotic activity by the host cell. Uptake of *Chlamydia psittaci*, the organism responsible for psittacosis, is not markedly sensitive to cytochalasin B, thus differing from typical phagocytosis.

205

*USES AND ABUSES
OF ENDOCYTOTIC
AND
HETEROPHAGIC
PATHWAYS*

Recently, capacities to invade cellular vacuoles have been conferred on *E. coli* by transferring a gene from a naturally invasive organism, *Yersinia pseudotuberculosis*. This experimental success presages advances in the molecular analysis of invasion. The gene is known to code for **invasin,** a protein of the bacterial outer membrane.

4.4.3.1. Microorganisms That Escape from Endosomes or Lysosomes

Low-pH-induced fusion of the viral surface with the endosomal membrane (Section 3.1.3.3) is utilized by many pathogenic viruses, including the oncogenic retroviruses and perhaps the AIDS organism.* Not all of these viruses are membrane-enveloped. Poliovirus, for example, has no membrane but its coat proteins undergo pH-dependent conformational changes analogous to those of the spike proteins at the surface of membrane-enveloped viruses. As with the latter viruses, entry of poliovirus into HeLa cells can be induced, at temperatures low enough to inhibit endocytosis, by reducing the pH of the medium to 5.5–6.5; under these conditions the virus penetrates directly across the plasma membrane (Section 3.1.3.3). At higher temperatures, where polioviruses are endocytosed, agents that increase endosomal or lysosomal pH inhibit infection.

Adenovirus enters the cytosol by lysing the endosomal membrane rather than by fusing with it. This, at least, is the simplest explanation for the fact that escape of this virus to the cytoplasm is accompanied by escape of other electron-dense macromolecular or particulate tracers when these tracers are endocytosed along with the virus.

Section 3.1.3.4 outlined experiments on Semliki Forest viruses indicating that certain mutants require especially low pHs to penetrate membranes and therefore are delayed in making their way out of the endocytotic apparatus

---

*This conclusion is currently under reexamination for several noteworthy viruses, including the HIV strains that cause AIDS, and Sendai virus. Recent experiments suggest that these viruses do not need a low pH to penetrate into cells and may even enter by crossing the plasma membrane.

FIG. 4.12. Sporozoites of the obligate intracellular parasite *Theileria parva* (the circular structure in the center of **a** and the large ovoid body occupying most of the field in **b**) invade bovine lymphocytes by forming very intimate associations with the lymphocyte surface (the association appears as a thickened membrane in a) and then zippering themselves into the host cells, apparently without active phagocytotic involvement by the host. The thickened, compound membrane (host plus parasite) that initially envelops the internalized parasite progressively resolves into a single membrane (as at the right in b—the three layers visible correspond to the expected appearance of a single membrane), which is interpreted as reflecting lysis of the host cell's vacuolar membrane. Microtubules (cut in transverse section at the right side of b), and then lysosomes, accumulate along the parasite's surface, but the lysosomes do not fuse with the parasite.

From Fawcett, D., Musoke, A., and Voigt, W. (1986) *Tissue Cell* **16**:873. a, × 60,000 (approx.); b, × 100,000 (approx.). (Copyright: Longman Group, United Kingdom.)

until they reach late endosomes (with BHK cells, the mutants take an average of 45 min to escape, whereas wild-type viruses need only 15 min). Can viruses escape even later, after entering definitive lysosomes? For most viruses, this question has not been studied in detail; it is assumed that even if there is some escape from the lysosomes, such escape would be quantitatively insignificant compared to penetration from earlier structures, and that most of the particles that get as far as secondary lysosomes are degraded.

For reoviruses, however, microscopic, biochemical, and cell fractionation evidence suggest strongly that viruses normally begin their replication in acid hydrolase-containing, degradative structures. The prevailing theory is that the reovirus enters cells via lysosomes [though, if acidified "prelysosomal" endosomes prove to possess appreciable proteolytic capacities (see Section 2.4.4.1), the theory may require revision]. This is surmised because low pH alone is not enough to foster productive viral entry into cells: for infection to occur, the viral core—double-stranded RNA and proteins—must be exposed by proteolysis of outer proteins. Thus, although placing potential host cells (e.g., L cells) in a medium containing ammonium ions at concentrations that raise intralysosomal pH does impede infection, this effect probably depends on inhibition of proteolysis rather than directly on the low pH, because viruses pretreated with proteases still are infectious under these circumstances. The outer proteins of the virus seem to be much more susceptible to acid protease than are the inner, and the double-stranded RNA may also be relatively resistant to lysosomal attack (if in fact RNase is present in the virus-containing compartments). But how do the viruses proliferate once the heterophagic system has prepared them to do so? One possibility is that the initial invaders do not actually free themselves into the cytosol; perhaps instead they stick in the membranes of the host-cell lytic compartments and there generate new, progeny, viruses that assemble in the cytosol.

Among prokaryotes, the Rickettsiae responsible for diseases like Rocky Mountain spotted fever are the best known of the organisms believed to escape from endocytotic structures to the cytoplasm. The trypanosomes that produce Chagas' disease are among the few eukaryotic microorganisms thought able to do this (the organism in Fig. 4.12 is another). Lytic secretions by the invasive organisms, or lipases present in their surface layers, have been invoked to explain such escapes but evidence is still scanty.

Transcytosis vehicles may be used by reoviruses to cross the gut epithelium at Peyer's patches (Section 4.3.2.2) and perhaps by gonococci traversing Fallopian tubes. The ability of some other organisms to travel along neurons has been tentatively attributed to the entry of these organisms into the endocytotic structures that form at nerve terminals and undergo "retrograde" transport to the cell bodies.

4.4.3.2. Protein Toxins

Diphtheria toxin is a bacterial product (actually, the protein is coded for by a bacteriophage infecting the bacterium *Corynebacterium diphtheriae*) that kills cells by inhibiting protein synthesis. The toxin ADP-ribosylates the elongation factors responsible for growth of nascent polypeptides. As initially translated, the protein is a single polypeptide chain, but the toxin is thought to undergo conversion into two chains, "A" and "B," linked by a disulfide bond and derived from the parent chain by proteolytic cleavage (see below). The A subunit is the enzyme that catalyzes the ribosylation of elongation factor.

207

USES AND ABUSES
OF ENDOCYTOTIC
AND
HETEROPHAGIC
PATHWAYS

Killing of cultured mammalian cells by the toxin normally requires endocytosis, involving cell surface "receptors" (probably glycoproteins) and the usual coated pit–endosome routes. Once inside the cells, much of the toxin is rapidly inactivated—probably by lysosomal digestion—but penetration of as little as one molecule into the cytosol is believed to suffice for the toxic effects.

Diphtheria toxin is relatively ineffective in the presence of ammonium chloride or the other agents used to raise the pH in the heterophagic pathway. On the other hand, the toxin can exert its effect without prior endocytosis when the pH of the culture medium is lowered to below 5. At low pH, the B subunit undergoes a conformational change that exposes a hydrophobic zone, permitting the subunit to insert in membranes (Fig. 4.13); the subunit apparently can serve as some sort of channel, for it increases the ionic permeability of membranes into which it is inserted. As with the membrane-enveloped viruses,* the findings suggest that under normal conditions the acidification of endocytotic structures promotes the insertion of some diphtheria toxin molecules into the membranes of the acidified compartments. This insertion, mediated by the B subunit, opens a route of access for the A subunit into the cytoplasm.

How is the linkage between the A and B subunits broken during these events? There obviously is no dearth of proteases that might help—the toxin is likely to encounter such enzymes during release from the bacteria, at the cell surface, in endosomes, in lysosomes, and perhaps even when portions of the molecule become exposed to the cytosol. Toxin "prenicked" by brief exposure to proteases so that only disulfide bonds hold the As and Bs together is much more effective than native toxin. Does this imply that the host cell can reduce the disulfides? A nice idea is that when inserted in the endosomal membrane, the conformation of a toxin molecule is such as to expose the key disulfide to the cytosol where it is accessible to the systems normally responsible for metabolic disulfide reductions, such as the glutathione system (see Section 3.4.1.5). But it has not even been ruled out fully that the A subunit can exert its effects upon becoming exposed on the cytoplasmic surface of pertinent membranes, to which it remains anchored through its linkage to the B subunit. (Elongation factors might be attacked as they "float by" in the cytosol though given the effects of nicking mentioned above, this seems less plausible than attacks by free-floating toxin subunits.)

The gradient in H^+ ions from endosome interior to cytoplasm is thought to provide the energy for successful crossing of membranes by diphtheria toxin. And the toxin requires anions such as Cl^- for successful entry, perhaps partly because such anions are needed for acidification of the entry vehicles (Section 3.1.5.3). A current hypothesis suggests that the toxin's "receptor" is a membrane constituent that is linked to anion transport proteins of the plasma membrane; if so, further study could help unravel the present uncertainties about the behavior of plasma membrane ion-transport systems during endocytosis.

Other protein toxins differ from diphtheria toxin in mode of action. Cholera toxin (Section 2.3.2.11), a bacterial product with high affinity for the G_{M1} gangliosides commonly present on the surfaces of animal cells, is one of several prokaryotic proteins that alter intracellular levels of cyclic nucleotides in eu-

---

*The similarities in entry routes between diphtheria toxin and the membrane-enveloped viruses have been exploited in the development of screening regimens for obtaining mutant cell lines defective in fundamental processes of endocytosis. Cells simultaneously affected in the entry of toxins and of viruses would be unlikely to be defective, for example, simply in particular receptors. Several of the cell lines exhibiting abnormal endosomal acidification mentioned earlier were obtained through such a strategy.

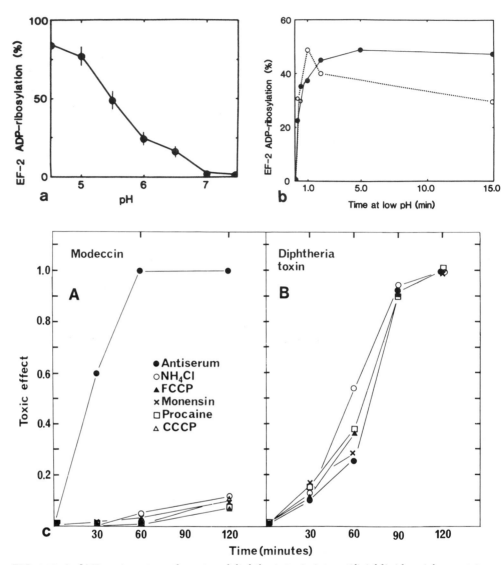

FIG. 4.13. (a, b) Experiments on the entry of diphtheria toxin into artificial lipid vesicles containing elongation factor (EF-2) and NAD in their interiors. (a) When the vesicles are exposed to proteolytically nicked toxin at the indicated pHs for 5 min and then are incubated at pH 8 for 30 min, the extent to which the enclosed elongation factor is ADP-ribosylated depends on the pH during the initial 5-min exposure. (b) An experiment showing that as little as 1 min at pH 4.7 (○) or 5.5 (●) suffices to promote near-maximal ADP-ribosylation during subsequent incubation at pH 8. From Montal, M., Simon, M. I., and Donovan, J. J. (1985) *J. Biol. Chem.* **260:**8817.

(c) Inhibition of the effects of the toxins, modeccin and diphtheria toxin, by weak bases and other agents. The ordinate indicates the toxic effect relative to the maximum observed effect and the abscissa, the time after toxin administration when the inhibitory agents were added. The data in B suggest that diphtheria toxin rapidly enters the cell (hence it is no longer accessible to antibodies). Quite soon thereafter the toxin escapes from endocytotic compartments to the cytoplasm (the various agents, such as ammonium chloride, that prevent its escape when present at time 0, no longer are effective inhibitors once the toxin molecules have penetrated the membranes of the endocytotic compartments). Modeccin (A) also enters the cell rapidly but it remains vulnerable to the various inhibitors for a considerable time, suggesting that it differs from diphtheria toxin in mode or site of escape. Data are in terms of "ID_{50}," the concentration of toxin needed to reduce protein synthesis in green monkey kidney cells to 50% of control, and are expressed as ID_{50} for control cells / ID_{50} with the inhibitor. From Olsnes, S., and Sandvig, K. (1986) In *Endocytosis* (I. Pastan and M. Willingham, eds.), Plenum Press, New York, p. 195.

209

USES AND ABUSES
OF ENDOCYTOTIC
AND
HETEROPHAGIC
PATHWAYS

karyotic targets. They do so either by affecting enzymatic mechanisms in the target cells or by the agent itself producing excess levels of the nucleotides. [For the whooping cough bacterium *Bordetella pertussis*, one (controversial) proposal asserts that increased levels of cyclic AMP impede the production of reactive oxygen metabolites by the professional phagocytes, thus helping the bacterium to survive.] Toxic lectins, such as ricin and abrin, produced by poisonous plants, affect protein synthesis by enzymatic alterations of the ribosomes (see Table 3.1). Ricin is an N-glycosidase, able to remove one or more essential adenines from ribosomal RNA. As with diphtheria toxin, endocytosis is required for this intervention, and access to the recipient cell's protein synthetic machinery depends on the crossing of membranes by the toxins, or at least of enzymatically active toxin subunits.

Despite the similarity just mentioned, several of the toxic plant lectins probably differ from diphtheria toxin in the mechanisms by which they gain access to the cytoplasm. In fact, mutant cell lines resistant to diphtheria toxin sometimes are hypersensitive to toxins like ricin (see Table 3.1). Ricin and certain others of the plant toxins require a neutral pH in the incubation medium for successful entry and the effectiveness of these toxins is enhanced by exposure of the cells to monensin or to weak bases, perhaps because monensin and the bases reduce lysosomal degradation of the toxins. The toxins continue to enter cells under ionic conditions that block formation of coated vesicles (Section 2.3.2.8), suggesting that the structures they use to enter cells are not the same as those used by diphtheria toxin; the plant toxins also are degraded less rapidly by cells than is diphtheria toxin. HRP conjugates of ricin (and ones of cholera toxin) have been found to accumulate largely in Golgi-associated sacs and tubules (see e.g., Section 5.3.2). The same toxins, when administered as complexes with colloidal gold, are confined principally to endosomes and lysosomes.

Further complicating the picture is one plant toxin, modeccin. Modeccin fails to intoxicate cells in which intracellular acidification has been counteracted by weak bases or ionophores. But, unlike the membrane-enveloped viruses or diphtheria toxin, it does not penetrate the plasma membrane when the pH of the medium is lowered (see Fig. 4.13c for additional differences).

4.4.3.3. Organisms That Survive within Lysosomes or Prelysosomal Structures (see Tables 4.3 and 4.4)

For some of the microorganisms that have adapted to life within the vacuoles of the heterophagic system, this adaption is facultative. For others, passage into the vacuoles is obligatory for sustained infection of host organisms and the conditions that prevail in the vacuoles promote proliferation or survival of the microbe. This situation probably accounts for some of the difficulties encountered in trying to culture certain pathogens under cell-free conditions and for many metabolic features of the pathogens. For example, the *Coxiella* forms responsible for Q fever grow best at low pH. *Prochloron*, a prokaryotic symbiont of ascidians, requires both a low pH and a source of tryptophan for survival outside its host. *Leishmania* reportedly takes up amino acids and glucose better at acid pH than at neutral pH.

"Endosymbionts" establish long-term residency in the cells they inhabit. Many of the pathogens eventually proliferate to the point where the host cell is killed and disrupted.

TABLE 4.4. Locations Where Intracellular Parasites Survive[a]

| | Virulent microorganisms that multiply | |
In phagolysosomes	In unfused phagosomes	In cytoplasm (after escape from phagosomes or phagolysosomes)
Listeria monocytogenes	*Toxoplasma gondii*	*Trypanosoma cruzi*
Leishmania mexicana	*Chlamydia psittaci*	*Rickettsia mooseri*
Salmonella typhimurium	*Legionella pneumophilia*	*Mycobacterium leprae*
Yersinia pestis	*Mycobacterium tuberculosis*	
Mycobacterium lepraemurium	*Mycobacterium microti*	

[a]Courtesy of P. D'Arcy-Hart. (See also Table 4.3.)

Of special interest for analysis of the controls of lysosomal fusion (Section 2.4.4) are those organisms that reduce or prevent fusions of lysosomes with the endocytotic vacuoles in which they enter cells. Several sorts of prokaryotic and eukaryotic pathogens do this. So does *Chlorella*, an alga that survives as an apparent symbiont within the digestive cells of *Hydra*. Most often, failure to fuse is demonstrated by prelabeling the host's lysosomes with electron-dense markers and establishing that these markers do not enter the vacuoles containing live, invading organisms (Fig. 2.4). Strictly speaking, this shows only that **secondary** lysosomes do not fuse with the vacuoles, but the lack of cytochemically demonstrable acid hydrolases in the vacuoles suggests that primary lysosomes do not fuse with them either.

How are fusions with lysosomes avoided or prevented? With plasmodia in erythrocytes, the absence of lysosomes in the host cell obviates the need for special devices. But in most situations, products of the invader seem to be required. In fact, when the invading cells are killed by heat treatment or other means before they are taken up, lysosomes often do fuse with the phagocytotic vacuoles, suggesting that continued activities of the invasive organisms are needed to inhibit the fusions. The same host cell can contain live organisms within vacuoles that exhibit few or no fusions with lysosomes, neighbored by vacuoles with dead organisms, which do fuse; thus the inhibitions probably are exerted locally rather than depending on global changes that would affect all the cell's heterophagic structures, such as in the overall cytoplasmic levels of regulatory ions or molecules.

For some invaders, coating their surface with antibodies or polycations relieves the inhibition of lysosomal fusions, suggesting that direct interactions of the microbe's surface with host cell membranes, which the coating of antibodies impedes, are part of the inhibitory mechanism. For *Mycobacterium tuberculosis*, sulfatide lipids from the bacterial walls are suspected to be responsible for inhibiting fusions, and schemes have been advanced in which the sulfatides alter fusion properties of host cell membranes, or affect the interactions of these membranes with the cytoskeleton so as to perturb organelle motility. As with other effects of polyanions, however, the mode of sulfatide action is currently under skeptical scrutiny (Section 2.4.7.2).

The protozoan *Toxoplasma* (Figs. 2.4 and 4.14) extrudes a network of membranous tubules as it invades. Some of these membranes may become incorporated in the surface of the intracellular vacuoles ("parasitophorus" vac-

211

USES AND ABUSES
OF ENDOCYTOTIC
AND
HETEROPHAGIC
PATHWAYS

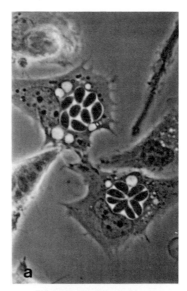

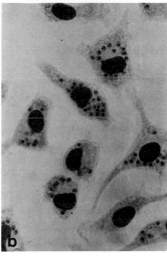

FIG. 4.14. Light micrographs of mouse resident peritoneal macrophages containing living, replicating *Toxoplasma gondii* **(a)** and *Leishmania donovani* **(b)**. The parasites are seen as the large ovoid bodies within vacuoles in a, and as the much smaller dark cytoplasmic bodies in b. Courtesy of H. W. Murray. a from Nathan, C. F., Murray, H. W., and Cohen, Z. A. (1980) *N. Engl. J. Med.* **303:**622. (Copyright: *N. Eng. J. Med.*)

uoles) in which *Toxoplasma* resides—antigens from the organisms are detectable in the membranes bounding the vacuoles. These vacuoles not only fail to fuse with lysosomes, but they also show much less acidification than is normal in the heterophagic pathway. This is important because the viability of *Toxoplasma* is reduced when the pH is lowered to 6 or below. [*Toxoplasma* survives much better in macrophages that have not been activated than in those that have, which verifies that activation of macrophages extends its effects to the "covert" world of conditions within the heterophagic system. (Activation usually is monitored by the release of materials to the extracellular medium or by changes in overt behavior of the macrophages, such as phagocytotic activity; its impact within the cell is harder to study.)]

Toxoplasma is one of the few cases for which there is any direct evidence that the organism modifies the membrane of the vacuole in which it resides. For a handful of other organisms, provocative differences have been noted between vacuoles containing surviving microorganisms and vacuoles containing other material, such as nonviable microorganisms of related strains. For example, the vacuoles (nonlysosomal) in which bacteria persist within *Amoeba proteus* reportedly release a bacterial protein to the ameba's cytosol. Malaria-causing plasmodia are claimed to insert antigens both into their vacuolar homes within the erythrocyte, and into the erythrocyte's plasma membrane. In the latter positions the plasmodial products might, it is speculated, promote retention of infected cells in capillary beds, thereby minimizing exposure to the splenic mechanisms capable of destroying the parasites (Section 4.6.1.1). *Listeria monocytogenes* secretes a hemolytic protein that is necessary for its growth in some of the cell types that it invades.

During the formation of parasitophorous vacuoles containing *Legionella*, there is a marked accumulation of host cell vesicles, and later of mitochondria, near the surfaces of the vacuoles. Subsequently, host cell ribosomes aggregate near the surface of the vacuole, forming a layer that remains at a distance of about 10 nm from the vacuolar membrane. (This distance is substantially greater than the normal separation of ER-bound ribosomes from the membranes of the ER.) Such observations, and similar ones with other invaders, bespeak intriguing interactions between host and parasite. Do the ribosomes actively generate proteins that accumulate in or near the vacuolar membrane, or that pass into the vacuoles to be used by the residents?

Certain types of microorganisms do not prevent fusion of lysosomes with the vacuoles in which they reside, but nonetheless survive. When coated with antibodies before uptake, *Mycobacterium tuberculosis* exhibits this behavior. So does *M. lepraemurium*, an organism causing leprosy-like diseases in rodents. The commonest mechanism for survival within lysosomes is thought to be resistance of walls or other coats surrounding the microbe to disruption by the enzymes and antimicrobial devices of the lysosomes. The lipopolysaccharides of bacterial cell walls inhibit certain lysosomal enzymes, at least under test-tube conditions. Some microorganisms may facilitate their survival by producing catalase or other "scavengers" of reactive oxygen metabolites. The amastigote forms of the protozoan *Leishmania*, thrive in very large lysosomal vacuoles in mammalian macrophages (Fig. 4.14 and Table 4.4).* The host cells can endure this occupation for prolonged periods and the vacuoles remain accessible to newly endocytosed material. The vacuoles also show labeling with markers preincorporated in the host cell's lysosomes before infection, but only low levels of acid hydrolase activities are demonstrable cytochemically, perhaps indicating that the parasite somehow binds or inactivates the hydrolases. Another speculation is that *Leishmania* produces ammonia, raising the vacuolar pH to levels more hospitable for the organism.

Low temperature reportedly affects relations between host cells and their intracellular denizens in ways that shadow the low-temperature effects repeatedly alluded to elsewhere in this book. For instance, when the temperature is dropped sufficiently, sea anemones extrude the dinoflagellate symbionts they customarily harbor.

---

*The promastigote form of *Leishmania* inhabits the digestive tract of insects. Thus, resistance to digestion is needed for survival in both of the hosts this parasite uses during its life cycle. Perhaps *Leishmania* first adapted to survival in the insect digestive tract and then became more and more of a virtuoso as its evolution responded to the demands of both habitats. *Leishmania* may even "use" opsonization by the C3 components of complement to promote its uptake by host phagocytes and by so doing minimize the respiratory burst and associated antimicrobial activities of the phagocytes.

213

*USES AND ABUSES
OF ENDOCYTOTIC
AND
HETEROPHAGIC
PATHWAYS*

Pathogenesis by some microorganisms that dwell within the heterophagic apparatus probably relates to impairment of lysosomal or endosomal activities important for defense, such as the processing of antigens. In contrast, endosymbionts and other long-term residents presumably pay for their upkeep by benefitting their hosts. The form of this payment no doubt varies considerably but it is hardly surprising that, as with *Chlorella* mentioned above, the microorganisms found in endosymbiont-like relations with other cells often are photosynthetic (even the ciliated protozoan members of planktonic communities are thought to harbor algae that survive for long times).

Section 3.1.2.1 mentioned that the usefulness of chloroquine as an antimalarial agent may follow from its impact on lysosomal degradation of hemoglobin by the plasmodial parasites. Aside from this, and a few studies demonstrating that the lysosomes of organisms such as the trypanosomes do contain lytic enzymes, including proteases, not much is known of the lysosomes of invasive eukaryotes or of how these lysosomes might participate in the invasion or its sequelae.

4.5. Clearance and Turnover

Macrophages clear away debris generated pathologically or during normal developmental and steady-state processes—the resorption of the tails of developing tadpoles for example, or the physiological cycles of mammary gland and uterus. In so doing, the macrophages utilize combinations of mechanisms comparable to those used to deal with invaders: phagocytosis; the release of regulatory molecules and other signaling agents; and the secretion of enzymes that help reduce dead cells and extracellular fibers and multimolecular complexes to manageable dimensions (see Sections 6.1.1 and 6.2.2).

4.5.1. Clearance of Biologically Active Molecules Released in Pathological Circumstances

Many of the biologically active molecules that accumulate in extracellular locales during inflammations or like processes, can be endocytosed by macrophages, fibroblasts, cells of the liver, or other cell types. Endocytosis of enzymes and of regulatory molecules—e.g., chemotactic peptides, complement proteins, interleukins—provides a means for reducing the extracellular levels of these materials. This can help turn off responses when they no longer are needed and to limit the spread of active molecules from sites of tissue insult or injury.* Some of the endocytotic receptors of the phagocytes may be "designed" partly to participate in such phenomena. For example, the macrophages' "mannose receptors" (Man/GlcNAc receptors; Section 2.2.1.2) bind various glycoconjugates and glycans possessing saccharide chains terminating in mannoses, N-acetylglucosamine, or fucoses. Almost certainly these receptors are multifunctional. For the present context, note that while they provide macrophages with the potential to recognize and phagocytose microorganisms with mannose-rich walls, such as yeast, they also might equip macrophages for

---

*On the other hand, by oxidizing or degrading useful materials in the extracellular environment, the reactive oxygen metabolites and enzymes released by the professional phagocytes may make some things worse while trying to make other things better.

clearance and scavenging of molecules such as lysosomal hydrolases released by their own organisms at sites of infection or other pathology (Sections 6.2 and 6.3).

4.5.1.1. Plasma-borne Inhibitors of Proteolysis ("Antiproteinases")

Certain plasma proteins of blood are designed to deal with potentially injurious materials. Haptoglobin, for example, rapidly complexes with hemoglobin molecules released by intravascular hemolysis and the complexes are then endocytosed in the liver and elsewhere. Scavenging of heme groups by hemopexin has already been mentioned.

A particularly prominent group of circulating proteins, accounting in total for as much as 20% of the plasma globulins, serve as "antiproteinases." These proteins limit the spread and the duration of action of proteases released into tissue spaces or the circulation by invasive microorganisms, by phagocytes, during fibrin activation and the resolution of thrombi, during activation of complement, or as a result of cell death. The antiproteinases bind to the proteases and the resultant complexes are endocytosed.*

The α_1-antiproteinase (often assayed as an "antitrypsin") is the most abundant of the blood-borne antiproteinases. Its importance is emphasized by the association of hereditary α_1-antitrypsin deficiencies with pulmonary emphysema and hepatic cirrhosis. The antiproteinase with broadest range of action is α_2-macroglobulin. This protein is secreted by fibroblasts of lung and other tissues, and by macrophages, among other cell types. It forms complexes with a very wide assortment of proteases. As it does so, the α_2-macroglobulin is hydrolyzed and undergoes conformational changes; for example, thiolester groups are exposed and in many cases, these groups participate in forming covalent links between the antiproteinase molecule and the protease with which it is complexed. When complexed with proteases, or treated experimentally in other ways to engender the proper conformational changes, α_2-macroglobulin is bound avidly by endocytotic receptors on macrophages and fibroblasts (see Fig. 3.1). The endocytosed complexes are degraded in lysosomes but the receptors recycle.

The antiproteinases probably have little effect on intralysosomal proteolysis because they are rapidly endocytosed only when already complexed with proteases and thus are not free to cause trouble. Moreover, some proteases retain enzymatic activity, especially against small substrates, even when bound in complexes with α_2-macroglobulin; for certain of the enzymes exhibiting this behavior, the activity is enhanced at acid pH.

4.5.2. Endocytosis of Circulating Molecules in Normal Physiology

4.5.2.1. Hormones and Growth Factors

After its secretion, a given molecule of polypeptide hormone such as insulin or epidermal growth factor (EGF), generally circulates in active form, for

---

*The antiproteinases in the circulation collaborate with other devices for containing the extracellular activities of proteolytic enzymes. For instance, endothelial cell surfaces are enriched in **thrombomodulin,** an agent that helps activate the proteolytic capacities of thrombin but that could, some believe, also mediate endocytotic clearance of this protease by the endothelial cells. Such clearance could be one of the cluster of devices that controls thrombus formation in capillaries.

only a few minutes. This situation affords the organism effective opportunity to adjust levels of circulating regulatory molecules rapidly in accordance with changing physiological circumstances. Some of the smaller hormones, such as somatostatin or the enkephalins, are inactivated principally by peptidases acting at the cell surface. In many cases, however, including insulin and EGF, endocytosis is an essential facet of the clearance mechanism.

Some of this endocytosis probably takes place in the kidney (Section 4.5.2.3). Most interest, however, has been devoted to endocytosis of hormones and growth factors by cells that are biological targets of these agents. Such endocytosis is often rapid and extensive. Usually it is mediated by the same receptors that produce the characteristic biological effects of the hormone or growth factor. For several hormones or growth factors, EGF included, uptake eventuates in delivery to and degradation in lysosomes. This is known: because (1) microscopically detectable forms of the endocytosed hormones or growth factors are seen in lysosomes; (2) inactivation and degradation of the molecules is inhibitable by chloroquine and other weak bases or by protease inhibitors such as leupeptin; under such inhibitory conditions, accumulation of detectable forms of the hormones and growth factors within lysosomes increases; (3) temperatures low enough to reduce rates of delivery of endocytosed material to definitive lysosomes also reduce the degradation of the hormones and growth factors.

Possibilities nonetheless remain open that compartments other than the lysosomes are importantly involved in the ultimate fate of endocytosed hormones and growth factors, at least in some cell types and particularly for certain agents, such as insulin (Section 5.2.1.2). Adipocytes begin to degrade insulin within 3 min of its endocytosis, raising again the likelihood of initial enzymatic attack in "prelysosomal" or nonlysosomal compartments. With hepatocytes some of the insulin that undergoes degradation is detectable in cell fractions of lower density than is typical for lysosomes. Occasionally, partial cleavages of insulin, insensitive to chloroquine, have been noted. And proteases or peptidases capable of attacking insulin have been found in cell fractions thought to derive from the "cytosol," leading a few researchers to argue that insulin somehow escapes to (or is exposed to) the cytoplasm outside the heterophagic system. Some of the older proposals for inactivation of insulin invoked reductions of disulfides as a very early step, taking place at the cell surface or very soon after the hormone departs to the cell interior. Recall that the cell may not always wait until lysosomal digestion has occurred before dealing with disulfide bonds in endocytosed molecules (see Sections 3.4.1.5 and 4.8.2).

Because endocytosis of hormones and growth factors is mediated by the same receptors through which these agents act on cells, and because some hormones or growth factors reportedly produce suggestive intracellular redistributions of endocytotic compartments or of lysosomes (e.g., migration to perinuclear locales), endocytosis has often been regarded not simply as a degradative mechanism but also as a necessary step in the action of hormones or growth factors. Or, endocytosis might enhance hormone effects, by translocating active hormone–receptor complexes from the cell surface to advantageous intracellular sites. Some investigators have even suggested that polypeptide hormones or steroids act directly via the lysosomes by promoting "leakage" of lysosomal contents into the nucleus or cytoplasm (Section 6.3.1).

Most such schemes for hormone action via lysosomes or endosomes are still minority viewpoints, timorously advanced or vigorously resisted. Arguments against the proposals are easy to come by. Investigators reporting apparent induction of redistributions of lysosomes or endosomes, from observations that microscopically labeled hormones or growth factors accumulate in particular cell regions, have not always ruled out that their labels are simply following normal routes for intracellular movement of endocytic organelles (Section 2.4) that operate whether or not the supposed inducing agents are present. pHs and other conditions in endocytotic compartments are likely not to be optimal for hormone–receptor affinities. (Even so, the concentrations of the hormones may be high enough to produce substantial receptor occupancy.) It is exceptionally hard to demonstrate the intracellular release of lysosomal hydrolases in viable cells unequivocally (see Section 6.3). Notions that polypeptide hormones or growth factors exert their effects after the hormones or growth factors themselves escape from endocytotic compartments have been discouraged by findings that a number of the early cellular responses to agents like insulin can be evoked in the absence of the agents themselves by antibodies that cross-link and redistribute the corresponding receptors in the plane of the plasma membrane (Section 2.3.2.10). Moreover, some major effects of key polypeptide hormones and growth factors almost certainly do not require internalization of the agents or their receptors by the cells: The changes these agents induce in ionic distributions, transmembrane transport, phosphorylation of proteins, or levels of cyclic nucleotides and other second messengers, occur very rapidly—often within seconds of hormonal administration, before much hormone could have gotten far into the cell. In addition, agents like insulin have been shown to affect target cells under circumstances in which endocytosis is unlikely, such as when the hormone is bound to large agarose beads. (Although it is always difficult to rule out the endocytosis of one or two molecules of hormone in such experiments, the observations usually are taken as demonstrating the possibility for mediating the observed biological effects solely through interactions at the cell surface.)

Despite all this, many intriguing findings on the intracellular locations and effects of hormones have yet to be adequately explained in terms of the agents acting solely at the cell surface. It could yet turn out that some of the longer-term effects do in fact depend on the agents or their receptors reaching the cell's interior. At the least it will be worth following up observations that receptors (e.g., those for EGF), having been phosphorylated or otherwise altered in response to binding of ligands, persist in this modified form within the cell for many minutes after endocytosis. Periodically, reports appear that peptide hormones entering the cell later pass even into the nucleus. Usually, such claims are instantly discounted on grounds of flaws or uncertainties in technique, but this dismissal is not always fully convincing.

The cycling and degradation of hormone **receptors** internalized from the plasma membrane in endocytotic vesicles are of considerable regulatory importance. However, as the receptors are endogenous to the cells that cycle and degrade them (and as authors always are faced with decisions that can only be resolved arbitrarily), I have categorized these processes as pertaining to autophagy rather than to heterophagy and will treat them in Section 5.2.

4.5.2.2. Steady-State Turnover of Serum Proteins: General Considerations

Haptoglobin, fibrinogen, or the immunoglobulins and antiproteinases are removed from the circulation rapidly when they participate in their specialized functions because these functions produce changes in the proteins that lead to

217

*USES AND ABUSES
OF ENDOCYTOTIC
AND
HETEROPHAGIC
PATHWAYS*

enhanced affinities for endocytotic receptors. Some otherwise normal intracellular proteins from cell interiors also are cleared at a fast pace when injected into the blood: most likely this signals the operation of devices that come into major play when cells are disrupted during pathology. But the proteins of the blood also undergo a normal, slow steady-state turnover in which synthesis and degradation are balanced; this is true both for the proteins mentioned at the beginning of the paragraph and for others, such as transferrin and albumin, whose functions do not yield increases in rates of their clearance from the circulation (Sections 4.1.2 and 4.3.3). For different serum proteins in humans and other mammals, it takes less than a day to more than a week for most of the molecules that were present in the circulation at a given moment to be replaced by new ones.

The factors bearing on the evolution of mechanisms for the turnover of proteins—those in the circulation and those within the cell (see Chapter 5)—include the following:

1. Proteins are subject to spontaneous denaturation or other changes in conformation, which occur at a finite rate even under conditions optimal for the proteins' functions. These changes can reduce affinities for other molecules, or render enzymes nonfunctional, and they sometimes promote precipitation of the proteins by exposing hydrophobic domains. Were there no mechanisms for getting rid of them, "defective" molecules would accumulate over time, with potentially serious consequences for physiology and metabolism, especially in cells and organisms that are no longer growing and so cannot simply dilute out molecular casualties.

2. Similar needs for disposal apply to abnormal proteins produced, for example, through errors in transcription or translation, as a result of mutations (e.g., grossly abnormal recessive genes in heterozygotes can produce quantities of nonfunctional molecules), or through "wear and tear" phenomena (like accidental encounters with unfriendly enzymes, extreme conditions of pH, or disruptive temperatures). Normal protein molecules may have to be disposed of when, for example, multicomponent protein complexes are being produced and the cells make more of one or another component than can be assembled into the complexes (see Section 5.5.5.1). It may also be necessary sometimes to dispose of proteins that have been missorted to incorrect locales in the cell but are otherwise normal.

3. To change metabolism or physiology, it often is necessary to alter the absolute amounts or the relative proportions of different species of proteins. Versatility in such modulations is increased by being able to control both synthesis of new molecules and degradation of old ones.

4. The degradation of some of its own proteins can enable an organism to survive hardship situations by providing needed low-molecular-weight metabolites that ordinarily are obtained from outside sources (Sections 5.1.5.1 and 5.6.1). Degradation must be selective so as, for example, to avoid disrupting essential biochemical pathways or the osmotic balances to which blood proteins contribute.

5. The processing and maturation of proteins (e.g., the removal of signal sequences and of other peptides during the genesis of secretory proteins or during the activation of digestive enzymes) can yield molecular fragments that must be disposed of.

For turnover in the steady state—when no net changes in amounts or types of molecules are occurring—our intellectual predilections for efficiency and tidiness lead us to expect preferential degradation of defective proteins. But, in fact, evolution could make do with a less discriminating system that might turn over both normal and abnormal proteins, trading the losses of the former for the advantages of possessing means for dealing with the latter. Related to this is an important detail of the steady-state turnover of serum proteins: the system responsible is not markedly selective for "old" molecules. This is indicated, for example, by findings that when serum proteins are isolated, radioactively labeled with iodine, and returned to the circulation, the subsequent decline in the label present in the circulation starts without appreciable lag and generally follows simple "exponential" kinetics.* Different serum proteins have different half lives but for each type of protein the kinetics of steady-state degradation approximate those expected for mechanisms that are indifferent to the age of the molecules—older and newer molecules seemingly face much the same probabilities of being destroyed.

4.5.2.3. Mechanisms

Heterophagy is the principal mechanism for the degradation of serum proteins. Steady-state turnover rates, therefore, must depend upon rates of formation of heterophagic structures and upon the affinities and abundances of participating receptors. Other factors impinge as well, such as the distribution of the proteins among the circulation, tissue fluids, and other sites that they enter. For instance, the tubule cells of the proximal portions of the nephron take up and degrade many of those proteins and peptides that pass the glomerular filter. The kidney is quite efficient in such endocytosis, and correspondingly,

---

*The pattern of decline resembles, mathematically, the decay of radioactive atoms and so is referred to as "first-order" or "pseudo-first-order." A similar pattern recurs in other important turnover phenomena (e.g., see Figs. 5.9, 5.21, and 5.23). When considering steady-state populations of molecules, such "simple" exponential loss of a cohort of labeled molecules implies a mechanism that removes a constant proportion of molecules from the population per unit time, independently of the "age" of the molecules. The "simple" exponential curve means, in essence, that the absolute rate of loss of label is maximal at time points immediately after labeling, when labeled molecules are most abundant in the population. As these molecules disappear and are replaced by newer (unlabeled) ones, the proportion of labeled molecules declines and so does their **absolute** rate of disappearance. But the **percentage** of the remaining labeled molecules lost per unit time remains constant. [For the uninitiated, it may be useful to think of a box initially containing only red ("labeled") balls, from which you remove a fixed number of balls every minute, replacing each ball you remove with a white one. At first there are many red balls in each removed sample, but as time progresses, there will be fewer and fewer red ones left, so that a random removal process will have less and less probability of encountering a red ball. It will take a relatively long time to remove the last few red balls.] Age-dependent mechanisms (see Sections 4.6.1 and 4.6.2) would yield a distinctly different pattern in which, rather than slowing, the rate of loss of a pulse of label should accelerate with time, as the labeled molecules get older.

Turnover data generally are displayed on a logarithmic scale, which gives straight lines for exponential data (see Figs. 5.9, 5.21, and 5.23). The time it takes for half the label to disappear is easily seen in such plots and is called the "half-life" of the population. [The half-life is shorter than the average (mean) life.]

The principal methods for studying turnover of circulating proteins are reintroduction of labeled proteins into the bloodstream, as mentioned in the text, and estimation of turnover rates from rates of synthesis in the steady state. The more diverse technical armory, described in Section 5.5.1.2 for intracellular turnover, occasionally is applied to circulating proteins. Section 5.5.1 will outline some of the methodological limitations in the estimation of turnover rates. (For example, the earliest time points often are difficult to study and deviate from simple expectations. And changes in the proteins due to iodination or other labeling procedures could alter turnover behavior.)

proteinuria is a common accompaniment of kidney disorders. But there are size limitations as to what can pass from the blood into the tubules. In humans, proteins with molecular weights much above 50–60 kD are precluded from such passage; in some other vertebrates, the cutoff points are a bit higher. Lysozyme, RNase, light chains of immunoglobulins, and several hormones can pass the filter when introduced into the bloodstream experimentally. Many such proteins, however, when present normally in the circulation, are complexed with larger materials either covalently (as with the immunoglobulin chains, which are parts of larger assemblies) or noncovalently (as with some of the hormones).

We lack detailed enough inventories of all the factors weighing on turnover, to allow judgment as to how each contributes. For example, to what extent is the endocytosis of the proteins adsorptive and to what extent is it by fluid phase uptake? Mechanisms solely based on fluid phase pinocytosis would seem to lack the specificity needed to account for the differences in half lives among different circulating proteins. But if adsorptive receptors are required, how selective need these be? And are they a special set devoted principally to turnover or is steady-state turnover just one of their functions? For example, do receptors like the ones exhibiting high affinities for α_2-macroglobulin complexed with proteases or those for IgG–antigen complexes, also have sufficient affinity for normal α_2-macroglobulin or immunoglobulin molecules, not in complexes, to support the gradual steady-state turnover of these serum proteins?

The theories currently taken most seriously suggest that circulating proteins are designated for uptake by specific changes, which then are recognized by corresponding adsorptive endocytotic systems. For steady-state turnover the changes in the proteins evidently must take place as readily for young molecules as for old ones, but this is not too hard to envisage. Spontaneous denaturation, for instance, is presently modeled as occurring when the normal thermally responsive "flexing," "breathing," and similar "random" conformational modulations of proteins accidentally cross thresholds from which they cannot return to normal. Such a process can have strongly age-independent aspects. Changes in proteins resulting from denaturation—the exposure of previously buried hydrophobic portions of the molecule is commonly cited as an example—can increase affinities of the molecules for membranes and therefore could promote endocytosis. Indeed, serum proteins denatured by exposure to formaldehyde and then reintroduced to the circulation, are very rapidly endocytosed, largely by the Kupffer cells of the liver.

4.5.2.4. The Asialoglycoprotein (ASGP) Receptor and Other "Lectins"

Encounters with enzymes can also predispose serum proteins to rapid degradation. For example, proteolysis can have effects similar to denaturation. The most detailed studies have concerned removal of sialic acids. Many of the serum proteins have oligosaccharide side chains terminating in sialic acids and it is well established that experimental desialylation, by exposure to neuraminidase, leads to the rapid endocytotic clearance of many species of serum proteins from the circulation (or from perfusion fluids and culture media used to sustain perfused livers or cultured hepatocytes). Endocytosis is followed by lysosomal degradation.

In mammals, this clearance is mediated chiefly by hepatocytes, using the

ASGP receptor, which I previously described as one of those that recycle repeatedly between coated pits and endocytotic compartments while delivering ligands to lysosomes. The receptor has been given the alternative name **hepatocyte Gal/GalNAc receptor** because it has high affinity for galactoses and N-acetylgalactosamines in terminal positions in oligosaccharides. Gal and GalNAc units are present in the next-to-terminal (penultimate) positions in the oligosaccharide side chains of many glycoproteins, including a variety of the ones in the circulation (see Section 7.1.1.1). The "exposure" of the units by desialylation explains why the ASGP receptor has high affinity for enzymatically desialylated serum proteins (asialo forms of fetuin, orosomucoid, ceruloplasmin, haptoglobin, α_2-macroglobulin and so forth), as well as for glycopeptides with appropriate oligosaccharides, and for "artificial" molecules ("neoglycoproteins"), such as serum albumin to which Gal residues have been enzymatically added.

The diversity of proteins that could be handled by the ASGP receptor system make the system an attractive candidate for participation in normal turnover.* Given that the professional phagocytes do secrete enzymes (Sections 4.4.1.2 and 6.2), and that some cell fractionators have long championed the existence of cell surface-associated neuraminidases, one might plausibly expect that circulating proteins normally encounter neuraminidases once in a while. Perhaps this happens in extracellular media where phagocytes are active, such as in regions of the spleen or liver. Or perhaps some of the proteins cycle into and back out of cells during endocytosis or transcytosis and encounter neuraminidases during this cycling.(Sections 5.5.4.3 will summarize evidence that glycoproteins of the plasma membrane do seem to lose terminal units of their oligosaccharides during their normal residence periods in the membranes, indicating the operation of mechanisms for desialylation of at least certain materials exposed to the extracellular world.)

Bits of evidence suggest that the ASGP receptor does play a role in turnover like the one envisioned here: Though normal levels of desialylated proteins detectable in the circulation are quite low, the levels increase when the liver is damaged in ways expected to reduce endocytosis by hepatocytes. And, in chickens, which lack the ASGP receptor, the blood contains considerably higher levels of proteins lacking terminal sialic acids than are found in mammals. Such evidence is circumstantial, however. If desialylation does govern normal, steady-state turnover of serum proteins, correlations would be predicted between rates of natural turnover of different glycoproteins and features of the proteins such as their susceptibility to loss of sialic acids or the rates of their binding to the ASGP receptor. Until these expectations have been assessed, or other evidence stronger than that now available is gathered, fans of the ASGP receptor system will probably remain prudent: So far they have carefully avoided overselling the system as the sole or even necessarily the most important turnover mechanism for blood-borne materials, or as the principal means by which cells deal with glycoproteins that become partially deglycosylated by enzymes released pathologically. Other devices might well cooperate with the

---

*The system might also handle materials entering the circulation under abnormal conditions. For example, preliminary studies suggest that a vitamin B-binding protein released from neutrophils during inflammations (Section 4.4.2.1) can be taken up via the ASGP receptor of hepatocytes and via analogous receptors on macrophages and other cell types. Uptake could ensure that this potentially dangerous protein does not accumulate excessively in tissue spaces or the circulation.

ASGP receptor, including other "lectin-like" receptors—so called because they recognize carbohydrate configurations in glycoconjugates. Indeed, in chickens a different receptor probably substitutes for the absent ASGP receptor: Chicken hepatocytes, and other cell types in birds, have a receptor-mediated endocytotic system based on affinity for terminal N-acetylglucosamines; such saccharides often are the "antepenultimate" units in oligosaccharide side chains, subject to exposure when sialic acids and galactoses are removed. Similar receptors have begun to turn up on various mammalian cells; these receptors may be related to the mannose receptors that have been studied most diligently in macrophages.

Conceivably, lectin-like receptors also take part in the slow breakdown of organized extracellular proteins of connective tissues—collagen, elastin, proteoglycans, and the like—which is detectable under normal conditions through most of the life span of a vertebrate organism. One series of concepts builds upon hints that these proteins are subject to a gradual covalent addition of glucose residues, probably by nonenzymatic reactions, which leads eventually to the presence of significant numbers of glucose derivatives on old molecules. Macrophages readily take up artificially glucosylated soluble proteins, giving grounds for speculation that these cells also attack glucosylated connective tissue proteins arising in nature.

4.6. Phagocytosis in the Steady-State Turnover of Cells and Cell Parts

Heterophagy disposes of cells and cell parts that die or are shed during normal development or turnover. The recognition devices and other mechanisms underlying such involvement must harmonize with developmental and homeostatic programs so that the phagocytes do not damage cells that should survive.

4.6.1. Destruction of Erythrocytes

Unlike the turnover of circulating proteins, the steady-state turnover of mammalian red blood cells is age dependent. An erythrocyte survives for a period characteristic of the particular animal species, and then disappears. The "average life spans" of the cells are defined as the times from passage of the cells out of marrow into the circulation, to the midpoints of the sigmoidal curves that describe this disappearance. (It takes several days for all the members of an age cohort of erythrocytes to depart.) In human males, erythrocyte life spans average 120 days; in rats, 60 days.

Aged ("senescent") erythrocytes are disposed of principally in the spleen, though the liver can make a substantial contribution particularly when the spleen is removed. As many as 10^9-10^{11} cells are destroyed per day in humans and other mammals. Destruction is chiefly through phagocytosis ("erythrophagocytosis") by macrophages and related cells. The erythrocytes' macromolecules are degraded, making iron and some other materials available for reuse (Section 3.5.2.2).

Other blood cells and cell derivatives turn over with their own characteristic life spans some of which are considerably shorter than those of erythrocytes. For instance, in humans, platelets—specialized cell fragments whose graveyard probably also is the spleen—last about 10 days if not used, and 10^8 white blood cells are destroyed per day.

4.6.1.1. Mechanical Screening

To account for the age-dependent turnover of circulating erythrocytes, most researchers propose that the cells undergo progressive changes as they age and that once these changes pass a threshold, the cells are recognized as aged and removed from the circulation ("sequestered") for destruction. Circulating erythrocytes do, in fact, change notably with time. Enzymatic activities, such as hexokinase and certain dehydrogenases and transaminases, exhibit slow declines; the cells lose a bit of their volume and surface area; they become less deformable, and less resistant to osmotic rupture. The fact that erythrocyte buoyant density increases with age has been a basis of techniques for separating younger and older red blood cells.

A fair amount is known about mechanisms that screen the circulation for abnormal erythrocytes and these mechanisms have served as starting points for ideas about the sequestering and degrading of normally senescent cells. For example, the arterial circulation of the red pulp of the spleen shunts some of the blood it carries into the "splenic cords"—a meshwork of reticular fibers, macrophages and other cells, from which the erythrocytes then pass into the venous circulation chiefly by squeezing through slits in the walls of venous sinuses.* Cytoskeletal arrangements and extracellular associations maintain these slits at widths as little as 1 μm or less so that the erythrocytes must distort considerably if they are to filter rapidly through the reticular meshwork and into the veins (Fig. 4.15). The biconcave disk shape of normal red cells—a shape established by the flexible network of spectrin and other proteins that underlies the plasma membrane—is favorable for such distortions; shape change is possible with less alteration in cellular surface area than would be required, say, for a sphere.

Sickled erythrocytes, or spherical ones like those present in hereditary spherocytosis, fail the test of passage through the spleen more often than do normal erythrocytes. Cells that fail are retained in the spleen and are phagocytosed by splenic macrophages. When such "mechanically abnormal erythrocytes" are present in too large numbers, anemias can result from their destruction. Anemias also can ensue when erythrocytes come to contain relatively nondeformable inclusions such as malarial parasites, or the aggregates of denatured or oxidized hemoglobin ("Heinz" bodies) that form in thalassemias and are produced experimentally by exposing erythrocytes to oxidants such as phenylhydrazine. Red blood cells containing abundant inclusions tend to be trapped by the spleen. When only a few inclusions are present, they sometimes can be removed from the cells by "pitting," the pinching off and phagocytosis of bits of cytoplasm containing the inclusions. This process takes place as the erythrocytes pass through and out of the splenic cords. Evidently the inclusions become stuck in narrow spaces, such as the slits of the venous sinuses, and are left behind as the rest of the cell squeezes through.

---

*Over the years, different views of the splenic circulation have been in vogue. Though the account given above has substantial support for species such as humans, some investigators still believe that much of the circulation in the cords is via more conventional capillary connections between arteries and veins, at least in certain species.

Erythrocytes that have undergone pitting can reenter the circulation, albeit with an altered surface area and shape (often spherical) that make them more prone to destruction the next time they enter the splenic cords.

A given human red blood cell passes through the splenic cords several thousand times in its normal life span, remaining each time for estimated periods of less than a minute to several minutes (sometimes longer). Oxygen levels and pH probably are low in the cords, and appreciable levels of extracellular hydrolases may be liberated by the macrophages. Even when normal, therefore, erythrocytes could be subjected to enzymatic modifications that might eventually pass thresholds, and the cells will also experience "stresses" that might push a marginal cell over the edge in osmotic terms or in its deformability. (Under experimental conditions at least, metabolically compromised red blood cells—ones with low supplies of ATP for example—can lose control of their shape if further stressed.)

223

USES AND ABUSES
OF ENDOCYTOTIC
AND
HETEROPHAGIC
PATHWAYS

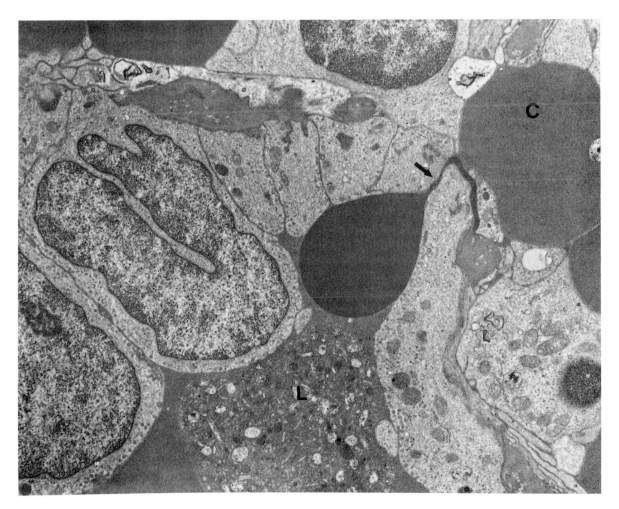

FIG. 4.15. From a human spleen, showing a region of the red pulp. At the arrow, an erythrocyte is seen squeezing through the wall of a sinus to pass from the splenic cords (C) to the lumen of a venule (L). Note the extreme distortion of the cell due, presumably, to its being forced through a narrow slit. From Chen, L. T., and Weiss, L. (1972) *Am. J. Anat.* **134:**425. × 7000 (approx.). (Copyright: Alan R. Liss, Inc.)

4.6.1.2. Screening of the Erythrocyte Surface: Antibodies to Senescent cells?

Do changes in the surface of the red blood cell during normal senescence govern the cell's fate? Erythrocytes obtained from the circulation can readily be labeled with ^{51}Cr and this label used to follow their behavior upon reintroduction of the cells into the bloodstream. When the cells are subjected to neuraminidase before being returned to the circulation, their sequestration is markedly speeded. These observations, plus findings that aged erythrocytes normally have fewer sialic acid residues at their surfaces than do younger cells, led initially to a spate of hypotheses that normal erythrocyte senescence involves progressive loss of the sialic acids from their surface glycoconjugates. Phagocytes might "sense" the degree of such desialylation either because of the likely reduction in the density of surface charge attendant upon selective loss of the negatively charged sialic acids, or through intervention of receptors akin to the ASGP receptor discussed in Section 4.5.2.4. (Macrophages and related mononuclear phagocytes do have such "galactose" receptors.) In tissue culture experiments, erythrocytes enzymatically shorn of a sufficient proportion of their sialic acids are phagocytosed at enhanced rates, by macrophages.

Some of the observed age-related decline in sialic acids at the red blood cell surface may, however, result from shedding of portions of the plasma membrane and its associated glycocalyx to the extracellular world, by vesiculation or other mechanisms. In contrast to the effects of differential enzymatic hydrolysis, this mode of loss would have little impact on the **density** of surface charge or of groups recognized by specific endocytotic receptors. In addition, macrophages can bind desialylated cells such as lymphocytes, without internalizing them, suggesting that the signals for erythrophagocytosis are more complex than simply depletion of sialic acids. Most importantly, recent findings could imply that the mechanisms for sequestering neuraminidase-treated erythrocytes are not identical to those than handle normally senescent cells. For example, in some experiments the sequestration of neuraminidase-treated young red blood cells appears to be only temporary so that, in contrast to older cells, which are destroyed, the younger cells can later return to the circulation. In addition, infusion of considerable numbers of neuraminidase-treated erythrocytes into the circulation seems to have little impact on the rate of destruction of normally aging cells that have not been exposed to neuraminidase; this suggests that the neuraminidase-treated cells do not compete for access to the normal turnover mechanism. (As would be expected, analogous infusion experiments do seem to indicate some form of competition of neuraminidase-exposed erythrocytes with one another.)

As alternatives to loss of sialic acids, some investigators have argued, for example, that the erythrocyte surface undergoes changes in its lipid composition, or perhaps in the distribution of phosphatidylserine or other phospholipid species between the two faces of the plasma membrane. Proposals about immune recognition currently are ascendant. There are specific IgGs present normally in the circulation that, under experimental conditions at least, bind preferentially to the surfaces of aging red blood cells, accumulating in numbers that may reach several hundred IgG molecules per erythrocyte (Fig. 4.16). These antibodies can opsonize erythrocytes for uptake by macrophages. One research group is convinced that the principal antigen recognized by the IgGs is a peptide arising through proteolysis of the "band 3 protein," an abun-

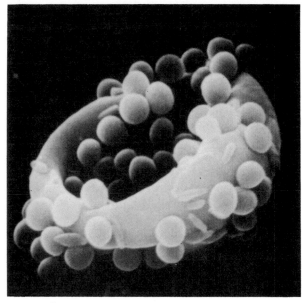

FIG. 4.16. Scanning electron micrograph of an erythrocyte demonstrating that an antibody (IgG) present in normal serum binds to senescent human erythrocytes obtained from normal blood by Percoll density gradient centrifugation. The antibody was first linked to killed bacteria (*Staphylococcus aureus*) so its binding to the erythrocyte could be visualized by the attachment of bacteria to the erythrocyte surface. From Galili, U., Rachmilewitz, A., Peleg, A., and Flechner, I. (1984) *J. Exp. Med.* **160**:1519. × 10,000 (Copyright: Rockefeller University Press.)

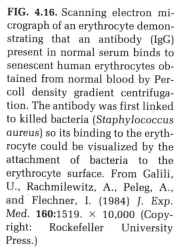

dant glycoprotein of the red blood cell surface, which normally serves as the "anion channel" for exchanges of Cl^- for HCO_3^-. Other groups favor altered glycophorins as the putative antigens. A third view is that $\alpha(1,3)$-linked galactosyl residues in glycoproteins or glycolipids become exposed by alterations in the senescing cell's surface and are key elements of the antigenic sites (the effects of neuraminidase might come in here). Still another suggestion is that macrophages recognize and phagocytose aged cells through a synergy of signals, some triggered by the antibodies and some by direct interaction of macrophage lectin-like receptors with altered glycoconjugates in the red cell surface.

4.6.2. Retinal Rod Cells

Phagocytes regularly dispose of portions of cells discarded during normal developmental processes. An example to which I have already alluded is the shedding of large "droplets" of cytoplasm ("residual bodies") enclosed in plasma membranes, by maturing sperm; this material is taken up by neighboring Sertoli cells. As mammalian erythroblasts mature in the marrow, they "extrude" their nuclei, surrounded by a thin rim of cytoplasm enclosed in plasma membrane; macrophages engulf the extruded material, apparently being able to recognize the nucleated cell fragment as different from the parent cell.

Vertebrate photoreceptors differ from these two examples in that the cells are very long-lived and the shedding of cell portions is an aspect of steady-state turnover, rather than of differentiation. Rod cells have been studied most extensively. The rod "outer segments" (Fig. 4.17) are packed with stacked mem-

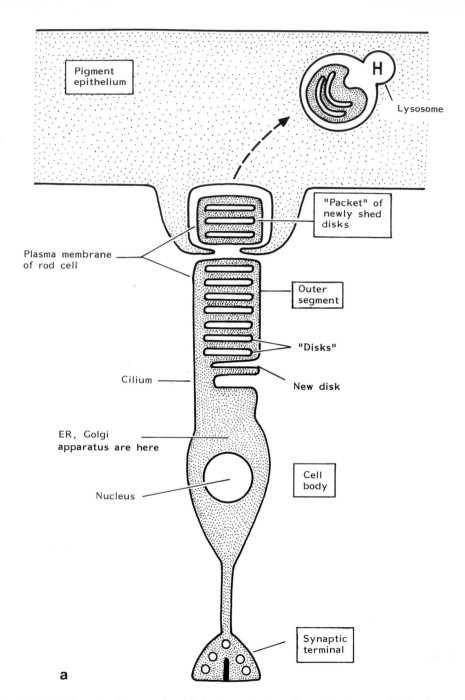

a

FIG. 4.17. (a) Schematic diagram of a rod photoreceptor from frog retina. During the development of these cells, the photoreceptive "outer segment" arises as a highly modified cilium and thereafter the outer segment remains attached to the cell body by a ciliary "stalk." The membranous disks of the outer segment arise by folding of the plasma membrane, migrate up the outer segment, and eventually are degraded when the cell sheds its "tip" as a packet of the oldest disks. The shed tips are digested in phagolysosomes formed by the pigment epithelial cells, which are closely associated with the outer segments.

(b) The upper two thirds of this micrograph is occupied by the cytoplasm of a pigment epithelial cell that sends processes down among several rod outer segments (R). P indicates packets

227

*USES AND ABUSES
OF ENDOCYTOTIC
AND
HETEROPHAGIC
PATHWAYS*

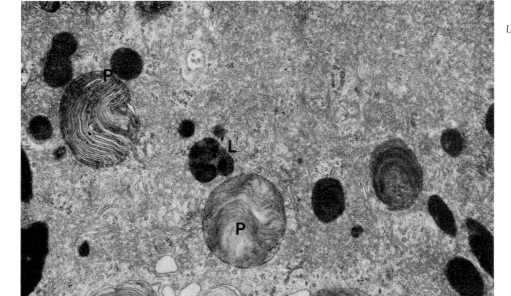

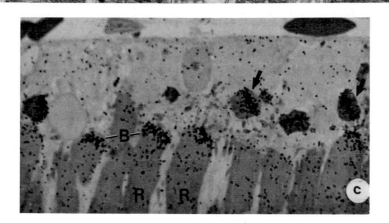

of disks phagocytosed by the pigment epithelial cells and L, a secondary lysosome (probably a residual body). From Young, R. W. (1971) *J. Cell Biol* **49**:303–318. × 16,000 (approx.).

(c) Region near the tips of photoreceptors (R indicates two rod outer segments) from a frog retina taken from an animal that had been injected with a pulse of radioactive amino acids 6 weeks earlier. In this autoradiogram, grains corresponding to radioactive proteins are seen clustered in bands (B) near the tips of the outer segment and in phagocytotic vacuoles in pigment epithelial cells (arrows). The bands represent the disks that had formed during the period shortly after the label was administered and by now have migrated the length of the outer segment and are undergoing shedding. From Young, R. W., and Bok, D. (1969) *J. Cell Biol.* **42**:392. × 1100.

branous disks composed principally of the usual phospholipids and of protein. Ninety percent of the protein is rhodopsin, a transmembrane glycoprotein, **opsin,** conjugated to **retinal**—a vitamin A derivative. The outer segment proteins and lipids are synthesized by the cell body and are transported to the plasma membrane of the outer segment. Disks arise at the base of the outer segment, by folding of the plasma membrane. In rods of most species, each disk eventually pinches off as a flattened intracellular sac.

Autoradiography shows that a pulse of label incorporated into the proteins of the outer segment, migrates as a band up the outer segment, reflecting the fact that the disks move progressively up the outer segment, as if being displaced from the base of the outer segment by newer disks. At the tip of the outer segment, "shedding" occurs: A packet of the oldest disks, enclosed in plasma membrane, separates from the tip and is engulfed in a vacuole formed within actin-rich pseudopod-like processes extended by the retinal pigment epithelial cells (Fig. 4.17). Aided by microtubules, this vacuole moves toward the base of the pigment epithelial cells and soon acquires acid hydrolases.

In frogs and tadpoles, a single shed packet may contain as many as 150 disks, representing on the order of 10% of the 50-μm length of an outer segment; at room temperature, a given photoreceptor sheds every third or fourth day. In rats, the packets contain about 100 disks each, and a given pigment epithelial cell handles the output from as many as 300 photoreceptors.

4.6.2.1. Proteins versus Lipids

Overall then, this sequence of events represents a steady-state turnover system that processes hundreds to thousands of square micrometers of membrane per photoreceptor per day, utilizing heterophagy as its central degradative mechanism. Normally, formation of new disks at the base of the outer segment is balanced with shedding at the tip so that the rods maintain a constant, species-specific length. Correspondingly, the proteins of the outer segment membrane are degraded in a strongly "age-dependent" fashion in the sense that most of the molecules have more or less the same life spans, fixed by geometry and rates of migration. The time it takes for a disk to move from one end to the other varies from a few days to a few weeks in different species, depending on the length of the outer segment and on the pace of migration and of shedding. The marching of bands of label up the outer segments described above, attests that most of the protein molecules present in a given disk at the outset of its journey are still there when that disk is shed.

The lipids behave quite differently from the proteins. When radioactive glycerol or other labels are introduced into the phospholipids and followed autoradiographically, radioactivity spreads rapidly throughout the outer segment. Evidently, disks exchange lipid molecules with one another and the result is that the label becomes more or less randomized among all the disks within a given cell. As would be predicted from such considerations, a pulse of label incorporated into the lipids is lost exponentially from the photoreceptors—the lipid molecules are removed from the outer segment at random with respect to their age. Presumably, lipid molecules are lost from the outer segment when they are unlucky enough to find themselves in a disk at the time it is shed, but mechanisms additional to this one are not ruled out. For frogs, the half lives of outer segment phospholipids are 3 weeks or so, whereas the outer segment proteins have life spans of about 40 days. The lipid molecules phagocytosed as components of shed disks are probably digested in lysosomes, but

229

USES AND ABUSES
OF ENDOCYTOTIC
AND
HETEROPHAGIC
PATHWAYS

there are noteworthy features of lipid turnover not explained by this degradation (see also sections 5.3.3.2. and 5.5.4). For one thing, the lipids seem to turn over a bit faster than would be expected were phagocytosis the only route operating. For another, different portions of given types of lipid molecules can turn over at different rates: "head" groups such as cholines can be removed and replaced without destroying the glycerol backbone or other parts of the molecule.

Pigment epithelial cells eventually accumulate considerable deposits of lipofuscin (Section 3.5.2.3) under normal conditions, and they are among several cell types that show notably enhanced lipofuscin deposition when supplies of vitamin E, an antioxidant, are inadequate. These features are as anticipated both from the intensity of the retinal pigment epithelium's participation in membrane turnover and from the fact that retinal lipids tend to be highly unsaturated. (Unsaturated fatty acids supposedly are more prone to peroxidations and other mishaps that render them indigestible or insoluble.)

4.6.2.2. Mechanisms and Rhythms

Rod cells of retinas denuded of their pigment epithelium do not shed their tips, suggesting that intervention of the phagocytotic cells is required for shedding. But the outer segments probably participate actively in detaching the packets of disks and in some way, the pigment epithelial cells and photoreceptors collaborate in setting the dimensions of the packet of disks shed—these dimensions are species specific, at least to an extent. One clue may be that as shedding is initiated, the disks near the point at which separation of the packet is to occur show vesiculations and other changes manifested even in the presence of agents (e.g., cytochalasin B) that prevent the final separation of the disk packet. The tips of rod outer segments also incorporate radioactive choline much more intensively than do other regions of the segment. And, early in the shedding sequence, the interiors of the rod tips near the points where separation of the disk packet from the outer segment is incipient, become accessible to Lucifer Yellow (Section 2.2.2.3) unlike other zones of the outer segment.

Ordinarily, the pigment epithelial cells remain in close proximity to the photoreceptor tips at all times, so it might be thought that phagocytosis of the shed disks does not require specific recognition devices or complex regulatory mechanisms. Indeed, the pigment epithelial cells are somewhat promiscuous in their phagocytosis, being able to take up latex beads and other particles as well as fragments of outer segments, both in culture and *in situ*. Still, adhesion of the pigment epithelium to the tips of the photoreceptors is strongest during the period of shedding and the pigment epithelial cells are not entirely unselective; the same pigment epithelial cells that can engulf latex beads will almost ignore heat-killed bacteria presented simultaneously, and the cells of explanted pigment epithelium bind and take up outer segment fragments much more avidly than they do latex. The pigment epithelial cells do have some lectin-like receptors, including one resembling the macrophage mannose receptor, but so far it has not proved possible to implicate specific sugars or oligosaccharides unambiguously in controlling the interactions between pigment epithelium and photoreceptors. Likewise, the pigment epithelial cells also have Fc receptors, and serum can encourage uptake of outer segment fragments in culture, but involvement of immune signals in disk shedding or phagocytosis has yet to be demonstrated clearly.

Disk shedding is temporally regulated. Typically, in animals maintained under conditions of alternating 12-hr periods of light and dark, the main burst of shedding occurs shortly after light onset, corresponding presumably to the natural "dawn." This behavior has both intrinsic circadian aspects and light-responsive ones; different species of animals vary somewhat in these regards. When maintained in constant light, rats or other animals show inhibition or even virtual abolition of shedding but when subjected to constant darkness, they continue shedding at more or less normal rhythms, for many days. Among the leading candidate mechanisms for regulating the timing of shedding are circadian cycles of production of melatonin or related hormone-like agents within the retina, and light-induced changes in the release of transmitters or other small regulatory molecules from the photoreceptors or interneurons.

4.6.2.3. Abnormal Phagocytosis in Retinal Dystrophies

In rats afflicted with an inherited retinal dystrophy that mimics human disorders—the *Royal College of Surgeons* rats—the pigment epithelial cells fail to engulf material from the photoreceptor outer segment. As the retinas develop, excess membrane accumulates at the tips of the photoreceptors; in time, the photoreceptors degenerate. The pigment epithelial cells from these rats, like those from normal animals, readily endocytose latex beads and other particles in tissue culture. But pigment epithelial cells from the dystrophic retinas, unlike those from normal animals, cannot take up fragments of outer segment from dystrophic animals or from normal ones, even though these fragments bind to the pigment epithelial cell surface. Apparently, the mutation alters the capacity of the pigment epithelial cells to respond to some signal, conveyed specifically by outer segment surfaces and necessary for the engulfment of the shed disks. One explanation proposes that in normal animals the pigment epithelial cells are somehow restrained from engulfing outer segment tips and that these restraints are relaxed temporarily during the shedding period; the dystrophy would owe its origin to a failure of the mechanisms that should lift the restraints. It is, however, premature to rule out the possiblity that the dystrophic retinas are deficient in a receptor system or another device that promotes uptake of the rod tips.

4.7. Do Yeast? Do Plants?

4.7.1. They Can

There are many reasons why yeast cells or plant cells might be expected not to engage in endocytosis. They do not depend upon phagocytosis for defense or to clear away dead cells (see Section 6.3.2.1). Turgor pressure (Section 3.4.2) and the presence of cell walls could limit possibilities for the cellular and compartmental contour changes needed for endocytotic internalization; vesicles filled with the dilute extracellular fluids in which plant cells frequently live should, for example, virtually collapse almost instantly upon budding into the cell because of the osmotic extraction of water. And large-scale endocytotic uptake of dilute media could cause problems for the cells' water balances. As for nutrition, though unicellular algae and yeast do encounter exogenous particles and macromolecules in their environment, the extracellular spaces adjacent to the plasma membranes of most cells of higher plants probably contain few such materials. The cell walls surrounding cells of higher plants (and those around yeast) generally are major impediments to the passage of particles or large molecules to the plasma membrane: the walls often severely limit or preclude access of molecules larger than polypeptides or small proteins.* [To the extent that macromolecules do circulate in higher plants, they probably do so chiefly through plasmodesmata—cell–cell continuities that provide direct cytoplasmic communication through the walls. The vascular system of higher plants does not contain a population of circulating cells. It is assumed to principally transport inorganic salts, small organic solutes (sugars and nitrogen compounds), and hormones and growth factors, mostly of relatively low molecular weight. The fluid in the xylem is usually pictured simply (sometimes oversimply) as a solution of such ions and molecules with few larger molecules present. Fluid in the phloem, however, is known to contain macromolecules and structures of microscopic dimension as components of the cytoplasm that persists in mature phloem elements (see Section 6.3.2.1).]

231

USES AND ABUSES
OF ENDOCYTOTIC
AND
HETEROPHAGIC
PATHWAYS

Nevertheless, both yeast cells and plant cells are thought to be capable of endocytosis. The chrysophycean protists—unicellular, chloroplast-containing, relatives of algae—can incorporate half-micrometer latex spheres and bacteria into cytoplasmic vacuoles. [These protists lack elaborate walls, although they do secrete thick coats ("loricae").] Endocytosis by yeast and higher plants is most readily demonstrated with "protoplasts"—cells enzymatically denuded of their walls (see Fig. 1.4) and placed in media with which they are in osmotic balance. Especially for plants, these preparations are found to take up traditional macromolecular pinocytotic tracers such as cationic ferritin and in some cases, they may engulf larger particles such as latex beads or membrane-enveloped viruses.

In algae and higher plants, some of this endocytosis has been shown to depend on coated vesicles resembling those of animal cells (Fig. 4.18). Endosome-like compartments, structures in the Golgi region, and MVBs also become labeled with endocytotic tracers. Some of the bodies in which the tracers accumulate also have cytochemically demonstrable acid phosphatase activity.

Vesicles and other structures like those mediating endocytosis in protoplasts are seen in plant cells with undisrupted walls, but it has been difficult to demonstrate unequivocally that they function in endocytosis, because appropriate tracers cannot penetrate to the cell surface. Under some osmotic conditions, plant cells with intact walls do seem to incorporate extracellular fluid containing microscopically detectable materials—fluorescent dyes or metal ions such as lead. But the studies are complicated both by possible abnormal effects of the tracers and conditions, and by experience with animal cells suggesting that metal ions might accumulate in intracellular compartment via paths other than endocytosis (e.g., during fixation of metal ion-exposed material for microscopy).

The extent to which endocytosis in plants delivers materials to lysosomes is still unclear. As already mentioned, endocytotic tracers have been found in MVBs, small vacuoles, Golgi-associated sacs, and other compartments and certain of these sites are suspected to be lysosomal. However, the evidence concerning entry into large central vacuoles—the types that occupy most of the

*The "pore size"—a measure of the maximum diameter of molecules or particles that can penetrate to an appreciable extent—is estimated to be less than 5 nm even in the simpler types of the walls surrounding cells of higher plants. Passage through the wall is conditioned not only by size, but also by factors such as the negative charges of pectins and other materials of the wall matrix.

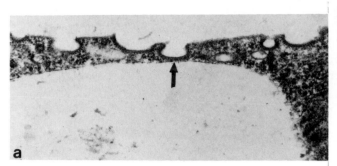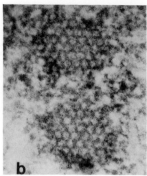

FIG. 4.18. Coated vesicles forming at the plasma membrane of the alga *Boergesenia forbesii* after the cells were mechanically wounded (punctured). **a** shows coated pits sectioned transversely (e.g., at arrow) and **b**, face views of two vesicles. That the vesicles are endocytotic was demonstrated by experiments with ferritin. From Laclaire, J. W., and O'Neil, P. (1984) *Science* **225**:331. a, × 25,000; b, × 100,000. (Copyright: American Association for the Advancement of Science.)

cell's volume—is still equivocal. Very recently, the vacuoles of soybean protoplasts have been shown to accumulate fluorescent tracers or small amounts of cationic ferritin, but most prior studies failed to demonstrate access of endocytosed materials to such vacuoles (perhaps because the accumulation is slow enough to be overlooked).*

For yeast, virtual unanimity has been reached, rather suddenly, about the abilities of cells with intact walls to endocytose: Uptake of Lucifer Yellow, of fluorescein-labeled dextrans, of viruses, and of bacterial amylase (M_r 55,000) has been reported for the yeast *Saccharomyces cerevisiae* by microscopy (Fig. 4.19) and cell fractionation. Normally, much of the tracer winds up in the vacuole—the yeast's principal lysosome—and there are some signs of endosome like "prevacuolar" structures as well. Thus, yeast seem to have a heterophagic pathway. There is, however, not much information yet about the uptake of electron microscopic tracers; the few remaining skeptics point to this and to the possible problems with tracers like fluorescent dextrans* and Lucifer Yellow (Section 3.4.1.2) as among the reasons they have not raced for the bandwagon. Believers counter with the argument that the presumption of endocytotic capacities already has helped explain important biological phenomena, as will be outlined shortly (Section 4.7.2.2).

Work on the morphology of endocytosis in yeast is still at early stages but preliminary findings suggest that the initial stages of endocytosis continue in

---

*Both for these studies, and for ones on yeast in the following paragraph, fluorescent dextrans were among the tracers used. (The fluorescent dextrans are also being used to estimate pHs in endocytotic compartments in plants and yeast.) As in Section 3.1.1.1, doubts have cropped up about the suitability of the dextrans because, especially when high concentrations of the tracers are used, as in many of the experiments on plants and yeast, relatively small percentages of low-molecular-weight contaminants can prove very troublesome by moving across membranes. For plant cells, control experiments and observations with electron microscopic tracers provide a degree of reassurance that endocytosed materials do enter the vacuoles but only a handful of such observations have yet been made and several studies with electron microscopic tracers have failed to find much entry into large central vacuoles. See also Section 5.2.1.3 for the possible use of vacuoles as "dumping" grounds for MVBs and other bodies that may be accessible to endocytosed materials.

233

USES AND ABUSES
OF ENDOCYTOTIC
AND
HETEROPHAGIC
PATHWAYS

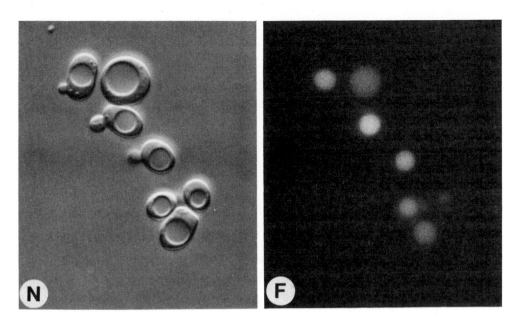

FIG. 4.19. Uptake of Lucifer Yellow into the vacuoles of yeast cells (*Saccharomyces cerevisiae*). N shows the cells as seen with Nomarski optics and F, the fluorescence of the dye. Courtesy of H. Riezman.

yeast whose ATP supplies are substantially depleted; under these conditions, tracers become trapped in compartments that may be acidified but seem to be "prevacuolar." Temperatures of 10–15°C substantially block endocytosis; ones between 15 and 20 °C inhibit transfers of already endocytosed materials to the vacuole. Clathrin coats may not be essential for endocytosis, if they are involved at all: endocytotic tracers are taken up (slowly) by mutant yeast lacking functional clathrin. These mutants are, however, sickly. Moreover, at least one group believes that specific "suppressor" mutations are needed, to somehow compensate for the absence of clathrin.

Several yeast mutants blocked in late steps of secretory pathways show reduced endocytotic capacities—evidently, endocytotic and exocytotic pathways share common components or mechanisms or otherwise intersect.

4.7.2. Why?

Uptake of bacteria by unicellular algae can provide a significant mode of nutrition. Endocytosis of bacteria has been observed in natural populations, as well as under experimental conditions. Some lake-dwelling chrysophyceans take up a mass of bacteria equal to one third of their own mass per day and it has been estimated that, under conditions of low illumination they are likely to encounter in nature, they can derive as much as one half of the carbon they need by degrading the bacteria.

In contrast, traversal of the endocytotic pathway to the vacuole in yeast is a low-volume affair; rates of uptake of Lucifer Yellow, seemingly as a fluid phase tracer, suggest that the equivalent of less than 0.1% of the cell's volume is taken into the vacuole per cell generation. And degradation of endocytosed proteins appears to be relatively slow. The cells would go hungry if they ate this way.

4.7.2.1. Nitrogen-Fixing Bacteria

The bacteria (*Rhizobium* species) that confer nitrogen-fixing capacities on the root nodules of leguminous plants dwell, as "bacteroids," in membrane-delimited compartments within the cytoplasm of root cells. The membrane bounding the compartments is related to the plasma membrane of the host cell in its antigens and enzymatic composition. Not every one of the bacteroid-containing compartments in a given host cell is sealed off from the extracellular space, but many are: the formation of the compartments thus is often analogized to phagocytosis, though the analogy is only a rough one. The bacteria gain entry to host tissues by active invasion so that a more precise analogy is to the situations in animal cells where invasive microorganisms assist their own entry into intracellular compartment (Section 4.4.3). Specific recognition devices couple particular bacterial strains to particular hosts.

The bacteroid-containing compartments have a low pH, which is attributed either to release of acid from the bacteria or, more often, to the presence in the compartment's bounding membrane, of proton pumps like those found in the plasma membrane. Hydrolytic enzymes have also been detected in the compartments but it is not established whether these are of bacterial or of host cell origin or how the bacteroids survive exposure to the enzymes. There is extensive metabolic traffic in oxygen, ammonia, carbon compounds (e.g., organic acids), and other materials between the bacteroids and the host cells' cytoplasm. This traffic must depend significantly upon the (mostly unknown) permeability and transport properties of the compartment's bounding membrane.

At late stages in the history of some nodules, and under other circumstances, the bacteroids eventually disintegrate, but it is not yet established whether this is truly a "heterophagic" process, dependent on entry of host-cell hydrolases into the vacuole. (It could, instead, reflect bacteroidal or, even sometimes, host-cell autolysis.) Similar uncertainty arises when interpreting cases in which heterotrophic plants, whose cells depend on intimately associated fungi for nutrition, exhibit eventual disruption of the fungi within membrane-bounded compartments in the plant cells' cytoplasm.

4.7.2.2. Cycling of Plasma Membrane or of Ligands Associated Therewith

When giant algal cells are wounded, they undergo rapid reorganization of their cytoplasm into forms with lower total surface area. While this is occurring, many coated vesicles bud from the plasma membrane, as would be expected if the vesicles were internalizing "excess" plasma membrane (Fig. 4.18). This is one of several observations suggesting that endocytotic-like processes in plants and in yeast participate in the cycling of membranes to and from the cell surface. [Given that endocytosed tracers do pass into Golgi-associated structures, it might even be that the formation of vesicles from the plant cell surface normally serves principally to balance the addition of membrane to the surface during phenomena such as secretion (Section 5.3.2).]

The two yeast mating types communicate with one another by means of mating factors ("pheromones"), which in *S. cerevisiae* are polypeptides 13 amino acids long. The mating factors released by one mating type bind to cell surface receptors on the responding type. The polypeptides are then endocytosed and degraded, apparently in the vacuole. Mutant yeast strains defective in endocytosis display markedly depressed sensitivity to the mating factors; uptake thus seems to be required for normal biological responses to the pheromones or at least is integral to the responses. Endocytosis of the mating factors can be accompanied by a decline in the number of receptors at the plasma membrane, a situation reminiscent of the endocytosis-mediated regulation of receptor abundance in the plasma membrane of animal cells (Section 5.2.1.2).

Endocytotic receptors now are being sought in plant cells. If found, they are expected to differ interestingly from those of animal cells. For instance, pHs in the extracellular spaces around plant cells are often low—5 to 6, or even less. Therefore, speculation has begun that plant receptors and endosomes might reverse the situations described in Chapter 3, by using **increases** in pH over that prevailing in the outside world, to promote dissociations of ligands.

235

USES AND ABUSES
OF ENDOCYTOTIC
AND
HETEROPHAGIC
PATHWAYS

4.8. Liposomes; Methyl Esters, Chimeric Toxins, and the Like

Experimental or therapeutic manipulation of lysosomes and endocytotic structures by the introduction of exogenous materials has a considerable history. Examples already cited include the endocytotic uptake of materials that alter organelle density, facilitating centrifugal separation (Section 1.3.1); the use of acidotropic agents, including antimalarial drugs, to modify lysosomal pH (Section 3.1.2); the introduction of enzyme inhibitors (Section 1.5.2)*; the selective osmotic rupture of lysosomes in cell fractions through use of methyl esters of amino acids (Sections 3.4.1.4) that penetrate into the lysosomes and are there hydrolyzed into products whose escape is slower than the entry (certain peptide derivatives like glycyl-phenylalanyl-β-naphthylamide can be similarly employed); and the influencing of LDL uptake by altering the abundance of LDL receptors in the plasma membrane (Section 4.1.1.3). Later sections will describe experiments in which lysosomal enzyme deficiencies have been "corrected" by introducing exogenous hydrolases (e.g., Section 6.4.1.4).

4.8.1. Liposomes

How can heterophagic processes be exploited to target drugs, antitumor agents, or experimental reagents to particular cell types while minimizing effects elsewhere? One group of procedures encloses the biologically active agent in an "inert capsule" designed to be endocytosed and then disrupted, freeing the agent intracellularly. The most used capsules are **liposomes,** small bodies constructed of artificial phospholipid membranes that can be degraded in lysosomes (Fig. 4.20). Such bodies have been used to introduce tracer molecules such as radioactive inulin, and enzymes or drugs, into lysosomes of cultured cells and into phagocytotic cell populations in intact animals.

Interactions of liposomes with cells and tissues depend on the sizes and composition of the bodies, their surface charges, the mobility of their membrane components, and many other factors. By adjusting composition and the conditions under which they are produced, liposomes can be made to enclose

---

*Protease inhibitors have been employed most often but, for example, castanospermine has been used to inhibit glucosidase activity, gentamicin seems to decrease lysosomal sphingomyelinase, and other inhibitors are being tried.

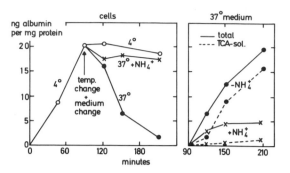

FIG. 4.20. Uptake of liposome-entrapped [125I] albumin by Kupffer cells. When liposomes are adsorbed to rat Kupffer cells in culture at 4°C and the cells are then warmed to 37°C, the liposomes are internalized and degraded, as is indicated by release of label from the cells (left) and accumulation of TCA-soluble label in the medium (right). Keeping the cells at 4°C inhibits internalization and degradation of the liposomes. The presence of ammonium chloride inhibits degradation. From Scherhopf, G. S. (1985) *Ann. N.Y. Acad. Sci.* **446**:375. (Copyright: New York Academy of Science.)

solutions of desired content within a single bounding membrane ("unilamellar" liposomes) or within multiple, concentric membranes ("multilamellar" liposomes). The bodies can be designed with diameters down to 50–100 nm, small enough to escape from some portions of the circulatory system and to fit within pinocytotic vesicles as well as phagocytotic vacuoles.

Sterols, proteins, and other materials can be incorporated into the membranes, or proteins can be attached to the surface of a liposome by cross-linking to lipid groups such as ethanolamines. Targeting to selected destinations has been essayed through attachment of proteins that enhance the binding of the bodies to the surfaces of specific cell types, such as anti-Thy antibodies or antibodies against histocompatibility antigens. Liposomes modified in such manner are being used to deliver drugs—for example, derivatives of methotrexate—that by themselves may be incapable of penetrating efficiently into cells.

FIG. 4.21. **Targeting of liposomes.** Small, unilamellar liposomes (SUV, where V = vesicles) containing a radioactive solute (^{111}In-labeled bleomycin) and with the serum glycoprotein orosomucoid (OR) or its asialo derivative (AOR) attached to their surfaces were injected into mice. Six hours later, the mice showed the indicated distribution of label. When AOR is attached, the liposomes are taken up by the liver much more rapidly, presumably reflecting endocytosis via the asialoglycoprotein receptor; the simultaneous presence of free asialoorosomucoid abolishes this effect, most likely owing to competition for the receptors.

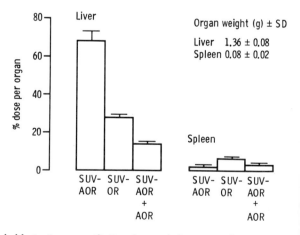

Uptake in the spleen—corresponding probably to "nonspecific" endocytosis by macrophages—is not similarly enhanced by the asialoglycoprotein because the spleen cells lack the appropriate receptor. From Gregoriadis, G., Senior, J., Wolff, B., and Kirby, C. (1984) In *Receptor-Mediated Targeting of Drugs* (G. Gregoriadis, G. Poste, J. Senior, and A. Trouet, eds.), Plenum Press, New York, p. 243. [See also *Ann. N.Y. Acad. Sci.* **446**:319 (1985) and *Biochem. Soc. Trans.* **12**:337 (1984).]

Most targeting studies have been done in tissue culture. The experiment on mice illustrated in Fig. 4.21 typifies work on intact organisms. In this study, attachment of desialylated glycoproteins to the surfaces of unilamellar liposomes promoted a marked enhancement of hepatic clearance of liposomes from the circulation owing, presumably, to interactions with the hepatic receptors discussed in Section 4.5.2.4.

Delivery does not always depend on heterophagy. Liposomes have been made that remain separate from the cell but nevertheless deliver lipid molecules to the plasma membrane. Other types of unilamellar liposomes, which fuse with the cell surface, are employed to introduce dyes, inorganic ions, and other materials directly into the cytoplasm of cells in culture. It may also prove possible to use liposomes whose fusion with natural membranes is promoted by low pH, for deliveries via endosomes.

4.8.2. Other Carriers and Other Strategies; Immunotoxins

Macromolecules—DNA, polysaccharides, or most often polypeptides or proteins—can also be made into carriers that mediate endocytosis of other molecules. Polylysine compounds have been used in this way for experiments with tissue culture cells, because their charge leads to effective endocytosis by a variety of cells (see Section 2.2.2.2) and because they contain numerous amino groups to which drugs or other small molecules can be conveniently linked. The delivered molecules are freed by lysosomal hydrolysis.

Results offering therapeutic promise have been obtained with chimeric "immunotoxins"—antibody molecules to which toxins, such as ricin (Section 4.4.3.2), are attached. Immunotoxins in which the antibodies are to cell surface proteins can be targetted to exert selective toxic effects on specific cell types in mixed culture populations, and also have been found to have somewhat selective antitumor activity in mice. Like the toxins themselves, the immunotoxins appear to be endocytosed by receptor-mediated mechanisms prior to affecting cells. Subsequent events are not well understood, but active portions of the immunotoxins are presumed eventually to cross the membrane bounding an endosome, lysosome, or other intracellular compartment, thereby entering the cytoplasm. Several of the most effective conjugates are of ricin subunits linked to antibodies by disulfide bonds; intracellular reduction of these bonds might free the toxin subunit to enter the cytoplasm (see Section 4.4.3.2).

Endocytosable, soluble, synthetic carriers have versatile potential. Therapeutic agents can be linked to polymers of methylacrylamide derivatives or to poly D-lysine through degradable linkers such as short peptides (or, it is hoped, through low pH-labile linkers) and the polymers can be targeted by incorporating oligosaccharides or other receptor-recognized materials onto them.

A brutal approach that has been used with some success is to load endocytic structures simultaneously with a biologically active agent and with high concentrations of sucrose or other impermeant solute and then to lyse the structures, intracellularly, by osmotic shock. The lysis of endosomes by adenovirus (Section 4.4.3.1) could prove useful along similar lines.

Long-term implants utilizing synthetic carrier polymers can sustain prolonged, slow, controlled delivery of steroids. Enzymes accumulating extracellularly during inflammatory responses to the implant gradually free the hormones and eventually degrade the carrier as well.

Acknowledgments

M. Chrispeels, L. Griffing, V. Herzog, L. Opresko, B. Pernis, J. Swanson, D. Wall, and R. Wallace read portions of this chapter (or listened patiently to outlines of it) and provided helpful ideas and comments.

For useful information, comments, and discussion on specific topics, as indicated, I am indebted to: G. Ashwell (ASGP receptor), J. Besharse (retina), S. Busson-Mabillot (yolk), S. Diament (parathyroid hormone), L. Eckhardt (immune responses), L. E. Ericson (thyroid), R. A. B. Ezekowitz (macrophages), D. W. Fawcett (protozoan parasites), B. Fields (reovirus), U. Galili (turnover of erythrocytes), S. Gordon (macrophages), G. Gregoriadis (liposomes), L. Griffing (endocytosis in plants), J. Harford (ASGP receptor), V. Herzog (thyroid), J. Hollyfield (retina), J.-P. Kraehenbuhl (transcytosis), M. Locke (insect organelles), A. Mahowald (insect organelles), M. Montal (toxins), J. W. Moulder (intracellular parasites), H. W. Murray (intracellular parasites), M. Neutra (transcytosis in Peyer's patches), S. Olsnes (toxins), L. Opresko (yolk), B. Pernis (immune processing), H. Reizman (yeast endocytosis), R. Rodewald (transcytosis of IgA), H. Ryser (drugs and toxins), K. Sandvig (toxins), R. Schenkman (yeast), G. L. Scherhopf (liposomes), S. Silverstein (macrophage properties), W. H. Telfer (insect yolk), C. G. Vallejo (*Artemia* yolk), D. Wall (yolk), R. Wallace (yolk), Z. Werb (macrophages), R. Young (retina).

Further Reading

Adams, D. O., and Hanna, M. G. (eds.) (1984) *Macrophage Activation*, Plenum Press, New York.

Adiyodi, K. G., and Adiyodi, R. G. (1983) *Reproductive Biology of Invertebrates*, Wiley, New York.

Adler, R., and Farber, D. (eds.) (1986) *The Retina: A Model for Cell Biology Studies*, Academic Press, New York. (See especially reviews by J. C. Besharse and V. M. Clark.)

Anderson, R. E., Kelleher, P. A., and Maude, M. B. (1980) Metabolism of phosphatidylethanolamine in the frog retina, *Biochim. Biophys. Acta* **62**:227–235.

Babior, B. M. (1987) The respiratory burst oxidase, *Trends Biochem. Sci.* **12**:241–243.

Bartles, J. R., and Hubbard, A. L. (1988) Plasma membrane protein sorting in epithelial cells. Do secretory pathways hold the key? *Trends Biochem. Sci.* **13**:181–184.

Beutler, B., and Cerami, A. (1986) Cachectin and tumor necrosis factor as two sides of the same biological coin, *Nature* **320**:584–588.

Bird, D. F., and Kalff, J. (1986) Bacterial grazing by planktonic lake algae, *Science* **231**:493–496.

Bok, D. (1985) Retinal photoreceptor–pigment epithelium interactions, *Invest. Ophthalmol.* **26**:1659–1694.

Boller, T., and Wiemken, A. (1987) Dynamics of lysosomal functions in plant vacuoles, in *Plant Vacuoles* (B. Marin, ed.) Plenum Press, New York, 361–368.

Boxer, G. J., Curnette, J. T., and Boxer, L. A. (1985) Hematologic disorders 1: Polymorphonuclear leukocyte function, *Hosp. Pract.* **20**(3):69. (See also *ibid.* p. 129.)

Carpentier, J. L., White, M. F., Orci, L., and Kahn, R. C. (1987) Direct visualization of the phosphorylated epidermal growth factor receptor during its internalization in A431 cells, *J. Cell Biol.* **105**:2751–2762.

Cohen, W. D. (ed.) (1985) *Blood Cells of Marine Invertebrates*, Liss, New York.

Curnutte, J. T., Kuver, R., and Scott, P. T. (1987) Activation of neutrophil NADPH oxidase in a cell free system, *J. Biol. Chem.* **262**:5563–5567.

D'Arcy-Hart, P., Young, M. R., Gordon, A. H., and Sullivan, K. H. (1987) Inhibition of phagosome–lysosome fusion in macrophages by certain mycobacteria can be explained by an inhibition of lysosomal movement observed after phagocytosis, *J. Exp. Med.* **166**:933–946.

Dean, R. L., Locke, M., and Collins, J. V. (1984) Structure of the fat body, in *Comprehensive Insect Physiology, Biochemistry and Pharmacology*, Pergamon Press, Elmsford, N.Y., 155–210.

Donovan, J. J., Simon, M. I., and Montal, M. (1985) Requirements for the translocation of diphtheria toxin fragment A across lipid membranes, *J. Biol. Chem.* **260**:8817–8823.

239

USES AND ABUSES
OF ENDOCYTOTIC
AND
HETEROPHAGIC
PATHWAYS

Downer, R. G. H., and Chino, H. (1985) Turnover of protein and diacylglycerol component of lipophorin in insect hemolymph, *Insect Biochem.* **15**:627–630.

Eaton, J. W., Konzen, D. K., and White, J. G. (eds.) (1985) *Cellular and Molecular Aspects of Aging: The Red Cell as a Model*, Liss, New York. (See especially the articles by Kay and Galili *et al.*)

Ericson, L. E. (1981) Exocytosis and endocytosis in the thyroid follicle cell, *Mol. Cell. Endocrinol.* **22**:1–24. (See also *J. Endocrinol. Invest.* **6**:311–324, 1983.)

Feinman, R. D. (ed.) (1983) Chemistry and biology of alpha-2 macroglobulin, *Ann. N.Y. Acad. Sci.,* **421**.

Fisher, A. B., and Chandler, C. (1985) Intracellular processing of surfactant lipids, *Annu. Rev. Physiol.* **47**:789–802. (See also *J. Biol. Chem.* **261**:6126–6131, 1986.)

Fitzgerald, D. J. P., Padmanabhan, R., Pastan, I., and Willingham, M. C. (1983) Adenovirus induced release of EGF and Pseudomonas toxin into the cytosol of KB cells during receptor mediated endocytosis, *Cell* **32**:607–617.

Golde, D. W., and Gasson, J. C. (1988) Hormones that stimulate the growth of blood cells, *Sci. Amer.* **259** (1):62–70.

Gordon, S. (1986) Biology of the macrophage, *J. Cell Sci. Suppl.* **4**:267–286.

Gupta, A. P. (1986) *Hemocytic and Humoral Immunity in Arthropods*, Wiley, New York.

Hafner, R. P. (1987) Thyroid hormone uptake into the cell and its subsequent localization to the mitochondria, *FEBS Lett.* **224**:251–257.

Havel, R. J. (1986) Functional activities of hepatic lipoprotein receptors, *Annu. Rev. Physiol.* **48**:119–134.

Herzog, V., Neumoller, N., and Holzmann, B. (1987) Thyroglobulin, the major and obligatory exportable protein of thyroid follicle cells carries the lysosomal recognition marker, M6P, *EMBO J.* **6**:555–560. (See also *J. Cell Biol.* **97**:607–617, 1983.)

Hilmer, S., Depta, H., and Robinson, D. G. (1986) Confirmation of endocytosis in higher plant protoplasts using lectin gold conjugates, *Eur. J. Cell Biol.* **41**:142–149.

Hohman, T. C., McNeil, P. L., and Muscatine, L. (1982) Phagosome–lysosome fusion inhibited by algal symbiont of *Hydra viridis*, *J. Cell Biol.* **94**:56–63.

Hook, G. E. R., and Gilmore, L. B. (1982) Hydrolases of pulmonary lysosomes and lamellar bodies, *J. Biol. Chem.* **257**:9211–9220.

Howard, J. C. (1985) Immunological help at last, *Nature* **314**:494–495.

Huebers, H. A., and Finch, C. A. (1987) The physiology of transferrin and transferrin receptor, *Physiol. Rev.* **67**:520–582.

Indrasith, L. S., Sasaki, T., and Yamashita, O. (1988) A unique protease responsible for selective degradation of a yolk protein in *Bombyx mori*: Purification, characterization and cleavage profile, *J. Biol. Chem.* **263**:1045–1061.

Isberg, R. R., and Falkow, S. (1985) A single genetic locus encoded by *Yersinia pseudotuberculosis* permits invasion of cultured animal cells by *Escherichia coli K12*, *Nature* **317**:262–264. (See also *Cell* **50**:769–778, 1987.)

Jeon, K. W. (ed.) (1983) *Intracellular Symbiosis*, Academic Press, New York.

Kawooya, J. K., and Law, J. H. (1988) Role of lipophorin in lipid transfer to the insect egg, *J. Biol. Chem.* **263**:8748–8753.

Krumins, S. A., and Roth, T. F. (1985) Interaction of very low density lipoproteins with chick oocyte membranes, *J. Cell Biochem.* **26**:281–288.

Lachmann, P. J. (1986) A common form of killing, *Nature* **321**:560–561.

Lackie, J. M. (1988) The behavioral repertoire of neutrophils requires multiple signal transduction pathways, *J. Cell Sci.* **89**:449–452.

Lee, K. T. (ed.) (1985) Atherosclerosis, *Ann. N.Y. Acad. Sci.* **454**.

Lemmon, S. K., and Jones, E. W. (1987) Clathrin requirements for normal growth of yeast, *Science* **238**:504–509. (See also Payne *et al.* below.)

Lewis, R. W., Harwood, J. L., and Richards, R. T. (1987) The fate of instilled pulmonary surfactant in normal and quartz treated rats, *Biochem. J.* **243**:679–685.

Lissitzky, S. (1985) La thyroglobuline et la biosyntheses des hormones thyroidiennes, *C.R. Soc. Biol.* **179**:157–167.

Locksley, R. M., and Klebanoff, S. J. (1983) Oxygen dependent microbicidal systems of phagocytes and host defenses against intracellular protozoa, *J. Cell. Biochem.* **22**:173–185.

Makarow, M. (1985) Endocytosis in *Saccharomyces cerevisiae*: Internalization of alpha-amylase and fluorescent dextran into cells, *EMBO J.* **4**:1861–1866. (See also *J. Cell Biol.* **104**:67–75, 1987, and Preson *et al.* below.)

McClure, M. D., Marsh, M., and Weiss, R. A. (1988) Human immunodeficiency virus infection of CD4 bearing cells occurs by a pH independent mechanism, *EMBO J.* **7**:513–518.

Metz, C. B., and Monroy, A. (1985) *Biology of Fertilization*, Academic Press, New York. (See especially reviews by M. Hoshi, L. Nelson, and H. Schuel.)

Middlebrook, J. C., and Dorland, R. B. (1984) Bacterial toxins: Cellular mechanisms of action, *Microbiol. Rev.* **48:**199–221.

Miller, D. K., Griffiths, E., Lenard, J., and Firestone, R. A. (1983) Cell killing by lysosomotropic detergents, *J. Cell Biol.* **97:**1841–1851.

Moehring, J. M., and Moehring, T. J. (1983) Strains of CHO-K1 cells resistant to Pseudomonas are cross resistant to diphtheria toxin and viruses, *Infect. Immun.* **41:**999–1009.

Mostov, K. E., and Simister, N. E. (1985) Transcytosis, *Cell* **43:**389–390. (See also *Cell* **46:**613–621, 1986.)

Moulder, J. W. (1985) Comparative biology of intracellular parasitism, *Microbiol. Rev.* **49:**298–337.

Moya, M., Dautry-Varsat, A., Goud, B., Louvard, D., and Boquet, P. (1985) Inhibition of coated pit formation in Hep2 cells blocks the cytotoxicity of diphtheria toxin but not that of ricin toxin, *J. Cell Biol.* **101:**548–559.

Munro, J. M., and Cotran, R. S. (1988) The pathogenesis of atherosclerosis: Atherogenesis and inflammation, *Lab. Invest.* **58:**249–261.

Neutra, M. R., Ciechanover, A., Owen, L. S., and Lodish, H. F. (1985) Intracellular transport of transferrin- and asialoorosomucoid–colloidal gold conjugates to lysosomes after receptor mediated endocytosis, *J. Histochem. Cytochem.* **33:**1134–1144.

Neutra, M., Phillips, T. L., Mayer, E. L., and Fishbind, D. T. (1986) Transport of membrane bound macromolecules by M cells in follicle associated endothelium of rabbit Peyer's patch, *Cell Tissue Res.* **247:**537–546.

Opresko, L. K., and Wiley, H. S. (1987) Receptor mediated endocytosis in Xenopus oocytes, *J. Biol. Chem.* **262:**4109–4116. (See also *Cell* **51:**557, 1987.)

Pastan, I., Willingham, M. C., and Fitzgerald, D. J. P. (1986) Immunotoxins, *Cell* **47:**641–648.

Pasteels, J.-J. (1973) Yolk and lysosomes, in *Lysosomes in Biology and Pathology* (J. T. Dingle, ed.), Vol. 7, North-Holland, Amsterdam, pp. 216–234.

Payne, G. S., Hasson, J. B., Hasson, M. W., and Schekman, R. (1987) Genetic and biochemical characteristics of clathrin deficient *S. cerevisiae*, *Mol. Cell. Biochem.* **7:**3888–3898. (See also *J. Cell. Biol.* **106:**1453, 1988 and Lemmon and Jones above.)

Pernis, B., Silverstein, S., and Vogel H. (eds.) (1987) *Antigen Processing and Presentation*, Academic Press, New York.

Peters, K.-R., Carley, W. W., and Palade, G. E. (1985) Endothelial plasmalemmal vesicles have a striped bipolar structure, *J. Cell Biol.* **101:**2233–2238.

Pierce, S. K., and Margoliash, E. (1988) Antigen processing: An interim report, *Trends Biochem. Sci.* **13:**27–29.

Postlethwait, J. H., and Giorgi, F. (1985) Vitellogenesis in insects, in *Developmental Biology* (L. W. Browder, ed.), Vol. 1, Plenum Press, New York, pp. 85–119.

Preson, R. A., Murphy, R. F., and Jones, E. W. (1987) Apparent endocytosis of fluorescein–isothiocyanate conjugate by *Saccharomyces cerevisiae* reflects uptake of low molecular weight impurities, not dextran, *J. Cell Biol.* **105:**1981–1988.

Rice-Evans, C. A., and Dunn, M. J. (1982) Erythrocyte deformability and disease, *Trends Biochem. Sci.* **7:**282–286.

Riezman, H., Chavtchko, Y., and Dulic, V. (1986) Endocytosis in yeast, *Trends Biochem. Sci.* **11:**325–328.

Rizki, T. M., and Rizki, R. M. (1984) The cellular defensive system of *Drosophila melanogaster*, in *Insect Ultrastructure* (R. C. King and H. Akai, eds.), Vol. 2, Plenum Press, New York, pp. 579–604.

Rodewald, R., and Kraehenbuhl, J.-P. (1984) Receptor mediated transport of IgG, *J. Cell Biol.* **99:**159S–164S.

Royer-Pokora, B., Kunkel, L. M., Monaco, A. P., Goff, S. L., Newberger, P. E., Boehner, R. L., Cole, F. S., Curnette, J. T., and Orkin, S. H. (1986) Cloning the gene for an inherited human disorder—chronic granulomatous disease—on the basis of its chromosomal location, *Nature* **322:**32–38.

Ryser, H. J. P., and Shen, W. C. (1978) Conjugation of methotrexate to poly (L-lysine) increases drug transport and overcomes drug resistance in cultured cells, *Proc. Natl. Acad. Sci. USA* **75:**3867–3870.

Sandvig, K., Tornessen. T. I., Sand, O., and Olsnes, S. (1986) Requirement of a transmembrane pH gradient for the entry into cells at low pH, *J. Biol. Chem.* **261:**11639–11644. (See also *J. Cell Biol.* **102:**37–47, 1986.)

Saxton, M. J., and Breidenbach, R. W. (1988) Receptor mediated endocytosis in plants is energetically possible, *Plant Physiol.* **86:**993–995.

241

USES AND ABUSES
OF ENDOCYTOTIC
AND
HETEROPHAGIC
PATHWAYS

Selsted, M. E., Brown, D. M., DeLange, R. J., Harwig, S. S. L., and Lehrer, R. I. (1985) Primary structure of six antimicrobial peptides of rabbit peritoneal neutrophils, *J. Biol. Chem.* **260**:4579–4584.

Sibley, L. D., Krahenbuhl, J. L., Adams, G. M., and Weidner, E. (1986) Toxoplasma modifies macrophage phagosomes by secretion of a vesicular network rich in surface proteins, *J. Cell Biol.* **103**:867–874.

Simionescu, M., and Simionescu, N. (1986) Functions of the endothelial cell surface, *Annu. Rev. Physiol.* **48**:279–293.

Smith, R. M., and Jarett, L. (1988) Biology of disease: Endocytosis and intracellular processing of insulin: Ultrastructural and biochemical evidence for cell-specific heterogeneity and distinction from nonhormonal ligands, *Lab. Invest.* **58**:613–629.

Solari, R., and Kraehenbuhl, J. P. (1984) Biosynthesis of the IgA-antibody receptor: A model for the transepithelial sorting of a membrane glycoprotein, *Cell* **36**:61–71.

Takeuichi, K., Woods, H., and Swank, R. T. (1986). Lysosomal elastase and cathepsin G in beige mouse: Neutrophils of beige (Chediak–Higashi) mice selectively lack elastase and cathepsin G, *J. Exp. Med.* **163**:665–677. (See also *J. Exp. Med.* **166**:1362, 1987.)

Tanchak, M. A., Griffing, L. R., Mersey, B. G., and Fowkes, C. C. (1984) Endocytosis of cationized ferritin by coated vesicles of soybean protoplasts, *Planta* **162**:481–486. (See also *Protoplasma* **138**:173–182, 1987.)

Tavassoli, M., Kishimoto, T., and Kataoka, M. (1986) Liver endothelium mediates the hepatocyte's uptake of ceruloplasmin, *J. Cell Biol.* **102**:1298–1303.

Telfer, W. H., Huebner, E., and Smith, H. S. (1982) The cell biology of vitellogenic follicles in Hyalophora and Rhodnius, in *Insect Ultrastructure* (R. C. King and H. Akai, eds.), Vol. 1, Plenum Press, New York, pp. 118–146.

Tirrell, D. A., Donaruma, G., and Turek, A. B. (eds.) (1985) Macromolecules as drugs and as carriers for biologically active molecules, *Ann. N.Y. Acad. Sci.* **446**.

Trager, W. (1987) *Living Together: The Biology of Animal Parasitism*, Plenum Press, New York.

Van Furth, R. (ed.) (1980, 1985) *Mononuclear Phagocytes*, Nijhoff, The Hague.

Vaysee, J., Gattegno, L., Bladier, D., and Aminoff, D. (1986) Adhesion and erythrophagocytosis of human senescent erythrocytes by autologous monocytes and their inhibition by β-galactosyl derivatives, *Proc. Natl. Acad. Sci. USA* **83**:1339–1343.

Vitetta, E. S. and Uhr, J. W. (1985) Immunotoxins, *Annu. Rev. Immunol.* **3**:197–212. (See also *Science* **238**:1098–1104, 1987.)

Vlassara, H., Brownlee, M., and Cerami, A. (1985) High-affinity, receptor-mediated uptake and degradation of glucose-modified proteins: A potential mechanism for the removal of senescent macromolecules, *Proc. Natl. Acad. Sci. USA* **82**:5588–5592.

Wall, D., and Maleka, I. (1985) An unusual lysosomal compartment involved in vitellogenin endocytosis by Xenopus oocytes, *J. Cell Biol.* **101**:1651–1664. (See also *Dev. Biol.* **119**:275–289 and *J. Biol. Chem.* **262**:14779, 1987.)

Wallace, R. A. (1985) Vitellogenesis and oocyte growth in nonmammalian vertebrates, in *Developmental Biology* (L. W. Browder, ed.), Vol. 1, Plenum Press, New York, pp. 127–177.

Weissmann, G. (ed.) (1980) *The Cell Biology of Inflammation*, North-Holland, Amsterdam.

Young, R. W. (1982) Biological renewal: Applications to the eye, *Trans. Ophthalmol. Soc.* **102**:42–75.

Youngdahl-Turner, P., Rosenberg, L. E., and Allen, R. H. (1978) Binding and uptake of transcobalamin II by human fibroblasts, *J. Clin. Invest.* **61**:133–141.

5

Autophagy and Related Phenomena

Autophagy—the lysosomal degradation of part of a cell's own cytoplasm—is a common phenomenon. Autodegradation of this sort was first clearly recognized in the early 1960s by Ashford and Porter, Essner and Novikoff, Swift and Hruban, and others. It is observable in many cell types under normal circumstances but is enhanced by stresses such as injury or nutritional deprivation.

Four major sorts of autophagic processes are distinguishable (Fig. 5.1): (1) "classical" autophagy, in which the cell sequesters and degrades a region of its cytoplasm within a membrane-delimited vacuole; (2) lysosomal digestion of membrane internalized from the cell surface; (3) **crinophagy,** in which secretory structures fuse with lysosomes rather than with the plasma membrane; (4) degradation of endogenous stores during embryonic development or at other times. The term "autophagy," when used without further specification, almost always refers to the "classical" variety.

Less is established about the details of autophagic processes than about heterophagy, in part because autophagy and its relatives have received less attention and in part because heterophagy has proved easier to manipulate experimentally in fruitful ways. Most studies of autophagy have been on mammalian cells, especially hepatocytes and cultured cell lines. Protozoa engage in classical autophagy and slime molds may as well, but very little is known of the situation in yeast and other experimentally important "simple" eukaryotes (see Section 7.5.2.1). Autophagic phenomena in plants have been captured only in outline, though certain plant tissues are known to engage in processes akin to classical autophagy (Sections 5.1.6.1 and 5.1.7.1) and others make extensive use of autophagic digestion of nutrient stores (Section 5.4).

5.1. Classical Autophagy

"Autophagic vacuoles" are identified, microscopically, by the presence of recognizable cytoplasmic organelles within a membrane-bounded compartment. Mitochondria, peroxisomes, ribosomes, granules of glycogen, or fragments of endoplasmic reticulum (ER) are the most frequent identifiable occupants. In liver, kidney, and other tissues of mammals, newly formed autophagic structures ordinarily are about 0.5–1 μm in diameter, although ones larger or smaller than this are not infrequent in some cell types. Fusions among the vacuoles, generating quite large structures, have occasionally been reported, especially in plants.

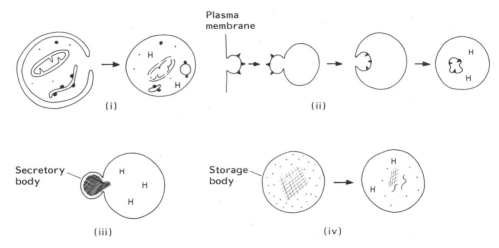

FIG. 5.1. The four principal types of autophagy considered in the text. (i) In "classical autophagy" the cell engulfs part of its cytoplasm within a membrane system that starts off as a double-walled saclike structure; this configuration evolves eventually into a closed compartment bounded by a single membrane, and acquires lysosomal hydrolases (H). (ii) Membrane from the cell surface can be degraded when, for example, the membrane is taken into the cell by endocytosis and becomes internalized within evolving endocytotic structures, such as MVBs, which become lysosomes. The black triangles are meant as diagrammatic "markers" for a patch of membrane and as indicators of the orientation of the side of the membrane that starts out facing the cytosol. However, the membrane may be altered considerably before reaching the lysosome interior. (iii) In **crinophagy**, secretion granules fuse with lysosomes (or prelysosomes) and the contents of the granules are broken down. (iv) Materials can be stored within membrane-bounded compartments for prolonged periods and then on appropriate triggering of the cell, undergo digestion by lysosomal hydrolases.

Cell fractions enriched in autophagic vacuoles can be prepared by Percoll or metrizamide gradient centrifugation, especially from tissues in which vacuole abundance has been experimentally enhanced (Section 5.1.5.3; see also footnote on p. 250). The fractions contain the usual panel of lysosomal enzymes and under appropriate incubation conditions they exhibit substantial *in vitro* proteolysis of their contents. The volume of cytoplasm enclosed within an average autophagic vacuole in cell fractions of rat liver is about 0.5 μm³.

Discrimination of vacuoles of autophagic origin from similar-looking phagocytotic vacuoles is by context and by observing stages in vacuole formation. Unlike phagosomes, which are bounded by a single membrane from the outset, many newly forming autophagic vacuoles are delimited by a pair of membranes (rarely more): As reconstructed from electron microscopy (e.g., Fig. 5.2), young vacuoles arise as regions of cytoplasm surrounded by a sac that eventually

FIG. 5.2. (a) Autophagic structures in cells of the fat body of a butterfly (*Calpodes*), during the period when peroxisomes are selectively autophagocytosed (see Fig. 5.7). A indicates a peroxisome-containing autophagic structure in which the two membranes of the sequestering sac (arrowheads) are still readily visible. The delimiting membrane of the peroxisome enclosed in the nascent vacuole is seen at m. At the arrow, a sac is seen partially surrounding a peroxisome: this probably is an autophagic structure caught early in the process of sequestration. From Locke, M. (1982). In *Insect Ultrastructure* (R. C. King and H. Akai, eds.), Vol. 2, Plenum Press, New York, p. 151.

(b) Autophagic vacuoles from root meristematic cells of the plant *Euphorbia characias* showing a sequestering sac surrounding a region of cytoplasm (b1) and the presence of cytochemically demonstrable acid esterase activity in a similar sac (b2). From Marty, F. (1978). *Proc. Natl. Acad. Sci. USA* **75:**852. × 35,000 (approx.).

fuses with itself to seal off ("sequester") the vacuole's contents from the rest of the cytoplasm. (Some authors refer to the sequestering sacs as "isolation membranes.") Later, one of the two membranes of the sac disappears, leaving the maturing vacuole with only a single delimiting membrane.

At their earliest stages most autophagic vacuoles lack cytochemically demonstrable acid phosphatase or other enzymes: To parallel the terminology

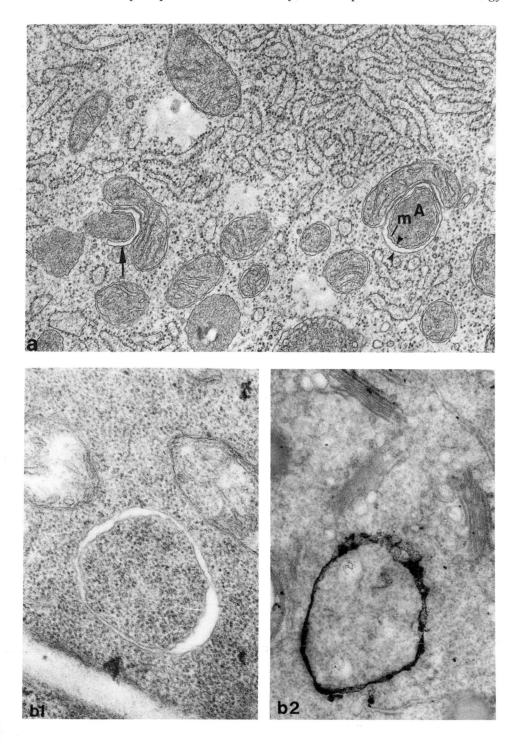

used for heterophagy, such vacuoles are sometimes called **autophagosomes** (or **cytosegresomes**) whereas vacuoles that have acquired hydrolases along with sequestered cytoplasm are called **autolysosomes.** For many situations, however, detailed information is lacking as to when the vacuoles being studied acquire hydrolases. So **autophagic vacuole** (AV) remains an accepted umbrella term covering both "nascent" AVs (those lacking hydrolases) and "degradative" AVs.

Eventually, the vacuoles evolve into smaller "dense bodies" similar to heterophagic residual bodies. During this evolution the organelles within the vacuoles are progressively disrupted and soon (Section 5.1.5.2) become unrecognizable microscopically.

In engulfing ER an AV takes in a small portion of what in many cells is an extensively interconnected network of "cisternae" (sacs) and tubules. Not established is whether the AV actually "bites off" a piece from the network or whether instead a zone of the ER separates off from the rest before it is sequestered. Similar questions arise with respect to other organelles. Mitochondria, for example, often are tens of micrometers in length and sometimes are extensively branched or even form elaborate, continuous networks. Peroxisomes in many cell types exist as clusters of interconnected organelles. Rare images, like that in Fig. 5.3, could mean that an AV nibbles off portions of a mitochondrion or other structure, but to demonstrate this persuasively, more must be learned about the three-dimensional structure of early AVs and about the factors that govern the sizes and shapes of the vacuoles.

It is not always easy to distinguish forming AVs from artifacts of tissue preparation or "accidents" of thin sectioning (curved sacs of nonautophagic nature can be cut so as to make it appear as if they have completely surrounded an island of cytoplasm; see Fig. 5.13).

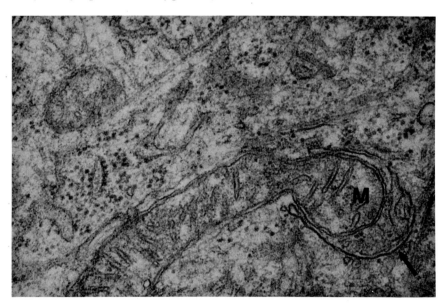

FIG. 5.3. From the cell body of a neuron in a rat ganglion 1 day after the cytoplasmic reorganization known as chromatolysis, which precedes regeneration, was initiated by cutting the axon; chromatolysis involves substantial autophagy in the cell body. At the arrow, a thickened membrane (see Fig. 5.5) partially surrounds the end of a mitochondrion (M); this seems to be a nascent autophagic vacuole that is engulfing a segment of the mitochondrion. From Holtzman, E. (1976) *Lysosomes: A Survey,* Springer-Verlag, Vienna.

5.1.1. The Membranes Surrounding Newly Forming AVs

Because the enveloping structure is a sac rather than a single membrane, and because some conditions produce sharp rises in the frequency of AVs within minutes, it is considered likely that the membranes responsible for classical autophagy arise directly from a preexisting cellular membrane system rather than being formed *de novo* from lipids and protein molecules. This idea is consonant with the prevailing agreement that membranes generally form by growth or modification of preexisting membranes.* Agreement as to which cellular system is the source of the vacuoles' membranes has been more elusive. Virtually every cellular membrane from the plasma membrane to the nuclear envelope has been accused. The ER and Golgi-associated systems, being composed largely of sacs, are the favorite suspects.

5.1.1.1. ER versus Golgi versus "None of the Above"

Backers of the ER note that although autophagic processes frequently are observable near the Golgi apparatus, they are by no means confined principally to this locale. The ER is much more widely distributed in the cytoplasm than are recognizable Golgi structures, and AVs often arise quite near recognizable elements of the ER. In fact, AVs are seen to form at sites, such as the distal regions of axons, where the Golgi apparatus is not in evidence but elements of ER are present.

On the other hand, from the inception of autophagic sequestration the membranes directly involved lack ribosomes. In addition, though very close **associations** of forming AVs with ribosome-studded ("rough") ER are common, direct **continuities** between the vacuolar membranes and rough membranes are exceedingly difficult to find, even with serial sections. Images suggesting such continuities generally are ambiguous—most could be close overlaps rather than true continuities. This is an important problem, because the presence of ribosomes is the only clear morphological criterion for identifying a membrane system as part of the ER. The rough ER might simply lose its ribosomes when it takes part in autophagy but this would be a difficult process to demonstrate, and has not been shown to take place. Extensive organized systems of **smooth** ER are present in some cells constitutively, as in axons where most of the ER is smooth, or in regions of the hepatocyte cytoplasm. Such smooth ER has repeatedly been ascribed roles in autophagy but, even in hepatocytes or neurons, individual sacs or tubules of smooth ER cannot unequivocally be distinguished, on morphological grounds, from other smooth membrane systems— sacs and tubules related to the Golgi apparatus for example, or certain endocytotic structures.

De novo origin of sequestering membranes continues to be posited by some investigators to explain situations where particles injected into the cytoplasm become surrounded by a membrane without the obvious intervention of preexisting membranes. Keep in mind also that although many autophagic vacuoles do start out with a pair of bounding membranes, the available evidence does not definitely demonstrate that all do. A "compromise" viewpoint, with a few adherents, holds that autophagic sequestration is mediated not by preexisting membranes but by newly made ones, which, however, form in intimate association with the ER, from lipids and proteins provided by the ER.

Beyond microscopic description, virtually nothing is known about how or why the membranes surround and sequester a region of cytoplasm. Perhaps current proposals that microtubules or other cytoskeletel elements control the organization and movements of Golgi and ER structures will eventually provide leads to pertinent mechanisms.

Those who think the Golgi apparatus is the chief culprit in autophagic sequestration point out that the functional and structural relations between the Golgi apparatus and the ER can be so intimate that it is difficult to be sure from microscopy, where one ends and the other begins (see also Sections 7.2.2.1 and 7.2.2.4). The two systems exchange membranes by vesicular vehicles but they also exhibit close structural associations that may even include direct continuities between ER cisternae and Golgi sacs (continuities are seen rarely, and their functional significance is uncertain). Though the distinctive-looking systems of stacked sacs that are morphologically diagnostic of the Golgi apparatus do have a restricted distribution in the cell, Golgi-related membranes can be more widely distributed. In many cell types, there are smooth-surfaced tubules, sacs, or networks, which show regions of very close association with the stacked Golgi sacs but also extend distances away from these regions (Section 7.2.2.3). Furthermore, the Golgi apparatus sometimes gives rise to small, discrete, tubular and vesicular elements that can migrate far. Such elements move, for instance, down the entire length of axons, with the result that Golgi derivatives and bona fide portions of the smooth ER intermingle in axons, forming mixtures very difficult to sort out visually.

In hepatocytes, newly forming AVs supposedly are bounded by membranes only 5–6 nm thick; this is similar to the thickness of the ER membranes

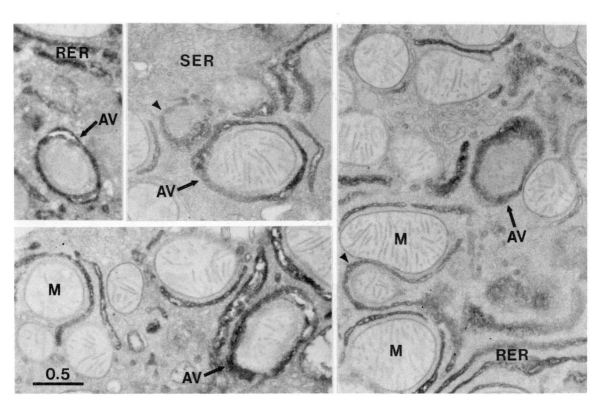

FIG. 5.4. Immunocytochemical staining (HRP technique) of the membranes bounding autophagic vacuoles forming in rat liver; autophagy was induced by perfusion with a medium depleted in amino acids and containing glucagon. In this set of micrographs, antisera against a purified membrane protein (ribophorin II) of the rough ER, or against a relatively pure ER membrane preparation, are seen to bind to the limiting membranes of autophagic vacuoles (AV). Staining is also observed in the rough ER (RER) but not in the smooth ER (SER). M indicates mitochondria. Whether the antigens detected in the AVs are identical to those in the ER requires further study of the antibodies' specificities. Courtesy of W. Dunn.

but is notably thinner than most Golgi membranes and other cellular membranes, including those of mature lysosomes. Efforts to buttress such evidence by cytochemical discrimination among membrane systems have, however, yielded disappointing results. Neither glucose-6-phosphatase, an enzyme present in the rough and smooth ER of hepatocytes and some other cell types, nor the diphosphatases that characterize certain Golgi elements, are reliably demonstrable in the sacs involved in autophagic sequestration. Acid phosphatase is sometimes present in these sacs but the significance of this localization is not understood (see Section 5.1.3). In some cell types, the sacs stain with methods based on the deposition of metal ions or of derivatives of osmium tetroxide, but these methods lack the specificity required for diagnosis of membrane origins. And, although immunocytochemical approaches should eventually prove powerful in helping determine the sources of the sacs, these approaches have only recently begun to be used systematically. (In the study illustrated in Fig. 5.4, the sacs responsible for the genesis of AVs in rat liver were found to bind antibodies from antisera that had been raised against proteins found principally, though not exclusively, in the rough ER. But in other studies, ER antigens were virtually undetectable in the surface membranes of the AVs, even though the antigens could be demonstrated in ER sequestered inside the same vacuoles.)

Freeze-fracture electron microscopy of AVs forming in mouse pancreas shows the patterns of intramembrane particles (IMPs) in the membranes of the sequestering sac to differ from the patterns in the ER or the Golgi apparatus. The outer membrane of the sac—the one facing the cytoplasm—has fewer IMPs than the ER or Golgi membranes, and the IMPs are more patchily distributed. The inner membrane is unusually poor in particles. These observations would seem to imply either that the attention to the ER and Golgi apparatus as likely sources of the membranes bounding AVs is misplaced, or that the membranes responsible for autophagic sequestration are specialized derivatives or transformed states of the ER or Golgi systems. Unfortunately, the earliest stages of vacuole formation cannot be readily identified in the freeze-fracture preparations, so that it is not known whether the pattern of IMPs described above for the sequestering membranes is present from the outset.

5.1.2. Changes in the Membranes during Evolution of AVs

The paucity of IMPs in the inner membrane of the sequestering sac might reflect a very rapid "deterioration" of this membrane. Because the inner membrane seems to be the one that disappears as the vacuoles mature, such changes are not surprising. But is the disappearance simply a degradative process, like the digestion of other material in the AV, or are more "interesting" processes involved?

Early AVs frequently show regions like those in Fig. 5.5 where the two membranes of the sequestering sac are very closely approximated to one another. For a while, the possibility was discussed that such "compacted" pairs of membranes somehow merge with one another, yielding the single membrane of the mature vacuole. This would, however, represent unusual behavior for lipoprotein membranes. Less disconcerting proposals suggest that the inner membrane is modified and then destroyed by the hydrolases that the vacuoles eventually acquire. (these suggestions rest on the still-unproved assumption that the hydrolases are delivered before the membrane is altered; Section 5.1.3.) If the inner membrane is degraded by lysosomal hydrolases, why is the outer

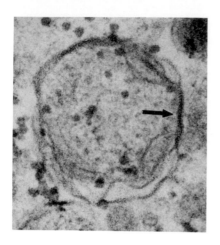

FIG. 5.5. AV from the rat adrenal medulla. At the arrow, the membrane delimiting the vacuole is seen to be quite thick and to be bisected by a dense line. Apparently in this region, the two membranes of the sequestering sac have become very closely apposed, obliterating the space that usually is present between them—above and below this region, the two membranes are more widely separated. From Holtzman, E. (1976) *Lysosomes: A Survey*, Springer-Verlag, Vienna. × 90,000.

membrane spared? Could it be that by the time the enzymes act, the outer membrane is somehow "protected" by having been provided with a set of lysosomal membrane proteins? In preliminary studies accompanying that illustrated in Fig. 5.4, an antigen characteristic of lysosomal membranes (Section 3.2.3.3) was found to be present in many AVs; the antigen was detectable in a few of the AVs at relatively early stages in their life history, when the inner membrane of the sequestering sac was still present. Fragmentary observations also hint that the "halo" that characteristically separates the contents of lysosomes from the delimiting membrane (Section 3.2.1) appears in AVs very soon after the inner membrane disappears and would therefore be available to carry out its supposed barrier functions. It would be instructive to know whether the material in this halo zone comes from within the sequestering sac (see Section 7.4.3.7).

Not only must the membrane persisting at the surface of a maturing AV survive, it must also serve functions that change as the AV evolves, fuses with lysosomes, degrades its contents, and then eventually transforms into a residual body. Proton-pumping capacities for example, and suitable transport and permeability pathways for digestive products must be provided and then maintained and regulated. Is it significant in these terms that the number of IMPs per unit area seems to increase as AVs mature? The AVs also sometimes show deep infoldings of their surfaces, which have been interpreted either as due to mechanisms that interiorize portions of the vacuole surface (see Section 5.2.1) or as devices providing an increased surface area for exchanges with the cytoplasm.

5.1.3. Acquisition of Hydrolases

AVs gain at least some of their hydrolases by fusion with preexisting secondary lysosomes. When autophagy is particularly intense, transient depletion of the cell's population of lysosomal "dense bodies" can result from these fusions: In rat hepatocytes responding to the hormone glucagon (Section 5.1.5.1), more than one dense body "disappears" for every AV formed.* When secondary lysosomes are prelabled by exposing cells to presumedly endo-

---

*These estimates come from microscopic determinations of organelle abundance in cell fractions trapped on Millipore filters; the studies were among the first efforts to use cell fractionation to follow the evolution of the lysosomal population during active autophagy. *In situ* estimates from microscopic morphometry (Section 1.5.3) lead to roughly the same conclusions.

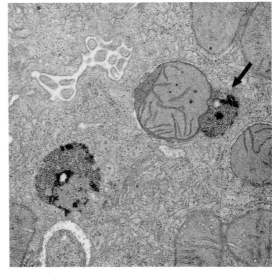

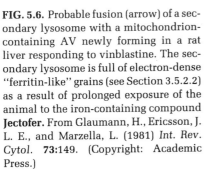

FIG. 5.6. Probable fusion (arrow) of a secondary lysosome with a mitochondrion-containing AV newly forming in a rat liver responding to vinblastine. The secondary lysosome is full of electron-dense "ferritin-like" grains (see Section 3.5.2.2) as a result of prolonged exposure of the animal to the iron-containing compound **Jectofer.** From Glaumann, H., Ericsson, J. L. E., and Marzella, L. (1981) *Int. Rev. Cytol.* **73**:149. (Copyright: Academic Press.)

cytotic tracers, the tracers later appear in newly forming autophagic lysosomes. These observations probably indicate both that "old" residual bodies fuse with AVs (see Fig. 5.6 and Section 2.1.3.3) and that autophagy and heterophagy are mutually intersecting pathways of lysosomal function rather than processes sharply segregated from one another. Systematic studies are needed to firm up the evidence for such intersection and to determine if it can take place at "prelysosomal" stages by fusions between endosomes and nascent AVs. If, as most researchers believe, prelysosomal structures persist as such only briefly, these fusions would probably be rare, if they occur at all; this expectation seems to be borne out by the few shreds of pertinent information available about distributions of endocytosed tracers and of membrane antigens.* If it turns out that endosomes and AVs do fuse to appreciable extents, it will be interesting to determine whether recycling mechanisms are tuned finely enough that endocytotic receptors are retrieved for reuse from the hybrid vacuoles.

Participation of primary lysosomes in delivering hydrolases to AVs has been little studied because autophagy is not prominent in the few cell types where primary lysosomes are readily identified. However, small hydrolase-containing vesicles like those discussed in Sections 2.1.3.1 and 7.3.2.1 have

---

*Preliminary reports have claimed success in "forcing" such fusions under artificial conditions. Cells that had sequestered lactose into presumptive early autophagic vacuoles (see footnote, p. 258) under conditions that the authors believe suppress fusion of new AVs with lysosomes (presence of asparagine) were able to digest the lactose when supplied with exogenous β-galactosidase, which the cells endocytose.

Reasonably pure AV fractions can be obtained by judicious choice of conditions to evoke autophagy, and of time after treatment at which to homogenize the cells. Attempts have been made, therefore, to follow the interactions of dense bodies and autophagic structures using behavior in centrifugation as the principal criterion for classifying the organelles. Such work can be dishearteningly difficult, however, especially because AVs often are obtained from material subjected to treatments (e.g., chloroquine, leupeptin) that modify the densities and sizes of many other lysosome-related bodies. In addition, it is not easy to predict the centrifugal behavior of the organelles formed by fusions of dense bodies with nascent AVs. In other words, some of the same problems we have been plagued with in deciding whether a body is a "lysosome" or an "endosome" arise in connection with autophagy as well.

been seen lying near AVs in cytochemical and immunocytochemical studies, and it is surmised that these vesicles do deliver enzymes to the vacuoles.

The most controversial observations concerning the delivery of enzymes to AVs are the cytochemical demonstrations of hydrolase activity in the membrane systems that carry out sequestration (Fig. 5.2). Such activity has often been seen, but usually only in a small minority of the vacuoles forming in a given preparation. One interpretation of these images is that sequestration can be carried out by sacs already charged with hydrolases, such as elongate lysosomes (Section 2.4.5) or the sacs located near the *trans* face of the Golgi apparatus (Sections 7.2.2.1 and 7.2.2.4). As always, acid phosphatase is the enzyme usually demonstrated in the configurations. But other enzymes have been demonstrated (Fig. 5.2), even if only rarely, and it could be that the sequestering sacs provide a battery of enzymes to the forming AVs. The enzymes would be released into the vacuole interior when the inner membrane of the sac disappears. An alternative proposal is that the hydrolases in the sacs get there when lysosomes happen to fuse with nascent AVs before the inner membrane goes. If this alternative is correct, then, presumably, the absence of hydrolase activity from most double-walled autophagic structures implies either that the inner membrane usually is gone before hydrolases are delivered or that the membrane is very rapidly destroyed upon enzyme delivery. (See also the comments on negative cytochemical results in Section 1.3.2.3.)

5.1.4. Selectivity?

At first blush, classical autophagy often gives the impression of being a more-or-less random process: Most classes of cytoplasmic organelles are subject to uptake and there is nothing special about the organelles being taken up, at least insofar as their microscopic appearances are concerned. (Soon after they are sequestered, however, the organelles within AVs often do reveal stigmata of deterioration—e.g., condensation of matrices, or disruption of structure.)

When a given AV is forming, it does not obviously discriminate among the structures present in the gulp of cytoplasm it is surrounding—a vacuole often contains a mixture of different types of membrane-delimited structures accompanied by "cytosolic" elements like granules of glycogen, or ferritin molecules. There do seem to be constraints related to the size of the bodies involved, so that, for example, when a mitochondrion is present in a nascent AV, not much room is left for other material. Still, it is by no means uncommon for a few profiles of ER sacs, ribosomes, and other structures to be seen along with the mitochondrion.

The impression of randomness is, however, at least somewhat misleading. Nuclei are not known to be autophagocytoses even in cases, such as the maturing erythrocyte, where they are superfluous (Section 4.6).* In addition, situations have been described in which cells selectively autophagocytose specific

---

*Autophagy of nuclei has been suggested for unusual situations such as the destruction of nuclei in multinucleate slime molds or during developmental tissue reorganization in insects. Membrane-bounded vacuole-like structures containing nuclei sometimes have in fact been seen in such situations, but phagocytotic origins of these structures have not been ruled out. In ciliated protozoa following conjugation, macronuclei fragment and eventually disappear, apparently by some sort of degradation within the cell. There have been claims of lysosomal involvement in this process too, based on acid phosphatase cytochemistry, but the details have not been worked out.

classes of cytoplasmic structures. For example, during metamorphosis in the butterfly *Calpodes*, the cells of the fat body exhibit sequential waves of autophagy (Fig. 5.7; see also Table 5.2): The first AVs to form contain chiefly peroxisomes. Later, vacuoles containing mitochondria predominate and still later, vacuoles arise that contain mainly ER. For hepatocytes responding to glucagon, morphometric analyses indicate that the frequency of peroxisomes in AVs is much lower than would be expected if the vacuoles are truly random "gulps" of cytoplasm. Similar studies suggest that the ER is selectively autophagocytosed in rodent liver recovering from phenobarbital (this drug induces a reversible hypertrophy of the smooth ER) and during the regression of the prostate gland induced by castration; hepatocytes in perfused liver are reported even to discriminate between smooth and rough ER, sequestering the former preferentially when the perfusion medium contains certain levels of amino acids and hormones. (See also Section 5.5.4.1 for selective loss of ER by mechanisms that appear to be nonautophagic.)

For more ordinary conditions as well, morphometric analyses sometimes indicate that given types of organelles are present in AVs at relative frequencies different from those in the cytoplasm. For instance, in normal liver and kidney, peroxisomes are relatively "overrepresented" in AVs as compared with mito-

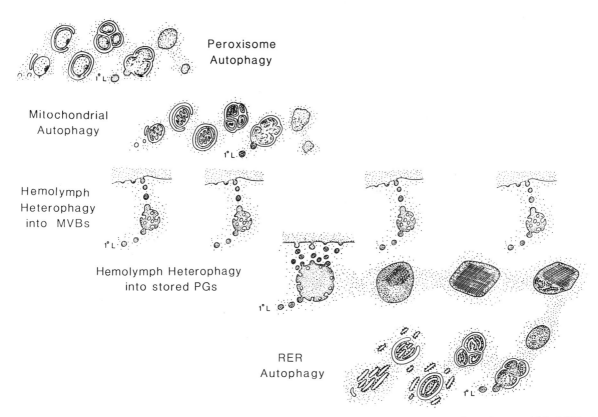

FIG. 5.7. Sequence of autophagic and heterophagic processes occurring in the fat body of the butterfly *Calpodes ethlius* during the 36 hr prior to pupation. Sequential waves of autophagy are selective for peroxisomes (Fig. 5.2a), mitochondria, and ER. The fate of endocytosed he-molymph proteins shifts from degradation within MVBs, to storage in large protein granules (see Fig. 4.5). From Locke, M. (1982), in *Insect Ultrastructure* (R. C. King and H. Akai, eds.), Vol. 2, Plenum Press, New York, p. 151.

chondria. In gland cells, secretory granules typically are greatly underrepresented (but see also Section 5.3).

These morphometric studies are interpreted on the assumption that the volumes of different types of engulfed structures are not differentially affected by engulfment. More importantly, the studies deal with the organelles that are **recognizable** in AVs. If, for example, after they are sequestered, different classes of structures were to be degraded into unrecognizable forms at markedly different rates, the enumerations of relative frequencies of the structures in AVs could give an incorrect indication of the relative frequencies with which the structures actually are autophagocytosed. Reassuringly, analyses in which data have been corrected for estimated rates of organelle deterioration in AVs still show that AVs are not strictly "random samples" of the cytoplasm.

5.1.4.1. Does Autophagic Selectivity Require Special Signals?

For cases like that of the *Calpodes* fat body, some signaling system must be posited to account for the autophagic targeting of particular classes of structures at particular times. For less extreme cases, special signals may not be necessary. For instance, there is no compelling reason to assume that AVs form with equal probability in all regions of the cell: as earlier indicated, many investigators have had the impression that autophagic structures are particularly abundant near the Golgi apparatus or in certain other zones. If these impressions do reflect the existence of preferred sites of vacuole origin, one would expect vacuole contents to be skewed in favor of organelles that abound near these sites.

Mechanical and size factors could also bias access of sequestering systems to different cytoplasmic structures. It is striking that organized cytoskeletal structures such as centrioles, basal bodies, microtubules, or filament arrays have only exceptionally been identified in structures that might be autophagic. Perhaps individual microtubules and small arrays of filaments are so labile structurally that they disappear when included in an AV. Larger filament sets—those of muscle, most obviously—tend to be closely knit into extensive configurations that might not readily be "attacked" by autophagic sequestration mechanisms (see also Section 5.5.5 but note also that the cytoskeleton almost certainly must be disrupted or deformed to some degree in order for sequestration of organelles to take place). That basal bodies generally are anchored to relatively large cilia and flagella, might afford similar "protection." Centrioles most often occupy focal positions in radiating arrays of microtubules and these arrays might impede sequestration of the centrioles. (Or perhaps centrioles actually can be autophagocytosed, but this takes place very rarely because they usually are present in only two copies per cell so that the likelihood of being caught in a random autophagic roundup is low.)

Lysosomes are recognized within AVs infrequently (see Section 5.1.7.2). This may mean that upon encountering nascent AVs, lysosomes fuse with them rather than persisting within the enclosed cytoplasm.

5.1.5. Rates; Perturbation

5.1.5.1. Influences of Nutrition and Hormones

In liver or kidney from animals kept under moderate conditions, 0.01 to 0.5% of the total cytoplasmic volume is present within recognizable AVs at a given moment. Other lysosome-related structures, such as dense bodies, are much more abundant (see Table 5.1). But the intensity of autophagy varies considerably with physiology; increases of vacuole frequency up to five- to tenfold can be induced by nutritional deprivation or other stimuli. This is seen, for example, when perfused livers are deprived of amino acids (Fig. 5.21) or in unicellular organisms such as *Euglena* or *Paramecium* under starvation conditions. The frequency of AVs in such material falls upon the return to more normal nutritional conditions, sometimes transiently approaching zero, as in livers that have been starved and then refed. Starvation does not reliably engender **sustained** increases of autophagy in tissues such as liver of intact animals, apparently because altered overall metabolic and hormonal balances lead eventually to counteracting influences, including the partial restoration of pools of circulating amino acids by degradation of muscle proteins (Section 5.5.5).

A different sort of nutritional effect is seen when yeast cells are shifted from a medium in which the cells need alcohol oxidase to use the available carbon compounds, to a glucose-rich medium, where the oxidase no longer is required. The cells' content of alcohol oxidase declines, in company with a decline in the abundance of the peroxisomes in which this enzyme is localized; one influential group of investigators believes the peroxisomes are degraded autophagically.

Certain amino acids are more important than others in their impact on autophagy in cultured cells or perfused organs. With liver, leucine is the single amino acid whose absence has the most dramatic effects. But, along with leucine, several other amino acids are needed to "suppress" autophagy in perfused liver to levels as low as they are with rich nutrition. Histidine is one of these amino acids; it also has especially marked influences in cell culture. For glutamine, the ability to help "suppress" autophagy-related phenomena has been speculatively laid to weak-base effects of ammonia generated, metabolically, from the amino acid.

Glucagon—a hormone whose levels change with nutritional state—was early found to increase the frequency of AVs in hepatocytes. Insulin has the opposite effect, diminishing the abundance of vacuoles in several cell types. Other hormones have been studied less, though several, including thyroid hormones and steroids, do seem to affect degradative phenomena dependent upon autophagy. Cultured cells maintained in serum-poor media often show enhanced frequency of AVs, probably in part because of the lack of serum-borne hormones and growth factors such as insulin-related polypeptides, fibroblast growth factors, and perhaps some of the steroids. With insects such as *Calpodes*, autophagy can be induced in the fat body by ecdysone, a steroid hormone that controls several facets of development.

In rats maintained under various laboratory circumstances, including *ad libitum* feeding, diurnal cycles of autophagy are observed in liver and kidney; the frequency of AVs declines to very low values at night and rises during the day (Fig. 5.8). Retinal photoreceptors also exhibit a diurnal cycle of autophagy in which vacuole frequency in rod cells peaks in the morning. Likely, these effects have their origins in combinations of circadian rhythms intrinsic to the cells, hormonal rhythms, and rises and declines in circulating nutrients.

5.1.5.2. Life Spans of AVs as Recognizable Entities

The frequency of vacuoles containing recognizable sequestered cytoplasm drops precipitately in hepatocytes of rats after insulin injection or in perfused

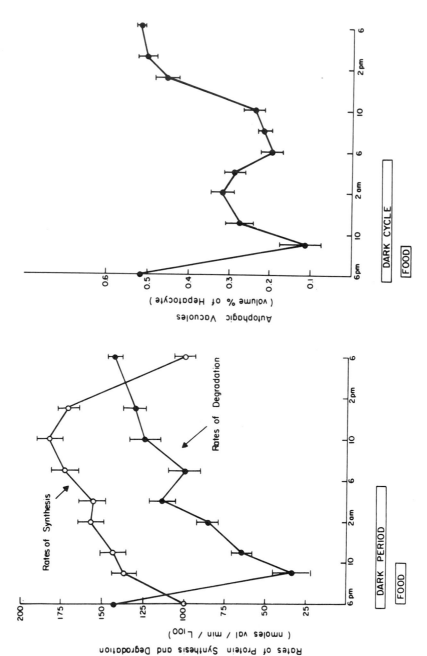

FIG. 5.8. Diurnal cycles in protein turnover and in the abundance of AVs within hepatocytes of male rats trained to eat during the indicated time interval. From Khairallah, E. A., Bartolone, J., Brown, P., Bruno, M. K., Makowski, G., and Wood, S. (1985). In *Intracellular Protein Catabolism* (E. A. Khairallah, J. S. Bond, and J. W. C. Bird, eds.), Liss, New York [see also *Acta Biol. Med. Germ.* **36:**1735, 1977; *Protein Turnover and Lysosomal Function* (H. L. Segal and D. Doyle, eds.), Academic Press, New York, 1978; *Biochem. J.* **162:**257, 1977].

liver deprived of amino acids and then resupplied with them. The kinetics of vacuole "disappearance" (decline in vacuole abundance) in such material are roughly "first-order" (see footnote on p. 218; Figs. 5.9, 5.19, and 5.21c). Accordingly, "half lives" for the vacuoles have been calculated from morphometric studies of the proportion of the total cytoplasmic volume that is included within the vacuoles at sequential time points. (The calculations assume that under the prevailing conditions, few if any new vacuoles are being formed.) The half lives are regarded as measures of the interval between the formation of a given vacuole and the transformation (degradation) of its contents into a state where their origins as engulfed cytoplasm can no longer be recognized microscopically.

In rodent hepatocytes, AVs containing different types of sequestered organelles mostly "disappear" with similar half lives of roughly 8–10 min (Fig. 5.9), though vacuoles in which glycogen predominates are longer-lived (half lives close to 20 min).

A possibly pregnant interpretation of the exponential kinetics of vacuole disappearance holds that fusion of lysosomes with new AVs is by random collision and that the rates of these fusions govern the rates of transformation of

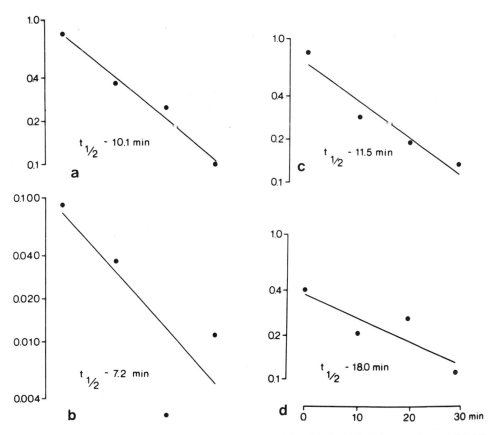

FIG. 5.9. "Disappearance" of AVs containing recognizable mitochondria **(a)**, peroxisomes **(b)**, ER **(c)**, or glycogen **(d)** as the predominant sequestered material. The data are presented as the volume of the AVs relative to the total cytoplasmic volume plotted versus time on a semilogarithmic scale. Half lives are calculated from the linear curves drawn to fit the data. Data were obtained by electron microscopy of hepatocytes from insulin-injected rats. From Pfeifer, U. (1978). *J. Cell Biol.* **78:**152.

the vacuoles into residual bodies. The absence of special guiding devices to bring nascent AVs and lysosomes together also could help explain why, when hepatocytes in perfused liver step up their rates of autophagy in response to amino acid deprivation, there are delays of several minutes before appreciable numbers of the newly forming AVs acquire cytochemically demonstrable hydrolases.

Analysis of the mechanisms underlying changes in the frequency of AVs is complicated by difficulties in deciding whether the changes are due to altered rates of autophagic sequestration, to altered rates of transformation of the vacuoles into residual bodies, or to both. From morphological studies, it does seem likely that changes in the rates of formation of new vacuoles underlie most of the increases and decreases in vacuole abundance described above. But quantitative approaches are needed for more detailed studies. In one such approach now being tried, the pace and extent of sequestration are estimated by introducing indigestible substances, such as sucrose or dextrans, directly into the "cytosol" and determining the rates at which the materials subsequently accumulate in lysosomes.*

5.1.5.3. Drugs; Temperature; Inhibitors; Weak Bases

That autophagy requires metabolic energy is suggested by the diminution of the frequency of AVs observed in explanted flounder kidney tubules exposed to cyanide, in liver depleted of ATP by treatment with ethionine, and in hepatocytes exposed to atractyloside. Both the formation of new AVs and the functioning of already existing lysosomes probably require energy. Sequestration of sucrose in the experiments to which I alluded in the last paragraph, also depends on adequate supplies of ATP. [On the other hand, the abundance of presumptive autophagic structures in chloroquine-treated macrophages (see below) is not much altered by concentrations of inhibitors of energy metabolism sufficient to block endocytosis.] Low temperature also reduces autophagy: With mouse seminal vesicle explants subjected to conditions that ordinarily induce increases in frequency of AVs, simultaneously lowering the temperature to 20–25°C partly prevents the increases. For hepatocytes in culture or in perfused liver, temperatures of 15–20°C can inhibit the protein degradation thought to take place in AVs (see Section 5.5.3); preliminary evidence suggests that these temperatures both reduce formation of new vacuoles and significantly suppress fusion of lysosomes with nascent vacuoles.

The induction of autophagy by glucagon and by some other hormones that operate via cyclic nucleotides might mean that cyclic AMP is one of the agents that controls autophagy. In a few experiments, exposure to cyclic AMP itself has seemed to enhance vacuole frequency. As ever, such observations must be looked at cautiously in light of the possibility of indirect and multiple effects; glucagon, for example, alters the flux of metabolites through gluconeogenetic

---

*Loading of the cells with the indigestible material is achieved by one or another of the bulk "microinjection" procedures now available (Section 5.5.6) or by techniques such as "electropermeabilization"—the reversible permeabilization of the plasma membrane by subjecting the cell to a brief period of high-voltage discharge. As evaluated by cell fractionation, when sucrose or lactose is loaded into tissue culture cells by the latter procedure, the sugar does become progressively associated with membrane-delimited compartments containing acid hydrolases. It would be dangerous, however, to extrapolate the data too rapidly to normal autophagy, as some sucrose also seems to be sequestered within mitochondria.

pathways and may also influence regulatory pathways dependent on inositol phosphates.

The literature on inhibitors of protein synthesis is difficult to make sense of, in part because there are differences in key details of different studies—in the timing of exposure, the concentrations of agents used, and the conditions of culture or nutrition. Cycloheximide was early reported not to have much effect on the waves of autophagy induced in hepatocytes by glucagon or in flounder kidney by explantation. It was inferred that new proteins are not needed for the formation of AVs. But cycloheximide subsequently was found to exert inhibitory influences—for example, to antagonize or reverse the increase in microscopically recognizable AVs and some associated changes in rates of protein degradation ordinarily observed in hepatocytes exposed to vinblastine or leupeptin (see below). Because cycloheximide's effects sometimes are evident relatively rapidly (within less than an hour), it may be that its crucial influence is mediated through increases in intracellular pools of amino acids or of tRNAs charged with amino acids rather than through effects on the supplies of specific proteins required for autophagy. Or, perhaps, cycloheximide alters associations of ribosomes and polysomes with the ER so as to affect sequestration, or perturbs levels of regulatory molecules such as cyclic nucleotides. In some studies with tissue culture preparations, cycloheximide's effects take several hours to develop and in these cases it could be that supplies of needed proteins run out. Prolonged inhibition of protein synthesis could even affect the availability of new hydrolases for resupplying the lysosomes (Chapter 7).

Chloroquine and other weak bases, and the protease inhibitor leupeptin, dramatically increase the abundance of AVs seen within macrophages or within hepatocytes in situ or in perfused livers. At first it was assumed that rates of vacuole formation were being enhanced. Now, however, more emphasis is placed on the likely slowing of transformation of AVs into residual structures due to inhibition of degradation within vacuoles.

In isolated hepatocytes or in explanted seminal vesicles, vinblastine and similar microtubule-altering agents sharply increase frequencies of AVs while depressing rates of lysosomal proteolysis. The conventional explanation is that depolymerization of microtubules diminishes the rates of fusion of lysosomes with nascent AVs, thus prolonging the average life span of the vacuoles as recognizable autophagic structures. However, AV-rich cell fractions obtained from livers exposed to vinblastine are not notably poor in acid hydrolases or in degradative capacities, seemingly implying that fusions with lysosomes may not always be much affected.

Figure 5.10 and Table 5.1 exemplify the especially promising observations made with the nucleotide **3-methyladenine** (3MA), which diminishes the abundance of AVs in cells such as isolated rat hepatocytes maintained under conditions of amino acid deprivation. Both morphological observations and sucrose-sequestration assays suggest that this agent specifically affects the sequestration step in vacuole formation. If this is borne out, 3MA should prove an exceptional tool for dissecting the life history of AVs. Note, however, that 3MA reportedly can also affect heterophagy.

5.1.6. Remodeling of Developing or Injured Cells

Sublethal injuries and developmental crises can notably affect the frequency of AVs. Section 5.1.4 summarized the waves of autophagy observed in the fat

FIG. 5.10. 3-Methyladenine depresses protein degradation in isolated rat hepatocytes incubated at 37°C. Illustrated are the effects of increasing concentrations of the agent on the release of [^{14}C]valine (between 30 and 90 min of incubation) from proteins labeled *in vivo* for 24 hr before the hepatocytes were isolated and exposed to methyladenine at 37°C in a culture medium. The open circles show the impact of methyladenine on release of label during the interval 30–90 min after the cells were isolated and placed, with methyladenine present, in a minimal medium (lacking amino acids). When the medium was supplemented with a mixture of amino acids (●), rates of autophagy diminished (see Table 5.1), and when the weak base, propylamine (10 mM), was added (△), lysosomal degradation was suppressed; under these circumstances (which themselves depress degradation), methyladenine has little effect, which is taken to confirm that its impact is principally on lysosome-related systems.

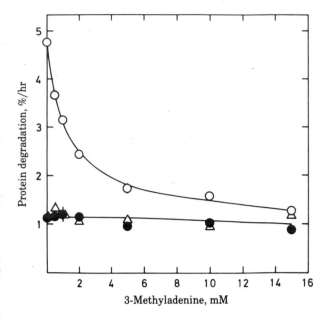

From Seglen, P. O., and Gordon, P. B. (1982) *Proc. Natl. Acad. Sci. USA* **79:**1889.

body of *Calpodes* during metamorphosis. Another example is seen when the axons of peripheral neurons are interrupted. The cells respond by marked reorganization preparatory to axonal regeneration; an early aspect of this reorganization is the appearance of numerous AVs in the cell bodies and in the stumps of the axons. Cells in aging cultures such as late-passage fibroblasts, or cells in confluent monolayers, contain more structures thought to be of recent autophagic origin than do cells in younger cultures. In contrast, cells that are growing and dividing rapidly, such as those of liver during the first days of regeneration after "partial hepatectomy" (removal of large parts of the liver), often show very few AVs.

TABLE 5.1. Reduction in Hepatocyte "Autophagosome" (early AV) Content by 3-Methyladenine and Amino Acids[a]

| | Cytoplasmic volume fraction (mm³/cm³) | |
Addition	Autophagosomes	Secondary lysosomes[b]
None	12.1 ± 0.7	39.3 ± 1.6
Amino acid mixture	1.6 ± 0.4	20.0 ± 5.5
3-Methyladenine	2.6 ± 0.7	29.0 ± 2.0

[a]Isolated rat hepatocytes were incubated for 2 hr at 37°C with no additions, with an amino acid mixture, or with 3-methyladenine (5 mM). Samples were fixed and prepared for electron microscopy, and analyzed morphometrically to determine the fraction of the total cytoplasmic volume occupied by the indicated structures. From Seglen, P. O., and Gordon, P. B. (1982) *Proc. Natl. Acad. Sci. USA* **79:**1889. [See Kovaks et al. (1981) *Exp. Cell Res.* **133:**431 for the methods used; see also Fig. 5.10.]
[b]Chiefly dense bodies, plus some late AVs.

Increases in the frequency of AVs in injured or unhealthy cells have been taken by a few investigators as signs of causative roles of autophagy in pathology or cell death. For the most part, however, autophagy after injury is thought of as promoting survival, though this is less a well-founded conclusion than it is an intuition, arising mainly from the assumption that autophagy induced by nutritional deprivation provides low-molecular-weight metabolites needed for survival.

5.1.6.1. Cellular Remodeling: Plant Cells; Red Blood Cells; Slime Molds; Yeast

The origins of the large vacuoles of mature plant cells have been hotly debated for decades. But for a number of plants, electron microscopy and enzyme cytochemistry of cells passing from the dividing meristematic condition into more differentiated states, suggest that the large vacuoles arise by processes resembling classical autophagy. This autophagy is initiated by networks of tubular sequestrational membrane systems (Fig. 5.11). Vacuoles that begin their formation in this way may subsequently grow by fusing with one another and with preexisting small vacuoles, by fusion of Golgi-derived vesi-

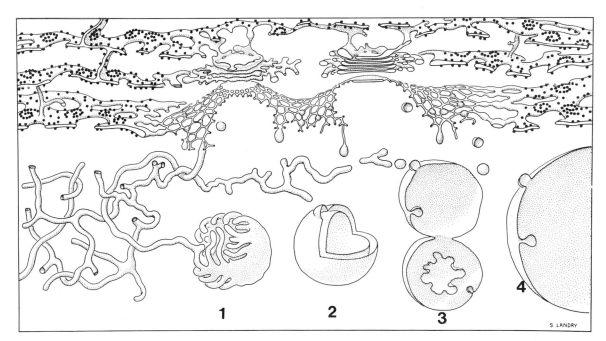

FIG. 5.11. Diagram illustrating the proposal that autophagy is responsible for the origin of the large vacuoles that arise during the differentiation of parenchymal cells of root meristems. The figures in the lower half of the diagram are successive stages (1–4) in a sequence inferred from electron microscopy: The vacuoles arise by the autophagic sequestration of cytoplasmic regions within cagelike networks of smooth-surfaced tubules; the tubules eventually fuse and close off to form double-walled AVs. Subsequently, the young vacuoles fuse with one another and with additional smaller vacuolar structures to enlarge into the mature vacuoles that occupy much of the cell. The invaginations of the vacuole surface seen in 3 and 4 refer to the apparent uptake of adjacent cytoplasm by mature vacuoles (Section 5.1.7.1 and Fig. 5.12).

The upper half of the diagram illustrates the presence in the meristematic cells, of extensive smooth-surfaced membrane systems. These systems are the likely source of the "provacuolar membranes" responsible for the autophagy diagrammed in the lower part of the diagram. They are associated closely both with the ER and with the *trans* face of the Golgi apparatus (Chapter 7).

From Marty, F. (1978) *Proc. Natl. Acad. Sci. USA* **75:**852.

cles, and by other mechanisms (see the discussions of nonclassical autophagy in Section 5.1.7.1). The sequestrational systems carry cytochemically demonstrable acid phosphatase and esterase activities (Fig. 5.2) and seem often to correspond to the "*trans* Golgi" or "GERL" systems of animal cells, about which much more will be said in Section 7.2.2.

As erythroblasts mature into red blood cells, they lose most of their cytoplasmic organelles [as well as the nucleus, in mammals (Section 4.6.2)]. Late erythroblasts and early anucleate stages ("reticulocytes") have numerous AV-like structures. Mitochondria are particularly evident in the vacuoles (Table 5.2), reflecting, perhaps, some selectivity in their engulfment, but ribosomes and ER can also be sequestered. "Expulsion" of mitochondria from maturing red cells has been seen frequently both *in situ* and with explanted cell preparations. Such expulsion is most readily explained by inclusion of the mitochondria within sequestering membrane systems, like those of AVs, followed by defecation through fusion of the vacuoles with the plasma membrane. The sources of the sequestering membranes are especially mysterious because reticulocytes are relatively depleted in ER and Golgi elements. Acid phosphatase has been demonstrated, cytochemically, in some of the configurations being discussed, but it seems unlikely that the vacuoles all acquire an extensive lysosomal hydrolase complement, as few lysosomes are present in late erythroblasts or reticulocytes. It may well be that the vacuole contents are extruded from the cells largely intact, to be degraded later through heterophagy. Comparable processes for the disposal of cell surface proteins are described in Section 5.2.1.3.

TABLE 5.2. Morphometric Data on Autophagy of Mitochondria in Rat Erythroblasts and Reticulocytes[a]

Parameters	EN− erythroblast	P[b]	EN+ erythroblast	P	Reticulocytes
No. of analyzed cell sections	47		88		76
% of sections with autophagosomes	4		25		34
% of sections with autophagocytosed mitochondria	0		10		12
Per 100 μm^3 cytoplasm					
Mitochondria (μm^3)	7.89 ± 3.80	0.000	5.18 ± 3.19	0.000	2.08 ± 1.69
Autophagosomes (μm^3)	0.09 ± 0.56	0.003	0.43 ± 0.95	0.402	0.37 ± 0.67
Autophagocytosed mitochondria (μm^3)	0.0 ± 0.0	0.034	0.11 ± 0.40	0.431	0.13 ± 0.36
Autophagocytosed cytoplasm (μm^3)	0.0 ± 0.0	0.097	0.06 ± 0.28	0.327	0.03 ± 0.16
Degraded material in autophagosomes (μm^3)	0.09 ± 0.56	0.024	0.26 ± 0.72	0.327	0.22 ± 0.52
Mean area of mitochondrial profiles (10^{-3} μm^2)	87 ± 46	0.068	73 ± 27	0.038	65 ± 30

[a]Values expressed as mean ± standard deviation. EN−, early erythroblast stages; EN+, later stages, just prior to nuclear extrusion. Note the disappearance of mitochondria during erythrocyte maturation (expressed in terms of the cytoplasmic volume occupied by mitochondria) and the selective autophagy of mitochondria during this period (although mitochondria occupy 8% or less of the total cytoplasmic volume, they occupy more than 25% of the volume of AVs). "Autophagocytosed cytoplasm" refers to recognizable nonmitochondrial material, such as ribosomes or ER. From Heynen, M. J., Tricot, G., and Verwilghen, R. L. (1985) *Cell Tissue Res.* **239**:235. (Copyright: Springer-Verlag, Berlin.)
[b]P = level of significance of difference between maturation stages tested with the one-sided Kruskal–Wallis test.

For rat bone marrow, electron microscopic morphometry (Table 5.2) indicates that autophagic-like sequestration of mitochondria could account for all of the disappearance of mitochondria from developing red blood cells, if, on the average, vacuoles persisted for less than 30 min after forming. Such life spans seem reasonable from the data on transformation of AVs outlined above, and from findings that vigorously feeding *Paramecium* expel the indigestible residues from a given food vacuole within 20 min after the vacuole forms. Still, there remain those who argue, from seeing swollen or disrupted mitochondria free in the cytoplasm of reticulocytes, that many mitochondria are destroyed within the cell by nonlysosomal means. One line of thought is that oxidative damage to the mitochondria initiates their destruction by making the organelles vulnerable to the proteases known to be present inside mitochondria (Section 5.6.4) or to other nonautophagic degradative systems known to be present in the reticulocyte's cytoplasm (Section 5.6.2.2). The latter enzymes might also contribute to the destruction of materials such as ribosomes and cytosolic proteins: the evidence that autophagy disposes of such materials is both much less detailed and much less extensive than is true for the removal of mitochondria.

The developmental transition to a multicellular state, induced by starvation in slime molds like *Dictyostelium discoideum*, includes extensive intracellular breakdown of endogenous proteins. Chloroquine arrests the transition, leaving starved slime molds in an aggregated but not yet differentiated state; this inhibition can be overcome by providing suitable levels of free amino acids. In yeast, sporulation is also accompanied by extensive intracellular proteolysis. That vacuolar proteases participate is concluded from the effects of mutational manipulations of these enzymes (see Section 7.4.1). Neither in yeast nor in slime molds is it known how substrates and proteases are brought together, but some form of autophagy might well be involved.

5.1.7. Sequestration of Cytoplasmic Components by "Semiclassical" Mechanisms? Microautophagy?

Considerable speculation has been put forth that cytoplasmic macromolecules are taken up directly by preexisting secondary lysosomes (or by certain prelysosomal structures, especially by MVBs; see Section 5.2.1). Often this uptake is envisaged as occurring in small "gulps" in which a few cytoplasmic molecules are transferred to the lysosome interior within a membrane-delimited vesicle derived from the surface of the lysosome, but the evidence for such "microautophagy" is equivocal.

5.1.7.1. Autophagy by the Vacuole in Mature Plant Cells

Mature vacuoles sometimes show membranous inclusions thought to arise by uptake of adjacent cytoplasm within pinched-off infoldings of the tonoplast. When these configurations were originally noted in conventionally prepared electron microscopic thin sections, the possibility was raised that they are preparative artifacts. This seems less likely now because corresponding images have been obtained with other approaches, notably in freeze-fracture preparations (Fig. 5.12). That vacuoles might have the capacity to incorporate adjacent cytoplasm is an attractive idea. It could, for example, help explain why vacuoles are supplied with such enzymes as RNase whose most likely substrates are in the extravacuolar cytoplasm. But vacuoles, of course, also participate in

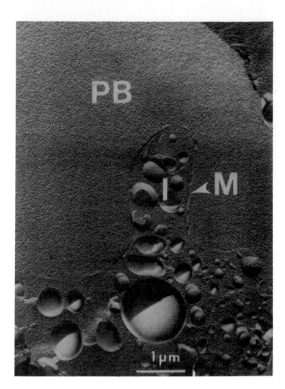

FIG. 5.12. Plant vacuoles and related structures like the protein body (PB; Section 5.4.2), shown here in a freeze-fracture electron micrograph, exhibit deep invaginations of their bounding membrane (M) in which cytoplasmic structures (I) are included. This has been interpreted as a form of auto-phagy—the invaginations are believed to pinch off into the protein bodies, subjecting the enclosed cytoplasm to digestion.

The body shown is from a 3-day seedling of mung bean (*Vigna radiata*). From Herman, E. M., Baumgartner, B., and Chrispeels, M. J. (1981) *Eur. J. Cell Biol.* **24:**226.

autolysis and other nonautophagic destruction of cytoplasm, and the enzymes might be present principally to serve in these processes (Section 6.3.2.1). There will, therefore, not be much more to say about the structures in question until more data are gathered on frequency and dynamics, and until more is learned about cytoplasmic turnover in plants.

Hypothetical mechanisms based on tonoplast infolding, or on direct penetration of the tonoplast have occasionally been proposed to account for the presence of lipid droplets or other inclusions in vacuoles of yeast or fungi.

5.1.7.2. Autophagic Sequestration by Preexisting Lysosomes in Animal Cells and *in Vitro*?

Mouse macrophages treated with the "antimicrotubule" agents, nocodazole or colchicine, exhibit little change in the abundance of classical AVs but they do show odd lysosomal configurations (Fig. 5.13): The cells show many sacs, with electron-dense content and cytochemically demonstrable acid phosphatase activity, arrayed in "cuplike," "U-shaped," or other configurations that partially surround other regions of cytoplasm. Because these configurations are reminiscent of conventional autophagy, they have been claimed to reveal capacities of lysosomes to take up cytoplasm directly. A more arresting conjecture is that the configurations represent specifically targeted autophagy of the lysosomes themselves. The orthodox alternative explanation is that the alterations in the cytoskeleton have simply resulted in reorganization of interconnected networks of lysosomes into convoluted forms that are not actively sequestering anything (see Section 7.2.2). Sometimes, in a given electron microscopic thin section the sacs appear as complete rings that, at first glance, look as if they have completely segregated a zone of cytoplasm. In many cases, however, sections along a different plane would have revealed, that the cytoplasm is only partially enclosed (see Fig. 5.13). As in our other encounters with such structures, the problem is in deciding how to interpret them—are they accidents, persistent long-term structures, or sacs caught in the act of surrounding cytoplasm that eventually will be sequestered entirely?

Profiles of lysosomal dense bodies occasionally give the impression that the lysosomes are extending membrane-bounded "flaps" that could be engulfing adjacent cytoplasm in a manner reminiscent of phagocytosis or micropinocytosis by cells. Correspondingly, vesicle-like inclusions are occasionally seen within dense bodies. Once again, however, only rarely have any of the observed inclusions been shown to be completely enclosed within the lysosomes rather than representing deep infoldings of the lysosomal surface that appear free owing to the plane of thin section. Particularly intriguing are observations that lysosomal dense bodies in cell fractions from hepatocytes can be induced to incorporate microscopically visible particles (e.g., the colloidal silica of Percoll gradients) into infoldings of their surface. Some of the infoldings seemingly do pinch off to lie free as membrane-bounded vesicles, in the lysosome interior. Because similar cell fractions are found to degrade proteins added to the suspension medium and this degradation is inhibitable by weak bases and by disruption of the lysosomes, it is tempting to conclude that iso-

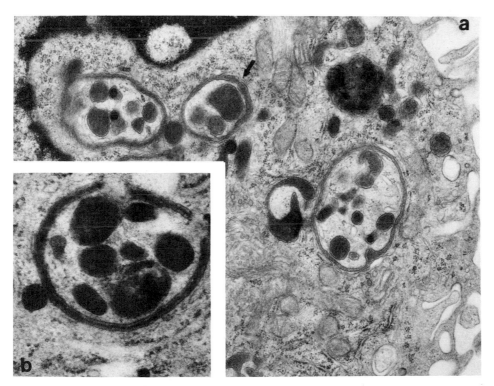

FIG. 5.13. Configurations seen when mouse macrophages are exposed to the "antimicrotubule" drug nocodazole. (a) Sacs with electron-dense content that, in suitable cytochemical preparations, exhibit acid phosphatase activity. The sacs apparently are surrounding other dense bodies as if in preparation for engulfing them autophagically, but in many of these configurations the area enclosed within the sacs is still in communication with the surrounding cytoplasm, indicating that sequestration is incomplete (arrow). (b) Endocytosed HRP demonstrated cytochemically in structures like those in a. From Thyberg, J., Blomgren, K., Hellgren, D., and Hedin, U. (1982) *Eur. J. Cell Biol.* **27:**279. a, × 50,000 (approx.); b, × 75,000 (approx.).

lated lysosomes can "microautophagocytose" and degrade materials from their surroundings. More information is needed, however, to confirm this picture and especially to decide whether it is relevant to the situation *in vivo*.

5.2. Degradation of Cell Surface Materials

5.2.1. MVBs as Autophagic Structures

Some of the MVB-like bodies in the region of the Golgi apparatus, in nerve terminals and elsewhere, form by mechanisms resembling classical auto-phagy—that is, by a sequestering sac surrounding a cluster of preexisting vesicles or a few smooth-surfaced sacs and tubules that vesiculate upon being taken up (Fig. 2.14). This mode of origin would account for the occasional observations of secretory granules still clad in their own delimiting membranes, or of vesicles containing endocytotic tracers, within MVB-like bodies.* It seems prudent at present to regard such bodies as distinct from the MVBs of endo-somal origin. But these latter, "endocytotic" MVBs seem also to have some autophagic functions, which is my next topic.

As a prelude, it will be useful to recall some of the details about endo-cytotic MVBs scattered in prior sections: (1) The majority of vesicles within typical MVBs are about 50–100 nm in diameter. (2) Some of the structures that appear as "vesicles" in a thin section actually are tubules and some of the tubules are continuous with the MVB's delimiting membrane. Serial section-ing, however, indicates that most of the vesicles in typical MVBs lack such connections and truly are interior structures. (3) The invaginations of the MVB surface, from which many of the internal vesicles are thought to arise as the MVB forms, range from a size similar to the vesicles, up to the relatively large infoldings characteristic of "cuplike" MVB precursors; these larger infoldings apparently fragment as they pinch off into the nascent MVB (Fig. 2.14).

The forces driving vesicle formation have not been studied, though the larger inpocketings of the surfaces of nascent MVBs look as if they could arise through local osmotic effects. A few preliminary findings suggest that weak bases slow vesiculation or alter the composition of the forming vesicles. One proposal envisages a shuttle system in which vesicles move repeatedly back and forth between the interior of the MVB and its surface, though what this would accomplish by way of transport is not certain. Most investigators believe rather that the vesicles are degraded within the lysosomes to which the MVBs give rise. As with other lysosomes, membrane within seems vulnerable to destruction whereas the surface membrane somehow is protected.

---

*These observations are interpreted in terms of autophagic sequestration because, for vesicles with endocytotic tracers inside them to reach the interior of an MVB, the entry process must involve one or another form of engulfment (presumably vesicles cannot penetrate the MVB's bounding membrane while themselves remaining intact). As already emphasized, typical MVBs participat-ing in endocytosis accumulate endocytosed tracers in the spaces **between** the vesicles (Fig. 2.15), the tracers having been delivered either by fusions of vesicles with the MVB or directly to the interior of an MVB precursor as it forms from the cell surface (Section 2.3.1.3). Similarly, secreto-ry granules lose their membranes when they fuse with MVBs (Section 5.3.1.2).

The frequency with which MVB-like bodies arise through sequestration of preexisting vesi-cles is difficult to estimate in part because particularly careful study is required to distinguish these bodies from other types of structures. Moreover, in MVBs with endocytotic tracers in the spaces between their vesicles, the tracers sometimes coat the outer surfaces of the vesicles, giving the impression, to the unwary, that the tracers are within the vesicles.

The production of internal vesicles by the MVBs involved in heterophagy, contributes to the regulation of the sizes of endocytotic structures and related lysosomes by providing a sort of microautophagic mechanism for disposing of "excess" membrane from the bodies' surfaces. (Each MVB vesicle represents on the order of $10^{-3}-10^{-2}$ μm^2 of membrane and a large MVB can contain dozens or, for exceptional cell types, hundreds of such vesicles.) Some of this membrane carries proteins of types that do not recycle extensively, seeming rather to be slated for destruction once they are internalized from the plasma membrane. Other pieces of membrane might get trapped simply by the accident of being present at the "wrong moment" along a stretch of the MVB surface that invaginates into the MVB.

As regards proposals that the vesicles budding into the MVBs carry adjacent cytoplasm to a degradative fate: A single vesicle includes a volume of roughly $10^{-5}-10^{-4}$ μm^3. This might seem like a negligible volume but the overall rates of vesicle formation in a cell could conceivably be quite high. Moreover, cytoplasmic uptake need not be via "fluid phase" internalization. It can be imagined that the MVB surface contains selective sites to adsorb cytoplasmic material and therefore that the vesicles resemble microautophagic analogues of the vesicles responsible for receptor-mediated endocytosis.

The cuplike precursors of MVBs could conceivably sequester larger gulps of cytoplasm within their inpocketings and such sequestration could account for the occasional observation of cytoplasmic structures (e.g., small secretory granules still surrounded by their own membranes) inside MVBs. Most often, however, the inpocketings look empty.

5.2.1.1. Degradation of Rhabdom Membranes of Invertebrate Photoreceptors

In the eyes of certain invertebrates, MVBs have quite prominent roles in disposing of membrane components internalized from the cell surface.

Photoreception in the eyes of most invertebrates takes place along specialized stretches of plasma membrane, organized as microvilli protruding from regions of the surfaces of **retinula** cells. Typically, neighboring retinula cells group their microvilli in closely packed arrays called **rhabdoms.** In many species the size of the rhabdom varies dramatically during the diurnal cycle; changes can occur at several time points, but often there is a considerable diminution shortly after dawn.

In various crustaceans, insects, and spiders, diminution of the rhabdom is accomplished largely by the withdrawal of photoreceptive plasma membrane directly into the same retinula cells that produced the membrane. This takes place by endocytotic-like formation of vesicles and tubules—including the usual coated structures—at the bases of the microvilli. The internalized membrane becomes incorporated into large MVBs that acquire acid hydrolases and eventually evolve into residual bodies. That the vesicles inside the MVBs are composed of membrane derived from the microvilli is strongly suggested by immunocytochemistry, and by freeze-fracture electron microscopy, which reveals similarities in patterns of IMPs between the vesicles and the microvillar plasma membrane.*

---

*In some species, quantities of rhabdom membrane are shed to extracellular spaces and then taken up by endocytosis. This situation is reminiscent of that in vertebrates (Section 4.6.2), though the cell types involved are quite different: for example, in certain invertebrates, macrophage-like hemocytes move in from the circulation to help engulf shed membranes.

5.2.1.2. Degradation of Receptors for Hormones and Growth Factors; "Down-Regulation"

Cultured cells, when exposed to appreciable concentrations of agents like EGF or insulin, often undergo a diminution of the abundance of the corresponding receptors at their surfaces. This "down-regulation" probably helps modulate cellular responsiveness—preventing overresponses to bursts of hormone release or, in the case of neutrophils, adjusting sensitivities to chemotactic agents. Though the phenomenon has not been studied intensively in intact organisms, there are hints that it is important. For instance, sustained obesity can lead to decreases in levels of insulin receptors as cell surfaces, owing, presumably, to chronic oversupplies of insulin in the circulation. In culture, at least, the process is readily reversible: the receptor population at the cell surface can be replenished within time frames that range from a few minutes to a few hours, depending on the cell type, the hormone, and the situation.

Simple models propose that in the absence of their ligands, the receptors subject to down-regulation are dispersed in the plasma membrane, occurring at densities that sometimes are as low as one or a few molecules per square micrometer. Rates of endocytotic internalization of the receptors are correspondingly low. The binding of ligands to the receptors both activates the receptors, producing intracellular effects, and leads to receptor clustering (these two phenomena may, of course, be causally related; see Section 2.3.2). Receptors come to group in coated pits and are internalized endocytotically; it is this that initially depletes the cell surface of receptors. What happens next depends on the paths taken by the receptors after they are internalized (see Section 3.3) and on the balances between internalization and the replenishment of receptors at the cell surface through recycling or new synthesis. The receptors that undergo prolonged down-regulation are ones for which replenishment does not keep up with internalization, for one reason or another.

There is experimental backing for these models from immunocytochemistry, photoaffinity labeling of receptors with suitable ligand derivatives, cell fractionation, and studies with weak bases or other inhibitors of lysosome function. Chapters 2 and 3 pointed out that clustering and internalization of certain receptors is indeed driven by the corresponding ligands, rather than occurring constitutively. The receptors for EGF in tissue culture cells and liver, and those for chorionic gonadotropin in Leydig cell tumor lines have been shown to accompany their ligands to the lysosomes, where they are digested. For EGF, microscopic tracer methods indicate that the receptors pass with endocytotic membranes into the delimiting membranes of MVBs and then remain with the membranes as they are interiorized within the MVBs; as a result the receptors can be demonstrated in the MVB vesicles.* Fc receptors in macrophages seem able to follow a similar route when bound to multivalent ligands. Less is known about the behavior of several of the other receptors recently claimed to undergo rapid lysosomal degradation rather than recycling repeatedly. (In addition to receptors already mentioned, such claims have been

---

*Electron microscopic counts of particulate labels bound to EGF receptors in MVBs indicate that the membranes of the interior vesicles are substantially richer in demonstrable EGF receptors per unit area, than are the membranes bounding the MVBs. Perhaps this means that the internalization processes that generate the vesicles are selective in terms of the membrane components taken in. Remember also that explanations must be found as to why the EGF receptors do not "concentrate" in the tubular extensions and recycling vehicles emanating from endosomes, as do the receptors on which I focused in Chapter 3.

made for certain tumor necrosis factor receptors and certain interferon and interleukin receptors, among others.) (See also number 11 in Section 7.6.)

In many studies, the destruction of receptors is monitored by assaying their biological activity—especially their capacities to bind ligands—rather than directly demonstrating their digestion. This approach has its pitfalls, as inactivation can result from diverse changes in the receptor proteins or their environments, but it can reveal early changes and other phenomena not readily detected otherwise.† In some cases, when insulin is being internalized by cells, the insulin receptor becomes inactivated even in the presence of weak bases at concentrations sufficient to block much of the degradation of the hormone, as if the cell deals with the receptor in a compartment different from that in which it digests insulin (see Section 4.5.2.1 and below). The inactivation of the ASGP receptor when hepatocytes are exposed to antireceptor antibodies also occasionally shows "anomalous" features, such as relative insensitivity to leupeptin (see Section 3.3.4.3). Inactivation of insulin receptors in cells internalizing insulin has sometimes been reported to be diminished by exposing the cells to cycloheximide, as if rapidly turning-over proteins were required for the process.

The constitutively recycling receptors discussed in Section 2.3.2.7 generally are not strongly "down-regulated" by their ligands. Still, rather than being different sorts of phenomena, ligand-driven internalization and degradation of hormone receptors, and the constitutive recycling of other receptors probably are extremes in a continuum of receptor-mediated endocytotic phenomena. Intermediates between these extremes, or variants, have been described: with neutrophils, chemotactic peptides can seemingly evoke insertion of receptors for C3b (Section 2.2.1.2) into the cell surface, followed by internalization and degradation of those receptors, in the absence of evident C3b-related ligands. In liver, a proportion of endocytotically internalized EGF receptors may return to the cell surface even though many of the receptors are degraded. For cultured cells expressing human EGF receptors, some reports claim that EGF receptors normally recycle on the average of a few times before being degraded, and mutant forms of the receptor, lacking the receptor's kinase activity, recycle repetitively while nonetheless delivering EGF for degradation. (Other reports on other systems conclude, however, that the kinase activity is necessary for the EGF receptor to function in endocytosis.) As should be evident from what has been mentioned previously, the behavior of the insulin receptor reportedly varies considerably in different cell types and in different experimental or physiological circumstances: after internalization, the receptors sometimes are

---

†As previously pointed out, much more must be learned about the early events of internalization to dispel our conspicuous ignorance about how the biochemical and physiological consequences of ligand binding to receptors—the phosphorylations, changes in ion distributions, and the like—fit into the stories being promulgated here. Also to keep in mind is that many receptors, including ones strongly down-regulated by their ligands, are subject to additional mechanisms that contribute to modulating cellular responsiveness without alteration of receptor numbers at the cell surface. The desensitization of neurotransmitter receptor systems by their agonists or by other agents, is a familiar example.

A related matter: If receptors internalized in activated forms continue to exert their biological effects while inside the cell (Section 4.5.2.1), then the mechanisms by which these receptors are handled in endosomes or lysosomes could be of considerable importance for controlling their influences on the cell. For instance, differences have been reported in the length of survival of receptors for EGF within different cell types; perhaps these differences correlate with variations in the duration of impact of given "doses" of EGF on given cell types.

degraded; sometimes they persist in seemingly nondegradative compartments; and in adipocytes, fibroblasts, and certain other cell types, they can undergo appreciable recycling to the cell surface. (The importance of these different behaviors for cell physiology *in situ* is unclear, especially since some of the seeming variation is almost certainly due to the use of different assays or to similar "nonbiological" factors.)

Autoimmune responses can induce pathological changes in receptor abundance at the cell surface. For example, features of the disease *myasthenia gravis* probably result from the enhanced rates of endocytosis and lysosomal degradation of acetylcholine (ACh) receptors at motor end-plates, driven by the anti-ACh receptor antibodies known to be present in the circulation of afflicted individuals. The disorder also involves blockade of normal functioning of the receptors, and inflammatory phenomena at the nerve terminals. Enhanced rates of turnover of ACh receptors also follow upon denervation of muscles, as though the presence of a nerve ending slows internalization of the receptors.

5.2.1.3. Variations on the Theme: Reticulocytes; Plants? Myelin?

Among the components lost by reticulocytes as they mature into erythrocytes are sets of cell surface proteins, including transferrin receptors and transport systems for amino acids. Transferrin receptors are readily detectable within MVBs in the maturing cells, having gotten there, presumably, as constituents of membranes participating in endocytosis. Few of these MVBs acquire detectable acid hydrolases, probably because of the paucity of lysosomes in reticulocytes. Instead, the MVB contents are defecated and the transferrin receptors wind up being shed to the extracellular space (Figs. 3.12 and 5.14). Other proteins and lipids (such as sphingomyelin) from the reticulocyte's plasma membrane may well follow the same route as the transferrin receptor. One guess about this atypical behavior is that the membrane molecules are redirected to the MVBs by hypothetical cross-linking agents able to bind to transferrin receptors and other selected molecules at the surfaces of cells once the cells have reached the appropriate maturational state. Under culture conditions, release of transferrin receptor-rich vesicles is inhibited by chloroquine, requires ATP, and is enhanced by transferrin. The shed receptors still can bind transferrin and are not otherwise grossly abnormal.

Whether the defecation of undegraded MVB contents takes place extensively in cells other than reticulocytes is not known. Such a process could endow cells with interesting capacities to externalize materials acquired from their own cytoplasm (see Section 5.5.6.2) as well as those picked up by endocytosis. It does seem likely that MVBs in cells other than reticulocytes participate in developmental regulation of cell surface composition. For instance, *Drosophila* eggs, at about the time of ovulation, form many MVBs that may be involved in disposing of receptors no longer required for uptake of yolk proteins.

For plant cells, fusion of MVBs with the large central vacuole has been posited to account for the observed passage of small amounts of endocytosed tracers, and membranous debris, into the vacuole.

A different variant of autophagy may help in the disposal of specialized plasma membrane by injured peripheral nerves. When myelinated axons are separated from continuity with the corresponding cell bodies, their myelin sheaths (these are layered arrays of plasma membrane laid down by Schwann

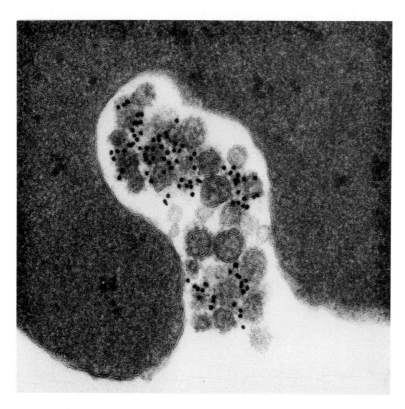

FIG. 5.14. Surface of a sheep reticulocyte that had endocytosed transferrin linked to gold particles, showing the exocytotic extrusion of the contents of an MVB including gold particles presumably still associated with transferrin receptors. From Pan, B. T., and Johnstone, R. M. (1985) *J. Cell Biol.* **101**:942. × 100,000 (approx.).

cells) are degraded ("Wallerian degeneration"). Normally, myelin is very long-lived—in rodents, the turnover of its constituents takes weeks or months. But in the injured state, it is rapidly destroyed in lysosomes. Some of this destruction is carried out by phagocytes, including macrophages, which invade the injured nerve. But some seems to take place within the Schwann cells themselves, which reinfiltrate their cytoplasm into the sheaths they had originally produced, fragment the myelin arrays into smaller chunks, and incorporate the fragments within large, acid hydrolase-containing vacuoles. The relative importance of this autophagic degradation as opposed to the heterophagic involvement of phagocytes is a subject of long-standing disagreement.

5.3. Lysosomes and the Fate of Secretory Structures: Crinophagy; Membrane Recycling in Secretory Cells

The production of thyroid hormones mixes a peculiar sort of autophagy with heterophagy. The gland secretes thyroglobulin molecules but then takes them back inside its cells to degrade the thyroglobulin in lysosomes, thereby liberating hormones (Section 4.2.4.1). This is an unusual example, but it dramatizes the fact that gland cells have adapted autophagy for interesting purposes.

5.3.1. Intracellular Degradation of Newly Made Secretory Products

In tissue culture systems, intracellular degradation of secretory products—collagen, insulin, casein, or immunoglobulin chains—can consume 10–50% of the cells' output of these molecules. More than one mechanism contributes to the digestion. Thus, fibroblasts in confluent cultures degrade about 10% of their recently made collagen. This degradation is largely unaffected by exposing the cells to chloroquine and hence is presumed not to depend on acidified compartments such as lysosomes (but see Section 6.1.5). In rapidly growing fibroblasts, however, as much as a third of the new collagen is broken down, and this degradation is largely sensitive to chloroquine. So is the extensive degradation of the abnormal collagen produced when proline analogues are present. Some abnormal viral proteins also go to lysosomes of the cells in which they are produced, rather than being inserted exocytotically in the cell surface as their normal counterparts are.

The last few observations are among the many showing that cells have screening devices to limit the buildup of nonfunctional or disruptive molecules in their cytoplasm (see also Section 5.6.2). Secretory materials altered severely by mutation, by genetic engineering, or by experimental perturbations of metabolism are often retarded in their passage through the secretory apparatus, owing, it is assumed, to changes in the proteins' folding, solubility, or interactions with membranes and intracellular receptors. The eventual degradation of the abnormal molecules presumably prevents their choking the compartments responsible for their production and transport. There are, however, profoundly changed proteins—chimeras produced by fusion of portions of different genes, glycoproteins lacking their oligosaccharides, and so forth—which traverse the secretory route more or less normally and are secreted, rather than degraded. Thus, the "criteria" for normalcy "employed" by the screening system must be of limited (or focused) stringency. Conversely, secretory molecules still are degraded by cells under conditions in which, as far as is known, the molecules are completely normal.

Some of the abnormal proteins whose passage through the ER is retarded are degraded by mechanisms less inhibited by chloroquine than expected for lysosomes. When chloroquine and other weak bases do reduce rates of degradation of newly made secretions, the conclusion usually drawn is that lysosomes are involved. This inference often is reasonable but it needs case-by-case examination in view of the fact that Golgi sacs and secretory bodies of various cell types are acidified and also contain enzymes capable of at least partial proteolysis of secretory molecules (Section 6.1.5). Similar caveats apply to experiments with endocytosed inhibitors of proteases: These compounds do seem to localize largely to lysosomes, but Golgi-associated compartments and secretory structures are accessible to membranes retrieved from the cell surface by endocytotic-like processes (Sections 5.3.2) and the extent of their accessibility to other endocytosed materials, like the inhibitors, has yet to be adequately explored (Section 7.2.2).

5.3.1.1. Autophagy of Secretory Materials

Secretion granules become extensively sequestered within AVs after injury or other stresses to gland cells (as, for instance, in explanted mammalian exocrine pancreas after several hours in culture or upon exposure to a nitrogen atmosphere, or when amphibian pituitary glands are transplanted from one animal to another). A potentially instructive case of selective autophagy of secretory material may be provided by the pancreas after exposure to alloxan *in situ*. This drug destroys the insulin-producing, endocrine, B cells of the pancreas. Reportedly, it also affects the "intermediate" cells, an unusual, relatively rare pancreatic cell type that, on morphological grounds, seems to possess both exocrine and endocrine granules. In these cells, alloxan produces differential autophagic engulfment of the granules that look like endocrine secretory granules.

Mammalian neurosecretory neurons utilize AVs in the terminals and varicosities of their axons to degrade part of their output of neurosecretion. It could be that autophagy here is helping to maintain balances among the production of neurosecretions, their transport down the axons, and their storage and release from the cell. [In one scheme, granules that have persisted for a time at secretory terminals without being released are supposedly "selected" for breakdown because such older granules eventually move (are displaced?) away from the sites of secretory release into nearby lysosome-rich regions of the axons (e.g., the axonal swellings known as "Herring bodies").] Other types of neurons also autophagocytose some of their synaptic vesicles (see Section 5.3.3.1).

5.3.1.2. Crinophagy

This process differs from classical autophagy in that, in crinophagy, secretion granules **fuse** with lysosomes or prelysosomal structures (Fig. 5.15) rather than undergoing engulfment. In much of the literature on intracellular degradation of secretions, this distinction is ignored or blurred.

The hallmark of crinophagic configurations is the loss of the secretion granule's own delimiting membrane as it enters the lysosome. To be sure, this criterion must be applied thoughtfully, given the possibility that granules autophagocytosed by the classical routes, may lose their membranes rapidly by degradation within the lysosomes.

Crinophagy was first described in detail for several cell types of the mammalian pituitary subjected to physiological manipulations that result in precipitate reduction of the rates at which secretions are released. For a while, the cells overproduce secretion granules, and the excess is disposed of by fusions with lysosomes. Crinophagic configurations were subsequently observed in other circumstances and other endocrine glands, such as in the endocrine pancreas under conditions involving altered rates of insulin secretion. Generalization is difficult because few glands and conditions have been investigated in detail. The process clearly is intensified under conditions of imbalance; under these conditions retargeting to the lysosomes, of secretion granules that would ordinarily be exocytosed, seems to take place. But crinophagy is also observed under apparently normal circumstances, such as in the adrenal medulla of rats subjected only to routine laboratory handling.

Both in the pituitary and in the adrenal medulla, the secretion granules disposed of by crinophagy often have "immature" morphologies. In the medulla, for example, the granules frequently look as if they have not yet fully "condensed" into the moderately electron-dense forms they normally assume (Fig. 5.15 insert). In the pituitary as well, many of the granules seem not yet to have undergone their final aggregation and condensation. On the face of it, then, crinophagy seems biased toward newly packaged granules. It has yet to be

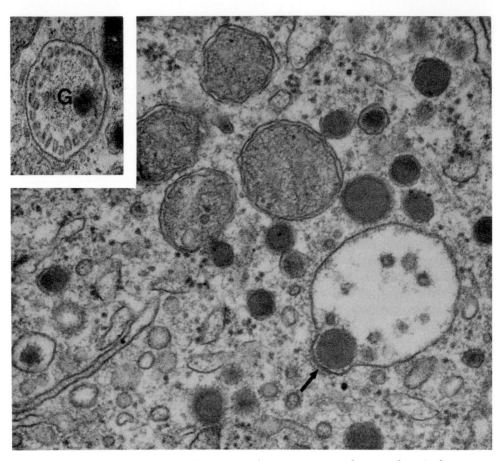

FIG. 5.15. Cell (gonadotropic) from the pituitary of an estrogen-treated castrated rat. At the arrow, a secretion granule has apparently just fused with a multivesicular structure. Insert: Structure with the morphology of an immature secretion granule (G) within an MVB in a cell of the rat adrenal medulla. Main picture from Farquhar, M. G. (1971) *Mem. Soc. Endocrinol.* **19**:79. × 60,000. (Copyright: Elsevier, Ireland.) Insert from Holtzman, E., and Dominitz, R. (1968) *J. Histochem. Cytochem.* **16**:320. × 75,000.

ruled out, however, that when older granules are taken up, they become modified so as to resemble younger ones: When it first enters a lysosome, the secretory material remains aggregated into a mass like that in the secretion granules (Fig. 5.15) and the hydrolases probably disperse this mass only gradually, chewing away first at the periphery; perhaps in consequence, as the granule is digested it "decondenses" slowly, producing appearances as in the insert to Fig. 5.15. A related explanation has been put forth for immunocytochemical demonstrations that early crinophagic configurations in the endocrine pancreas, though containing detectable insulin, largely lack the peptides that become excised from insulin molecules as they mature into functional hormones (see Section 6.1.5.1); the peptides might be particularly vulnerable to degradation because they remain at the margins of the dense aggregate of hormone molecules that forms within the secretion granules.

The lysosome-related structures involved most often in crinophagy are of the multivesicular variety. This might be simply because crinophagy is centered in the Golgi region, where secretion granules form and MVBs are numerous. Or perhaps the MVBs "fool" the secretion granules because, being endocytotic depots, the MVBs contain plasma membrane components of the sorts with which the secretion granules ordinarily interact during exocytosis. For example, secretion granules might actually fuse with early endosomes at stages when the endosomal membrane still carries an appreciable load of plasma membrane proteins. (When the endosomes then develop an MVB morphology, they might generate the internal MVB vesicles in part from the membrane added to the endosomes' surfaces by the fusion with secretory granules.)

Most morphological observations on the degradation of secretions come from systems involved in "regulated" secretion—cells that form distinctive storage structures for secretions and release the secretions intermittently, upon appropriate stimulus. There is much current attention to possible differences between such secretion, and "constitutive" secretory processes in which release is more continual. Whether the vehicles that carry constitutive secretions to the plasma membrane are also subject to crinophagy is difficult to judge because the vehicles seem often to be (coated?) vesicles without unique content or other distinguishing morphological features, rather than the distinctive-looking electron-dense structures used by gland cells to store and carry regulated secretions. Additionally, though microscopic configurations suggestive of crinophagy have been noted in exocrine glands, much less is known about the extent of crinophagic-like processes in such glands than for endocrine tissues. Cells of the exocrine pancreas and of lacrimal glands increase their incorporation of secretion granules into lysosomes when secretion is perturbed by colchicine or other drugs but more work is needed to determine whether this is a crinophagic or an autophagic response. In the case of liver, vinblastine-induced increases in the degradation of lipids have been interpreted as due to increased crinophagy of lipoproteins. However, microscopic identification of possible crinophagic structures containing lipoprotein particles in hepatocytes is not easy because hepatocytes can endocytose lipoproteins into MVBs and other lysosome-related structures, and the cells also package lipoproteins for secretion via Golgi-associated systems that can somewhat resemble degradative bodies morphologically and cytochemically (Section 6.1.5).

Pigment granules are sometimes seen within lysosomes in cells that produce or store melanin, and in retinular cells of invertebrate retinas. Consequently, it has been proposed that crinophagy (or classical autophagy) helps regulate intracellular levels of such granules. Here again it is often difficult to distinguish among crinophagic, autophagic, and heterophagic structures in the cells in question (Section 4.3.1).

5.3.2. Retrieval of the Membranes of Secretory Structures

During exocytotic release of secretions, the membranes bounding secretory structures become incorporated in the cell surface but this does not lead to long-term changes in cell surface area. A few secretory cell types—goblet cells that have released large quantities of their mucus stores rapidly, for example—shed much of the "excess" membrane to the extracellular space. More usually, membrane is withdrawn back into the cell, in the form of vesicles and tubules. Because these vesicles and tubules arise by budding from the cell surface and can be labeled with endocytotic tracers, their formation is frequently treated as a variety of endocytosis. But as far as is known, they are devoted mainly to retrieving membrane and do not have particular interest in acquiring materials from the extracellular space.

Certain cell types can change the transport or permeability properties of their plasma membrane reversibly by cycles of exocytotic insertion and endocytotic removal of membrane. The toad bladder is thought to use this device to adjust permeability properties of its epithelial lining, adding membrane that is permeable to water when stimulated to do so by antidiuretic hormones, and removing the membrane, endocytotically, to restore the permeability barrier. The cycling of proton pumps in the kidney and turtle bladder was outlined in Section 3.1.5.5.

5.3.2.1. Selective Retrieval

The withdrawal of membrane from the cell surface after secretion, is conceived of as a selective retrieval of components added during exocytosis rather than as the compensatory internalization of an equivalent area of preexisting plasma membrane. This explains how the plasma membrane of secretory cells remains different in its composition and other properties from the membrane of the secretory bodies.* Not infrequently, secretory cells fixed for microscopy during periods of active secretion exhibit profiles in which vesicles appear as if budding back into the cytoplasm directly from the exocytotic figures (Fig. 5.16). And for several glands, patterns of IMPs or of immunocytochemically demonstrable antigens indicate that the "patch" of membrane added to the cell surface during exocytosis of a secretion granule, retains its integrity as a discrete "microdomain" until it is retrieved into the cell. Perhaps dispersal of membrane constituents after exocytosis, is retarded by lateral interactions among the molecules in the membranes of secretory bodies, or by influences of the cytoskeleton, or by yet-to-be-identified barriers at the points where the plasma membrane and the secretory body membrane are confluent.

Nerve terminals were among the earliest material in which membrane recycling during secretion was convincingly demonstrated. Such cycling was suspected early because the abundance of the small "synaptic vesicles" that are thought to mediate transmitter release from the terminals changes little even under conditions where massive amounts of transmitter are released rapidly. Some groups envisage a very rapid retrieval of the vesicles after exocytosis, so that the membrane of a synaptic vesicle that has just released neurotransmitter is retrieved intact having spent less than a second in the cell surface. The more widely accepted models for membrane retrieval in nerve terminals, especially during and after prolonged periods of intense neurotransmission, propose that the components added to the cell surface by exocytotic fusions of the synaptic vesicles move considerable distances in the plane of the plasma membrane, to be retrieved at sites distant from the sites of transmitter release. If so, regroup-

---

*These compositional differences are usually reported from comparisons of secretory bodies with the plasma membrane as a whole. Data are sparse as to the composition of the local domains of the plasma membrane where secretions are released, and where the recycling being considered takes place. Furthermore, most detailed studies have been on "regulated" secretion (Section 5.3.1.2) though membrane retrieval is known to be extensive also in cells whose secretion is largely "constitutive," such as immunocytes releasing immunoglobulins. And in light of the discussions below on involvements of the Golgi apparatus and of lysosomes in membrane cycling, it will be important to determine whether cells that are both retrieving membrane after exocytosis and engaging in other forms of endocytosis, move the membranes internalized in both processes along the same intracellular pathways. Is it relevant that some cell types seem to focus conventional endocytosis at the opposite cell pole from secretion-related recycling?

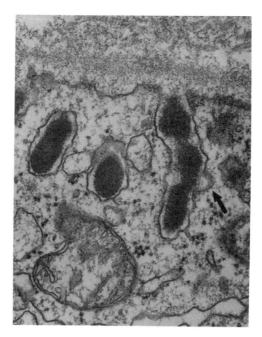

FIG. 5.16. Edge of a prolactin cell from rat pituitary showing an exocytotic configuration (a secretion granule that has fused with the plasma membrane); at the arrow, a coated vesicle seems to be forming from the membrane delimiting this configuration. Courtesy of M. G. Farquhar. × 23,000.

ing devices might come into play to collect membrane constituents selectively into appropriate clusters for retrieval.

Scattered evidence (some of it contradictory) suggests that membrane retrieval after secretion is energy dependent and requires divalent cations. "Coated" structures resembling those involved in receptor-mediated endocytosis frequently are observed at sites of retrieval, both in gland cells and in nerve terminals. But there is an ongoing ferment over whether a clathrin coat is obligatory for retrieval. In many secretory cell types, especially during intense secretory activity, retrieval seems to be carried out partly by vacuoles or tubules larger than coated vesicles. On the other hand, from looking, with quick-freeze and freeze-fracture technology, at IMP distributions during neurotransmission, some microscopists concluded that the coated vesicles of nerve terminals not only internalize much of the recycling synaptic vesicle membrane but also can even gather together synaptic vesicle components that have dispersed somewhat in the plane of the plasma membrane during migration away from the sites of exocytosis.

Be mindful that a small but insistent minority of investigators doubts that the synaptic vesicles of nerve terminals are important for neurotransmission, proposing instead that most transmitter molecules pass out of the terminals through gated channels. To me, their case against direct vesicle involvement is vastly weaker than the case for such involvement, but in any event there can be little doubt that synaptic vesicles do cycle by coupled exocytosis and endocytosis whatever the significance of the cycle may be.

5.3.3. Fate of Retrieved Membrane

Gland cells and nerve terminals possess appreciable populations of MVBs, and when endocytotic tracers such as HRP are internalized during secretion-linked membrane retrieval, the tracers soon become evident in the MVBs. Does this mean that membrane retrieved after secretion is scheduled for degradation by MVB-mediated autophagy? To some extent the answer is yes, but most of the retrieved membrane seems instead to be preserved and reutilized to package secretions.

5.3.3.1. Local Reuse in Nerve Terminals

Lysosomes are relatively scarce in axons and presynaptic terminals of neurons in vertebrates, so the fact that materials endocytosed at terminals often end up in lysosomes in the cell body usually is attributed to transport of MVBs and other "prelysosomal," endocytotic compartments up the axon to the cell body, where fusions with lysosomes take place. (This probably is an oversimplified account; acid hydrolases and lysosomes are not entirely absent from axons.) In frog retinal photoreceptors, the MVBs show the presence of immunocytochemically demonstrable membrane antigens characteristic of the cells' synaptic vesicles, indicating that the MVBs probably do help degrade membranes of the synaptic vesicles. But the frequency of MVBs in nerve terminals is low and is not markedly responsive to altered physiology; the bodies are likely therefore to be more important for the relatively slow steady-state turnover of synaptic components than as part of the secretory cycle.

That synaptic vesicle membrane retrieved by nerve terminals is reused extensively to reconstruct synaptic vesicles, was postulated from calculations showing that nerve terminals sustaining prolonged bouts of transmission would require implausible rates of supply of fresh membrane components were each vesicle to be used only once and then destroyed. These calculations were buttressed by demonstrations in my laboratory and others that endocytotic tracers such as HRP, taken up by actively secreting terminals, accumulate in vesicles appearing identical to most other synaptic vesicles. The HRP can then be depleted from the terminals by stimulating renewed release of transmitter; the tracer apparently is exocytosed along with other contents of the vesicles.

In the systems studied most in these respects, the neurotransmitters are small molecules—acetylcholine, catecholamines, and the like—that can be synthesized, or reaccumulated after release, by machinery in the nerve terminals. Transport systems and enzymes in the membranes of the synaptic vesicles participate actively in storing or synthesizing the transmitters as well as some other components the vesicles contain, especially inorganic ions and nucleotides.* To reconstitute functional vesicles from retrieved membrane, the cell's recycling process thus must have sufficient fidelity to provide the vesicles with rather specific necessities. It will be important to determine the minimum requirements for a functional vesicle and thereby to find out how permissive recycling can be.

Nerve terminals stimulated to extremes of transmitter release, or terminals in which membrane retrieval has been perturbed by exposure to low temperature, accumulate large, membrane-delimited structures ("cisternae") aris-

---

*Neurosecretory neurons, whose principal secretions are peptides and proteins that cannot be synthesized locally in nerve terminals, retrieve their vesicles' membranes endocytotically, but seem not to reutilize them directly, at least not within the secretory terminals. Several other types of neurons package large molecules such as proteins or glycosaminoglycans in synaptic vesicles, along with smaller neurotransmitters. Some of these large molecules are released during neurotransmission but others are claimed to recycle (through mechanisms that have yet to be analyzed).

ing from endocytosed membrane. These structures are held to be intermediates in recycling because they can subsequently generate synaptic vesicles, probably by budding. The cisternae vary in appearance in different preparations but often have the form of vacuoles, sacs, and tubules, calling to mind the endosomes involved in receptor-mediated endocytosis. There is no direct evidence whether the cisternae are acidified (some synaptic vesicles are) or whether they play membrane-sorting roles analogous to those of endosomes (Section 3.3.4.4). (With frog photoreceptors, however, my laboratory finds that ammonium chloride induces accumulation of cisternae as might be expected were they acidified and endosome-like.) Very little is known about the molecules the cisternae or the synaptic vesicles normally acquire from the extracellular space during membrane retrieval in nerve terminals. Are any of these molecules ligands for specific receptors in the cycling membranes? (Nerve terminals do take up toxins and growth factors, apparently by receptor-mediated endocytosis, but the relations of this uptake to recycling of synaptic vesicles are uncertain.) Can the presence of exogenous molecules in vesicles arising from recycled membrane affect the physiology of the terminals?

Temperature-sensitive mutants recently identified in *Drosophila*, are defective in membrane recycling in nerve terminals and other secretory cells. They should prove very useful in analyses of the processes just discussed. Provision of adequate Ca^{2+} in the medium can overcome the effects of the mutations.

5.3.3.2. Membrane Reuse in Gland Cells: Golgi Involvement; Lysosomal Involvement

That gland cells reuse the same membrane components repeatedly for packaging and release of secretion was suspected from findings that the proteins of the membranes bounding secretory granules have average life spans much longer than would be expected were they to be used only for one round of secretion. However, in the initial microscopic studies of the membrane retrieval by gland cells, actively secreting gland cells never showed passage of presumed fluid phase tracers such as HRP into the ER and only rarely did the tracers reach the Golgi apparatus or nascent secretion granules. Rather, the tracers first were endocytosed into vesicles, tubules, and small vacuoles and later were found in lysosomes. In other words, the vehicles of endocytotic retrieval seemed to be fusing with degradative organelles rather than with the compartments that could reuse membrane directly. Did this imply that membrane components are actually not reused or perhaps that reuse depends on the dismantling of the membranes into their constituent molecules, which only then become available for packaging secretions?

Such speculation largely died away with the utilization of other endocytotic tracers, including dextrans and especially tracers that adhere strongly to membranes: cationic forms of ferritin (which bind to membrane anions; Section 2.2.2.3) or HRP–cholera toxin conjugates (which bind to gangliosides). Secretory cells retrieving membranes after secretion, do accumulate these tracers in sacs associated with the Golgi apparatus (Fig. 5.17) and in secretion granules as well as in lysosomes and "prelysosomes." Endocytosed cationic ferritin, and some dextrans, can enter most or all of the sacs of a given Golgi stack in an active gland cell, suggesting that membrane recycling can distribute components throughout the Golgi apparatus (see Section 7.2). Recycling of membrane

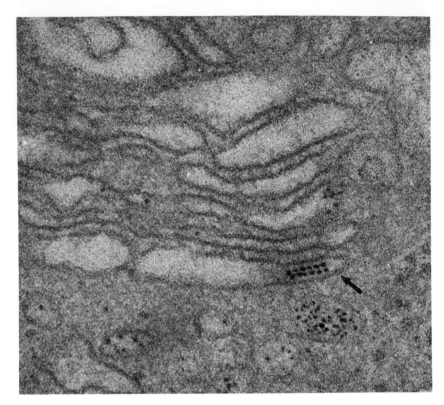

FIG. 5.17. Golgi region of a thyroid cell from a preparation of "inside-out" pig thyroid follicles 30 min after administration of cationic ferritin. Ferritin grains are evident in a Golgi sac (arrow). From Herzog, V. (1981) *Trends Biochem. Sci.* **6**:319. × 150,000 (approx.).

directly to the ER has still not, however, been demonstrated convincingly with any endocytotic tracer and thus presumably does not take place.

How should the difference in fate between the different tracers be interpreted? An appealing possibility is that after reentering the cell, the endocytotic structures bearing retrieved membrane go first to lysosomal or prelysosomal compartments where "undesired" contents, such as material from the medium surrounding the cells, are dumped; perhaps the membrane itself also undergoes sorting, to select components for reuse. From these lysosomal way stations the membrane would then progress to the Golgi apparatus and fuse with newly forming secretory bodies or with compartments that generate such bodies.

There is still too little evidence with which to choose between this idea and the alternate possibility that multiple routes coexist from the cell surface to lysosomes or to the Golgi apparatus, with different routes being favored by different tracers. All that is clear is that cells can send materials of a given endocytotic compartment to divergent fates—witness the sorting of endocytosed materials during transcytosis (Section 4.3.2) and the sorting of ligands and receptors after receptor-mediated endocytosis. Similarly, in *Paramecium*, membranes retrieved at the cytoproct after lysosomal defecation, recycle to the oral region for reuse in feeding, whereas membranes retrieved at sites where trichocysts are released move elsewhere; these two types of membranes differ in microscopic appearance and seemingly can be recognized as different by cellular transport systems.

If the recycling membrane does visit more than one depot before being reused, it must be able to make the circuit rapidly; when radioactively labeled amylase, a cationic, membrane-adherent pancreatic secretory protein, is taken up during secretion by the pancreas, it can reach the Golgi apparatus within a very few minutes after its uptake.

The thyroid gland focuses questions about recycling with particular directness (see Fig. 4.7): Thyroglobulin is both secreted and endocytosed at the same cell pole. The endocytosed protein is transported to the lysosomes in vesicles formed at that pole, but membranes retrieved along this same cell surface during secretion of the protein for storage, recycle to the Golgi apparatus. Do the thyroid cells use the same membrane components for both routes or do they have a subtler sorting system? Is the anionic character of thyroglobulin, or some more specific tag (e.g., the M6P groups mentioned in Section 4.2.4.1) responsible for targeting the vehicles carrying endocytosed thyroglobulin to the lysosomes?

As with endocytotic receptors, a proportion of the membrane that recycles after secretion may find itself trapped and degraded within lysosomes by accidents of rate, such as the budding off of a vesicle into a nascent MVB that had recently received retrieved membrane at its surface and had not yet had time to send all of this membrane on its way elsewhere. Degradation of retrieved membrane can be made intense; my laboratory has found, for example, that when adrenal medullary cells are stimulated to extremes of release of their epinephrine, the cells subsequently incorporate much retrieved membrane into lysosomal MVBs and apparently degrade it. Perhaps more modest phenomena of a similar sort account for normal turnover of secretion granule membranes (see Section 5.5.4.3).

Little attention has been paid to the behavior of membrane lipids during cycling in secretory cells. Behavior different from the proteins might be expected given that individual lipid molecules can move from one membrane to another through the agency of exchange proteins, that the lipids can diffuse rapidly in the plane of the membrane, and in view of the findings on other types of cycling (e.g., Sections 4.6.2.1 and 5.5.4).

5.3.3.3. More on Golgi Involvement in Membrane Cycling

The observations on secretory cells considered in the last few sections foreshadowed the discovery of extensive endocytotic traffic between the cell surface and the sacs and tubules in the Golgi region, in many cell types. Now, for instance, even certain of the endocytotic receptors such as those for transferrin are known to pass into Golgi-associated systems (see Sections 4.1.2.2 and 7.2).

Researchers into this traffic are presently embroiled over the functional and structural boundaries and interactions between the Golgi apparatus and the lysosome-related structures that accumulate nearby. For example, several cultured cell types can reattach sialic acids to recycling ligands like transferrin and to plasma membrane proteins, when sites suitable for such addition are created experimentally by enzymatic removal of preexisting sialic acids (or by adding galactoses). Such "resialylation" would be simply explained if, during their transit through the cell, recycling proteins visit the sacs of the Golgi system that are responsible for the addition of sialic acids to oligosaccharides (see Chapter 7). These events could permit the cell to "repair" glycoproteins that have been shorn of portions of their oligosaccharides during their exposure to the extracellular world. (The ASGP receptor, for example, functions abnormally if its oligosaccharides are desialylated or otherwise altered.) But resialylation rates for proteins cycling via endocytosis are slow compared to rates of recycling. One proposed explanation is that although the proteins follow much the same itinerary each time they pass through the cell (e.g., endosome or lysosome to Golgi apparatus to cell surface), the attachment machinery for sialic acid is so inefficient or always so near saturation that many molecules cycle past it without being affected. Alternatively, a given protein molecule cycling repeatedly in and out of the cell might traverse the path into and out of the relevant Golgi structures only occasionally, most often using some other route that does not expose it to the resialylation enzymes. I will return to these muddy waters in Section 7.2.

Membrane traffic to and within the Golgi region is not limited to endocytotic and exocytotic structures. Membrane-bounded vehicles carry proteins from the ER to the Golgi sacs and from the Golgi apparatus to nonsecretory structures, including lysosomes. Transport from one sac of the apparatus to another may depend on movements of vesicles. And the maturation of secretion granules includes considerable increases in concentration of their contents, generating an "excess" of delimiting membrane to be removed, presumably by the budding off of vesicles or tubules from the surface of the secretory body. Most of the traffic in the Golgi region is currently conceptualized in terms of two-way vesicular shuttles. It may, however, also be that the numerous lysosome-related structures in the Golgi region capture some of the membrane, autophagically, and destroy it. There even are cell types, such as in the fat body of mosquitoes at the time when the cells cease massive secretion of yolk proteins, in which processes resembling classical autophagy seemingly come into play to reduce the extent of the Golgi apparatus itself.

5.4. Digestion of Stored Materials

Storage and digestion of yolk in animals is largely through modified heterophagic mechanisms (Section 4.2.2). Other stores are of endogenous provenance, and autophagy is involved in their digestion.

5.4.1. Storage Polymers in the Vacuoles of Plants and Fungi

Plant cells and fungi utilize their vacuoles for storage of materials that can be hydrolyzed to mobilize nutrients for development or survival, and to alter osmotic balances. The large central vacuoles of plant tubers, for example, are rich in carbohydrate polymers [e.g., stachyose (galactose-galactose-sucrose; Table 5.3) or fructans (chains of fructoses with sucrose at one end)] whose constituent sugars are utilized during sprouting. Acid hydrolases capable of degrading such polymers have been localized to the vacuoles of tubers and increases in the activities of these enzymes precede or accompany mobilization of the stores.* Even sucrose may sometimes be digested in vacuoles, by acid invertases.

---

*Vacuole-containing cell fractions are somewhat enriched in certain of the enzymes for synthesizing the storage polymers as well, though others of these enzymes behave, in cell fractionation, as if localized in the cytoplasm outside the vacuoles. The processes of storage have not been described fully; nor has the governance of the balances between storage and degradation.

TABLE 5.3. Distribution of Major Saccharides and of Certain Enzymes between the Vacuoles and the Remainder of the Cytoplasm in Preparations of Tubers of the Plant *Stachys sieboldii*[a,b]

Saccharide or enzyme	Content or activity in protoplast lysate (μg or mU/ml)	% recovery in density gradient (protoplast lysate = 100%)	% in vacuoles
Fructose	20.3	69	63
Galactose	38.6	94	71
Glucose	12.5	66	51
Myoinositol	7.27	82	65
Sucrose	88.2	103	30
Galactinol	17.6	94	17
Raffinose	36.4	101	67
Stachyose	926	97	100
α-Galactosidase	62.2	86	99
Galactokinase	165	86	6

[a]See Fig. 1.4 for fractionation of the protoplasts. From Keller, F., and Matile, P. (1985) *J. Plant Physiol.* **119**:369. (Copyright: Gustav Fischer-Verlag.)
[b]The storage polymer stachyose is seen to be localized entirely within the vacuoles along with α-glactosidase, one of the enzymes involved in its digestion.

Neurospora stores polyphosphate compounds in its vacuoles (see Section 3.4.2.1) and can digest them, enzymatically, to provide phosphate when exogenous supplies are not available.

5.4.2. Protein Bodies of Seeds

The cotyledons of seeds in legumes (peas and beans) and many other plants, possess protein bodies (aliases: "aleurone grains" or "protein storage vacuoles"). These are electron-dense cytoplasmic structures (Fig. 5.18), bounded by a single membrane and containing concentrated stores of proteins and of "phytic acid" (salts of myoinositolhexaphosphoric acid with cations like Ca^{2+}). The stores serve as reserves for the development of seedlings, analogous in some ways to yolk, but differing from yolk in that endocytosis is not known to play a significant part in their accumulation.

The proteins in protein bodies include familiar lectins (e.g., concanavalin A or phytohemagglutinin) along with abundant "storage proteins." Many, though not all, are glycoproteins that are synthesized in the ER and glycosylated by the ER and Golgi apparatus (see Sections 7.1.1, 7.3.3.3., and 7.4) before delivery to the forming protein bodies. Subsequent to this delivery, additional modifications often take place, including limited proteolytic cleavages (Fig. 5.18); removal of terminal N-acetylglucosamines or other groups from the oligosaccharides; and more complex molecular rearrangements including oligomerization (and for concanavalin A, perhaps the ligation of polypeptides).

Mature protein bodies generally are spherical to ovoid with diameters ranging from 0.1 μm to 20 μm in different species of plants (diameters in the plants most studied, tend to lie between one and a few micrometers). The routes by which the bodies form vary somewhat among plant species and during development. Many of the bodies, however, are known to arise directly from large central vacuoles. In peas, for example, where over 10^5 protein bodies form per cotyledon cell, the central vacuole gradually fills with the electron-dense content characteristic of the protein bodies. As it fills, the vacuole contorts and fragments into separate small structures. A cardinal requirement for this process is the provision of membrane, for at the outset, the vacuole is delimited by a tonoplast with an area of a few thousand square micrometers, whereas the aggregate surface area of the protein bodies that eventually form is hundreds of times greater. Presumably, the Golgi apparatus provides this membrane along with major components of the stored materials.

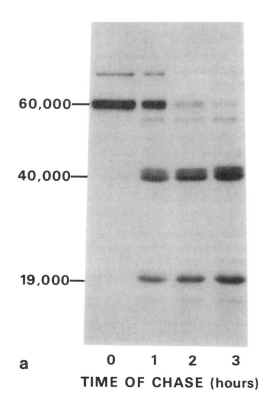

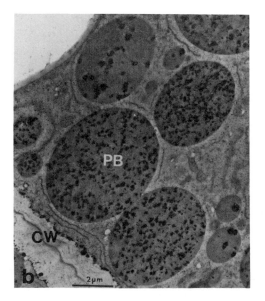

FIG. 5.18. (a) Fluorographs of SDS–PAGE gels showing forms of the storage protein **legumin** isolated by antibody–Sepharose columns, from protein bodies of cotyledons of developing pea seeds (*Pisum sativum*) at intervals after a 1-hr pulse of radioactive amino acids. The protein arrives in the bodies chiefly in the form of a 60-kD polypeptide that is processed, proteolytically, into smaller polypeptides over the ensuing few hours. From Chrispeels, M. J., Higgins, T. J. V., and Spencer, D. (1982) *J. Cell Biol.* **93:**306.

(b) Protein bodies in a cotyledon cell of a 3-day-old mung bean (*Vigna*) seedling showing the presence of acid phosphatase, demonstrated cytochemically. The reaction product for acid phosphatase seen at the cell surface [between the plasma membrane and the cell wall (CW)] may originate from extracellular enzyme activity. From Herman, E. M., Baumgartner, B., and Chrispeels, M. J. (1981) *Eur. J. Cell Biol.* **24:**226.

5.4.2.1. Digestion of the Contents of Protein Bodies

Mature protein bodies contain acid hydrolases, including phosphatases, acid glycosidases (e.g., mannosidase and β-N-acetylhexosaminidase), and some proteases. These enzymes may help in the final processing of materials before storage, but they seem incapable of extensive degradation. As they do not digest the stored protein, the proteinases appear to be only weakly active or to have quite restricted specificities. The progressive dehydration of the seed as it matures would also be expected to curtail enzymatic activities in the protein bodies.

In contrast, after the seeds imbibe water and start to germinate, the contents of the protein bodies are hydrolyzed relatively rapidly: active proteases, phytases, RNase, and even phospholipases are among the enzymes prominent in the bodies during this digestion (Table 1.2 and Fig. 5.18). Some of these enzymes are ones that were present in the maturing protein bodies, but additional enzymes, including potent "acid endopeptidases" and RNase, are added to the protein bodies after germination begins. The mechanisms of addition are not known, but are assumed to include fusion of hydrolase-transport vehicles with the protein bodies.

As the contents of the protein bodies become depleted, the bodies can fuse with one another. For cells of castor bean cotyledons, large, central vacuoles form in this way to participate in the subsequent life of the cells.*

Storage and degradation in seeds of barley and some other cereals is more an autolytic process than the autophagic phenomenon described here (see Sections 6.1.4.2 and 6.3.2). However, recent speculation about barley protein bodies suggests that the bodies merge their membranes with lipid droplets, thereby providing the lipases that degrade the lipid.

5.5. Intracellular Turnover of Proteins; Related Processes†

Section 4.5 surveyed the extensive involvement of heterophagic lysosomes in normal turnover, and in the degradation of extracellular materials attendant upon physiological or pathological change (see especially the background discussion in Section 4.5.2.2). To what extent are autophagic lysosomes similarly responsible for the capacities of cells to degrade their own macromolecules and organelles under steady-state conditions and in responses to metabolic stress or disease? For a while, soon after autophagy was discovered and when lysosomes were the only well-known sites of enzymes of intracellular digestion, it ap-

---

*Protein bodies that have degraded a portion of their content sometimes exhibit configurations like the ones considered in Section 5.1.7.1, as indicating autophagic uptake of cytoplasm by plant vacuoles. Perhaps autophagy of this sort is stepped up in embryonic cells as a mechanism by which the cells transform depleted protein bodies into large long-lived vacuoles (see Section 5.1.6.1). The proteases, RNase, and phospholipases in the protein bodies might participate in disposing of cytoplasmic materials entering the growing vacuoles.

†The term **turnover** is often used with a connotation of balance between synthesis and degradation, as in the steady-state replacement of molecules or organelles. But a looser usage is increasingly common, so that, for example, cells that increase their degradation of proteins even without compensatory increases in synthesis are often spoken of as showing enhanced turnover of proteins. It is impossible to be a semantic purist in discussing the turnover literature but in the following sections, where it is important, turnover phenomena taking place in the steady state will be explicitly identified as such.

peared that these questions had simple answers. Now, though it still seems that lysosomes make major contributions to degradation of endogenous macromolecules, other systems are known to participate as well; the relative importance of lysosomal and nonlysosomal mechanisms varies for different components that are turning over and for different cell types and physiological or pathological states.

The breakdown of proteins ("intracellular proteolysis") dominates discussion of turnover because much less extensive attention has been paid to degradation of nucleic acids, carbohydrate polymers, lipids, or other molecules. It does appear, for example, that when a rat is starved, its liver loses lipids and RNA as well as proteins. But the imbalance of attention in favor of proteins has left large gaps in the understanding of how the cell normally turns over its membrane systems and other organelles, and has also hindered analysis of the cellular etiology of the numerous lysosome-related diseases in which abnormal quantities of lipids, glycosaminoglycans, or related compounds accumulate (see Section 6.4).

5.5.1. Methodology

The operational compromises adopted to study turnover have been refined over time as more of the pitfalls have become evident.

5.5.1.1. Degradative Rates Vary with the State of the Cell

Which circumstances are to be taken as "normal" and what is the time period appropriate for deciding whether a cell or tissue is in a steady state? In rodent liver, the biochemists' favorite organ, the degradative rates of hepatocyte proteins alter dramatically under different conditions (Table 5.4). In starvation, degradation of some liver proteins is enhanced, as least for the first few days, providing amino acids for the hepatocyte's own use and for circulation to other tissues. During rapid growth, as, for example, in liver regeneration, degradation is substantially suppressed, which accounts for much of the rapid restoration of the regenerating liver's mass. The pace and nature of protein breakdown in hepatocytes change under more ordinary conditions as well, as the cells react to levels of circulating amino acids, to hormones such as insulin, and to other

TABLE 5.4. Protein Turnover in Mouse Liver under Steady-State Conditions (Adult) and in Several Circumstances of Net Accumulation of Proteins[a]

	Adult	Rapid liver growth		
		Regenerating	Refed[b]	Newborn
Total synthesis	66	76	78	92
Stable proteins				
Synthesis	32	30	45	36
Accumulation	0	20	48	18
Degradation	32	10	0	18

[a]From Scornik, O. A. (1984) Fed. Proc. **43**:1283. Data are in mg protein/day per 100 mg liver protein. Stable proteins are those that survive in the liver for at least many hours.
[b]Recovering from starvation.

influences expected to fluctuate during a normal day, depending on feeding schedules, circadian cycles, and the like.

Rats trained to eat on a fixed schedule show variations of as much as fivefold in the tempos of degradation of their liver proteins in the course of a day (see also Fig. 5.8). After a rich meal, amino acids from the food come to the liver through the bloodstream and are stored as labile proteins; later, amino acids are released to the circulation from the "reservoir" represented by these proteins. In other words, though the cells maintain an overall, daily steady-state balance, at different times during the day they show net increases or decreases in protein content. In this, and in many other situations, the time courses of changes in protein synthesis differ from those of degradation; often, changes of the synthetic rates are less marked than those of breakdown.

Tissue culture cells growing rapidly show lower total rates of internal protein degradation than do the same cells when, for example, they reach confluence and net growth slows or stops. Depriving tissue culture cells of serum hormones and growth factors, or of amino acids, steps up the rates of protein degradation by twofold or more in many cases. There are, however, cultured cell types that seem to respond to such changes more by altering rates of protein synthesis than by modulating degradative rates.

Both *in situ* and in culture, changes in rates of protein degradation often prove to be transient on a time scale of hours or days, even if the conditions that initially induced the changes persist. This probably reflects the organism's or the cell's ability to restore levels of key molecules by, for example, degrading muscle proteins so as to maintain circulating pools of amino acids (Section 5.5.5). Factors of this sort, along with subtle differences in experimental design, account for many of the seemingly contradictory findings that dot the literature on turnover.

5.5.1.2. Methods for Estimating Degradative Rates

With enzymes, the rate of decline in cellular enzymatic activity seen when protein synthesis is inhibited has been taken as a measure of the rates of breakdown of the enzymes. This approach has obvious drawbacks: Conditions are hardly normal (e.g., hydrolase synthesis is blocked and some cell cultures fail to exhibit relevant responses such as the increases in degradation often inducible by withdrawing serum). In addition, enzymes can be inactivated without necessarily being degraded. Nevertheless, there are circumstances under which this type of method is the only one available. Similarly, rates of turnover occasionally must be inferred from measuring rates of synthesis under conditions believed to be steady-state.

Most often, estimates of degradative rates for cytoplasmic proteins come from incorporating radioactive labels into cells and determining the time course with which these labels subsequently disappear from the cells, or from subcellular protein populations of interest. Sometimes, as with peroxidase-mediated iodination of cell surface proteins (Section 2.2.2.3), the labels are attached to preexisting proteins, but generally, labeling is by exposure of the cells to radioactive amino acids.* Care must be taken to discriminate between

---

*Only rarely, in such studies, has much information been gathered about the rates of entry of the labels into pertinent cellular precursor pools, such as the pool of tRNAs charged with amino acids, or about the duration of labeling of the pools. This information would be helpful for

the decline of label due to degradation and losses due, for example, to secretion. Hepatocytes secrete as much as 25–45% of the proteins they produce and many cells whose functions are not principally secretory nonetheless do release some proteins or other polymers.

The particular label used also is important, both because proteins differ in amino acid composition and because different labels can generate different artifacts. Particularly vexing has been **reutilization**—the reincorporation into new proteins, of radioactive amino acids released in the cell by degradation of the proteins labeled during the original exposure to radioactive precursors. By slowing the net rate of loss of label, this effect lengthens the apparent life spans of proteins. Leucine, the amino acid most frequently employed for turnover studies, is extensively reutilized; in some research this has led to overestimates of protein life spans of 1.5-fold or even much more.

Reutilization problems are minimized in some cell cultures by flooding the cells with an excess of unlabeled amino acid, after the pulse of label. For studies on hepatocytes, arginine labeled in its guanidino group can be employed as the radioactive precursor; the high levels of arginase in liver militate against reuse of this label. (*In situ*, however, the liver reutilizes arginine released to the circulation from other tissues with low arginase activity.) The guanidino groups of arginine also can be labeled in liver by exposure to labeled carbonate or bicarbonate compounds (Fig. 5.19 and Table 5.5), which enter the urea cycle.† Tritiated water has occasionally been used as a metabolic source of label for amino acids because it can be extensively incorporated during transaminations and seems not to be extensively reutilized. Iodinated tyrosines released by digestion of iodinated proteins generally are not reused by cells (though some cell types reputedly deiodinate proteins to an extent); nor seemingly are amino acids labeled by reductive methylation.

"Double label" procedures make use of two distinguishable forms of a given amino acid, most often leucine labeled with ^{14}C and leucine labeled with

---

thorough analysis of pulse–chase results, for interpreting observations under non-steady-state conditions, and for work on proteins that turn over especially rapidly. For example, though most steady-state turnover of endogenous proteins is interpreted in terms of exponential kinetics (see footnote on p. 218), the actual data often are more complex, particularly for time points soon after administration of label.

Also important is that many studies monitor degradation simply by determining rates at which radioactive labels are converted into TCA-soluble forms—no effort is made to distinguish between digestion to the level of small peptides and the final liberation of amino acids. Reassuringly, large intermediates in the intracellular degradation of proteins have proved hard to find when looked for, suggesting that breakdown rapidly goes to near completion. Peptidase inhibitors like bestatin are occasionally employed to accumulate the smaller types of intermediates and thereby to facilitate searches for the sites where various steps in breakdown are located. But nothing particularly surprising has emerged thus far, except for the possibility that some lysosomally generated small peptides are digested to amino acids in the cytosol.

Investigations of the turnover of specific molecules or groups of molecules obviously depend crucially upon methods for isolating, separating, and identifying the species of interest. In usual studies of degradation of endogenous proteins, gel electrophoresis is used to separate polypeptides on the basis of molecular size and little effort is made to identify specific proteins further. As might be expected, recent studies with two-dimensional separation methods have sometimes led to different conclusions than did work on the same tissue, using older methods.

†Pools of other groups potentially derived from carbonate, such as carboxyls of acidic amino acids, apparently are not labeled in liver to a degree that severely compromises the method. With other tissues, however, carbonate has been used deliberately to label acidic amino acids.

^{3}H. The cells are exposed to one label—say the ^{14}C form—for a while and then "chased" with nonradioactive medium for relatively long periods (often a day or more). During the chase, proteins with faster rates of turnover will lose a greater percentage of their radioactivity than will more stable proteins. The cells are then exposed to the other label—in our example, the ^{3}H form—for a brief period (often an hour or less). During this exposure, proteins that turn over faster will incorporate relatively more of the label. Thus, overall, the more "metabolically labile" (short-lived) proteins will have higher ^{3}H/^{14}C ratios than will more stable proteins (see Table 5.5 and Fig. 5.20). These methods are sensitive to the choice of time intervals used, and are not immune to reutilization problems, but they have proved valuable particularly for comparisons of the relative rates of degradation of different proteins in a given cell type.

In interpreting turnover data, keep in mind that what most frequently is measured is the amount of label present in structures or molecules isolated from samples of the cells at successive intervals, and not the actual rate at which label is liberated by degradation of the molecules. To the extent that organelle destruction depends on sequestration by another cellular system, such as lysosomes or prelysosomes, it is the act of sequestration rather than the process of digestion that will govern the apparent rates of loss of label from the organelle population. This is so because sequestered organelles will usually no longer be isolated in the same cell fractions as they are when free in the cytoplasm. Analogous possibilities arise when immunological methods or assays of enzyme activity are used to identify or quantify particular proteins being studied, as when determinations of specific activities are being made: Care must be taken to ensure that the disappearance of antigenicity or of enzymatic capacity actually reflects degradation, rather than some less drastic change.

5.5.2. Overview of Intracellular Turnover

Though methodological limitations have not paralyzed the analysis of intracellular turnover, most available data are reliable more as estimates of relative rates, than as precise measures of the absolute rates at which particular types of proteins are degraded. This understood, several kinds of findings have been particularly influential in shaping present views of degradative mechanisms. Of these, the only observations unanimously accepted are those that have relied on the usual batteries of inhibitors to demonstrate that intracellular digestion of endogenous proteins requires metabolic energy, provided presumably by ATP. The breaking of peptide bonds by proteolytic enzymes generally is not itself energy dependent. Therefore the energy requirement is an indication that the degradative phase of intracellular turnover is more complex than simply the attack on proteins by proteases.

5.5.2.1. Rates; Rapidly versus Slowly Turning Over Proteins

With the usual exceptions—the sperm of vertebrates, and mature mammalian erythrocytes—all eukaryotic cells that have been studied carefully show evidence of intracellular degradation that reaches most of their molecules, except for most nuclear DNA. "Typical" cultured cells, provided with adequate nutrients and serum, degrade their proteins at average rates of roughly

1–1.5% per hour. Comparable rates—0.5–1.5%—are observed in systems such as liver of well-nourished rats *in situ* or in perfused liver preparations under optimal conditions of nutrition and hormone levels. Thus, rates on the order of 1–1.5% are taken as "basal," though, for example, livers of mice that have been starved and refed, transiently show even lower tempos of degradation. Under "step-up" conditions such as starvation or serum deprivation, the pace of protein degradation increases to several percent per hour, or more.

Plants, yeast, and fungi have been studied much less than have mammalian cells, but these organisms seem also to turn over their proteins under normal conditions, and to increase rates of degradation in response to deprivation of nutrients and to other physiological stresses. Growing yeast break down less than 1% of their protein per hour but the same cells, when starved, increase these rates to 3–5%.

Under optimal growth conditions, bacteria degrade 1% or less of their proteins per hour. In hard times, such as when the cells are starved for nitrogen, the rates increase to as much as 5–10%.

For most cytoplasmic proteins in eukaryotes—those of organized structures as well as those thought to be free in the "cytosol"—the kinetics of disappearance of pulses of label plot approximately as simple exponential curves (Fig. 5.19; but see footnote on p. 292). Therefore, on grounds like those gone over in Section 4.5.2.2, intracellular turnover is treated as a process that is not dependent on molecular age and protein survival times are expressed as half lives. The estimated half lives for different cytoplasmic proteins in rodent hepatocytes range from 15 min or less, to several weeks; values of several hours to several days are taken as typical of the majority of proteins.

It has become the fashion to conceive of cytoplasmic proteins as falling into two discrete—rapidly turning over and slowly turning over—populations. This distinction has emerged especially from findings that the rates of degradation of proteins with half lifes of a few hours or less in cultured cells respond differently to various experimental manipulations than does the turnover of longer-lived species. For instance, when serum is provided to or withheld from cultured fibroblasts, the observed changes in rates of degradation affect principally the longer-lived proteins. In some experimental systems at least, there also seem to be temperature thresholds that apply more to longer-lived proteins than they do to shorter-lived ones (Section 5.5.6.3). As Sections 5.5.3 and 5.5.4 will detail further, such observations probably relate to differences in the mechanisms by which the cell degrades different types of proteins.

These observations notwithstanding, distinctions between "short-lived" and "long-lived" are rough and operational: both classes are heterogeneous and the boundaries are not precise. Customarily, half lives of a few minutes to an hour or two are taken as typical of short-lived species, but researchers on some systems have included proteins with half lives of a day or more in their "short-lived" category.

In liver, proteins with half lives of less than an hour account for 10–20% of protein synthesis, though they contribute only one to a few percent of the mass of protein present in the cells at a given time.

Enzymes such as tyrosine aminotransferase and RNA polymerase I in hepatocytes, certain nucleases in yeast, and certain bacteriophage proteins, fall in the short-lived category. Because a number of these proteins occupy nodal or critical positions in metabolic or transcriptional pathways, the argument has been advanced that fast turnover is a regulatory device that provides the cell

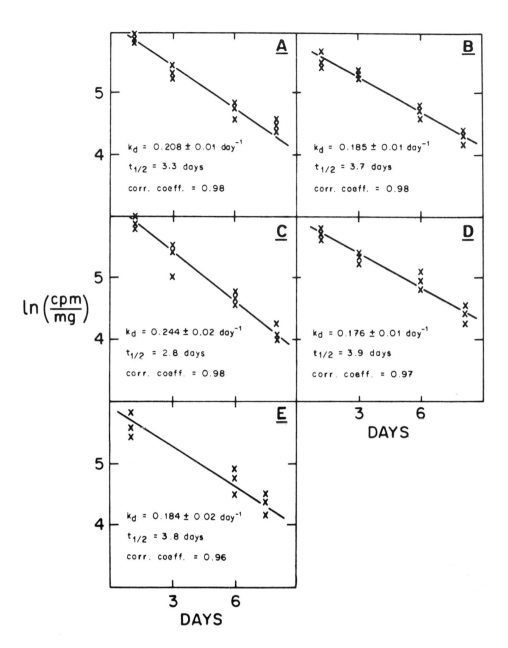

$\ln\left(\dfrac{cpm}{mg}\right)$

A

$k_d = 0.208 \pm 0.01 \text{ day}^{-1}$

$t_{1/2} = 3.3 \text{ days}$

corr. coeff. = 0.98

B

$k_d = 0.185 \pm 0.01 \text{ day}^{-1}$

$t_{1/2} = 3.7 \text{ days}$

corr. coeff. = 0.98

C

$k_d = 0.244 \pm 0.02 \text{ day}^{-1}$

$t_{1/2} = 2.8 \text{ days}$

corr. coeff. = 0.98

D

$k_d = 0.176 \pm 0.01 \text{ day}^{-1}$

$t_{1/2} = 3.9 \text{ days}$

corr. coeff. = 0.97

E

$k_d = 0.184 \pm 0.02 \text{ day}^{-1}$

$t_{1/2} = 3.8 \text{ days}$

corr. coeff. = 0.96

DAYS

FIG. 5.19. Turnover data for mitochondria in rat liver from a study using injection of [^{14}C]bicarbonate to label the proteins. The authors of the study, from which the data were taken, are among those who believe that the inner compartment (inner membrane plus matrix) of mitochondria is degraded largely as a unit (Section 5.5.4), but find also that the outer compartment (outer membrane and material between the two membranes) turns over faster than the inner compartment. Declines in specific radioactivity for fractions isolated at intervals after the injection are shown for: **A**, liver homogenate; **B**, intact mitochondria; **C**, outer mitochondrial compartment (material solubilized or released when the outer membrane is solubilized in digitonin); **D**, inner mitochondrial membrane; **E**, mitochondrial matrix (i.e., material enclosed in inner compartment). Note: the near-linear fall in label (data are plotted on a semilogarithmic scale); the differences between the half lives calculated for C versus those for D and E (outer compartment versus inner compartment); the similarities in turnover between D and E (inner membrane and the matrix it surrounds). Because of the design of the experiment, proteins with very rapid rates of turnover would make little contribution to these data. (See also Table 5.5.) From Lipsky, N. G., and Peterson, P. L. (1981) *J. Biol. Chem.* **256**:8652.

with the capacity to alter its gene expression or metabolism rapidly.* Some abnormal proteins also survive for only short times (Section 5.6.2), indicating, perhaps, a screening function for rapid degradation. The "fast" set might be expected to include as well by-products of protein maturation that serve no known function after their cleavage from the proteins, such as the NH_2-terminal "signal" peptides excised from secretory proteins after they traverse the membranes of the ER. Little explicit effort has been devoted to determining how much of steady-state turnover corresponds to the digestion of such fragments of proteins.

Prejudice favors the expectation that long-lived proteins will be ones serving the steady metabolic and structural "housekeeping" processes needed to sustain the biochemistry and physiology common to most cells under most circumstances. Longer-lived species may also tend to reside in membrane-delimited organelles or in other well-defined structures such as ribosomes or elements of the cytoskeleton. However, there are long-lived proteins in the cytosol. And some of the rapidly turning over enzymes are associated with membranous organelles: examples in liver are HMG-CoA reductase, an ER enzyme with a reported half-life of a few hours (Section 5.5.4.1), and D-aminolevulinate synthetase, a mitochondrial enzyme whose half-life has been estimated at 30–90 min. Turnover of the enzymes of the "glycosomes" (structures of the peroxisomal family) in trypanosomes is also unusually rapid for a membrane-delimited organelle; as judged by immunological assays, the proteins have average half lives of only 30 min. (This story is becoming more complex, however; it now appears that such rapid turnover may characterize only certain phases of the trypanosome's life cycle—at other times, the turnover may slow quite considerably.)

5.5.2.2. Correlations between Degradative Rates and Features of the Proteins

Intracellular degradation of endogenous proteins obviously is not by a uniform or random process affecting all species of cytoplasmic proteins at the same rates or with the same probability. But are there simple regularities underlying the observed diversity in protein half lives? From extensive double-label comparisons of the turnover rates of different proteins of given compartments (e.g., the cytosol) in liver and other tissues, it appeared for a time that on the average, larger polypeptides, proteins that are acidic, and glycoproteins, turn over faster than do smaller polypeptides, more basic proteins, or proteins lacking oligosaccharide side chains (Fig. 5.20). In addition, those proteins that turn over more rapidly *in vivo* were found to be ones that are particularly suscepti-

---

*Ornithine decarboxylase is one of the "metabolic enzymes" with the shortest reported life spans. Half lives of 15 min or less have been calculated for this enzyme in rodent hepatocytes on the basis of polyamine-responsive changes in enzyme activity. Because studies with antibodies to the protein suggest that the enzyme loses activity before it is degraded, one view is that an "antizyme" turns off the enzyme and targets it for destruction, which, though rapid, is not quite as fast as the loss of enzyme activity would suggest. Comparable complexities arise for some of the other enzymes whose half lives seem very short; tyrosine aminotransferase, for example, can be inactivated *in vitro* by a "microsomal" enzyme system that is not degradative. It may be that for these enzymes and even for those other enzymes that do show very rapid rises or falls in actual number of molecules present, the initial regulation is generally by nonproteolytic activations and inhibitions, with proteolytic breakdown coming into play subsequently.

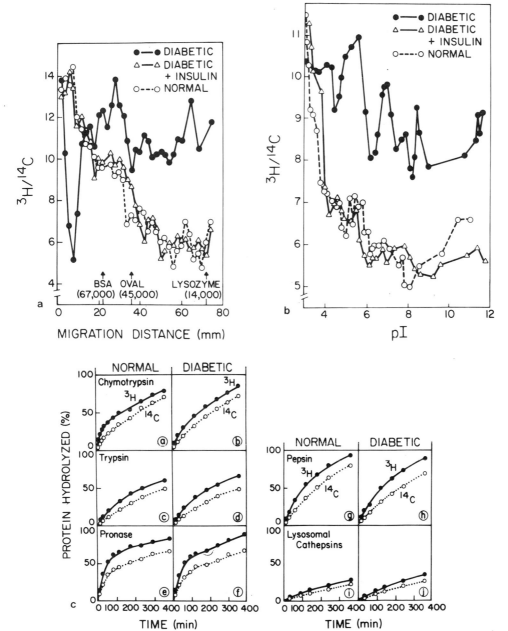

FIG. 5.20. Apparent correlations (and lacks thereof) between molecular properties and turnover rates. Rat liver cytosolic proteins were studied by a double leucine-label protocol (Section 5.5.1.2) such that high $^3H/^{14}C$ ratios indicate rapid turnover: Livers were labeled *in situ* and then cell fractions were prepared and proteins were separated by SDS–PAGE **(a)** or isoelectric focusing **(b)** In normal rats, more rapid turnover seems to be correlated with larger polypeptide size (a) and lower isoelectric point (b); the correlations are not seen in diabetic rats unless the animals are given insulin.

(c) The proteins that turn over more rapidly *in situ* are more readily hydrolyzed when the cytosolic proteins are exposed to purified proteases *in vitro*: In this panel, ^3H is released at a more rapid pace than is ^{14}C because the proteins with high $^3H/^{14}C$ ratios are more susceptible to proteolysis. This relationship is not altered in the diabetic material.

From Dice, J. F., and Walker, C. D. (1980). In Ciba Foundation Symposium 75 (new series): *Protein Degradation in Health and Disease*, Elsevier, Amsterdam, p. 331.

ble to proteolysis when tested *in vitro* by mixing them with purified proteolytic enzymes. Historically, these correlations focused attention on the likelihood that specific features of the "substrates"—the proteins undergoing turnover—help govern their relative susceptibilities to intracellular degradation. The known correlations cannot be pushed too far, however. They certainly do not apply to all cells, all proteins, and all circumstances, and some investigators argue that the methods used exaggerate them. For example, most of the comparisons were of bands on polyacrylamide gels rather than of specifically identified proteins. A relatively small number of prominent bands in the gels can dominate the overall pattern, masking "deviant" behavior of more numerous, "minor" bands. For cytosolic proteins of liver, relations of degradative rates to isoelectric point and polypeptide size seem to hold better under basal conditions than they do when animals are deprived of insulin (Fig. 5.20). And the apparent correlations are not good predictors of the rates of degradation of specific proteins introduced experimentally into the cytoplasm (see Section 5.5.6).

The literature on factors influencing proteolysis *in vitro* is rich. Proteins in their native conformations vary markedly in their susceptibility to attack by purified proteases but the differences often are diminished by denaturation, which notably increases the rates at which most proteins are degraded. "Oxidative" damage—amino acid modifications, cross-linking, or fragmentation of polypeptides—by free radicals and peroxides also tends to increase the rates at which proteins are hydrolyzed. In contrast, the presence of substrates, cofactors, or other ligands, can help protect enzyme molecules against proteolysis. These and many other observations conform to the expectation that the interactions between proteases and their substrates would depend on conformational effects that influence the accessibility of susceptible linkages in the substrates.

Extrapolating from the *in vitro* observations to the cell, theorists of turnover have sought metabolically mediated modifications of proteins that might govern rates of degradation. Phosphorylations, deamidations of glutamine and asparagine residues, carbamylations, and changes affecting disulfide links each have their advocates. Rates of intracellular degradation of certain sets of proteins sometimes have indeed seemed to correlate with the state of the glutathione system or other oxidation–reduction systems; with metabolic situations that influence levels of cofactors or rates of phosphorylation of particular proteins; or with the presence or absence of substrates (e.g., cultured cells alter their rates of degradation of glutamine synthetase in response to changes in the levels of glutamine in the medium). But the evidence is scattered and circumstantial and cannot reliably be interpreted or generalized.

The most striking findings of these sorts (see Section 5.6.2) are that bacteria and eukaryotic cells such as reticulocytes, selectively and rapidly degrade proteins made severely abnormal as a result of genetic changes, the presence of amino acid analogues, or the interruption of translation by puromycin (see Figs. 5.24 and 5.25). Gross changes in the conformation of the affected proteins presumably are involved. But even here the "rules" are not simple: Many sorts of genetically engineered chimeric proteins survive quite nicely within the cell (see Section 5.3.1). And degradative rates are relatively leisurely for some proteins afflicted with seemingly gross abnormalities, even ones in which folding or other features are awry to the extent that passage of the proteins through the ER and Golgi systems is retarded.

Currently, correlations between features of proteins' amino acid sequences and their turnover behavior are generating the most excitement though none of the generalizations has yet proved universal. For instance, many of the shorter-lived proteins have in common the presence of sequences rich in proline, glutamate, serine, and, to a lesser extent, threonine (these sequences are known as PEST sequences from the single-letter code for amino acids). Other such correlations will be summarized in Section 5.6.

5.5.3. The Case for Autophagy as a Principal Mechanism in Turnover

To what extent does autophagy account for the intracellular degradation of endogenous proteins? The strong divisions of opinion on this matter arise partly from differences in the cell types and phenomena studied by different groups. The conclusions most readily supported by the evidence next to be considered are: (1) Autophagic degradation is of considerable magnitude in cells experiencing physiological or pathological malaise. (2) Autophagy probably also governs much of steady-state turnover in organs like the liver. (3) Even in hepatocytes, however, a great deal of the degradation of short-lived endogenous proteins cannot comfortably be explained from what we know of lysosomes. In tissue culture cells growing rapidly, nonlysosomal proteolysis may overshadow the participation of autophagy.

5.5.3.1. Correlations; Quantitation

Morphometric estimates of the amounts of cytoplasm taken into AVs (Section 5.1.5) are in line with the quantities needed to explain most of the degradation of cytoplasmic proteins in nutritionally stressed liver. Likewise, autophagy in cell cultures responding to the absence of serum factors or the removal of amino acids, often seems sufficient to account for most of the increased degradation of proteins (see also Fig. 5.21). Conversely, decreases in the abundance of AVs sometimes are accompanied by declines in degradative rates of the order expected if autophagy were principally responsible for the breakdown. And some of the interesting details of normal turnover rates are congruent with details of autophagy: e.g., the average half lives of peroxisomal proteins are 1–2 days whereas mitochondrial proteins average several days, paralleling in qualitative terms at least, differences in the relative rates of autophagy of these two types of organelles (Section 5.1.4). Calculations based on observations like those summarized in this paragraph suggest, for instance, that autophagy is responsible for breaking down 50–70% or more of the mass of protein degraded within hepatocytes *in situ* in the course of a "normal" day.

What about "basal" turnover, as is observed in hepatocytes of perfused liver when they are well nourished and well supplied with insulin, or in cultured cells given adequate levels of amino acids and serum? Under these conditions, AVs containing readily identifiable cytoplasmic structures can be quite scarce but appreciable degradation of endogenous proteins still takes place. One possibility is that in basal situations, autophagic processes other than classical autophagy—crinophagy, microautophagy, and the like—become proportionally more important. This is impossible to evaluate directly at present because most of such processes are difficult to recognize unambiguously

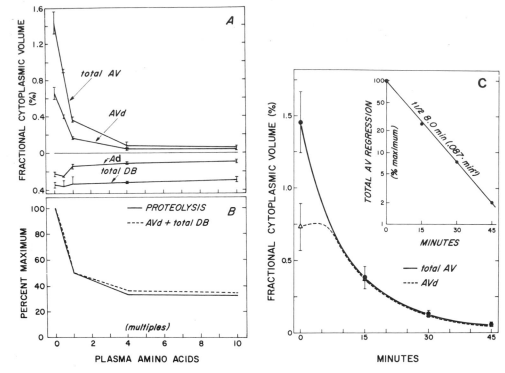

FIG. 5.21. Autophagy in hepatocytes of perfused rat liver.

(A) The abundance of autophagic vacuoles (AV), expressed here in morphometric terms (percent of cytoplasmic volume occupied by the structure in question), diminishes with raised concentrations of amino acids in the perfusion medium (expressed here in terms of multiples of the normal plasma levels of amino acids). AVd are autophagic vacuoles delimited by a single membrane and thought to have acquired hydrolases. The lower curves indicate the abundance of dense bodies in the same cells (Ad is a type of dense body containing glycogen or certain indigestible residues).

(B) Parallelism between lysosome abundance and rates of proteolysis in the perfused livers used for A.

(C) Changes in abundance of autophagic vacuoles during the period immediately following the perfusion of high concentrations of amino acids through a liver that previously was being perfused with no amino acids in the medium. The exponential falloff of vacuole abundance is plotted on a semilogarithmic scale in the insert.

in the microscope, confounding morphometric analyses of their contributions (see Section 5.1.7.2). Indirect approaches are summarized below (e.g., Section 5.5.3.2).

5.5.3.2. Effects of Weak Bases and of Inhibitors of Proteases

Nowadays, estimates of the extent of lysosomal involvement in cellular degradation of proteins are most often based simply on the degree to which degradation in suitably labeled cells is inhibited by the agents assumed to exert their relevant effects chiefly on lysosomes: chloroquine, ammonium chloride, protease inhibitors (especially leupeptin or pepstatin), vinblastine, 3-methyladenine, or low temperature. Unfortunately, important assumptions about how such agents distribute in the cell and how they impact on cellular metabolism have rarely been checked critically enough to quell suspicions that they sometimes affect nonlysosomal systems (see Section 1.5.2). Moreover, the agents, even when used in combinations, cannot be relied upon to abolish lysosomal proteolytic activities completely: As judged from enzyme assays *in vitro* or from studies on proteolysis in lysosomes isolated from cells exposed to the inhibitors, lysosomal proteases like cathepsin L retain significant activity in the pH ranges produced by the weak bases, and the protease inhibitors often leave the lysosomes capable of significant activity.

Inhibitor studies nevertheless have been very heavily relied upon to help determine roughly how important autophagy may be. With intact liver under various conditions, the presumed lysosome inhibitors depress breakdown of endogenous proteins by 50–75%. The effects are stronger on longer-lived proteins than on shorter-lived ones, but even liver proteins in the fast-turnover category show decreases in degradative rates of as much as 30–40% after exposure to the inhibitors.

Estimates are much lower—down to 10–30% of overall degradation—for the usual mammalian cultured cell lines, when the cells are growing rapidly and are supplied with adequate levels of amino acids and serum factors. With such cells, some investigators find the intracellular proteolysis of the longer-lived proteins to be significantly more sensitive to the bases and protease inhibitors than is breakdown of short-lived ones. Other researchers regard these differences as small or negligible. There is broader agreement that when cell cultures are induced to step up their degradation of endogenous proteins by depleting the medium of serum or amino acids, longer-lived proteins are targeted for destruction to a disproportionate extent and much or all of the increase in degradative rate can be countered by the lysosome inhibitors. Customary generalizations drawn from these findings are that cultured cell lines use autophagy mainly in situations of nutritional or hormonal inadequacy and that when they do use autophagy, they select long-lived proteins to degrade. A corollary (still-contentious) implication is that these same cells, confronting more optimal conditions, use nonlysosomal mechanisms for most of their digestion of endogenous proteins, whether short-lived or long-lived.

Hepatocytes of perfused livers exposed to leupeptin while being maintained under conditions that reduce digestion of endogenous proteins to basal levels sequester 1–2% of their cytoplasm in AVs during the first hour of leupeptin treatment. On the assumption—not universally accepted—that short-term leupeptin exposure simply stops the tranformation of AVs into residual bodies, with no added induction of autophagy, these findings have been used to argue that under basal conditions, 1–2% per hour of the cell's cytoplasm is sequestered into lysosomes. This intensity would account for the known rates of basal protein turnover.*

Lysosome-rich cell fractions obtained from leupeptin-injected rats contain active cytosolic enzymes, notably lactic dehydrogenase (LDH), aldolase, and tyrosine aminotransferase (TAT). If one accepts the logic of the last paragraph and assumes that leupeptin is merely preventing proteolysis and thereby revealing the normal locations of enzymes that ordinarily undergo degradation, it would appear that cytosolic proteins enter autophagic lysosomes at substantial rates (and in forms that have not undergone presequestrational inactivation).

---

*Supportive evidence is that lysosomes isolated from livers kept under basal degradation conditions contain proteins that undergo digestion if the lysosomes are now suspended in an appropriate medium. These proteins have no obvious heterophagic source, their presence is accounted for by postulating that some form of autophagy persists at significant rates under basal conditions, despite the extreme sparseness of classical images of autophagy in the livers.

This holds both for short-lived enzymes (TAT; see Section 5.5.2.1) and for long-lived ones (aldolases and LDH have half lives of 2–3 days). A rough estimate from the percentages of total cellular activities that are recovered in AV-rich fractions from leupeptin preparations, puts autophagic sequestration of the three enzymes at about 2.5% of the liver's content per hour. These rates would account for all of the normal steady-state turnover of LDH or aldolase. And, as might be expected, were autophagy the principal turnover mechanism for LDH, levels of LDH in lysosomes of leupeptin-treated rats increase when dexamethasone induces increases in total cytoplasmic content of the enzyme. For TAT, however, degradative mechanisms additional to lysosomal ones must come into play to bring about its rapid turnover.

5.5.4. Are Organelles Degraded as Units?

The uptake of a mitochondrion or of a ribosome into an AV leads to the simultaneous disappearance of all of the engulfed structure's component molecules from the cell's population of mitochondria or ribosomes (Section 5.5.1.2). Therefore, if autophagy is the sole or the dominant mechanism for turnover of cytoplasmic structures, very strong correlations would be expected among the half lives of the different constituents of given classes of organelles; all should turn over at pretty much the same rates.

At first, studies on turnover in liver and other material seemed to demonstrate such clustering of turnover rates for the different proteins of mitochondria, for the RNAs and proteins of ribosomes, for peroxisomal proteins, and for some proteins of the plasma membrane. The fact that even the DNA of mitochondria turns over suggested that the organelles are destroyed as wholes (or in very large chunks; Section 5.1). The lack of obvious age dependency in the kinetics of turnover of organellar components was consistent with the apparently nonselective facets of autophagic sequestration (Sections 4.5.2.2 and 5.1.4), and the fact that sequestration can be somewhat selective could be used to explain differences in turnover rates of different organelles.

Matters are much more complicated now that more data have been collected; the simple correlations have broken down. For membrane-delimited organelles, turnover of lipid components is found to be at a different rate from that of the membrane proteins and evidence also has accumulated for substantial heterogeneity in half lives among the proteins of a given type of organelle. The absence of age dependency, which seemed to suggest that older and newer organelles of a given type faced equal probabilities of destruction, may attest more to ambiguities in the concept of organelle "age."*

---

*That is, concepts of "older" and "newer" may apply poorly to many cellular structures. Examples particularly pertinent to our concerns: (1) As they grow, membranes of the ER or chloroplast thylakoids are thought to acquire new lipids and proteins throughout their extent rather than segregating into older and newer expanses. (2) Diffusion in the plane of the membrane and the extensive recycling of plasma membrane components are likely to randomize the distribution of plasma membrane molecules of different ages, at least within some cell surface domains. (3) Peroxisomes seem often to maintain extensive connections with one another, raising the possibility that older and newer constituents equilibrate among many or all of the members of a cell's peroxisome population. (4) Similarly, extensive and stable interconnections among what had been thought to be separate mitochondria have been reported for yeast and several other cell types, and in a number of other cell types, mitochondria seem to fuse with one another or to separate off from larger mitochondria. (The normal rates of these fusion and fission phenomena have been little studied but the fact that mitochondrial and chloroplast DNAs can engage in

TABLE 5.5. Data from a Double-Label Experiment in Which [^{14}C]-Bicarbonate Was Injected into the Rats First and Then, at the Indicated Interval, the Animals Were Injected with [^{3}H]-Leucine 30 min before the Livers Were Homogenized[a]

	ln (^{3}H/^{14}C)					
	5 days		6 days		9 days	
Liver homogenate	1.49	1.46	2.12	1.97	1.74	1.91
Mitochondria	0.99	0.96	1.16	0.95	0.78	0.92
Outer compartment	1.32	1.30	1.53	1.29	1.28	1.28
Inner membrane	0.71	0.69	0.92	0.82	0.52	0.64
Matrix	0.70	0.68	0.82	0.69	0.57	0.64

[a]Note the differences between inner and outer mitochondrial compartments and the similarities between the inner membrane and the mitochondrial matrix. From Lipsky, N. G., and Peterson, P. L. (1981) *J. Biol. Chem.* **256**:8652. (See also Fig. 5.19.)

None of this need be fatal to the idea that autophagy is important in normal turnover. The retinal rod (Section 4.6.2.1) provides an excellent precedent for steady-state destruction of membrane systems dependent on bulk incorporation within lysosomes; there is nonetheless substantial heterogeneity in half lives and in age-related behavior among membrane components (especially between lipids and proteins). For that matter, there still are investigators who insist that the heterogeneity of turnover rates for the constituents of given types of organelles has been overstated and who emphasize instead seeming correlations in behavior of cytoplasmic constituents. For instance, several students of mitochondrial turnover (see Table 5.5 and Fig. 5.19) are convinced that many of the proteins of the inner mitochondrial membrane and of the matrix this membrane surrounds, show similar half lives (which are somewhat longer than the half lives of outer membrane components).

Still, even on the most "optimistic" of assumptions, autophagy cannot be the whole story. Additional processes must be operating to account for the fact that some proteins in organelles like mitochondria do turn over much faster than others, and to produce differences in half lives between organellar proteins and lipids. Early hypotheses proposed that individual molecules can both enter and leave the organelles, superimposing molecule-by-molecule turnover on "bulk" degradation via autophagy. This still seems realistic for lipids, given the existence of lipid exchange proteins and similar devices that can move individual lipid molecules from one membrane system to another. Certain of the protein components of non-membrane-delimited structures such as ribosomes or multienzyme complexes might also be so situated as to enter or leave without requiring gross dismantling of the rest of the structure. Post-translational **entry** of certain integral proteins into membranes and of newly made proteins into the interiors of preexisting peroxisomes, mitochondria, and chloroplasts is known to be extensive. Nonexocytotic **departure** of intact proteins from inside the cytoplasm's membrane-delimited structures is more diffi-

genetic recombination also suggests that the inviolability of the individual organelle cannot be taken for granted.) (5) Some of the cell's microtubules and actin microfilaments are relatively short-lived as organized structures, assembling from soluble pools of their constituent proteins, exchanging constituents with these pools and soon disassembling. (6) Lysosomal fusions with endosomes and AVs (with one another too?) mix lysosomal constituents of different ages.

cult to envisage, however. True, peroxisomes have persistently been reputed to be relatively "leaky" organelles from whose interior proteins might conceivably exit. But for most of the other organelles, attention is swinging toward the likelihood that proteases present in the organelles themselves break down some proteins locally (See Section 5.6.4 and Fig. 5.25 for comments on proteases in mitochondria and chloroplasts).

5.5.4.1. Modulation or "Atrophy" of Organelles?

Connected to these considerations are claims that mitochondria, peroxisomes, or other organelles can degenerate in the cytoplasm of otherwise healthy cells, without undergoing autophagy (Section 5.1.6.1), or that individual organelles can survive intact while modulating their internal enzymatic capacities drastically by losing some enzymes while gaining others. Evidence adduced centers on microscopic observations of "sick-looking" organelles that seem to be degenerating in place. These images are, however, difficult to interpret, both because of the possibility of artifact during preparation of the samples for microscopy and because ordinarily appearances do not tell much about the health of an organelle until matters have reached an advanced state of malady. But there is some stronger evidence. Chloroplasts, for instance, undergo degradation ranging from the rapid turnover of a few of their proteins (Section 5.6.4) to extensive breakdown of their membrane systems, but to my knowledge, admittedly incomplete, autophagy of chloroplasts has rarely if ever been demonstrated beyond the possibility of contravention. Even when "excess" chloroplasts are lost completely, as in the zygotes of algae, the process looks more like disintegration than autophagy (though perhaps fragments of the degenerated organelles are taken up by vacuoles).

Selective degradation of particular organellar macromolecules without autophagy of the organelle may actually be widespread. ER, for example, though clearly subject to autophagy, also seems able to lose enzymatic activities or immunocytochemically detectable proteins without concomitant destruction of the membranes. A case in point is encountered when cultured cells of a CHO-derived line are induced to hypertrophy their smooth ER by exposure to the drug **compactin,** an inhibitor of the enzyme HMG-CoA reductase (Section 4.1.1.1.), and then are provided with sterols; levels of ER-associated HMG-CoA reductase decline after this regimen much more rapidly than does the extent of the ER itself. To accomplish this autophagically, the cell presumably would have to aggregate molecules of the enzyme in the plane of the ER membrane, producing membrane specially enriched in HMG-CoA reductase, which then would have to be degraded selectively. In the absence of evidence for such a process, a plausible alternative is that HMG-CoA reductase molecules, or at least fragments of these molecules bearing their active sites, are enzymatically cleaved from the ER, to be disposed of as individual molecules. Proteolytic attacks by endogenous enzymes on HMG-CoA reductase have recently been induced by permeabilizing cells and then subjecting them to treatments that include the provision of ATP. Puzzlingly, however, sterol enhancement of degradation of HMG-CoA reductase is abolished in cells genetically engineered to produce truncated forms of the reductase lacking the domains that ordinarily insert in the ER membrane.

5.5.4.2. Cytosolic Proteins; Nuclear Proteins

The heterogeneity in turnover rates of cytosolic proteins is difficult to reconcile with beliefs in the primacy of lysosomes, without introducing important assumptions that are difficult to test. Efforts at such reconciliation thus have been eclipsed by the current attention to degradation of cytosolic proteins by nonlysosomal proteases (Section 5.6), but it is not yet time to abandon altogether the ideas that were generated before these proteases were discovered: Some observers speculated about targeting or recognition mechanisms whereby autophagic structures could sequester different proteins at different rates (see Section 5.5.6.4). This thinking was encouraged by observations that some of the proteins once thought to float free in the cytosol now seem likely to be structured in labile cytoplasmic arrays (see Section 1.5.1.1). Perhaps rates at which given types of proteins associate and dissociate from these arrays influence the probabilities of their encounters with autophagic bodies (see Section 5.5.5.1). Less commonly, proposals were made about lysosomal hydrolases having active sites on the cytoplasmic surface of the lysosomal membrane, which would enable the lysosomes to attack cytosolic proteins without internalizing them. Through this mechanism, the proteins would undergo degradation at rates dependent largely on their susceptibility to attack by the lysosomal hydrolases. Another posited mechanism with similar outcome was reversible autophagy—hypothetical processes through which random samples of cytoplasm could be exposed transiently to the lysosome interior and then returned to the cytosol.

Degradation of nuclear proteins poses problems similar to the breakdown of cytosolic materials in that the rates of breakdown vary substantially for different proteins. This degradation has not yet been studied systematically. Some of the proteins that contribute to the fundamental structure of the chromosome turn over slowly: For growing cells, it often is claimed that most of the histones are very long-lived. Yet during spermiogenesis in fish and other animals, the histones are entirely replaced by protamines. Certain oncogene-coded proteins that localize to the nucleus are reputed to last only a relatively brief time. And when *Xenopus* oocytes are injected with ribosomal proteins under circumstances in which there are not enough rRNAs to associate with the proteins, the excess protein molecules are broken down, probably within the nucleus. Are yet-to-be-discovered nuclear proteases responsible for most of the turnover of nuclear proteins? Or are the proteins attacked by cytoplasmic machinery after passing through nuclear pores, or when the nuclear envelope breaks down in cell division?

5.5.4.3. Turnover of the Plasma Membrane

The blebbing off of vesicles or other shedding of material to the extracellular space may sometimes contribute to loss of plasma membrane components from cells (see Section 5.2.1.3). But the endocytotic and recycling processes described in Section 5.2 are the most obvious steady-state phenomena through which plasma membrane proteins would normally be put in jeopardy of destruction. By suitably choosing among the processes known to occur in one or another of the endocytotic, heterophagic, and autophagic events already described, most of the data on turnover of plasma membrane constituents can be explained in terms of lysosomal degradative mechanisms. Such reconciliation, as outlined next, is still chiefly at the level of plausible hypothesis, however, and there are significant discordant details.

Both *in situ* and in cell cultures, many proteins of the plasma membrane turn over with the same sort of apparently age-independent kinetics as other cytoplasmic proteins; typical half lives range from several hours to several days. In many reports, half lives for different plasma membrane proteins of a given cell vary considerably. In a few double-label studies on cell cultures, however—especially on transformed cells lacking well-differentiated cell surface subdomains—heterogeneity is less marked and the half lives for different protein species seem to fall into a few groups. Even for liver cells and other differentiated cell types, similarities among the turnover rates of different major plasma membrane proteins have been reported from studies in which the cell surface was radioactively labeled (glycosylated or iodinated) by enzymes added to the extracellular medium. Lipids have been studied in less detail, but differences from the proteins are to be anticipated given the divergences already discussed in considering the turnover of other membranes. [As with other membranes, newly made or recycling lipids can move into and out of the plasma membrane by mechanisms and routes different from those used by most of the proteins. In addition, certain plasma membrane lipids (notoriously, the inositol lipids and those containing arachidonic acid) undergo special physiologically important cycles of partial breakdown and resynthesis.]

It would be quite surprising were all the proteins of a cell's plasma membrane to exhibit the same turnover behavior. For one thing, a structurally polarized cell usually possesses several differentiated cell surface domains (e.g., Figs. 4.9, 5.16, and 5.22) differing in the rates and nature of endocytotic and exocytotic membrane movements.* Even for "undifferentiated" cells in culture, the substantial differences known to exist in the internalization behavior and intracellular routing of different receptors (see Sections 3.3 and 5.2) should be, and are, reflected in different rates of degradation. Less expected was the discovery that the oligosaccharide chains of plasma membrane glycoproteins in several cultured cell lines lose certain residues at rates considerably faster than the rates at which the polypeptide chains are degraded (Fig. 5.23). Unlike the degradation of the polypeptides, the loss of these residues from the oligosaccharides is not inhibited by chloroquine or other lysosome inhibitors, suggesting that it takes place either extracellularly, or in some non-lysosomal intracellular compartment encountered during cycling of the pro-

---

*Occasional microscopic images suggest selective endocytotic internalization of specialized plasma membrane domains: For example, configurations strongly resembling gap junctions seem to be taken into the cells at enhanced rates when junctional relations between cells are undergoing change. Desmosomes, and even tight junctions also have been claimed to undergo endocytotic internalization (see Fig. 5.22). If these configurations are stages in degradative paths, as their discoverers sometimes have advertised (on reasonable grounds), it would be valuable to determine whether their abundance correlates with rates of degradation of relevant proteins, such as those of the gap junction connexons. The data that do exist on turnover of the proteins of cell–cell junctions are similar to those on turnover of other plasma membrane proteins in that the rates fall in the hours-to-days range and seem to vary even for different proteins of a given type of junction. But detailed, systematic analyses oriented toward the issues of interest to us here have not been published. When cells desquamate from striated epithelia or are released from matrices, relatively large digestion products of the molecules that had previously anchored them turn up in extracellular media, so it seems clear, at a minimum, that not all the release of cells from bonds to adjacent structures is based on endocytosis. [Extracellular enzymes (Chapter 6) are obviously among the agents that could participate.]

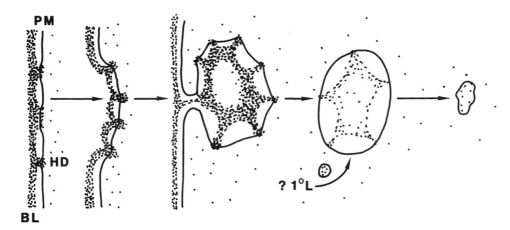

FIG. 5.22. During at least two periods in the developmental sequence of the butterfly *Calpodes* (Fig. 5.7), the fat body cells form endocytotic vacuoles in which they take up what seems to be regions of their basal laminae (BL), attached to portions of the plasma membrane (PM) by "hemidesmosomes" (HD). The contents of these vacuoles presumably are degraded by lysosomal hydrolases though whether the enzymes are acquired from primary (1°L) or secondary lysosomes is not certain. From Locke, M. (1982). In *Insect Ultrastructure* (R. C. King and H. Akai, eds.), Plenum Press, New York, p. 151.

teins. To some extent the losses can be made up by "reglycosylation" mechanisms like those I mentioned when discussing the involvement of the Golgi apparatus recycling (Sections 5.3.3.2 and 5.3.3.3; see also Section 7.2). It may even turn out that a typical glycoprotein molecule of the plasma membrane undergoes several cycles of partial deglycosylation and reglycosylation in the course of its life.

Life spans of the cell surface receptors that undergo repeated recycling often must be interpreted with special care (cf. Section 5.2.1.2). In some cases

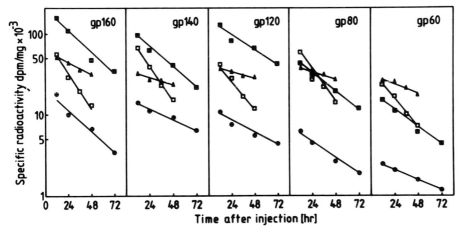

FIG. 5.23. Turnover data for several glycoproteins (gp) of the plasma membrane of rat liver cells demonstrating the differences between turnover rates for oligosaccharide components that are added late during glycoprotein synthesis, versus "core portions" (polypeptide and core oligosaccharide moieties) of the same proteins. The data represent the radioactivity in proteins prepared from cell fractions enriched in plasma membranes; the fractions were prepared at the indicated times after a 2-hr pulse of label. ●, mannose; ■, N-acetylmannosamine; ▲, leucine; □, fucose. Semilogarithmic plot. From Tauber, R., Park, C. S., and Reutter, W. (1983). *Proc. Natl. Acad. Sci. USA* **80:**4026.

the receivers' cycling is specially sensitive to "environmental variables." In others, receptors behave as though more than one pool exists at the cell surface or show odd behavior within the cell. Half lives of less than 10 hr have been estimated for the α_2-macroglobulin receptors of macrophages, the transferrin receptor of erythroleukemia cells, and the transcobalamin receptor of fibroblasts. Why should these receptors turn over faster than many of the other plasma membrane proteins of the same cells? Perhaps for the receptors that recycle repeatedly into endosomes and back out to coated pits, part of the answer lies in the recycling. Maybe, on each of its cycles, a given receptor molecule faces a finite probability of a degradative encounter. Because the receptors cycle several times per hour, such diversion of only a few percent per cycle, say by accidental "missorting" to the lysosome interior, could account for relatively rapid turnover. [Minor changes in the probability of missorting, induced for example by antibodies (see Section 3.3.4.3) or by endosomal proteolysis, could have large effects on the degradative rate.] But why then do not all recycling receptors turn over fairly rapidly? The LDL or ASGP receptors of some cell types have reasonably long half lives—of a day or more. And why is the rate of turnover of the LDL receptor not vastly slowed by mutations that abolish clustering of this receptor into coated pits and therefore should reduce its rates of internalization into endosomes? Could it be that recycling receptors are not degraded as an offshoot of their recycling but rather that some other endocytotic process (see Sections 2.2.2.1 and 2.3.2.11) gathers a few of the molecules every once in a while and delivers them to their doom?

5.5.5. Striated Muscle

The size and protein content of vertebrate muscles change dramatically with nutrition, exercise, and disease. These changes depend upon coordinated regulation both of rates of protein synthesis and of rates of breakdown of endogenous proteins. (Striated muscle fibers are not very active endocytotically and heterophagy probably plays a very minor role in their economy, at least quantitatively.) In normal rats the principal contractile proteins and other constituents of the myofilament apparatus turn over with half lives of several days to several weeks. Half lives for the contractile proteins in muscle cultures are a week or two. These rates increase in starved animals, which degrade muscle proteins to maintain levels of circulating amino acids. Enhanced intracellular proteolysis is also observed in responses to fever; inactivity; molting or metamorphosis of invertebrates; exposure to cachectin or to certain glucocorticoids, prostaglandins, or interleukins; lack of insulin; provision of thyroid hormones to preparations previously deprived of these hormones; muscular dystrophies; and many other situations.

Assessing the roles of lysosomes in such degradation has proved particularly frustrating. Muscle fibers do contain conventional lysosomes, including AVs whose frequency varies with physiological and pathological condition. In the heart, for example, autophagy is intensified after oxygen deprivation. But relative to the enormous mass of the contractile apparatus, lysosomal bodies are very sparse.* This, together with mechanical problems in organelle isola-

---

*Cytochemists have occasionally reported acid hydrolase activity in one of the extensive membrane systems that lie close to the contractile structures (the T system or the sarcoplasmic reticulum) but there is no general agreement as to what these findings signify. Sometimes the reaction product is artifactual, but this "explanation" probably cannot account for all of the findings.

tion, has hindered biochemical studies of muscle lysosomes, and limits microscopic work as well. Inferences that lysosomes do participate importantly in degrading muscle are usually drawn from indirect evidence, such as the increases in lysosomal hydrolases often seen in situations of enhanced breakdown. Sometimes, as in the resorbing tadpole tail where muscles are completely destroyed, or in pathological changes, these hydrolase increases are due mainly to influxes of phagocytes or mast cells. But there are cases where enhanced degradation of endogenous proteins takes place within intact muscle fibers and in such situations acid hydrolase activities can augment without influxes of phagocytes, or are already noteworthy by the time the phagocytes arrive.

Weak bases like methylamine substantially antagonize the stepped-up intracellular proteolysis seen in rat muscles when insulin or amino acids are lacking. These bases have much less effect on the lower ("basal") rates of protein breakdown characteristic of muscles kept under better conditions (e.g., explants maintained under some mechanical tension, in a rich medium). Perhaps as in other cells and tissues, autophagy is more important for stepping up degradation than for basal turnover. In some experiments, however, in which intramuscular proteolysis was increased by exposure to ionophores or other manipulations thought to raise levels of Ca^{2+} in the fibers, the enhanced degradation was relatively unresponsive to weak bases. This might imply that lysosomes play little role in the reactions to ionic changes, except that the enhanced proteolysis is sensitive to leupeptin, which, for many of the phenomena discussed in prior sections, has been taken as a sign of lysosomal involvement. One possibility is that in muscle, leupeptin affects nonlysosomal proteolytic mechanisms. This would be consonant with the ability of leupeptin to inhibit many proteases in the test tube, but how does the inhibitor get access to nonlysosomal proteases (see Section 1.5.2)? An alternative possibility is that muscle lysosomes possess proteases that are relatively insensitive to weak bases because they have high or broad pH optima.

5.5.5.1. Protection by Association?

The geometry of muscle's organized contractile apparatus would pose a formidable problem for autophagy unless some dismantling occurs before components of the filaments are engulfed (see Section 5.6.4). At the least, nonlysosomal attacks might be needed to prepare the contractile material for lysosomal degradation. Many investigators go further and doubt that lysosomes actually do degrade the proteins of the myofilaments extensively, even if lysosomes do digest other muscle proteins. Along with the considerations above, these doubts arise, for example, from observations that when muscles are dismantled during insect metamorphosis or tadpole tail resorption, the filament systems disorganize by processes described as "fragmentation" or "dissolution"; filaments are not often seen in AVs, even in cases in which the same types of muscle fibers do autophagocytose recognizable mitochondria. What is more, when muscles proteolyze their proteins in responses to starvation or denervation, the release of methylhistidine shows little sensitivity to presumed inhibitors of lysosomal function even though the inhibitors do reduce the release of other amino acids such as tyrosine in the same preparations. (Methylhistidine, which is methylated posttranslationally and not reutilized for protein synthesis, is popularly regarded as almost unique to actin and myosin. However, reliance on release of this amino acid from muscle as an index of proteolysis of the filaments is not yet universally agreed upon.)

As suggested earlier, turnover of other cytoskeletal proteins, additional to the "muscle proteins," might also be conditioned by association and cellular organization. For example, proteins organized in the cytoskeleton of the extended processes of neurons or glial cells might face different degradative threats than do similar proteins in more typical cytoplasm—this has been pointed out in efforts to explain why some neuronal or glial proteins show multiphasic turnover kinetics. Many of the cilia of ciliated cells persist for quite prolonged times even when there is extensive upheaval in other microtubule systems in the same cell; this is readily evident for organized ciliary arrays, such as those along the oral regions of ciliated protozoa.

Erythrocytes differentiating in chick embryos degrade the cytoskeletal protein α-**spectrin** by mechanisms responsive to the usual lysosome inhibitors. Breakdown of a related cytoskeletal protein, β-**spectrin,** in the same cells is much less affected by the inhibitors. The proteins in question come from the pool of molecules that are not assembled into structural aggregates and the digestion is thought to be important in regulating the levels of unassembled cytoskeletal materials in the developing cell. The same species of proteins present in assembled cytoskeletal arrays in the same cells, are degraded very slowly if at all during the cells' lifetimes.

Possibly analogous observations are beginning to be made on plasma membrane systems. For several multisubunit receptors, cultured cells make a large excess of the subunits, assembling some into receptors and degrading, rapidly, those that do not get assembled. For the T-cell receptor, degradation of excess chains in T cells seems to occur in lysosomes, but in fibroblasts transfected with T-cell-receptor genes, the degradation occurs prelysosomally, perhaps even in the ER (see also Section 5.3.1). Degradation may also take care of tubulins made in excess, ribosomal proteins produced with not enough rRNA around (Section 5.5.4.2), and myelin proteins synthesized in systems prevented from myelinating.

Are nuclear proteins protected by their association with chromatin or with the organized "matrix" thought by some investigators to pervade the nucleus? There has been speculation about selective proteolysis of proteins bound to DNA contributing to the regulation of genetic activity but it has also been suggested that breakdown of nuclear proteins depends upon passage of the proteins to the cytoplasm (see Section 5.5.4.2), a passage that would require prior dissociation of the molecules from intranuclear structure.

5.5.6. Fates of Microinjected Molecules

The experimental introduction of quantities of specific molecules into the cytoplasm is a powerful way to study intracellular degradation. Many methods have been employed: transplantation of cytoplasm between amebas; microneedle injection of oocytes or cultured cells; liberation of the contents of pinocytotic vesicles by osmotic shock (Section 4.8.2); transient compromising of the plasma membrane barrier electrically (see footnote on p. 258) or by exposure of macrophages to extracellular ATP (Section 3.4.1.2); fusions of liposomes with the cell surface (Section 4.8.1). Most current work uses viruses or polyethylene glycol to induce fusions of cultured cells with erythrocyte ghosts previously loaded with materials of interest by osmotic treatments.*

---

*Microinjection thus far has aimed mainly at introducing soluble molecules into the cytosol. Much less has been done with membranous material though there have been a few uses of fluorescent

Upon microinjection, Lucifer Yellow, radioactively labeled sucrose or dextrans, or proteins such as BSA tagged with fluorescent dyes, are sequestered into lysosomes or related membrane-bounded compartments. Rates of sequestration vary considerably for different materials but it is impressive that some of the fluorescent materials can be taken up rapidly enough to clear the rest of the cytoplasm within hours after injection. Sucrose is taken into sedimentable structures believed to be lysosome-related (see footnote on p. 258), at rates faster than expected from the rates at which cells normally degrade cytosolic components. On the other hand, cultured BHK fibroblasts accumulate microinjected dextrans into sedimentable structures only to the extent of 0.5% of the cells' dextran content per hour, even under conditions where autophagy has been stepped up by effects of high culture density. Perhaps these last values approximate rates of "fluid phase" autophagy under near-normal conditions whereas the higher rates with sucrose reflect inductive or selective phenomena due to the abnormal nature of the material taken up or of the exerimental situation. For Lucifer Yellow the ability to cross membranes under some circumstances may permit specially rapid sequestration (see Section 3.4.1.2); if so, which of the cells' own small molecules follow similar routes?

5.5.6.2. Degradation; Release of Intact Proteins

For a time after injection, the behavior of microinjected proteins often differs considerably from their normal counterparts. This hinders work on rapidly turning over molecules, though things do settle down after a few hours. Limited proteolytic changes or other alterations in the proteins to be injected sometimes occur during loading of the erythrocytes. Quite frequently, a proportion of the microinjected material, as high as 1% per hour, is released from the recipient cells, in high-molecular-weight form. Part of this release may be by cell death or by the shedding of cytoplasm from the fused cells. Could it also be that some steps of autophagy are reversible or are mediated by recycling carrier compartments resembling those involved in heterophagy at least to the extent that materials taken into nacent AVs can depart from the vacuoles, eventually to reach the cell surface? Musing along this last tack is prompted by the observations on expulsion of receptors from reticulocytes (Section 5.2.1.3) and the fact that autophagy and heterophagy use the same lysosomes and otherwise intersect.

When suitable account is taken of factors like abnormal behavior during the first hours, systematic estimates of degradative rates for different proteins or peptides injected by erythrocyte fusion procedures into mammalian cell cultures often give half lives similar to the half life of corresponding endogenous species (Table 5.6). Degradation of the erythrocyte-injected proteins also shows responses to serum deprivation and other manipulations comparable to the responses seen with endogenous proteins.

5.5.6.3. Lysosomal Involvement

Degradation of erythrocyte-injected proteins takes place both in the lysosomes and extralysosomally, with different proteins showing different be-

---

lipid derivatives (Section 3.3.4.5) or of the fusion of isolated organelles or membrane-enveloped viruses with the cell surface.

TABLE 5.6. Comparison of Half Lives Obtained by Microinjection
and Biosynthetic Labeling[a]

Protein	$t_{1/2}$ injected	$t_{1/2}$ biosynthetic	Reference
Ubiquitin	11	9	J. Biol. Chem. **256**:5916
Ferritin	47	36	J. Biol. Chem. **247**:5234
AAT	66	72	Science **127**:287
Aldolase	77	22–101	Curr. Top. Biol. Med. Res. **1**:125
G3PDH	84	24–74	"
LDH(H)	172	85–168	"
Pyruvate kinase	187	120–146	Biochem. J. **204**:89

[a]The injections were done by osmotic loading of erythrocyte ghosts, which were then fused with the cultured cells by exposure to Sendai virus. The data reveal that proteins microinjected into HeLa cells exhibit half lives similar to their half lives *in situ*. From Rechsteiner, M., Hough, R., Rogers, S., and Rote, K. (1987) In *Lysosomes: Their Role in Protein Breakdown* (H. Glaumann and F. J. Ballard, eds.), Academic Press, New York, p. 487.

havior.* In cultures of growing HeLa cells (Table 5.6), most of the degradation of BSA and much of the breakdown of microinjected cytosolic proteins, takes place outside the lysosomes. But digestion of those cytosolic enzymes whose natural half lives are long (2 days or more) does show significant sensitivity to the usual "lysosomal inhibitors."

The responsiveness of proteolysis of microinjected proteins to the lysosomal inhibitors increases under conditions of serum deprivation. But the serum-responsive component of degradation largely disappears in cells cooled to below 20°C, whereas the component that does not respond to serum shows a more continuous temperature curve, without sharp breaks. Such observations suggest the existence of at least two distinct intracellular proteolytic mechanisms only one of which depends on lysosomes, confirming the conclusions about turnover drawn above from other types of evidence.

Microinjected proteins fail to show simple correlations between rates of destruction and size or isoelectric point (see Section 5.5.2.2). Curiously, with proteins injected into HeLa cells, half life tends to be correlated with the degree to which the injected proteins resist extraction from the recipient cells by the detergent Triton X-100. Because the injected proteins are not membrane proteins, this finding is taken possibly to reflect the establishment of associations between injected proteins and cytoskeletal or other cellular structures. Similar implications have been drawn from observations that, in certain cases, sequestration in the lysosomal system seems to be preceded by segregation of the proteins into progressively restricted cytoplasmic regions, as if the proteins were associating with organized cytoplasmic materials before reaching definitive lysosomes. Injected ferritin can clump before it is autophagocytosed.

5.5.6.4. Signals Governing Degradation?

The degradation of erythrocyte-injected RNase A by confluent human fibroblasts has opened particularly exciting perspectives with respect to possible

---

*This has been established by cell fractionation and by use of the usual inhibitors of lysosome function. Some of these experiments investigated the fate of proteins coupled covalently to radioactive sucrose or raffinose; when these proteins are degraded, the sucrose- or raffinose-linked degradation products, being unable to cross membranes, should remain in the compartments where they are generated (or in compartments and extracellular spaces they subsequently reach by processes involving membrane fusions).

autophagic controls. The enzyme is degraded in lysosomes, and the rate of degradation is enhanced in serum-deficient media. A truncated form of the enzyme, lacking the 20 amino acids of the NH_2 terminus, is also degraded largely in lysosomes, as established both by inhibitor studies and by isolation of lysosomes. But degradation of the truncated form does not respond to changes in serum levels. A first tentative conclusion then is that, at least under conditions of culture confluence (which themselves step up autophagic degradation), the absence of effect of serum-deprivation does not necessarily imply a nonlysosomal degradative route.

A second conclusion is that the peptide missing from the truncated form confers serum sensitivity on degradation of the enzyme. The peptide seemingly is not required for autophagy to pick up the enzyme but there must be something about it that interacts with cellular sites so as to trigger augmented or more efficient autophagy when serum levels are low. The findings have been extended to show that the peptide missing from the truncated form is itself degraded by a serum-sensitive path when microinjected into cells, and that it can confer such sensitivity on proteins to which it is artificially linked. A pentapeptide segment (Lys-Phe-Glu-Arg-Gln) of the longer peptide provides the key signal. This sequence *per se* is present in only a few of the other proteins that have degradation characteristics like those of RNase A. But many of these other proteins are recognized by an antibody to the pentapeptide even though they lack the sequence itself—presumably, the proteins possess a conformational feature similar to that produced by the pentapeptide, or there is a family of interchangeable sequences. There have also been hints of the existence of cytosolic proteins that can bind selectively to the proteins in question.

These observations have been greeted enthusiastically because they promise molecular-level insights into what have been mysterious mechanisms. For instance, older proposals as to how autophagy might operate differentially to produce the varied rates of turnover of different cytosolic components (see Section 5.5.4.2) frequently suggested that molecules or organelles are targeted selectively for autophagy by modifications—enzymatic attacks, exposure of hydrophobic regions through denaturation, and so forth—to which the autophagic system responds. Different organelle classes and cytoplasmic proteins might vary in susceptibility to being modified. The approaches outlined in this section promise to lead from such general, inevitably vague models, toward testable schemes for the regulation and specificity of turnover processes.

5.6. A Brief Visit to Cytoplasmic, Nonlysosomal Degradation

5.6.1. General Considerations; Bacteria

It became obvious that lysosomes could not be responsible for all important types of intracellular proteolysis when it was realized that bacteria, which lack lysosomes altogether, and mammalian reticulocytes, which have very few, nonetheless can degrade "cytoplasmic" proteins extensively. Bacteria, for example, turn up intracellular proteolysis when starved for nitrogen, thereby making amino acids available for metabolic use, including the synthesis of new induced enzymes. They may also utilize proteolysis in transcriptional control mechanisms by which DNA sequences are freed from regulatory proteins.*

As more has become known about the proteolytic systems existing in the cytoplasm outside the lysosomes, many of the fissures in the portrait of intracellular turnover presented above have begun to fill in. However, only the barest outlines are yet evident as to how lysosomal and nonlysosomal mechanisms weave together in eukaryotic cells. For instance, it seems likely from Section 5.5.2.1 that nonlysosomal mechanisms are relatively more important for breaking down short-lived proteins than they are for the longer-lived proteins. To a degree this is due to compartmentalization of substrates—autophagy would have limited access to those proteins that are degraded rapidly within organelles like mitochondria (Section 5.6.4). Is it generally the case that turnover involves a kind of "competition" in which few of the shortest-lived molecules are around long enough for the lysosomes to have a crack at them? One important unknown in these contests is whether oligopeptide fragments arising by partial proteolysis outside the lysosomes can subsequently be sequestered and digested autophagically.

Nonlysosomal degradative systems must be regulated and selective, else the cell would not survive their presence. To what extent are the proteases governed simply by factors like the intrinsic susceptibility of proteins to proteolytic attack or to denaturation (Section 5.5.2.2) and to what extent are potential substrates protected by associations with membranes or other organized structures (see Section 5.5.5.1)? For bacteria, the regulatory molecule **ppGpp** has been implicated in mechanisms by which proteolytic activities are altered in response to changes in the levels to which tRNAs are charged with amino acids. What are the comparable signaling devices in eukaryotic systems? "Cystatins," which inhibit thiol-proteases, and other inhibitors of proteolytic enzymes are now being detected in the cytoplasm of many cells. Is regulation of nonlysosomal degradation one of the roles of such inhibitors? What, if any, is the relevance of supposed autodegradative activities of genetic repressor proteins or of the programmed proteolytic processing of viral polyproteins into separate functional proteins, to the degradative phenomena of concern to us here?

Nonlysosomal lipases, glycosidases (e.g., mannosidases), and other degradative enzymes additional to the proteases are encountered from time to time. Attention to these enzymes has almost exclusively concerned their participation in the production of macromolecules (e.g., in glycosylation of glycoproteins; Section 7.1.1.1), but maybe they also have degradative roles in turnover.

5.6.2. Abnormal Proteins; Abnormal Conditions

Nonlysosomal cytoplasmic proteolysis achieves degradation of severely abnormal proteins such as are produced with amino acid analogues, puromycin, drastic mutations, or oxidation (Fig. 5.24). Both in bacteria and in

---

*Nonlysosomal degradation of cytosolic enzymes or regulatory proteins often figures in proposals for metabolic control mechanisms but supportive evidence is not firm. In yeast, for instance, catabolite repression of enzymes can be paralleled by proteolysis of the enzymes, but the repression of activity is not blocked in mutants whose vacuolar proteases are inactive. On the basis of such observations, the opinion of some researchers on catabolite repression is therefore shifting away from the view that proteolysis is directly responsible for inactivating the enzymes and toward notions that the enzymes are first repressed by nondegradative changes (e.g., phosphorylations) with proteolysis playing a later, temporarily dispensable role.

reticulocytes, nonlysosomal, ATP-dependent proteolytic systems held responsible for much of the degradation of such proteins have been identified and partially purified. Similar systems probably exist in other eukaryotic cells, along with the lysosomal mechanisms that seem to pick up certain types of abnormal proteins (Section 5.3.1). One of the mammalian systems reportedly has the form of a large (19 S) multicomponent cytoplasmic particle.

5.6.2.1. Bacteria

Though intracellular proteolysis in bacteria may involve more than one system, the most intense interest concerns a large protease (M_r near 500,000) coded for by the *lon* gene of *E. coli* and exhibiting vanadate-inhibitable ATPase activity. How ATP participates in bacterial proteolysis is not understood. [Notice that we have avoided this issue in discussing autophagy as well, for, aside from the few findings mentioned in Section 5.1.5.3, there is no clear, direct, information as to which step(s) in autophagy help account for related energy requirements of protein turnover.] Current models suggest that ATP modifies the protease rather than its substrates and that once this has happened, abnormal proteins can activate the protease allosterically. How the system discriminates between molecules to be degraded and ones to be spared is a matter of importance not only for the bacterium's health but also for production of genetically engineered chimeric proteins or other useful molecules the bacterium might regard as abnormal. Bacteria made to overproduce the "lon protease" are sickly—they grow slowly and sometimes even die—so the system's selectivity may not be perfect.

lon gene transcription can be induced simply by the presence of abnormal proteins but the gene also is one of the heat shock genes, suggesting either that the proteolytic system has wide regulatory functions or that the stresses to which the heat shock genes respond also tend to generate abnormal proteins.

5.6.2.2. Reticulocytes; Oocytes

As well as breaking down abnormal proteins rapidly, reticulocytes degrade seemingly normal globin chains, when these chains are present in excess of the amounts that can be assembled into hemoglobin. (Excesses are notable in disorders like β-thalassemia where the balances of production of different classes of chains are severely askew.)

Frog oocytes probably use a nonlysosomal system to break down microinjected secretory proteins when these proteins are introduced into the cytosol in the form of "primary translation products" (i.e., as they are translated by ribosomes before undergoing the proteolytic removal of signal sequences and other modifications normally carried out by the ER and Golgi apparatus). Under ordinary circumstances, or in oocytes provided with foreign mRNAs, primary translation products for secretory proteins move cotranslationally from ribosomes into the interior of the ER where they seem secure from the attacks just described. Are the cytosolic proteolytic mechanisms tuned to destroy those secretory protein molecules that are accidentally missorted to the cytosol during translation, as a few molecules probably are even in normal cells? (See also Section 5.3.1; does any of the breakdown of abnormal proteins that do enter the reticulum cotranslationally depend on proteases in the ER?)

5.6.2.3. Aging?

Theories of aging often build around ideas that screening mechanisms for detecting and dealing with defective materials eventually become inadequate, leading to deterioration of the quality of organellar and macromolecular populations. The literature is full of conflicting and hard-to-interpret findings. For example, whereas some senescent cultures retain appreciable ability to degrade abnormal proteins, in certain studies senescent cell cultures deprived of serum or grown to high culture density have shown less increase in intracellular proteolysis than do younger cultures.

5.6.3. Ubiquitin: A Signal for Proteolytic Destruction?

Among the several nonlysosomal proteolytic systems identified so far in eukaryotes, special attention is being paid to those involving the 76-amino-acid polypeptide **ubiquitin.** In the most studied of the ubiquitin-dependent proteolytic mechanisms, before the substrate proteins are hydrolyzed, ubiquitin becomes conjugated to them via isopeptide linkages (these links are to the side chains of lysine residues and to the amino groups at the NH_2 termini of proteins). This "ubiquitinylation" requires ATP and is carried out by sequential action of three distinguishable enzymes. Ubiquitinylated proteins are favored substrates for proteolysis via a pathway that also requires ATP. The proteolysis

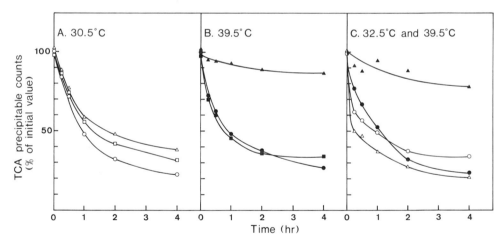

FIG. 5.24. A temperature-sensitive mutant mouse cell line, defective in the ability to conjugate ubiquitin to proteins, fails to degrade short-lived proteins at the nonpermissive temperature. In the experiments illustrated in **A** and **B**, the cells were grown for 15 min in amino acid analogues [Threo-α-amino chlorobutyrate and thialysine (S-aminoethyl L-cysteine)], labeled with a brief pulse of radioactive methionine, and then placed in normal medium. Protein degradation was monitored by the decrease in radioactive methionine present in molecules (proteins and large peptides) precipitable by trichloroacetic acid (TCA). (A) At the permissive temperature, wild-type cells (○), mutant cells (△), and a revertant cell line (□) all degrade the proteins at much the same rate. (B) At a higher (nonpermissive) temperature, the mutant cells are differentially affected.

(C) A similar experiment using exposure to puromycin, rather than to amino acid analogues, to generate the abnormal proteins. Circles are wild-type and triangles the mutant. The solid symbols refer to the higher temperature. Normal short-lived proteins (pulse-labeled with methionine under ordinary conditions) show patterns of behavior similar to those illustrated here, and weak bases have no effect at either the permissive or the nonpermissive temperature.

From Ciechanover, A., Finley, D., and Varshavsky, A. (1984). *Cell* **37:**57. (Copyright: *Cell Press.*)

generates peptides that subsequently are degraded to amino acids by cytoplasmic peptidases. In the course of these events, ubiquitin is liberated for reuse in conjugation. Overall then, this pathway accounts for the participation of ATP in breakdown of proteins, and it involves a discrete, identified change in the substrate molecules that targets them for destruction. Both *in vivo* and in cell-free preparations, the system is much more effective in attacking some proteins than others, perhaps in part because the ubiquitinylating enzymes show disparate efficiencies in conjugating ubiquitin to different proteins. Cataloging is now under way of features of proteins that may govern their handling by the ubiquitin system—under some experimental conditions, for example, the nature and state of the NH_2-terminal amino acid seems highly influential— but it is still too early for confident generalization. For a number of protein substrates, tRNAs, including histidyl and arginyl tRNAs, are needed for the ubiquitin system to function efficiently *in vitro*.

Ubiquitin is widely distributed in animal cells, being found both in cytoplasmic locations and attached to nuclear proteins, including some histone molecules. Capacities for ubiquitin-dependent proteolysis also are widespread, and have been implicated, at least circumstantially, in phenomena of biological importance. In some organisms or cell cultures, ubiquitin is among the proteins that increase in heat shock responses. In reticulocytes, the rapid, selective degradation of abnormal proteins may proceed in part by a system involving ubiquitin and this system may be responsible as well for the decline of glycolytic enzymes that normally accompanies erythrocyte maturation. A mouse mammary gland tumor cell expressing a temperature-sensitive mutation affecting one of the enzymes of ubiquitin conjugation has been found to be defective in degradation of short-lived proteins (Fig. 5.24). This cell line also shows temperature-sensitive deficiencies in heat shock responses and arrests in the G2 stage of the cell division cycle, raising thoughts about ubiquitin-mediated proteolysis of cytoplasmic regulators or nuclear proteins. Ubiquitin and ubiquitin-dependent proteolysis has recently been detected in yeast and in plants; among the plant proteins with which ubiquitin may form conjugates is phytochrome, a regulatory, light-sensitive protein whose turnover in some experiments is markedly influenced by light. In certain cell cultures, ubiquitin turns up at difficult-to-explain locations, such as attached to the extracellular portions of plasma membrane proteins.

The welter of information exemplified in the last paragraph has yet to be forced into a coherent picture. Suspicions are growing that ubiquitin plays several distinct roles in cells, even insofar as intracellular proteolysis is concerned. With some cell lines, extensive methylation of proteins before they are microinjected into the cells has little effect on subsequent degradation of the proteins; this is disquieting because methylation should abolish most of the sites to which ubiquitin can be bound. In a few experiments, ubiquitin has seemed to have direct regulatory effects on cytoplasmic proteases, that do not depend on its being conjugated to potential substrates and there has even been a report that ubiquitin itself is a protease.

5.6.4. Other Nonlysosomal Proteolytic Systems; Ca^{2+}-Dependent Proteases; Organellar Proteolysis

Of the eukaryotic nonlysosomal proteolytic systems that seem not to involve ubiquitin, those responsive to increases in cytoplasmic levels of free

FIG. 5.25. Preparations of isolated mitochondria degrade some of their own polypeptides. In the experiment illustrated, isolated rat liver mitochondria labeled for 20 min in radioactive methionine in the absence **(A)** or the presence **(B)** of puromycin were disrupted and incubated in the presence or absence of ATP and an ATP-regenerating system. Degradation of labeled proteins was monitored in terms of the percent of label that is still acid-precipitable at the indicated times. Puromycin increases the rate of degradation. From Goldberg, A. L. (1987). In *Lysosomes: Their Role in Protein Breakdown* (H. Glaumann and F. J. Ballard, eds.), Academic Press, New York.

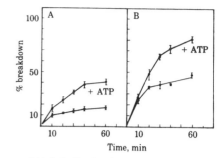

Ca^{2+} have been detected in the greatest variety of cell types. Several such calcium-regulated proteases have been identified, including, for example, the **calpains,** calcium-activated thiol-proteases somewhat resembling papain. Attacks by calcium-dependent proteolysis have been proposed as key to the dismantling of cytoplasmic filament systems—the intermediate filaments of neurons and other cell types, and the contractile filaments of muscle.

Most other cytoplasmic proteases have been less studied. An interesting one recently claimed to exist in erythrocytes catalyzes an ATP-independent degradation of oxidized hemoglobin, suggesting that even mature red blood cells possess useful, selective proteolytic capacities.

Mitochondria and chloroplasts have intrinsic proteolytic systems that are energy dependent but independent of ubiquitin (Fig. 5.25). "Organellar" proteolysis shares some features with the bacterial proteolysis described above, including the capacity to hydrolyze abnormal proteins especially rapidly. As much as 20% of the proteins made by isolated mitochondria seem to be degraded within the organelles. *In situ*, the usual targets of the mitochondrial and chloroplast systems probably include those proteins that are degraded very rapidly either in the steady state or as physiological conditions change. The possible functions of the proteolytic systems are legion: screening of translation products made within the organelles; breakdown of damaged proteins such as ones affected by light-induced oxygen radicals in chloroplasts; maintenance of balances between amounts of specific proteins and amounts of other materials such as chlorophyll or enzyme cofactors; balancing the amounts of proteins made inside the organelles with importation from the cytoplasm outside; digestion of polypeptide fragments released during such importation or during processing of recently synthesized organellar proteins; modulation of enzymatic capacities as occurs for example in yeast mitochondria responding to changes in availability of oxygen.

5.6.5. Degradation of Nucleic Acids

Information about how cells degrade their nucleic acids is surprisingly skeletal (see also footnote on p. 252). Only for rRNAs and mitochondrial DNAs is there strong reason to posit appreciable participation of the lysosomes; in hepatocytes these molecules turn over with half lives of a few days, rates comparable to those of the proteins of the same organelles (see Section 5.5.2.1). To a degree, degradation of RNA in cell cultures responds to weak bases or serum deprivation as expected for lysosomal processes. When insulin or amino acids are lacking, breakdown of rRNAs in perfused liver accelerates in parallel to the acceleration of autophagic degradation of proteins; when glucagon is used to induce autophagy, however, proteolysis seemingly is enhanced more than the degradation of rRNAs.

It seems inevitable that at least once in a while, an mRNA or tRNA will be caught, accidentally, by a nascent AV. But there clearly are effective non-lysosomal enzymes for breaking down RNAs: prokaryotes can degrade their RNA, sometimes quite rapidly and extensively; the RNAs within membrane-bounded cytoplasmic organelles are thought to turn over much faster than expected from rates of autophagic organelle destruction (mitochondria and chloroplasts seem to have their own nucleases); and the nuclei of eukaryotes are presumed to have systems with which to degrade RNA molecules, including the "spacers" and intron sequences released during the processing of nascent rRNAs and mRNAs. One problem in making sense of the available information is that cells have quite a variety of enzymes that can cut RNAs to some extent. And RNAs undergo several sorts of cleavages during their genesis as well as during their destruction.

Schemes invented to explain differences in turnover rates among mRNA molecules or between different classes of RNAs have attributed special importance to: the presence or absence of caps and poly(A) tails at the ends of mRNAs; differences in intrinsic susceptibility of different RNAs to attacks by nucleases; competitions, wherein the polysomal association of ribosomes with mRNAs protects the mRNAs from nucleases, but moments of vulnerability occur as ribosomes cycle onto or off the mRNAs; and analogous competitions between nucleases and membranes of the ER (or the plasma membrane in prokaryotes) with which polysomes associate during cotranslational translocation of nascent polypeptides across the membranes.

Acknowledgments

M. Chrispeels, J. F. Dice, W. Dunn, and J. Etlinger read and commented acutely upon extensive segments of the chapter. H. Glaumann and F. J. Ballard helped me obtain preprints of several chapters from their then-unpublished book (see below). In addition, for specific pieces of useful information, for references, or for discussions, I am particularly grateful to F. J. Ballard (turnover), M. Chrispeels (protein bodies in plants), A. Ciechanover (ubiquitin system), J. F. Dice (molecular signals for turnover; RNase turnover), D. Doyle (turnover of plasma membrane), W. Dunn (formation of AVs), H. Glaumann (autophagy), A. Goldberg (nonlysosomal systems; bacteria; muscle), D. F. Goldspink (muscle), A. Hershko (ubiquitin), E. A. Khairallah (turnover), R. M. Johnstone (reticulocyte maturation), M. Locke (insect storage and autophagy), L. Marzalla (autophagy), G. Mortimore (autophagy in perfused liver), L. Orci (crinophagy in pancreas), U. Pfeifer (kinetics of autophagy), M. Rechsteiner (microinjection experiments), W. Reutter (turnover of plasma membrane glycoproteins), O. Scornik (degradation in the biochemical economy of mammals), P. O. Seglen (autophagy).

Amenta, J. S., and Brocher, S. C. (1981) Mechanisms of protein turnover in cultured cells, *Life Sci.* **28**:1195–1208.

Annu. Rev. Physiol. **47** (1985) has several useful articles on insulin receptors by Czech, Bergeron, and others. (See also Further Reading, Chapter 4.)

Arrigo, A. P., Tanaka, K., Goldberg, A. L., and Welch, W. J. (1988) Identity of the 19S prosome particle with the large, multifunctional complex of mammalian cells (the protasome), *Science* **231**:192–194.

Ascoli, M. (1984) Lysosomal accumulation of the hormone–receptor complex during receptor-mediated endocytosis of human choriogonadotropin, *J. Cell Biol.* **99**:1242–1250.

Ballard, F. J. (1982) Regulation of protein accumulation in cultured cells, *Biochem. J.* **208**:275–287.

Berg, R. A., Schwartz, M. L., and Crystal, R. G. (1980) Regulation of the production of secretory proteins: Degradation of newly synthesized defective collagen, *Proc. Natl. Acad. Sci. USA* **77**:4746–4750.

Berhanu, P. (1988) Internalized insulin receptor complexes are unidirectionally translocated to chloroquine sensitive degradative sites, *J. Biol. Chem.* **263**:5961–5969.

Beynon, R. J., and Kay, J. (eds.) (1985) Cellular proteolysis. *Biochem. Soc. Trans.* **13**:1005–1026.

Bienkowski, R. S., Curran, S. F., and Berg, R. A. (1986) Kinetics of intracellular degradation of newly synthesized collagen, *Biochemistry* **25**:2455–2458. (See also *Biochem. J.* **214**:1–10, 1983 for a review.)

Brunden, K. R., and Podulso, J. F. (1987) Lysosomal delivery of the major myelin glycoprotein in the absence of membrane assembly: Posttranslational regulation of the level of expression by Schwann cells, *J. Cell Biol.* **104**:661–669.

Chertow, B. S. (1981) The role of lysosomes and proteases in hormone secretion and degradation, *Endocr. Rev.* **2**:137–173.

Chrispeels, M. J. (1984) Biosynthesis, processing and transport of storage proteins and lectins in cotyledons of developing legume seeds, *Philos. Trans. R. Soc. London Ser. B* **304**:309–322.

Chu, F.-F., and Doyle, D. (1985) Turnover of plasma membrane proteins in rat hepatoma cells and primary cultures of rat hepatocytes, *J. Biol. Chem.* **260**:3097–3107.

Desautels, M., and Goldberg, A. L. (1985) The ATP-dependent breakdown of proteins in mammalian mitochondria. *Biochem. Soc. Trans.* **13**:290–293.

Deutscher, M. P. (1988) The metabolic role of RNases, *Trends in Biochem. Sci.* **13**:136–139.

Dice, J. F. (1987) Molecular determinants of protein half-lives in eucaryotic cells, *FASEB J.* **1**:349–357.

Dice, J. F., Chiang, H. L., Spencer, E. P., and Becker, J. M. (1986) Regulation of catabolism of microinjected ribonuclease: Identification of residues 7–11 as the essential pentapeptide, *J. Biol. Chem.* **261**:6853–6859.

Eguchi, E., and Waterman, T. M. (1976) Freeze-etch and histochemical evidence for cycling in crayfish photoreceptor membranes, *Cell Tissue Res.* **169**:418–434.

Finley, D., and Varshavsky, A. (1985) The ubiquitin system: Function and mechanism, *Trends Biochem. Sci.* **10**:343–347. (See also *Science* **234**:179–186, 1986).

Frehner, M., Keller, F., and Wiemken, A. (1984) Localization of fructan metabolism in the vacuoles isolated from protoplasts of Jerusalem artichoke tubers (*Helianthus tuberosus*), *J. Plant Physiol.* **116**:197–208.

Glaumann, H., and Ballard, F. J. (eds.) (1987) *Lysosomes: Their Role in Protein Breakdown*, Academic press, New York.

Goldberg, A. L., Baracos, V., Roderman, P., Waxman, L., and Dinarello, C. (1984) Control of protein degradation in muscle by prostaglandins, Ca^{2+} and leukocytic pyrogen (interleukin 2), *Fed. Proc.* **43**:1301–1306.

Gonatas, N. K., Kim, S., Stieber, A., and Avrameas, S. (1977) Internalization of lectins in neuronal GERL, *J. Cell Biol.* **73**:1–13.

Gordon, P. B., and Seglen, P. O. (1988) Prelysosomal convergence of autophagic and endocytic pathways, *Biochem Biophys Res Comm.* **151**:40–47.

Grimes, G. W., and Gavin, R. H. (1987) Ciliary protein conservation during development in the cilated protozoan *Oxytrichia*, *J. Cell Biol.* **105**:2855–2859.

Hakanson, R., and Thorell, J. (eds.) (1985) *Biogenetics of Neurohormonal Peptides*, Academic Press, New York.

Harding, C., Heuser, J., and Stahl, P. (1983) Receptor mediated endocytosis of transferrin and recycling of transferrin receptor in rat reticulocytes, *J. Cell Biol.* **97**:329–339.

Hare, J. F. (1988) Dissection of membrane protein degradation mechanisms by reversible inhibitors, *J. Biol Chem.* **263**:8759–8764.

Herzog, V. (1981) Pathways of endocytosis in secretory cells, *Trends Biochem. Sci.* **6**:319–322.

Jarett, L., and Smith, R. M. (1985) Receptor regulation: Problems and perspectives, in *Mechanisms of Receptor Regulation* (G. Poste and S. T. Crooke, eds.), Plenum Press, New York, pp. 1–11.

Jingami, H., Brown, M. S., Goldstein, J. L., Anderson, R. G. W., and Luskey, K. L. (1987) Partial deletion of membrane bound domain of HMG-CoA reductase eliminates sterol enhanced degradation and prevents formation of crystalloid endoplasmic reticulum, *J. Cell Biol.* **104**:1693–1704.

Keller, F., and Matile, P. (1985) The role of the vacuole in storage and mobilization of stachyose in tubers of *Stachys sieboldii, J. Plant Physiol.* **119**:369–380.

Khairallah, E. A., Bond, J. S., and Bird, J. W. C. (eds.) (1985) *Intracellular Protein Catabolism,* Academic Press, New York.

Knutson, V. P., Ronnett, G. V., and Lane, M. D. (1985) The effects of cycloheximide and chloroquine on insulin receptor metabolism, *J. Biol. Chem.* **260**:14180–14188.

Levy, J. R., and Olevsky, J. M. (1988) Intracellular insulin receptor dissociation and segregation in a rat fibroblast cell line transfected with a human insulin receptor gene, *J. Biol. Chem.* **263**:6101–6108.

Lippincott-Schwartz, J., Bonifacino, J. S., Yuan, L. C., and Klausner, R. D. (1988) Degradation from the endoplasmic reticulum: Disposing of newly synthesized proteins, *Cell* **54**:209–220.

Locke, M. (1984) The structure and development of the vacuolar system in the fat body of insects, in *Insect Ultrastructure* (R. C. King and H. Akai, eds.), Vol. 2, Plenum Press, New York, pp. 151–194.

Lott, J. N. A. (1980) Protein bodies, in *The Biochemistry of Plants* (P. K. Stumpf and E. E. Conn, eds.), Vol. 1, Academic Press, New York, pp. 589–623.

Lowell, B. B., Ruderman, N. B., and Goodman, M. N. (1986) Evidence that lysosomes are not involved in the degradation of myofibrillar proteins in rat skeletal muscle, *Biochem. J.* **234**:237–240.

Marshall, S. (1985) Degradative processing of internalized insulin in isolated adipocytes: Dual pathway for intracellular processing of insulin, *J. Biol. Chem.* **260**:13517–13523, 13524–13531.

Marin, B. (ed.) (1987) *Plant Vacuoles,* Plenum, New York.

Marty, F., Branton, D., and Leigh, R. A. (1980) Plant vacuoles, in *The Biochemistry of Plants* (P. K. Stumpf and E. Conn, eds.), Vol. 1, Academic Press, New York, pp. 625–658.

Matile, P. (1978) Biochemistry and function of vacuoles, *Annu. Rev. Plant Physiol.* **29**:193–213.

Mayer, R. J., and Doherty, F. (1986) Intracellular protein catabolism: State of the art, *FEBS Lett.* **198**:181–193.

Mortimore, G. E. (1984) Regulation of intracellular proteolysis, *Fed. Proc.* **43**:1281–1282. (For parallel work on RNA, see also *J. Biol. Chem.* **262**:14514, 1987.)

Mortimore, G. E., Lardeux, B. R., and Adams, C. E. (1988) Regulation of microautophagy and basal protein turnover in rat liver, *J. Biol. Chem.* **263**:2506–2512.

Ohkuma, S., Chudzik, J., and Poole, B. (1986) The effects of basic substances and acidic ionophores on the digestion of exogenous and endogenous proteins in mouse peritoneal macrophages, *J. Cell Biol.* **102**:959–966.

Orci, L. (1985) The insulin factory: A tour of the plant surroundings and a visit to the assembly line, *Diabetologica* **28**:528–546.

Orci, L., Ravazzola, M., Amherdt, M., Brown, D., and Perrelet, A. (1986) Transport of horseradish peroxidase from the cell surface to the Golgi in insulin-secreting cells: Preferential labeling of cisternae located in an intermediate position in the stack, *EMBO J.* **5**:2097–2101.

Pan, B. T., Teng, K., Wu, C., Adam, M., and Johnstone, R. M. (1985) Electron microscopic evidence for externalization of the transferrin receptor in vesicular form in sheep reticulocytes, *J. Cell Biol.* **101**:942–948.

Pfeifer, U. (1982) Kinetic and subcellular aspects of hypertrophy and atrophy, *Int. Rev. Exp. Pathol.* **23**:1–45. (See also *J. Cell Biol.* **64**:698–621, 1975.)

Pontremoli, S., and Malloni, E. (1986) Extralysosomal protein degradation, *Annu. Rev. Biochem.* **55**:455–487.

Poste, G., and Crooke, S. T. (eds.) (1985) *Mechanisms of Receptor Regulation,* Plenum Press, New York.

Rechsteiner, M. (ed.) (1988) *Ubiquitin,* Plenum, New York.

Rechsteiner, M., Rogers, S., and Rote, K. (1987) Protein structure and intracellular stability, *Trends Biochem. Sci.* **12**:390–394. (See also *Annu. Rev. Cell Biol.* **3**:1, 1987.)

Reed, B. C., Glasted, K., and Miller, B. (1984) Direct comparison of the rates of internalization and degradation of covalent receptor–insulin complexes in 3T3-L1 adipocytes: Internalization is not the limiting step in receptor–hormone complex degradation, *J. Biol. Chem.* **259:**8134–8143.

Rez, G., and Meldolesi, J. (1980) Freeze-fracture of drug-induced autophagocytosis in the mouse exocrine pancreas, *Lab. Invest.* **43:**269–277.

Robinson, D. G. (1985) *Plant Membranes: Endo- and Plasma Membranes of Plant Cells,* Wiley, New York.

Sawano, F., Ravazzola, M., Amherdt, M., Perrelet, A., and Orci, L. (1986) Horseradish peroxidase uptake and crinophagy in insulin secreting cell, *Exp. Cell Res.* **164:**174–182.

Scornik, O. A. (1984) Role of protein degradation in the regulation of cellular protein content and amino acid pools, *Fed. Proc.* **43:**1283–1288.

Schellens, J. P. M., Vreeling-Sindelacova, H., Plomp, P. J. A. M., and Meijer, A. J. (1988) Hepatic autophagy and intracellular ATP: A morphometric study, *Exp. Cell Res.* **177:**103–108.

Schlessinger, J. (1988) The epidermal growth factor receptor as a multifunctional allosteric protein, *Biochemistry* **27:**3119–3123.

Segal, H. L., and Doyle, D. J. (eds.) (1978) *Protein Turnover and Lysosome Function,* Academic Press, New York.

Seglen, P. O., and Gordon, P. B. (1982) 3-Methyl adenine: Specific inhibitor of autophagic/lysosomal protein degradation in isolated rat hepatocytes, *Proc. Natl. Acad. Sci. USA* **79:**1889–1892.

Smith, R. E., and Farquhar, M. G. (1966) Lysosome function in the regulation of the secretory process in cells of the anterior pituitary, *J. Cell Biol.* **31:**319–336.

Snider, M. D., and Rogers, O. L. (1986) Membrane traffic in animal cells: Cellular glycoproteins return to the site of Golgi mannosidase I, *J. Biol.* **103:**265–275. (But see also Neefjes *et al., J. Cell Biol.* **107:**79, 1988.)

Szego, C., and Pietras, R. J. (1984) Lysosomal functions in cellular activation: Propagation of the action of hormones and other effectors, *Int. Rev. Cytol.* **88:**1–302.

Tauber, R., Park, C. S., and Reutter, W. (1983) Intramolecular heterogeneity of degradation in plasma membrane glycoproteins: Evidence for a general characteristic, *Proc. Natl. Acad. Sci. USA* **80:**4026–4029.

Tsurugi, K., Motizuki, M., Mitsui, K., Endo, Y., and Shiokawa, K. (1988) The metabolism of ribosomal proteins microinjected into Xenopus oocytes, *Exp. Cell Res.* **174:**177–187.

Turner, J. M., Tartakoff, A. M., and Berger, M. (1988) Intracellular degradation of the complement C3b/C4b receptor in the absence of ligand, *J. Biol. Chem.* **263:**4914–4920.

Viestra, R. D., and Sullivan, M. L. (1988) Hemin inhibits ubiquitin-dependent proteolysis in both higher plant and yeast, *Biochemistry* **27:**3290–3295.

Wolf, D. H. (1985) Proteinases, proteolysis and regulation in yeast, *Biochem. Soc. Trans.* **13:**279–283.

Zeman, R. J., Bernstein, P. L., Leidemann, R., and Etlinger, J. P. (1986) Regulation of Ca^{2+} dependent protein turnover by thyroxine, *Biochem. J.* **240:**269–272.

6

Extensive Release. Excessive Storage

This chapter concerns important "exaggerations" of lysosomal function observed in specialized cell types or in particular developmental or pathological states. Cases will be considered in which the contents of lysosomes are released in quantity from cells or within cells, or in which lysosomal accumulations build up as a consequence of the failure of digestion.

6.1. Some Cell Types Secrete Lysosomal Materials Extensively as a Normal Function

6.1.1. Osteoclasts Create Extracellular, Lysosome-like Digestive Compartments

Osteoclasts are multinucleated giant cells that arise by the fusion of monocyte-related precursors. Under the influence of parathyroid hormone, vitamin D, calcitonin, and other regulators, they mediate the reabsorption and remodeling of bone during normal growth and in pathological states. This the osteoclasts do by adhering closely to the surface of the bone, secreting lysosomal hydrolases and other degradative agents and then endocytosing materials quarried from the bone for further degradation within heterophagic compartments. Both secretion and endocytosis are focused at a specialized domain the osteoclast establishes where it adheres to bone (Fig. 6.1). At this zone, the osteoclast's membrane is extensively folded and invaginated—"ruffled." The extracellular space between the ruffled membrane and the bone—the **resorption chamber**—has a markedly lower pH than adjacent extracellular regions. This is detectable by microscopic procedures like those used to study lysosomes and by direct microelectrode measurements, which indicate that the pH is well below five. The osteoclast's cytoplasm at the rim of the chamber is rich in actin filaments and other cytoskeletal elements and its plasma membrane is closely appressed to the bone's surface. This organized border evidently minimizes diffusional exchanges between the resorption chamber and adjacent extracellular spaces; some such sealing off of the chamber would seem necessary for the maintenance of the low pH and to prevent enzymes secreted by the osteoclast from diffusing away. The seal also minimizes the influx of substances that could operate against resorption, such as the protease inhibitors present in extracellular media (Section 4.5.1.1).

319

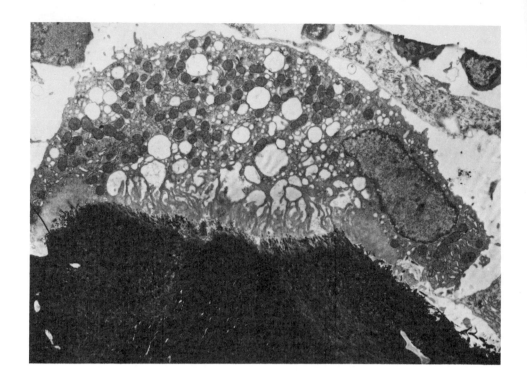

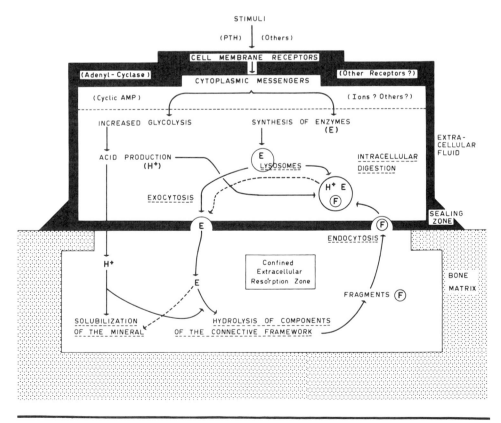

Although cytochemical methods for demonstrating lysosomal hydrolases generally show weak activity, if any, in the resorption chamber, the current consensus is that osteoclasts secrete sufficient lysosomal enzymes into their chambers for the enzymes to degrade the collagen, glycosaminoglycans, and other constituents of bone's organic matrix.* This breakdown continues in the endocytotic structures formed at the ruffled border, which pick up hydrolases as well as partially digested matrix materials. The lysosomal enzymes are aided in their attack on the bone by osteoclast-secreted collagenases and other non-lysosomal enzymes like those released by macrophages (see Sections 4.4.1.2 and 6.2.2).

The low pH in the resorption chamber sustains the activities of the secreted lysosomal hydrolases and also contributes directly to resorption by solubilizing mineral components of bone. The osteoclast maintains the acid conditions by means of "proton pumps" localized to the ruffled membrane bounding the acidified chamber. Several of the antibodies to lysosomal membrane proteins discussed in Section 3.2.3.3 have been found, immunocytochemically, to bind selectively to this region of the chick osteoclast cell surface. These antibodies include the one against a 100-kD component of rat lysosomes that cross-reacts with an H^+, K^+-ATPase from guinea pig gastric mucosa—the corresponding antigen at the osteoclast surface is thought to be part of the lysosomal acidification system. The proteins needed to maintain the low pH in the chamber probably are inserted into the ruffled region of the osteoclast surface by localized exocytosis and are prevented from diffusing laterally out of this domain by the specializations of the plasma membrane at the chamber rim, where the membrane adheres tightly to the bone.

Carbonic anhydrase activity in the osteoclast cytoplasm is a possible supplier of the protons needed to acidify the extracellular chamber.

---

*Cytochemical methods do readily demonstrate lysosomal enzymes in the osteoclasts' ER, Golgi apparatus, and presumed secretory vesicles; some of the latter structures are coated. [The "large" mannose-6-phosphate receptor (Section 7.3), thought to be involved in hydrolase sorting, can also be demonstrated immunocytochemically in the ER, Golgi apparatus, and vesicles, as well as in some endocytotic structures.] The customary explanations for the feebleness of demonstrable activity in the resorption chamber are: that the enzymes are too dilute to be readily detected by the insensitive methods used in cytochemistry; that endogenous substrates compete with those added in the cytochemical media; and that bone poses special technical problems that impair cytochemical methods. These explanations are essentially ad hoc so it is fortunate that direct biochemical evidence confirms the ability of osteoclasts to secrete lytic enzymes, including acid hydrolases, profusely. In addition, the presence of cathepsins in the resorption chambers has been demonstrated by immunocytochemistry.

←

FIG. 6.1. Osteoclasts. The electron micrograph is of an osteoclast from the metaphysis of a rat tibia, *in situ* upon the surface of the bone (approx. × 5000). The lower panel summarizes an interpretation of resorption mechanisms in such configurations. PTH, parathyroid hormone.

From Vaes, G., Eeckhout, Y., Delaisse, J.-M. Francois-Gillet, C., and Lenaers-Claeys, G. (1982). In *Collagen Degradation and Mammalian Collagenases* (M. Tsuchiya, R. Perez-Tamayo, I. Okazaki, and K. Maruyama, eds.), Excerpta Medica, Amsterdam, p. 223. (Copyright: Elsevier, Amsterdam.) The micrograph was provided by R. Schenk. [See *Verh. Dtsch. Ges. Pathol.* **58**:72, 1974 (Copyright: Gustav Fischer Verlag.).] Also see *J. Cell Biol.* **39**:676, 1968, and *Lysosomes in Biology and Pathology* (J. T. Dingle and H. B. Fell, eds.), Vol. 1, 1969, North-Holland, Amsterdam.

6.1.1.1. Do Other Cell Types Create Comparable Extracellular Compartments?

Elements of the mechanisms used by osteoclasts to resorb bone are available to other cells for other purposes.

When macrophages adhere by their Fc receptors to large surfaces coated with immune complexes, they create extracellular chambers whose communication with the rest of the extracellular milieu is restricted at least to the extent that the space between the macrophage and the surface becomes inaccessible to tracer proteins such as fluorescein-labeled immunoglobulins. Similar chambers established by macrophages adhering to collagen-covered surfaces are found to be markedly acidified. In addition to promoting localized digestion of extracellular materials, the ability to establish such adhesive sites might help account for phenomena in which macrophages, lymphocytes, or eosinophils kill other cells (to which the immune system has targeted them), by releasing toxic agents. The killer cells seem to have several different mechanisms at their disposal but among the putative "cytocidal" agents are reactive oxygen metabolites (Section 4.4.2.2), secreted proteases and other enzymes, and secreted proteins that form porelike holes in the surfaces of the target cells.

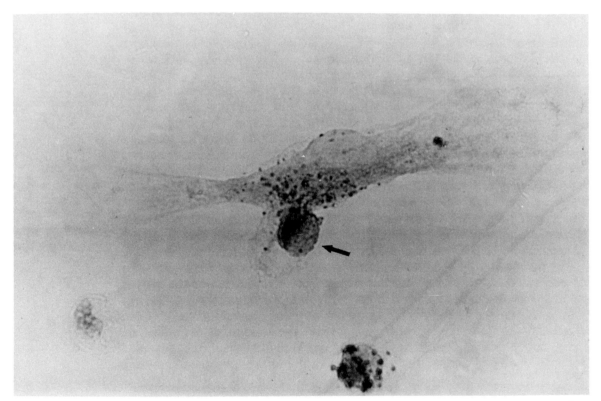

FIG. 6.2. Probable direct, cell-contact-dependent, transfer of the lysosomal enzyme β-glucuronidase from a mouse macrophage (arrow) to a human fibroblast. The cells were cocultured on coverslips [with M6P added to suppress long-distance transfer via secretion and endocytosis (see Section 6.4.1.4)]. The fibroblast cannot make its own β-glucuronidase because it is from a patient with a genetic enzyme deficiency (mucopolysaccharidosis, type VII) but it has acquired glucuronidase activity, demonstrated here by cytochemical staining of the numerous granule-like structures present near the zone of adhesion between the two cells. From McNamara, A., Jenne, B. M., and Dean, M. F. (1985) *Exp. Cell Res.* **160:**150. (Copyright: Academic Press.)

As with osteoclasts, releasing the agents into sealed-off compartments formed between the killer cells and their targets, could be advantageous for producing elevated local extracellular concentrations, for avoiding interference by extracellular inhibitors, and for restricting damage to neighboring cells and tissues.

Macrophages and lymphocytes can transfer lysosomal enzymes to membrane-bounded compartments within cultured fibroblasts through processes requiring adhesion of the donor cells to the recipients (Fig. 6.2). It may be that the adherent partners establish a sealed-off compartment into which the donor secretes enzymes, which then are endocytosed by the fibroblast.

6.1.2. Surfactant Secretion in the Lung

The type II alveolar cells ("granular pneumocytes") of the alveolar epithelium secrete **surfactant,** a coating material crucial to the maintenance of normal airway structure and air exchange. Surfactant includes specific proteins but is especially rich in phospholipids, typified by the disaturated phospholipid, dipalmitoyl phosphatidylcholine (DPPC). Within the type II cells, the phospholipids are packaged mostly in large, membrane-bounded "granules" often called lamellar bodies because their lipid content gives them a distinctive lamellated ultrastructure (Fig. 6.3). The bodies also contain immunocytochemically demonstrable surfactant proteins. Some of the lamellar bodies arise by transformation of MVBs, and many appear to be accessible to exogenous endocytotic tracers (Fig. 6.4). Cytochemical studies and biochemical analyses of

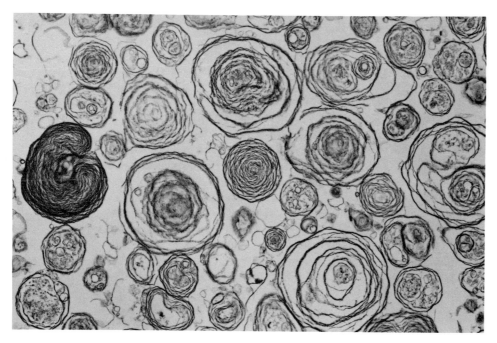

FIG. 6.3. Electron micrograph of a fraction highly enriched in lamellar bodies, isolated from rabbit lung on a discontinuous density gradient. As compared to intact lamellar bodies (Fig. 6.4), these seem somewhat swollen and they lack bounding membranes, but they show the lamellar structure of the bodies to advantage. The fraction shows little contamination with other recognizable structures (though it would be difficult to assess its content of extracellular surfactant and other materials from pulmonary extracellular spaces). × 10,000 (approx.). From Hook, G. E. R., and Gilmore, L. B. (1982) *J. Biol. Chem.* **257:**9211.

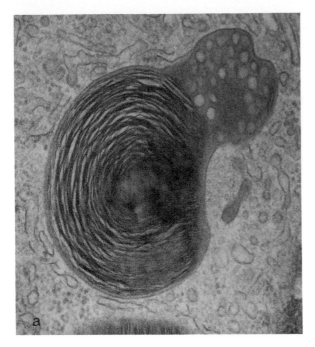

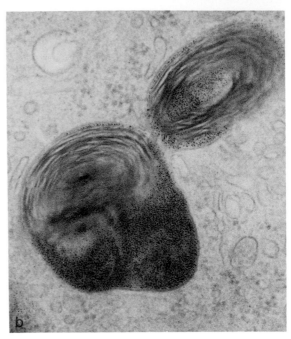

FIG. 6.4. From preparations of rat lung into which cationic ferritin (Section 2.2.2.3) had been introduced by catheter. **(a)** The occurrence of this sort of composite body, part MVB and part lamellar body, is taken as morphological evidence that lamellar bodies are related to lysosomes and endosomes. **(b)** The lamellar bodies shown here have accumulated cationic ferritin endocytosed by the cells. (The tracer is visible as small electron-dense grains scattered within the bodies.) After several hours of uptake, virtually all the lamellar bodies of the cell are labeled. From Williams, M. C. (1984) *Proc. Natl. Acad. Sci. USA* **81**:6054. × 45,000 (approx.).

isolated lamellar bodies indicate that the bodies possess a spectrum of lysosomal hydrolases (Table 6.1). They also accumulate the usual weak bases by an ATP-stimulated mechanism. Thus, within the limits of available methodology, the lamellar bodies qualify as lysosomes by the usual criteria.* Release of lamellar body contents to the lining of the alveolus is by exocytosis.

Is the lysosomal nature of lamellar bodies important for surfactant metabolism? In theory, lysosomal phosphatases could contribute to the biosynthesis of disaturated phospholipids by dephosphorylating phosphatidic acid as a prelude to the production of diacylglycerols elsewhere in the cytoplasm. Or lysosomal enzymes might participate in the "remodeling" of phospholipids, a process wherein the type II cell uses phospholipases of the "A" variety, and acyltransferases, to replace unsaturated fatty acids with saturated ones. For

---

*Room for doubt remains as to whether all the lamellar bodies in a cell so qualify. The cytochemical data establish only that many of the bodies contain at least one hydrolase (acid phosphatase) and that some contain other enzymes such as aryl sulfatase; even with methods for acid phosphatase, it is difficult to show that a majority of the bodies contain hydrolases. The biochemical data demonstrating the presence of lysosomal enzymes were gathered on reasonably pure cell fractions, but such data by their nature can only provide average values for the isolated fraction as a whole and are prone to distortion by the presence of contaminating organelles. It is therefore not yet ruled out that some of a cell's lamellar bodies are nonlysosomal secretory organelles or that some are endosomes rather than "full-status" secondary lysosomes. Nor is it known how degradation of surfactant within those lamellar bodies that are lysosomal is limited or controlled, if in fact it is.

TABLE 6.1. Content of Acid Hydrolases in Fractions Like That in Fig. 6.3, Compared with a Companion Lung Fraction Enriched in Conventional Lysosomes[a]

	Unfractionated homogenized lung	Lysosomes	Lamellar bodies	Relative distribution[b]
Acid phosphatase (p-nitrophenylphosphate)[c]	9.1 ± 1.1[d]	38.9 ± 3.5[d]	73.5 ± 13.2[d]	2.2
Acid phosphatase (β-glycerophosphate)[c]	8.4 ± 0.3	47.4 ± 5.7	83.1 ± 12.5	2.0
Acid pyrophosphatase[e]	7.9 ± 2.1	26.2 ± 4.8	13.0 ± 7.2	0.6
Acid phosphodiesterase I[c]	0.2 ± 0.02	2.4 ± 0.4	3.2 ± 1.6	1.6
β-N-Acetylglucosaminidase[c]	26.8 ± 2.6	115.5 ± 19.6	100.8 ± 24.7	1.0
α-N-Acetylglucosaminidase[c]	0.09 ± 0.02	1.2 ± 0.2	1.0 ± 1.0	1.0
α-Mannosidase[c]	6.9 ± 1.3	20.3 ± 2.2	75.4 ± 9.6	4.2
α-Galactosidase[c]	0.6 ± 0.1	3.4 ± 0.4	1.7 ± 1.5	0.6
β-Galactosidase[c]	1.9 ± 0.2	8.3 ± 1.7	8.3 ± 3.1	1.1
α-Glucosidase[c]	0.16 ± 0.01	1.3 ± 0.2	2.8 ± 4.0	2.5
β-Glucosidase[c]	0.23 ± 0.02	3.5 ± 2.2	7.6 ± 5.4	2.5
β-Xylosidase[c]	0.18 ± 0.05	3.7 ± 1.3	13.4 ± 15.8	4.1
α-Fucosidase[c]	1.2 ± 0.2	4.8 ± 0.7	8.0 ± 2.7	1.9
β-Fucosidase[c]	0.24 ± 0.08	1.7 ± 0.5	3.7 ± 3.7	2.5
β-Glucuronidase[c]	NM[f]	1.0 ± 0.8	0.3 ± 0.5	0.3
Arylsulfatase[c]	0.8 ± 0.3	17.2 ± 3.6	6.6 ± 1.6	0.4
Sphingomyelinase[c]	0.27 ± 0.24	0.63 ± 0.27	2.5 ± 1.9	4.6
Elastase[g]	0.26 ± 0.06	0.47 ± 0.19	0.09 ± 0.03	0.2
Fatty acid esterase[c]	2.5 ± 0.4	14.0 ± 7.6	5.1 ± 3.5	0.4
Nonspecific esterase[c]	192.7 ± 23.8	293.6 ± 55.5	31.9 ± 5.0	0.1
Alkaline phosphatase[c]	54.4 ± 12.9	220.8 ± 12.4	134.4 ± 20.7	0.7
Alkaline phosphodiesterase[c]	1.5 ± 0.4	7.0 ± 1.6	6.9 ± 5.3	1.1
Phospholipid/protein (mg/mg)	0.19 ± 0.03	0.34 ± 0.03	5.01 ± 0.41	

[a]Appropriate control experiments with liposomes made from lamellar body components, the differences in relative activities of the various hydrolases between the lamellar bodies and the lysosomes, and the cytochemical literature, have convinced investigators of the lung that adsorption of hydrolases during isolation is unlikely to account for the presence of the lytic enzymes in lamellar body fractions. From Hook, G. E. R., and Gilmore, L. B. (1982), *J. Biol. Chem.* **257:**9211.

$$[b]\text{Relative distribution} = \frac{\dfrac{\beta\text{-}N\text{-acetylglucosaminidase (lysosomes)}}{\text{hydrolase (lysosomes)}}}{\dfrac{\beta\text{-}N\text{-acetylglucosaminidase (lamellar bodies)}}{\text{hydrolase (lamellar bodies)}}}$$

[c]Nanomoles of substrate hydrolyzed/min per mg protein.
[d]Mean ± S.D. ($N = 4$).
[e]Nanomoles of PO_4^{2-} liberated/min per mg protein.
[f]NM, not measurable.
[g]Absorbance units/min per mg protein.

neither of these pathways, however, is it clear that the role of lamellar body hydrolases is quantitatively important, as compared to the roles of similar enzymes in other places. Strictly speaking, it has yet to be shown unequivocally even that **newly synthesized** surfactant lipids must pass through lysosomal bodies on their way to the cell surface (see footnote on p. 324, and Section 6.1.5) or that most surfactant protein molecules use such a route. (Autoradiography, however, has shown that the lamellar bodies do acquire choline- and leucine-labeled macromolecules soon after the molecules are synthesized.) Acyltransferases and other synthetic enzymes have occasionally been reported in lamellar body-rich cell fractions but the reliability of the methods used and the significance of the findings are in dispute.

Quantitative studies on the rates of radioactive labeling of surfactant and on its metabolic turnover, strongly suggest that components of extracellular surfactant are recaptured by the type II cells and reused to produce new surfactant. Some surfactant normally is engulfed by alveolar macrophages and might thereupon be degraded into forms that can be passed back to the type II cells. But retrieval directly by the type II cells almost certainly accounts for most of the metabolic recycling. One could conceive that surfactant is digested extracellularly into relatively small molecules and that it is these products that reenter the type II cells, but experimental support for this possibility is not strong. Endocytotic uptake of surfactant seems more likely especially given that the lamellar bodies are accessible to endocytosed tracers. Do the lysosomal hydrolases in the lamellar bodies degrade retrieved surfactant, making its components available to the cell's synthetic machinery? Or perhaps the enzymes somehow prepare endocytosed surfactant for direct return from the lamellar bodies to the cell surface (e.g., the hydrolases might "cleanse" the surfactant of undesirable materials acquired from the air).

If lysosomal hydrolases are secreted from the lamellar bodies along with surfactant lipids, as seems likely, the enzymes might collaborate with activities of the alveolar macrophages in establishing an unpleasant environment for foreign organisms. The intraalveolar extracellular space where surfactant is present is sometimes represented as having a relatively low pH, although the measured values are not impressive by lysosomal standards—most do not fall much below 7—so that only a few acid hydrolases would be strongly active. The surfactant layer is an impediment to diffusion of ions to or from the cell surface, and it may be that the degree of acidification that does obtain reflects the secretion of acidified lamellar body contents into the space below this layer. Or, when the lamellar bodies undergo exocytosis, proton pumps might be inserted into the cell surface along with the other components of the bodies' delimiting membranes.

6.1.3. The Acrosomes of Sperm

The sperm of many animals secrete hydrolytic enzymes that help create pathways to the egg cell surface during fertilization. Typically, eggs are enveloped by layers rich in extracellular glycosaminoglycans and proteins. The enzymes participate in chewing holes in these envelopes and may also help disperse the additional, cellular, layers that surround eggs of some organisms. The enzymes are released from the acrosome, a specialized Golgi-derived secretory organelle that fuses exocytotically with the sperm's plasma membrane early in fertilization. This fusion is part of the "acrosome reaction," triggered by contact with the egg's envelope and mediated in part by changes in the distribution of inorganic ions—Ca^{2+}, K^+, Na^+, and H^+—across the sperm's plasma membrane.

The membrane of the acrosome is not retrieved after exocytosis. Some of it may be shed, along with bits of the plasma membrane, in the small vesicles that separate from the cell surface of activated sperm of mice and some other animals. But much of the acrosomal membrane remains inserted as part of the cell surface. In many invertebrates, the acrosome reaction includes the formation of one or more cytoplasmic processes that extend from the sperm toward the egg and ultimately fuse with the egg's plasma membrane to establish cytoplasmic continuity between the cells. The processes are clad in the membrane created by the fusion of the acrosome with the plasma membrane. In mammals such as the mouse, fusion of egg and sperm involves posterior regions of the sperm head, rather than the anterior zones where the acrosome exocytoses. But in these cases too, the acrosome reaction is a prerequisite for normal fertilization and many investigators surmise that acrosomal components modify the plasma membrane in ways that assist fusion with the egg.

As just exemplified, acrosomal morphology, behavior, and enzymology vary considerably among different species of animals. It is common, however, for the enzymes within the acrosome to include proteases and also hyaluronidases, neuraminidases, sulfatases, and other enzymes that digest glycosaminoglycans and oligosaccharides. **Acrosins,** serine proteases active at neutral pH, are particularly prominent in mammalian sperm: these enzymes are stored in the acrosome and apparently require proteolytic activation during fertilization. Other acrosomal proteins ("bindins"; "lysins"), studied primarily in invertebrates, may disperse components of egg envelopes nonenzymatically or promote attachment of sperm to egg. Some of the enzymes—the acrosins included—are believed to remain bound to the acrosomal membrane after the acrosome reaction has taken place. Binding could permit focusing of enzymatic action in the direct vicinity of the sperm head, as has been postulated to account for the fact that the tunnel the sperm makes as it passes through the egg envelope often is not much wider than the sperm itself.

Acrosomes are of interest here because some of the hydrolases they contain closely resemble typical lysosomal enzymes. In addition to the neuraminidases and hyaluronidases already mentioned, acid phosphatases, esterases, aryl sulfatases A and B, glucosaminidases, phospholipases, dipeptidyl peptidases (Fig. 6.5), and other hydrolases with acid pH optima have been found in the sperm of

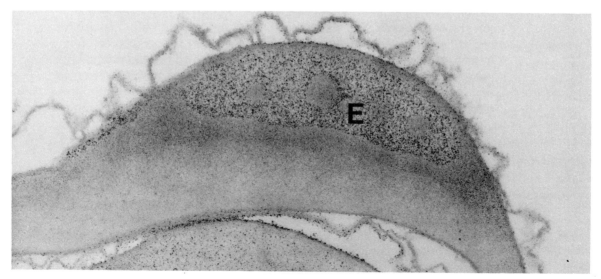

FIG. 6.5. Part of the large acrosome capping a guinea pig sperm, reacted cytochemically to demonstrate the exopeptidase **dipetidyl peptidase II.** The moderate sprinkling of electron-dense material at E is reaction product localized to a subregion of the acrosome, which might reflect a similar localization of the enzyme (but see Section 1.3.2). From Talbot, P., and Dicarlantonio, G. (1985) *J. Histochem. Cytochem.* **33:**1169. × 50,000 (approx.).

one or another species by biochemical or cytochemical assays. (Care must be taken in the biochemical work to rule out contributions from nonacrosomal sources, such as the lysosome-rich cytoplasmic droplets shed during sperm maturation.) Acid proteases like the lysosomal cathepsins are less frequently demonstrated, though they have occasionally been reported. In hamster, rat, and a few other species, acrosomes of living sperm can accumulate aminoacridine dyes or methylamine and therefore are considered to maintain a low internal pH. The mechanisms of this acidification have been studied only sketchily, but findings with the usual battery of inhibitors are reasonably consistent with the presence of an active proton pump.

For such reasons, acrosomes are frequently thought of as specialized or modified lysosomes in which the spermatid packages some of the hydrolases that cells ordinarily provide to lysosomes, along with other materials that pertain specifically to sperm. The acrosomal membrane is assumed to be specialized too. But it is not known how many of those acrosomal enzymes that resemble typical lysosomal hydrolases, are products of the same genes as their counterparts in ordinary lysosomes. And what is the nature of the extracellular milieu in the immediate vicinity of the reacting acrosome? Is the pH proper for lysosomal activity? Do inhibitors, like the acrosin inhibitors detectable in seminal plasma, limit hydrolysis by acrosomal enzymes so as to prevent overenthusiastic attacks on the egg envelopes?

In the eggs of some species, fertilization provokes a wave of exocytotic release of enzymes (e.g., peroxidases or proteases) from granules located just below the cell surface. These enzymes participate in modifying extracellular coats, detaching extra sperm, and other functions. Occasionally, the cortical granules have been analogized to acrosomes or even to conventional lysosomes, but there is little reason yet to press the analogies.

6.1.4. Glands, Plants, and Microorganisms

6.1.4.1. Prostate and Preputial Glands

The mammalian prostate gland and the rodent preputial gland are among the several glands, additional to elements of the digestive system, that secrete acid hydrolases in vertebrates. Prostatic secretions, along with enzymes from the testis and elsewhere, contribute to the relatively high acid hydrolase content of seminal fluid. The enzymes may help govern the cycle of coagulation and liquefaction undergone by the semen of many animals. The preputial gland has been used extensively as a source of β-glucuronidase for biochemical study but the metabolic importance of the enzyme is still uncertain; it could participate in the genesis of rodent pheromones.

The preputial gland and sebaceous glands employ holocrine secretory mechanisms so their release of acid hydrolases is as a facet of their general discharge of cytoplasm. In contrast, the prostate gland releases acid phosphatase and other hydrolases by exocytosis, from membrane-delimited secretory bodies packaged in the Golgi region. These bodies are thought to contain a more limited spectrum of enzymes than is present in typical lysosomes because the fluids into which prostatic secretions feed are rich in certain acid hydrolases, such as acid phosphatase, but not in others. The prostatic cells also contain ordinary lysosomes fulfilling the usual, nonsecretory functions.

Differences have been noted in immunocytochemical localization, enzymatic activities, antigenic properties, oligosaccharides, and responses to inhibitors between secreted prostatic acid phosphatase and comparable enzymes from known lysosomal sources. But the prostatic type of enzyme is not unique to the gland—similar (identical?) phosphatases are found in certain other cells. Overall, although it remains tenable to posit close evolutionary relations between the prostate's secretory bodies and "ordinary" lysosomes, as many authors have done, the closeness of the relationship cannot be assessed in the absence of amino-acid-sequence analyses of the enzymes, and genetic data.

6.1.4.2. Plants, Carnivorous and Otherwise

Cautions of similar genres apply to the miscellany of observations available on secretion of acid hydrolases in plant cells.

In the development of seeds of barley and several other major food plants, starchy endosperm tissues containing stored nutrients are digested by hydrolases—proteases, RNase, amylase, and others—secreted by adjacent, specialized layers of cells (aleurone layers).* Gibberellic acid is among the regulatory agents that induce production of these hydrolases.

Carnivorous plants secrete hydrolases into elaborate digestive chambers bounded by specialized cell walls. The chambers in some species are acidified and the hydrolases have correspondingly low pH optima.

The regulation of cell growth and differentiation by auxins and other plant growth regulators is popularly believed to involve the maintenance of a low pH in cell walls and extracellular spaces, and the secretion of lytic enzymes with acid pH optima, whose actions influence the plasticity of cell walls (see Fig. 5.18b). Secreted hydrolases, whose production is controlled in part by ethylene, participate also in leaf abscission and in other developmental phenomena, including the ripening of fruits.

In several of these examples, the hydrolases secreted by plant cells will probably prove to be released, exocytotically or by holocrine or autolytic mechanisms, from vacuoles or other lysosome-like compartments similar to those discussed in earlier parts of this book. For example, the secretion of digestive enzymes by certain carnivorous plants has been described simply as the exocytosis of lysosomal or vacuolar contents. But this description is based chiefly on observations that the relevant intracellular compartments and the extracellular ones both show reaction product for cytochemically demonstrable enzymes such as acid phosphatase and esterases. Suggestive as such observations are, cytochemical methods lack the discriminatory precision to rule out that different enzymes with overlapping properties might be present at the different locales. Keep in mind also that not every digestive process occurring in an acidified space is lysosomal; witness, for example, the digestion of foods in the stomachs of vertebrates.

6.1.4.3. Eukaryotic Microorganisms; Fungi

Invasive organisms such as fungi secrete cellulases and other hydrolases to assist their penetration of host structures. Many microorganisms that ordinarily are not invasive, including yeast and fungi, slime mold amebas, and protozoa,

---

*The structures in which endosperm of plants like barley or corn store nutrients during seed formation differ from those of legumes discussed in Section 5.4.2. The starch-rich endosperm of barley is laid down by cells that have died by the time digestion of the endosperm begins. And the protein bodies used for some of the storage in cereal seeds arise by accumulation of proteins directly within the ER. These bodies are unlikely to be lysosomal, which may explain why they require digestion from without in contrast to the digestion from within demonstrated for legumes.

also release proteases, glycosidases, phosphatases, and the like, to surrounding media, intensifying this release under conditions of general starvation or of deprivation of particular necessities. The slime mold *Dictyostelium discoideum* secretes hydrolases extensively at particular points in its developmental program. The yeast *Saccharomyces cerevisiae* synthesizes and secretes an inducible acid phosphatase when sources of inorganic phosphate run out. *Chlamydomonas reinhardtii*, a unicellular alga, secretes sulfatase when sulfate is low in its medium.

The secreted enzymes can be used to obtain or seek nutrients. *Pseudomicrothorax*, the ciliated protozoan that eats cyanobacteria (see Section 2.2.1.1), is believed to secrete lytic enzymes into the extracellular chambers where the filaments of its food become fragmented before being taken up. Other protozoa degrade portions of the exoskeleton of crustaceans to obtain food. More difficult to envisage is how secreted hydrolases could help when hungry cells release them into the surrounding medium, rather than into compartments or onto digestible surfaces; the enzymes would rapidly be diluted as they float into the extracellular world, not to mention matters of pH and the like. One view is that the enzymes released during starvation are less important in actually mobilizing nutrients than they are in generating trace amounts of products that signal the cell that useful nutrients are nearby. Another is that the cells limit the extent to which the enzymes become diluted. For example, some protozoa retain acid phosphatase and other hydrolases attached to their extracellular surfaces rather than releasing all of the enzyme molecules to the surrounding medium. In yeast and *Chlamydomonas*, the cell walls may retard diffusion of secreted enzymes away from the cell surface.

Various of the enzymes released by microorganisms exhibit no evident relationship to lysosomes—the invertase secreted by yeast is a well-known example. But some of the hydrolases are lysosomal. Release of lysosomal enzymes can be an incidental or accidental accompaniment of endocytotic membrane recycling (see Section 3.3.3) or of defecation of residues. Frequently, however, the levels of such release are uncomfortably high for "accidental" mechanisms. Endocytosing *Acanthamoeba* organisms can secrete 20% of their content of some lysosomal hydrolases per hour. Starving amebas of *Dictyostelium* secrete more than half of their content of many acid hydrolases over the course of a few hours. In this last example, preexisting lysosomes contribute heavily to starvation-induced release, but the secretion overall is more complicated than just the exocytosis of secondary lysosomal contents: The rates at which acid phosphatase is secreted differ from those for some of the other enzymes, and enzyme release does not parallel perfectly the release of indigestible materials like latex taken up before release was induced. What is more, some of the enzyme molecules are secreted without having undergone the maturation characteristic of the hydrolases in functioning lysosomes (Chapter 7), indicating that newly made lysosomal enzymes are being "diverted" from their travel to the lysosomes into a directly secretory route.*

Subpopulations of lysosomes specialized for use in hydrolase secretion have been postulated for *Tetrahymena pyriformis* and other protozoa because when the levels of activity of RNase, phosphatases, glycosidases, and other

*In many cell types, comparable diversion can be intensified by relatively simple experimental manipulations, such as exposure to weak bases, or by mutation. These effects have been exploited in experiments on the packaging of newly made hydrolases into lysosomes (Section 7.4.3).

hydrolases are compared to one another, the relative activities of the different enzymes in the set of hydrolases secreted to the medium diverge markedly from the ratios among the intracellular hydrolases. (In addition, overall rates of secretion of acid hydrolases from *Tetrahymena thermophila* correlate poorly with rates of defecation of indigestible tracers.) Subsets of lysosomes in which the ratios of some hydrolase activities resemble the secreted population have been tentatively identified in *Tetrahymena pyriformis* by density gradient centrifugation. But the fit is far from perfect and the conclusion that there are specialized "secretory" lysosomes must be further tempered by the many considerations that afflict analysis of the heterogeneity of lysosomes in a given cell (Section 7.5.1.2; one problem is that measurements of enzyme **activities** need not accurately portray the extent of release of enzyme molecules—e.g., conditions in the extracellular medium may differentially inhibit some enzymes).

6.1.5. Hydrolases in Secretion Granules. Processing of Polypeptides for Secretions

A frequent, but puzzling cytochemical finding on many glands, both exocrine and endocrine, is the presence of acid phosphatase activity in secretion granules, particularly in some of the granules whose appearance and location suggest they are among the ones most recently formed by the cell. Activity is also found in the Golgi sacs responsible for packaging the secretions into the granules. Cells producing melanin granules show a similar distribution of acid phosphatase (Section 4.3.1).

Other acid hydrolases mostly have not been demonstrable in the acid phosphatase-containing secretion granules, but proteases having antigenic properties in common with lysosomal proteases are sometimes detectable. And there is immunocytochemical evidence that secretory materials and various lysosomal hydrolases share some common locations within the dispute-ridden networks of membranes present near the "*trans*" face of the Golgi apparatus (Section 7.2.2.2).

Two major schools of thought about these observations have crystallized. One argues that the acid phosphatase in secretory structures is nonlysosomal even though in some properties, such as its inhibition by fluoride ions, it resembles the acid phosphatase(s) demonstrable cytochemically in lysosomes. By analogy with likely roles of diphosphatases in the Golgi apparatus, nonlysosomal acid phosphatases in Golgi sacs and new secretion granules might be regulatory agents, hydrolyzing the nucleotides or other by-products of the glycosylation of nascent glycoproteins (Section 7.1.1.1).

The opposing position is that the acid phosphatase is a genuine lysosomal enzyme, which somehow has become included in the secretion granules. The difficulty in demonstrating the presence of other lysosomal enzymes could be a matter of deficiencies in cytochemical techniques (e.g., their incapacity to detect immature forms of the hydrolases; see Section 7.4.1.1). Or it might mean that the secretion granules actually do contain only a few of the enzymes characteristic of definitive lysosomes.* The acid phosphatase could stray in by

---

*The mechanisms by which lysosomal hydrolases are routed and packaged probably differ somewhat for different enzymes (Section 7.3.3) and it is not hard to conceive of selective "diversion" of molecules of particular hydrolases to a nonlysosomal destination. One lysosomal enzyme—β-glucuronidase—is, in fact, differentially retained in the ER of rodent liver (Section 7.3.3.1).

accident—because, in using the same Golgi apparatus to handle lysosomal enzymes, secretions, and other materials, the cell cannot successfully sort or segregate every molecule to its proper compartment. Or lysosomal enzymes might be put into the secretory apparatus "deliberately," to function in the genesis of secretion granules.

6.1.5.1. Proteolytic Processing of Nascent Secretory Molecules

Discord over these alternatives has bubbled on back burners for more than a decade, awaiting either a better understanding of how the Golgi apparatus transports secretory materials and lysosomal enzymes, or impasse-breaking technical advances: cell fractionation methods good enough to resolve the phosphatase-containing granules and Golgi sacs from contaminating structures; immunocytochemical procedures sensitive enough to enable a more definitive decision about the presence or absence of hydrolases other than acid phosphatase in secretory bodies; or cell strains in which one might determine whether mutations affecting lysosomal phosphatases also affect the phosphatases in secretion granules. Meanwhile, other closely parallel disputes have broken out over findings that many polypeptide secretions—insulin, neurosecretory materials, albumin, the α-factor of yeast, and others—undergo maturational proteolysis during or soon after Golgi packaging of the secretions into membrane-delimited secretory bodies.

The processing of these polypeptides frequently involves cleavage by "convertase" enzymes at lysine–arginine sequences or other pairs of basic amino acids. In many cases, cleavage is followed by carboxypeptidase trimming and then by nonproteolytic modifications such as the NH_2-terminal acetylations and COOH-terminal amidations of neurosecretory products (Fig. 6.6).

The "convertases" are proteases, some described as trypsin-like serine proteases because they favor paired basic residues as sites of attack, and some

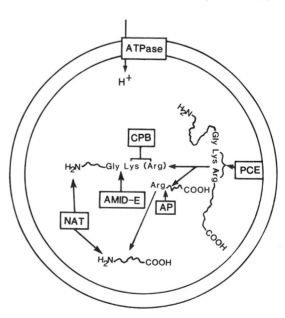

FIG. 6.6. Newly packaged neurosecretory polypeptides of the mammalian pituitary gland are processed within the secretion granules, one of which is schematized here, to generate smaller biologically active peptides. Details vary in different cell types and organisms. The initial hydrolysis by a "trypsin-like" converting enzyme (PCE) can yield precursors of two (or more) active final products. Hydrolysis by the converting enzyme is followed by steps that can include peptidase "trimming" (CPB, carboxypeptidase; AP, aminopeptidase), α-amidations at COOH termini of the peptides (AMID-E), and N-acetylations at NH_2 termini (NAT). From work by Y.-P. Loh and H. Gainer at the NIH; after Loh, Y.-P. (1985). In *Biogenetics of Neurohormonal Peptides* (R. Hakanson and J. Thorell, eds.), Academic Press, New York.

called "cathepsin-like" from their immunological properties or other features. These enzymes, and the carboxypeptidases used for trimming, have been demonstrated in cell fractions rich in secretory bodies. They differ from enzymes typical of lysosomes. For example: (1) The cathepsins that predominate in most lysosomes (see Section 1.1) are not serine proteases or trypsin-like. [Cathepsin B, however, was once thought to be trypsin-like, acrosin (Section 6.1.3) is a serine protease with some trypsin-like properties, and the elastase and cathepsin G of neutrophils are serine proteases.] (2) The convertases carry out much more limited cleavages of their substrates than do lysosomal enzymes. (3) The secretory-body peptidases are cobalt dependent unlike the comparable exopeptidases in lysosomes. From these differences, many investigators have concluded simply that the secretory-body enzymes bear no more than evolutionary relationships to lysosomal hydrolases. The convertases are, however, present in such minute amounts within secretory granules that progress in purifying them and determining their amino acid sequences has been slow and there is much else yet to be learned about conversions of secretions (and about lysosomal enzymes). Champions thus still exist for the notion that the observed differences in enzyme characteristics between convertases and lysosomal hydrolases are misleading. Perhaps, some assert, certain of the lytic enzymes packaged in secretory granules actually are lysosomal enzymes whose characteristics are atypical because the enzymes are in immature form, or because they have been modified during transport to the granules or affected by conditions within the granules. These last lines of thought have gained a bit of force with the appreciation that some of the enzymes that process secretions have acid pH optima, and that the pH is 5.5–6 within several types (not all) of secretion granules in which cells package proteolytically processed proteins. [Also, Golgi sacs involved in the production of secretions may also be moderately acidified (Section 7.2); this is one of the bases for speculation that proteolytic conversions of secretions may be initiated in Golgi compartments even when the conversions are completed later, in secretory packages.]

Finally, the possibility has been bruited, without much evidence one way or the other, that hydrolases of the Golgi apparatus or in secretion granules serve some frankly degradative functions—in crinophagy or in the disposal of some of the inactive protein fragments generated during processing of secretions.

6.2. Hydrolases in Extracellular Spaces in Animal Tissues

A secretory product that plagues its students with uncertainties like those just outlined is **renin**, a cathepsin D-like "aspartyl protease" and a participant in the cascade of proteolytic conversions that produce the vasoconstrictor angiotensin.* Renin is produced, probably as a proenzyme, by cells such as the "juxtaglomerular cells" of the kidney and is stored in and secreted from abundant granules in the cells. Some of the granules react cytochemically for acid phosphatase and some of the granules bind antibodies raised against cathepsin

---

*Traditionally, angiotensin was supposed to be generated chiefly in the circulation, from suitable precursors. Insurgent minority opinion now suggests, however, that much may actually be generated within tissues (and perhaps even within cells that either both synthesize angiotensin and then proteolyze it before it is secreted, or acquire the protein endocytotically).

B as well as antibodies against renin (and probably antibodies against α-glucosidase). Moreover, certain of the structures containing renin have all the earmarks of AVs. Renin is enough like a lysosomal enzyme that when cDNAs specifying this protease are injected into L cells or *Xenopus* oocytes, the renin molecules produced possess mannose-6-phosphates on their oligosaccharides, a hallmark of lysosomal hydrolases (see Chapter 7).

Renin is one of the surprising variety of hydrolases normally present in serum, synovial fluid, tears, urine, or other extracellular locations. The interstitial fluid that "bathes" the cell must be presumed to have some of these enzymes and even the cerebrospinal fluid is claimed to contain several. Prior sections have alluded to these hydrolases (e.g., Section 6.1.4) and certain ones are quite familiar—those of the gut and seminal fluid, for example, and the ones responsible for blood clotting and complement reactions (Section 4.4.1). The aim of this section, and the following few, is to examine normal and pathological phenomena involving hydrolases released to tissue fluids and spaces, with an eye especially toward understanding how lysosomes fit in.

6.2.1. Protection; Turnover; Activation and Inactivation of Biologically Active Molecules

We have already encountered examples of the three most obvious classes of roles the hydrolases are likely to play: (1) Sections 4.4.1.2, 4.4.2, and other sections alluded to their **defensive** potentials in attacking foreign organisms and materials. (2) The enzymes are often blamed (perhaps not always justly) for modifying other proteins, or cells, so as to trigger degradative **turnover** mechanisms (Sections 4.5.2.3 and 4.6.1.2) or necessitate repair (Sections 5.3.3.3 and 5.5.4.3). (3) They **activate** or **inactivate** other biologically active molecules.

This last set of roles includes participation not only in the complement and fibrin cascades that activate blood proteins but also in processes such as the normal assembly of collagen fibers, which depends on extracellular proteolytic modifications of the procollagen secreted by fibroblasts. On the other hand, many of the biologically active extracellular molecules that influence physiology or pathology—hormones, growth factors, proteins secreted by lymphocytes, and so forth—can be **inactivated** by lysosomal proteases or other secreted enzymes, at least in the test tube.* Because of such activations and inactivations, the concentrations and effects of regulatory or mediating molecules at loci of tissue damage or inflammation and at other sites where extracellular hydrolases abound, will depend not only on the rates of supply of the molecules but also on their interactions with the enzymes they meet.

6.2.2. Degradation of Connective Tissue Matrices; Arthritis and Autoimmune Disorders; Tumor Cells

Though the constituents of highly organized connective tissue fibers and matrices are very long lived, collagen, glycosaminoglycans (GAGs), and other connective tissue constituents do turn over. (Whether this is to be regarded as a

---

*Some of the inactivating and degradative enzymes that act extracellularly are attached to cell surfaces. This is true of lipases and of several of the peptidases that inactivate small peptide hormones or neuroeffectors. Cell surface locations have been suggested as well for neuraminidases.

true steady-state turnover is not established; see Section 4.5.2.4.) Normal half lives are measured in weeks or months for the GAGs of nonmineralized tissues and in months or years for some of the collagens; given their differences in location, organization, and other properties, one would expect different subpopulations of collagen molecules or of GAGs to turn over at different rates, even in a given tissue, and the bits of available data sustain this expectation. Quite modest levels of extracellular hydrolases would be needed to account for the observed normal turnover, but attention has been paid mainly to the elevated enzyme levels noted in certain developmental situations and especially in pathological conditions. Of particular interest have been disorders like rheumatoid arthritis, or periodontal diseases, where extracellular hydrolases contribute to dramatic erosion of cartilage, bone, and softer connective tissues. The disorders involve chronic inflammatory responses with associated influxes of blood-derived phagocytes, but these cells account for only part of the elevation of extracellular hydrolases. For example, in arthritis, hydrolases also come from cells proliferating in the synovial lining and probably from chrondrocytes as well.

Many explanations of the disorders rely on postulated or observed immune phenomena, often of types that could result directly in increases in extracellular hydrolases. For instance, in certain types of arthritis, antigen–antibody complexes resulting from autoimmune responses come to coat the surfaces of joints, and in some autoimmune diseases involving nephritis, comparable piling up of immune complexes takes place at the glomerular filters in the kidney. Hydrolase release at these sites is likely to attend the "frustrated phagocytotic" efforts of macrophages and neutrophils trying to engulf the immune complexes; the phagocytes are thought to fuse lysosomes and other enzyme containing bodies with the plasma membrane regions facing the surfaces on which the cells spread.

Interleukins and other agents secreted by lymphocytes and leukocytes also stimulate cartilage erosion, though it is not yet obvious just how this comes about, or why, in addition, synovial cells secrete proteins that stimulate release of collagenolytic enzymes from cultured chondrocytes. Many such effects are becoming known, especially from work on tissue cultures and explants, and they are beginning to help in picturing the pathogenesis of connective tissue damage, at least at the phenomenological level.

Connective tissues are extensively resculpted in developmental and regenerative reorganizations—wound healing, the reconstruction of the postpartum uterus, and so forth. Extracellular degradative enzymes probably participate in this remodeling by cutting up the protein fibers and the proteoglycan matrices of the connective tissues. Evidently, some of the resulting pieces can move away in extracellular fluids; relatively large segments of GAGs and proteins even wind up in the urine and are excreted. But the macrophages and other cells that accumulate in reorganizing tissues probably phagocytose many fragments of the fibers and matrices locally, and degrade them. Lysosomal enzymes such as dipeptidyl peptidase II, which is active against prolyl peptides, seem "designed" in part for dealing with products of the partial breakdown of collagen.

Cells from natural tumors, and transformed cultured cells release much higher levels of proteases to culture media than do most of their normal counterparts. These enzymes could exert many effects relevant to the properties and behavior of tumors. For example, they may alter the cells' surfaces, and their adhesion to one another or to extracellular materials. (Are the mitogenic effects and the influences on cell differentiation seen when tissue culture cells are exposed to proteases manifested in intact organisms as well?) The enzymes' impact on local concentrations of hormones, growth regulators, or angiogenic factors could also be significant, and they might also interfere with the processing of antigens (Section 4.2.4.2) or other immune phenomena. Relevant to this section of the book, the enzymes might aid in the disrupting of connective tissues and cellular layers as is required for effective metastasis (e.g., for the penetration of tumor cells through basal laminae). Do normal migratory cells, such as macrophages moving through connective tissues, also use secreted hydrolases to help forge their pathways?

6.2.3. Where Do the Extracellular Hydrolases in Tissues Come From? (Even Professional Phagocytes Secrete Nonlysosomal Enzymes)

First guesses often attributed much of the release of lytic enzymes to lysosomes. The presence of small amounts of extracellular hydrolases under normal circumstances was presumed to reflect steady-state lysosomal defecation. When enzymes accumulated in high concentrations, release mechanisms were sought in the death of neutrophils or macrophages, in frustrated phagocytosis, and in the leakage ("drooling") that occurs when phagocytes tackle large particles and engulf them slowly enough that lysosomes fuse with forming phagocytotic vacuoles not yet closed off from the exterior world.

Current portraits are more variegated. For one thing, it is now quite evident that many of the lytic enzymes secreted even by lysosome-laden professional phagocytes (Table 6.2) do not have acid pH optima, come from nonlysosomal compartments, and can be secreted without simultaneous release of acid hydrolases. Neutrophils release enzymes from their specific granules (Section 4.4.1.2) and they secrete a few other proteins, such as a "gelatinase" active against denatured collagen, whose intracellular sources are uncertain but seem not to be lysosomal. The lysozyme- and the proteoglycan-degrading enzymes secreted by macrophages, and most of the collagenases released by macrophages, neutrophils, and fibroblasts are not from lysosomes. (The collagenases are among the most important secretions for the concerns of this part of the book because one or another of them can initiate attacks on native forms of the various species of collagen found in fibers, basal laminae, and elsewhere. Native collagens resist attack by most other proteases.)

Complementing the lysosomal elastase of neutrophils (a serine protease), macrophages secrete a nonlysosomal metalloprotease that can act on elastin, as well as on other proteins. Many cell types secrete nonlysosomal proteases that, for convenience, are assayed as "plasminogen activators," angiotensin-converting enzymes, and the like, though their biological roles often are not identified and their ranges of action probably are much broader than their designation as "activators" or "converters" implies.

Uterine epithelial ("endometrial") cells secrete **uteroferrin,** a glycoprotein that transports iron to the fetus, perhaps by a transcytotic mechanism. This protein resembles lysosomal hydrolases in having acid phosphatase activity not inhibited by tartarate (see Section 1.3.2.3) and in that its oligosaccharides contain M6P residues (Section 7.4.2.1). Similar proteins have been detected in liver, and in the spleen a protein of this type is found in lysosome-like granules.

TABLE 6.2. Materials Known or Strongly Suspected to Be Secreted by Macrophages under Various Conditions[a]

Enzymes	Complement components
Neutral proteinases	C1
Plasminogen activator	C2
Metal-dependent elastase	C3
Collagenase, specific for interstitial col-	C4
lagens (types I, II, III)	C5
Collagenase, specific for basement mem-	Properdin
brane collagen (type IV)	Factor B
Collagenase (gelatinase), specific for	Factor D
pericellular collagen (type V)	Factor I (C3b inactivator)
Stromelysin	Factor H (β_1H, C3b inactivator acceler-
Cytolytic proteinase	ator)
Arginase	Reactive metabolites of oxygen
Lysozyme	Superoxide anion
Lipoprotein lipase	Hydrogen peroxide
Angiotensin-converting enzyme	Others
Acid hydrolases	Bioactive lipids
Proteinases and peptidases	Prostaglandin E_2
Glycosidases	6-Ketoprostaglandin $F_{1\alpha}$
Phosphatases	Thromboxane B_2
Lipases	Leukotriene C (slow-reacting substance
Others	of anaphylaxis)
Plasma proteins	12-Hydroxyeicosatetranoic acid
α_2-Macroglobulin	Others
α_1-Proteinase inhibitor	Nucleotide metabolites
Tissue inhibitor of metalloproteinases	cAMP
Plasminogen activator inhibitor	Thymidine
Fibronectin	Uracil
Transcobalamin II	Uric acid
Apolipoprotein E	Factors regulating cellular functions
Coagulation proteins	Interleukin-1 (endogenous pyrogen)
Tissue thromboplastin	Angiogenesis factor
Factor V	Interferon-α
Factor VII	Transforming growth factor-α
Factor IX	Transforming growth factor-β
Factor X	Platelet-derived growth factor
	Basic fibroblast-derived growth factor
	GM-CSF, G-CSF, M-CSF[b]
	Tumor necrosis factor α
	Erythropoietin

[a]Courtesy of Z. Werb. [Derived from Werb, Z. (1987) In *Basic and Clinical Immunology* (6th ed.) (D. P. Sittes, J. D. Stobo, and J. V. Wells, eds.), Appleton & Lange, Los Angeles.]
[b]CSFs (colony-stimulating factors) are proteins that help control the proliferation, maturation, and activities of macrophages (M) and granulocytes (G).

    Uteroferrin, like renin, is one of the enigmatic secretory proteins that seem related to lysosomal enzymes in major properties but have been studied chiefly in nonlysosomal sites and roles. Another example is a protease resembling the lysosomal cathepsin L, secreted by mammary gland cells in culture. The "major excreted protein" of mouse fibroblasts is also a cathepsin or a close relative. [It has M6P residues and may actually be cathepsin L in an enzymatically active, precursor form (Section 7.4.1).] Secretion of this protein is increased as much as 100-fold when the cells undergo oncogenic transformation.

6.2.3.1. Details of Hydrolase Release

Augmented release of hydrolases to tissue fluids and spaces need not depend on cell death or on "accidents" during endocytosis; exocytotic secretory pathways are available. This is clear from demonstrations that even when not dying or trying to endocytose large structures, unicellular organisms (see Section 6.1.4.3) or cultured mammalian cells can be induced to release much of their content of lysosomal hydrolases or other sets of enzymes in the course of one or a few hours. Mouse macrophages obtained by thioglycollate infusion release over 50% of their lysosomal enzymes during 3 hr. of exposure to NH_4Cl (Section 7.3.1.2). Extensive release of lysosomal enzymes from macrophages also can be triggered by particles such as opsonized zymosan, acetylated lipoproteins (Section 4.1.1.3), or asbestos fibers. Cultured fibroblasts under ordinary conditions release 1–10% of their content of lysosomal hydrolases to growth media per day and like the macrophages, augment this secretion on exposure to weak bases. Rodent kidney tubule cells *in situ* release 20% or more of their β-glucuronidase and β-galactosidase per day with no special inducement. [This release is much reduced in mice affected by various "pigment" mutations (Section 4.4.2.4).] Secretion of β-glucuronidase from the kidney can be enhanced selectively by testosterone. Platelets secrete acid hydrolases upon encounters with thrombin or collagen. When stimulated to release adrenaline, the adrenal medulla simultaneously increases its secretion of acid hydrolases.

Most of the information about the processes and controls of hydrolase secretion was gathered from explants and cell cultures and though many cell types have been studied—phagocytes, chondrocytes, fibroblasts, platelets, synovial cells, and so forth—the findings should not uncautiously be extrapolated to situations *in situ*. For example, macrophage-like cell lines seem to secrete lysosomal hydrolases at a much more rapid pace than do macrophages freshly obtained from animals, which tend to secrete such enzymes only if provoked as above. Nevertheless, some helpful outlines are emerging. For instance, release of hydrolases often responds to perturbations of the cytoskeleton or to influences affecting intracellular concentrations of inorganic ions, in much the same ways as does exocytosis of other material. In a few cases (e.g., in *Dictyostelium* amebas), release has been shown to require metabolic energy. And, as noted earlier, lysosomal enzymes secreted by macrophages and other animal cells can come from the population of preexisting lysosomes, by diversion of newly synthesized enzymes into a secretory route (see Section 7.3.1.1 and 7.3.1.2), or from both of these sources, depending upon the cell type and the circumstances.

Some hydrolase secretion is constitutive, in the sense that no specific triggering is needed under "ordinary" conditions. For instance, macrophages steadily release lysozyme, at least in culture. But examples already cited illustrate the diversity of experimental stimuli—immune complexes, lymphokines, phorbol esters, sucrose, bacterial lipopolysaccharides—that can alter the rates of secretion of various enzymes. There are abundant signs that comparable natural stimuli help integrate important biological phenomena. For example, inflammatory or activated macrophages secrete notably larger quantities of proteases than do macrophages in more quiescent states. Synovial fibroblasts cultured from normal joints, without special activation, secrete little collagenase, but considerable amounts of this enzyme are released for prolonged periods from cultures of arthritic material, or when normal cells are exposed to

phorbol esters or to interleukin-like molecules released by lymphokine-stimu-
lated macrophages.

339

EXTENSIVE
RELEASE.
EXCESSIVE
STORAGE

6.2.3.2. What's It Like Out There?

Even though it is now possible to specify many of the enzymes likely to be present extracellularly under given conditions, it still is difficult to predict precisely how they will behave once there. For example, normal, overall, extracellular pHs seem unlikely to favor activity of most of the acid hydrolases but some observers believe that the pH at sites of inflammation or in the immediate vicinity of sulfated GAGs may be low enough, at least transiently, for lysosomal enzymes to act.

The enzymes released in a given locale may interact with one another as well as with molecules from the blood. For instance, collagenases are secreted in inactive forms, which are activated proteolytically outside the cell—some of the other secreted proteases (or cell surface peptidases?) are held responsible. Macrophages and fibroblasts also secrete protease inhibitors, which could cooperate with the better known inhibitors from the serum to limit the effects of secreted enzymes. Is it only when levels of hydrolase release are too extensive for the inhibitors to control the enzymes that damage is done? Perhaps, for example, the synovial fluids of joints are so poorly accessible to serum inhibitors that when there is a massive influx of phagocytes responding to inflammatory signals, intraarticular surfaces find themselves poorly defended against degradative attack.* Is the central nervous system, with its paucity of extracellular proteins from the serum, also at special risk in these regards, and if so, is this risk contained largely by keeping release of hydrolases to a minimum (see Section 2.1.3.4)?

Many cell types possess receptors for M6P or mannose at their surfaces and with them can bind and endocytose lysosomal hydrolases (Section 6.4.1.4 and Chapter 7). In culture, at least, endocytosis via such receptors can effect the transfer of active enzymes from fibroblasts that secrete them, to the lysosomes of endocytosing recipient fibroblasts (Section 6.4.1.4). As already mentioned, such transfer of enzymes from one cell to another is occasionally invoked as a biologically important supply mechanism. But perhaps endocytosis in situ, by fibroblasts or by macrophages and other scavengers, is more widely important in helping regulate extracellular degradation, by diminishing the amounts of enzymes in extracellular spaces.

6.3. Release of Lysosomal Hydrolases within Cells

Over the years, the suggestion has repeatedly been made that leakage of lysosomal enzymes into the cytoplasm or into the nucleus explains pathology or accounts for normal phenomena such as effects of hormones (Section 4.5.2.1) or even the "breakdown" of the nuclear envelope during the cell division cycle. The literature on these proposals contains many provocative findings but methodological murkiness has diminished their impact.

---

*Culture media from normal synovial cells contain agents that minimize production of collagenase
  by the synovial cells. If, as some researchers assert, these are autoregulatory agents, they might
  prove useful for therapeutic strategies against joint diseases.

6.3.1. How to Tell Cause from Effect; Suicide Sacs?

There is little doubt that lysosomes disrupt when cells die. However, even in situations where lysosomes release much of their content within dying cells, it is extraordinarily difficult to decide whether the acid hydrolases play primary, causal roles in killing the cells, or whether the lysosomal changes are secondary responses to mortal injury or death.

Hydrolase release within cells has generally been assessed biochemically, by determining the relative proportions of "free" enzymes versus particle-bound enzymes in homogenates.* But when one observes changes in these proportions for cells responding to a given treatment, does this indicate release of enzymes *in situ* or does it result rather from alterations in the size, fragility, or other properties of the lysosomes that influence their susceptibility to disruption during homogenization and centrifugation? This question is presently unanswerable for the studies where only modest changes in the intracellular distribution of hydrolases are claimed, as in nonpathological material or sublethal cellular injuries. In drastic pathological circumstances, the situation often is even more complicated because cell death and secretion of enzymes increase the levels of free hydrolases in the homogenates, and invasions by neutrophils, macrophages, or other "inflammatory" cells add new lysosomal populations.

Cytochemical methods for localizing enzyme activities are prone to their own disabling ambiguities: The classical procedures for localizing acid phosphatase are notorious for yielding spurious, diffuse deposits of reaction product in nuclei and in the extralysosomal cytoplasm. Even with immunocytochemical methods, technical inadequacies can generate high enough artifactual labeling throughout the cell to create problems.

There also are conceptual issues to contend with. For instance, if the intracellular leakage of lysosomal hydrolases is to produce a discrete, sublethal change, such as the dissolution of a particular membrane system or the destruction of certain enzymes or nuclear proteins, what would limit the hydrolases to such effects? Either the enzyme release would have to be slight and the targets especially susceptible, or the cell would need regulators. Perhaps the near-neutral pH in the extralysosomal cytoplasm and the cytoplasmic protease inhibitors mentioned in Section 5.6.1 create such an unfavorable environment for the action of most lysosomal enzymes that if hydrolases somehow do escape, they can have only very limited effects, except when release is too extensive. (Evolutionary "scenarios" are imaginable in which the inhibitors arose to deal with innate imperfections in the cell's ability to keep all its hydrolase molecules physically confined; biological uses might then have been found for incomplete inhibition.)

6.3.1.1. Labilizers and Stabilizers

For a decade or so, effects of widely used therapeutic materials such as cortisones, and of other biologically significant molecules, were frequently ascribed to their impact on release of lysosomal enzymes. For instance, investigators of arthritis were struck by observations that vitamin A and its derivatives

---

* "Free" enzymes are those that are nonsedimentable under conditions of centrifugation that bring down the known membranous structures in the homogenate. They are also nonlatent (see Section 1.2), indicating that they no longer are within a compartment bounded by an intact membrane.

hasten, whereas cortisones slow, the autolytic destruction of cartilage that ensues when pieces of cartilage are explanted to organ culture. Because autodegradation is accompanied by release of the explant's lysosomal hydrolases to the medium, disruption of the lysosomes was held to be responsible for destroying the tissue. Vitamin A was viewed as a **labilizer** of the lysosomes and cortisone was considered a **stabilizer.**

Given the lipid solubilities and other properties of vitamin A derivatives, corticosteroids, and comparable "labilizers" or "stabilizers," it could indeed be that these molecules affect the delimiting membranes of the lysosomes. For a time, evidence for such effects was sought by measuring alterations in the ratios of free to bound hydrolases in homogenates or in the fragility of isolated lysosomes. Efforts were also made to standardize cytochemical incubation conditions so as to compare cells exposed to labilizing or stabilizing materials with control cells, with regard to the rapidity of hydrolysis of lysosomal substrates. (The intention was to devise in situ measures of changes in the accessibility of the lysosome interior, for use as indices of alterations in the delimiting membrane that might affect hydrolase leakage.)

But the agents and conditions investigated in these studies have effects, known or potential, on many cellular membranes, not just those of the lysosomes. What is more, the prejudice that the governing effects are on leakage of hydrolases or on the resistance of lysosomes to disruption was easily disputable: Why not seek instead, influences on the fusions of membranes, the exocytotic secretion of enzymes or regulatory molecules, the budding of vesicles, or the osmotic and permeability properties of the altered membranes? And why explain antiarthritic effects of antiinflammatory drugs in terms of lysosomal leakiness within cartilage cells when the drugs are known to affect also the professional phagocytes so prominent in inflammations? Such questions, and most of the work on labilizers or stabilizers, have been left in limbo awaiting the lifting of the methodological clouds.

6.3.2. Developmentally Programmed Cell Death

6.3.2.1. Autolysis in Plants

Autolysis, orchestrated by responses to environmental change and by hormonal growth regulators, is a prominent feature of development and of normal cyclical senescence of organs in higher plants. It is almost as if plants, lacking phagocytes, have come to use autolysis for projects of remodeling and clearance (and defense) of sorts that macrophages and other motile cells handle in animals. In the development and expansion of the vascular system's xylem, autolysis removes the remains of dead cells from their cell wall-bounded compartments, leaving interconnected spaces that evolve into "vessels" the plant uses to move fluid among its roots, leaves, and other organs (see Section 4.7.1). When flowers or leaves age and wither, autolysis of their cells liberates materials for withdrawal into the surviving plant before the senescing organ falls off. Such scavenging achieves conservational retrieval of molecules that were accumulated originally through considerable metabolic investment by the plants. Enzymes released by autolysis may also take part in morphogenetic remodeling of cell walls; this has been asserted, for instances, for some of the pectinases.

The vacuoles of the autolyzing cells are the obvious, plausible sources of autolytic enzymes. Early in autolysis the vacuoles appear disrupted and hydrolases become demonstrable, cytochemically, in diffuse distribution throughout the cell. These observations have led to a broadly accepted premise that changes in tonoplast integrity are among the cellular factors that **initiate** autolysis and that lysosomal hydrolases help kill the cell. Dissenters argue that nonvacuolar changes—in the plasma membrane, the nucleus, or the chloroplasts—are responsible for initiating cell death and a few observers are more impressed by autophagic use of the lysosomes in dying plant cells than by tonoplast disruptions.

When the phloem of the vascular system differentiates, the vacuoles of its cells disappear, along with some other cytoplasmic elements; eventually, the nuclei also go. But many macromolecules, the plasma membrane, and sometimes even mitochondria, plastids, and ER-like sacs, persist in the phloem. Does this persistence mean that the phloem vacuoles disappear by a mechanism that does not actually liberate degradative enzymes—perhaps by vesiculation rather than by tonoplast dissolution or disruption? Or is the bath in vacuolar enzymes when the tonoplast opens brief enough, or well enough controlled by inhibitors, by dilution of the enzymes, or by pH, that various phloem components survive the encounter until the systems that maintain mature phloem restore a more salubrious environment? [In many plants, long-term upkeep of the phloem is assisted by "companion" cells that make their contributions through cytoplasmic continuities (plasmodesmata) with the phloem.]

6.3.2.2. Animals

Animal cell necrologists distinguish two major sorts of cell death on the basis of the morphology of the affected cells: In **necrosis,** as seen often in pathology, cell swelling and considerable disruption of the cytoplasm are among the early hallmarks of impending demise; the cell looks as though its plasma membrane barriers have been compromised so that controls of osmosis or permeability no longer function well. In **apoptosis,** claimed by some researchers to typify normal turnover of cell populations, changes in nuclear morphology are among the initial signs; rather than swelling, the cytoplasm contracts or fragments. Apoptotic cells are phagocytosed by their normal neighbors and by professional phagocytes, or sometimes are shed into the lumina of organs. (Necrotic tissue also is cleared partly by phagocytes, but when necrosis is extensive, considerable autolytic degradation probably occurs before and during such clearing.)

Lysosomal involvement? The prominence of autolysis relative to autophagy and heterophagy probably varies considerably in different cell death phenomena, but quantitative relations have not been studied precisely. Dead animal cells do release large amounts of lysosomal hydrolases at sites of extensive pathological damage and in normal developmental situations, such as the massive destruction of tissues during metamorphoses; presumably the enzymes help in removing debris.

Lysosomes have often been supposed to initiate the programmed death of cells, for example, in the normal development of the nervous system, or in insect metamorphosis, but much of the concrete evidence is to the contrary: For instance, when muscles break down during insect metamorphosis, changes in plasma membrane permeability and other features of the cells precede detectable alterations in the lysosomes.

6.3.3. Pathology and Therapeutics

Drugs, hormones, carcinogens, photosensitizing agents, and crystals and fibers are among the materials known to accumulate in lysosomes. The primary actions of many such materials on cells have, therefore, been "explained" by supposing their induction of lysosomal hydrolase release into the cytoplasm. Not all such explanations are arbitrary concoctions. Some arise from hard-won circumstantial evidence, such as findings that photosensitizing agents can engender release of hydrolases from isolated lysosomes, or that hormones known to influence the nucleus seem to set off migrations of lysosomes to the perinuclear region (Section 4.5.2.1).

6.3.3.1. Silicosis

The most compelling evidence for direct effects on lysosomes has been garnered for certain crystals and fibers, especially the silica fibers that cause lung orders, and the uric acid crystals responsible for gouty inflammations. Silica particles of phagocytosable dimensions can kill cultured alveolar macrophages and other phagocytotically active cultured cells. In the microscope, the cells seem to phagocytose the particles before they show signs of distress. Soon after this uptake, lysosomal contents leak into the cytoplasm: endogenous hydrolases become diffusely distributed throughout the dying cell and so do HRP or fluorescent dyes that had been taken into the lysosomes before the exposure to silica.

The "perforation of the lysosomes from within" inferred from these findings could result from silica's membrane disruptive capacities. Such capacities are evident from in vitro hemolytic effects; they are believed to arise partly from hydrogen bonding by silicic acid with membranes. Silica particles coated with aluminum, or coadministered with polyvinyl-pyridine-N-oxide (a polymer that reduces the available hydrogen bonding sites at the particles' surfaces) are much less effective in killing cells than are naked particles.

Proposals integrating this information suggest that upon entering the lung (or on placement in protein-rich culture media), silica particles acquire a coating of denatured proteins or similar materials that minimizes damage to cells as long as the coating persists. Upon phagocytosis, the coating is degraded, uncovering the lytic silica surfaces, which compromise lysosomal integrity. The resulting injuries—lethal or sublethal—to macrophages and other lung cells trigger inflammatory reactions; among the sequelae are the release of agents that enhance deposition of collagen by fibroblasts and produce other characteristics of the later stages of silicosis.

Reasonable and strongly founded as this scheme is, skeptics point to findings that clash with it. Most startlingly, when culture media with little or no Ca^{2+} are used in place of usual media, silica's lethality for cultured cells is substantially mitigated even though the particles still produce the redistributions of microscopically detectable tracers that had been taken to reflect lysosomal perforation. Could this observation mean that the primary, mortal lesion depends directly on a silica-enhanced influx of Ca^{2+} through the plasma membrane? Does the finding also carry the disconcerting implication that lysosomal leakage need not be fatal to cells, even when leakage is extensive? This last possibility would enthuse those who believe that leakage is not a disaster and therefore could mediate nonlethal biological effects. But it might disappoint designers of therapeutic agents who hope to custom-make cytotoxic compounds that will be targetable to lysosomes of particular cell types, there to be converted to disruptive forms that would kill the cells by rupturing the lysosomes (see Section 4.8).

Interestingly, synthetic "lysosomotropic" detergents constructed of hydrocarbon chains attached to amines take several hours to kill cells even though they rapidly accumulate in lysosomes (as weak bases) and quickly induce leakage of Lucifer Yellow that had previously been incorporated into the lysosomes.

6.3.3.2. Damage Control in Plants

Vacuolar components are used by some plants in defenses against predators. Chitinases, for example, released as cells die during fungal attacks, may limit further fungal growth by disrupting the invaders' walls. Mustard oils and other compounds in plants such as horseradish deter grazing by making the experience of eating the plant unpleasant (see Section 3.5.1). The vacuoles of tomato plants contain protease inhibitors that are much more effective against proteases of animal origin than against plant enzymes. These inhibitors accumulate in increased amounts after damage, and it has been suggested that when the plants are subject to repeated attack by insects, the inhibitors cushion the impact or even eventually discourage the attackers.

6.4. Storage Diseases

Of the several dozen human disorders, and the various animal diseases known to be due to lysosomal dysfunctions, most are lysosomal **storage diseases** of genetic origin. Each such storage disease has its most marked effects on a characteristic set of cell types in which the lysosomes become engorged with undigested ("stored") content and often are enlarged (Fig. 6.7). Frequently, the lysosomes exhibit a disease-specific ultrastructure owing to the accumulation of abnormally large amounts of specific sorts of undigested materials: in a number of the diseases, stored lipids assemble into lamellar or quasi-crystalline forms within the lysosomes, whereas in others the lysosomes are packed with glycogen granules or less well-structured masses of carbohydrate-rich material.

The underlying defect in most of the diseases is a deficiency in the activities of one or another lysosomal enzyme (Table 6.3). Recessive mutations are responsible. Most of the mutations are rare, but some of the disorders are found with elevated incidence in particular ethnic groups—frequencies as high as one per several thousand individuals have been estimated for Tay–Sachs alleles in Ashkenazi Jews. These frequencies have generally been regarded essentially as matters of chance and of the genetics of small, isolated populations, but subtler effects such as selective advantages for heterozygotes may also contribute; e.g., from epidemiological evidence, the Tay–Sachs allele has sometimes been claimed to afford a degree of protection against tuberculosis. The majority of the mutations are autosomal though a few of the diseases are X-linked (e.g., α-galactosidase deficiency; Hunter's syndrome—a sulfatase deficiency disorder).

Almost all the known lysosomal storage diseases affect pathways for breakdown of lipids or of GAGs or oligosaccharides: historically, therefore, most

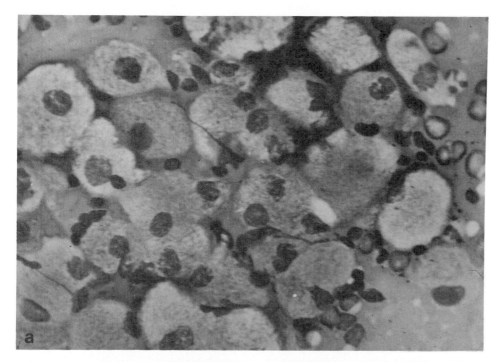

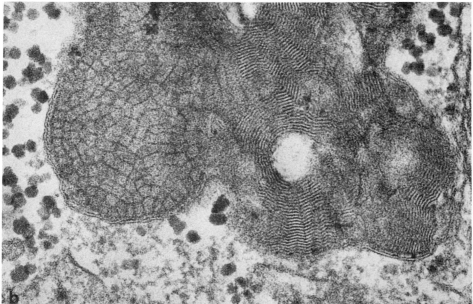

FIG. 6.7. (a) Light micrograph of "Gaucher cells" from the spleen of a patient with Gaucher's disease; the cells are derived from macrophages whose cytoplasm assumes a characteristic appearance (described as "wrinkled tissue paper" or "crumpled silk") owing to the accumulation of rodlike inclusions of glucosyl ceramide lipids. From Brady, R. O., and Barranger, J. A. (1983). In *Metabolic Basis of Inherited Disease* (5th ed.) (J. B. Stanbury, J. B. Wyngaarden, D. S. Frederickson, J. L. Goldstein, and M. S. Brown, eds.), McGraw-Hill, New York, pp. 842–856.

(b) Storage body from a fibroblast cultured from tissue of a patient with metachromatic leukodystrophy (Table 6.3) and grown in the presence of exogenous sulfated lipids. The body shows a distinctive morphology corresponding to arrays of lipids. Original from Rutsaert, J., Menu, R., and Resibois, A. (1973) *Lab. Invest.* **29**:527. Reprinted from Holtzman, E. (1976) *Lysosomes: A Survey*, Springer-Verlag, Vienna. × 100,000 (approx.).

TABLE 6.3. The Defects in Human Lysosomal Storage Diseases[a]

Disease	Deficient hydrolase(s)	Nature of defect[b]
Tay–Sachs disease, A variant	β-Hexosaminidase A	1. mRNA for α-precursor absent 2. Reduced synthesis of α-precursor 3. Insoluble α-precursor 4. Labile α-precursor 5. Mature α-chain larger than normal and yields inactive enzyme 6. Defective association of α-precursor with β-precursor 7. Reduced α-precursor synthesis and defective association with β-precursor
Sandhoff's disease	β-Hexosaminidases A and B	No synthesis of β-precursor; reduced maturation of α-precursor
Tay–Sachs disease, AB variant	β-Hexosaminidase activity with G_{M2} ganglioside as substrate	Absence of factor required for hydrolysis of G_{M2} ganglioside
Pompe's disease	α-Glucosidase	1. Precursor formed and phosphorylated but conversion to mature form impaired 2. Precursor formed; no phosphorylation and conversion to mature form impaired 3. Reduced synthesis of precursor; maturation unimpaired 4. No precursor formed 5. Precursor very rapidly degraded in prelysosomal compartment 6. Enzymatically inactive protein synthesized
Metachromatic leukodystrophy, late-infantile form	Arylsulfatase A	1. No precursor of arylsulfatase A formed 2. Arylsulfatase A rapidly degraded
Metachromatic leukodystrophy, late-onset form	Arylsulfatase A	Arylsulfatase A rapidly degraded in lysosomes
Pseudodeficiency of arylsulfatase A	Arylsulfatase A	Synthesis of a precursor reduced in size and with altered properties
Metachromatic leukodystrophy, variant form	Arylsulfatase A with sulfatide as substrate	Absence of factor required for hydrolysis of sulfatide
Multiple sulfatase deficiency	At least six lysosomal sulfatases and a microsomal sulfatase	Probably due to deficiency of factor required for stabilization of sulfatases
Mannosidosis	α-Mannosidase	Precursor of α-mannosidase not synthesized
Marateaux–Lamy syndrome, severe form	Arylsulfatase B	Mature arylsulfatase B (form I) is inactive and is rapidly degraded
G_{M1} gangliosidosis, infantile form	β-Galactosidase	Mature β-galactosidase polypeptide reduced in amount, inactive, and does not aggregate

TABLE 6.3. (*Continued*)

Disease	Deficient hydrolase(s)	Nature of defect
G_{M1} gangliosidosis, adult form	β-Galactosidase	Mature β-galactosidase reduced in amount and aggregates poorly
Morquio B syndrome	β-Galactosidase	Mature β-galactosidase has reduced activity and aggregates poorly
Galactosialidosis	β-Galactosidase and membrane-bound neuraminidase	Protective protein (subunit of neuraminidase) not synthesized; β-galactosidase degraded in lysosomes and neuraminidase inactive
Gaucher's disease, Type 1	β-Glucocerebrosidase	β-Glucocerebrosidase with decreased stability, enhanced degradation in lysosomes
Gaucher's disease, Types 2 and 3	β-Glucocerebrosidase	Maturation of β-glucocerebrosidase impaired
Hurler's disease	α-Iduronidase	Precursor of α-iduronidase not synthesized
Mucolipidosis II	Several	N-Acetylglucosaminylphosphotransferase deficient
Mucolipidosis III ("I-cell" disease)	Several	1. N-Acetylglucosaminylphosphotransferase partially deficient 2. N-Acetylglucosaminylphosphotransferase with reduced activity toward lysosomal enzymes

[a]From Tager, J. M. (1985) *Trends Biochem. Sci.* **10**:324. (Copyright: Elsevier, Cambridge.)
[b]The third column indicates the defects in synthesis, processing, or stability of hydrolases, or other proteins, responsible for different forms of the diseases. The table omits a number of suspected lysosomal disorders whose bases are not yet well understood.

have been referred to as lipidoses, sphingolipidoses, or mucopolysaccharidoses, depending on the predominant storage material. Sometimes, effects on the breakdown of the protein moieties of glycoproteins or proteoglycans are noted but these seem to be secondary offshoots of defective handling of the oligosaccharides or GAGs attached to the proteins.

Conventionally, it is assumed that defects with primary impact on the metabolism of polypeptides or nucleic acids have such dire consequences that affected fetuses do not survive, accounting for the absence of corresponding known diseases. The chief alternative possibility—that cells have substantial redundancy in degradative pathways for proteins and nucleic acids so that loss of one enzyme activity need not be fatal—runs counter to most tastes. Yet, yeast cells do survive the genetic ablation of vacuole protease activities, if the cells are not called upon to sporulate. The worm *Caenorhabditis elegans* survives mutations that reduce its levels of a cathepsin D-related protein to below 10% of normal; the only evident abnormality is some vacuolation of intestinal cells (due to slowing of rates of degradation of endocytosed material?). (*Caenorhabditis* also seems to tolerate mutations that lead to virtually complete loss of its principal DNase, making it unable either to digest the nucleoids of the bacteria it eats or to clear away the nuclei of cells that die during its developmental program.) In Chediak–Higashi disease, mature neutrophils may lack two of the

TABLE 6.5. Chromosomal Locations of Hydrolase Genes in Humans, as Determined Chiefly from Storage Disease Material[a]

Enzyme	Chromosome	Region[b]
Acid phosphatase-1	2	p23
Acid phosphatase-2	11	p12 → cen.
Arylsulfatase A	22	
Arylsulfatase B	5	
Cathepsin D	11	p ter. → q12
DNase (lysosomal)	19	
Esterase A4	11	cen. → q22
Esterase B3	16	
Esterase D	13	q14
Esterase activator	14	
α-L-Fucosidase	1	p32 → p34
α-Galactosidase A	X	q22 → q24
α-Galactosidase B (N-acetyl-α-D-galactosaminidase)	22	q13 → q ter.
β-Galactosidase	3	p ter. → q13
β-Galactosidase (processing factor?)	22	q13 → q ter.
α-Glucosidase (acid)	17	
β-Glucosidase (acid)	1	p11 → q ter.
β-Glucuronidase	7	
β-Glucuronidase modifier	19	
Hexosaminidase α-subunit	15	q22 → q ter.
Hexosaminidase β-subunit	5	cen. → q13
Lipase A (acid, lysosomal)	10	
Lipase B (acid, lysosomal)	16	
α-D-Mannosidase (lysosomal)	19	p ter. → q13
Sulfoiduronide sulfatase	X	

[a]From Galjaard, H., and Reuser, A. J. J. (1984) In *Lysosomes in Biology and Pathology* (J. T. Dingle, R. T. Dean, and W. Sly, eds.), Vol. 7, Elsevier, Amsterdam, p. 315. Data mainly derived from Human Gene Mapping Conference 6 (1981). *Cytogenet. Cell Genet.* (1982) Vol. 32/1.
[b]cen., centromere; ter., terminal.

mediates were lysosomal capacities normal. Paths of enzymatic attack thus can be inferred by determining which intermediates are abundant when particular enzymes are absent.*

Lipid-rich structures are the materials most slowly destroyed by normal lysosomes probably in part because the initial hydrolase attacks can affect peripheral molecules or moieties, but leave the core organization of membranes or micelles intact. These assemblies must be dismantled to complete the destruction.

Section 6.2.2 pointed out that extracellular proteases and endo-glycosidases such as hyaluronidase frequently are utilized to fragment the large connective tissue glycoconjugates into manageable form. After the fragments are endocytosed, lysosomes take over. When the GAGs contain sulfates, their

---

*The assumption that the intermediates piling up represent the substrates of the missing enzyme usually seems to work pretty well. There are a few complications, such as overlaps in the specificities and actions of certain lysosomal hydrolases providing alternative means for attack on particular bonds. For many steps, however, in the breakdown of complex lipids, oligosaccharides, GAGs, and proteoglycans, only one enzyme will do. And some of the alternative routes that seem possible from studies on the properties of purified enzymes probably are much less effective *in vivo* than the normal degradative sequences (see the comments on hexosaminidases in Section 6.4.1.3) and therefore cannot stanch the accumulation of abnormal stores.

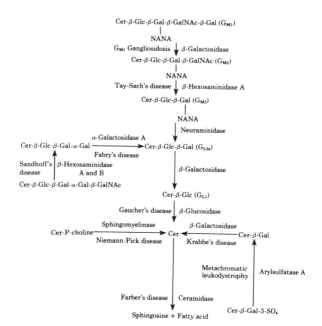

FIG. 6.8. Lysosomal degradation of sphingolipids, and diseases known to interrupt the pathway. Cer, ceramide; Gal, galactose; Glc, glucose; GalNAc, N-acetylgalactosamine; NANA, N-acetylneuraminic acid (a sialic acid). From Glew, R. H., Basu, A., Prence, E. M., and Remaley, A. T. (1985) *Lab. Invest.* **53**:250. (Copyright: U.S.–Canadian Academy of Pathologists.)

removal by lysosomal sulfatases often is a prerequisite for subsequent actions of glycosidases. The polysaccharide chains themselves usually are degraded from one end, by removal of one residue at a time, because the lysosomal glycosidases are mostly exoglycosidases that act at the nonreducing end of polymers.

Some of the lysosomal degradation sequences contain "surprises." For instance, in the digestion of the sulfated GAG **heparan sulfate,** prior acetylation of the GAG's glucosamines was found to be mandatory for hydrolysis of the bonds linking the glucosamines to the chain (Fig. 6.9). This requirement was first signaled by the existence of a mucopolysaccharidosis (Sanfilippo type C) in which acetylation is defective. Follow-up biochemical studies discovered acetylation machinery in lysosomes in the form of an enzyme complex built into the delimiting membrane and capable of transferring acetyl groups it acquires from outside the lysosomes, to substrates inside. Conjecture has opened that this machinery may be versatile. In addition to being important for degradation of GAGs, perhaps it also serves functions such as in minimizing the retention of diffusible amines by the lysosomes—acetylated amines are less basic than the unacetylated forms and thus should be less "acidotropic" (Section 3.1.2.1). Relationships also are conceivable between lysosomal acetylations and the acetylations occurring in other acidified compartments, such as neurosecretory granules (see Section 6.1.5).

6.4.1.2. One Gene, One Enzyme?

In many storage diseases the culpable mutation affects the structure and activity of a single type of enzyme. Sometimes, however, single mutations have broader effects. For example, lysosomes of mammalian cells contain two principal β-hexosaminidase activities distinguishable on the basis of their activities against a range of substrates, natural and artificial. From studies on Tay–Sachs

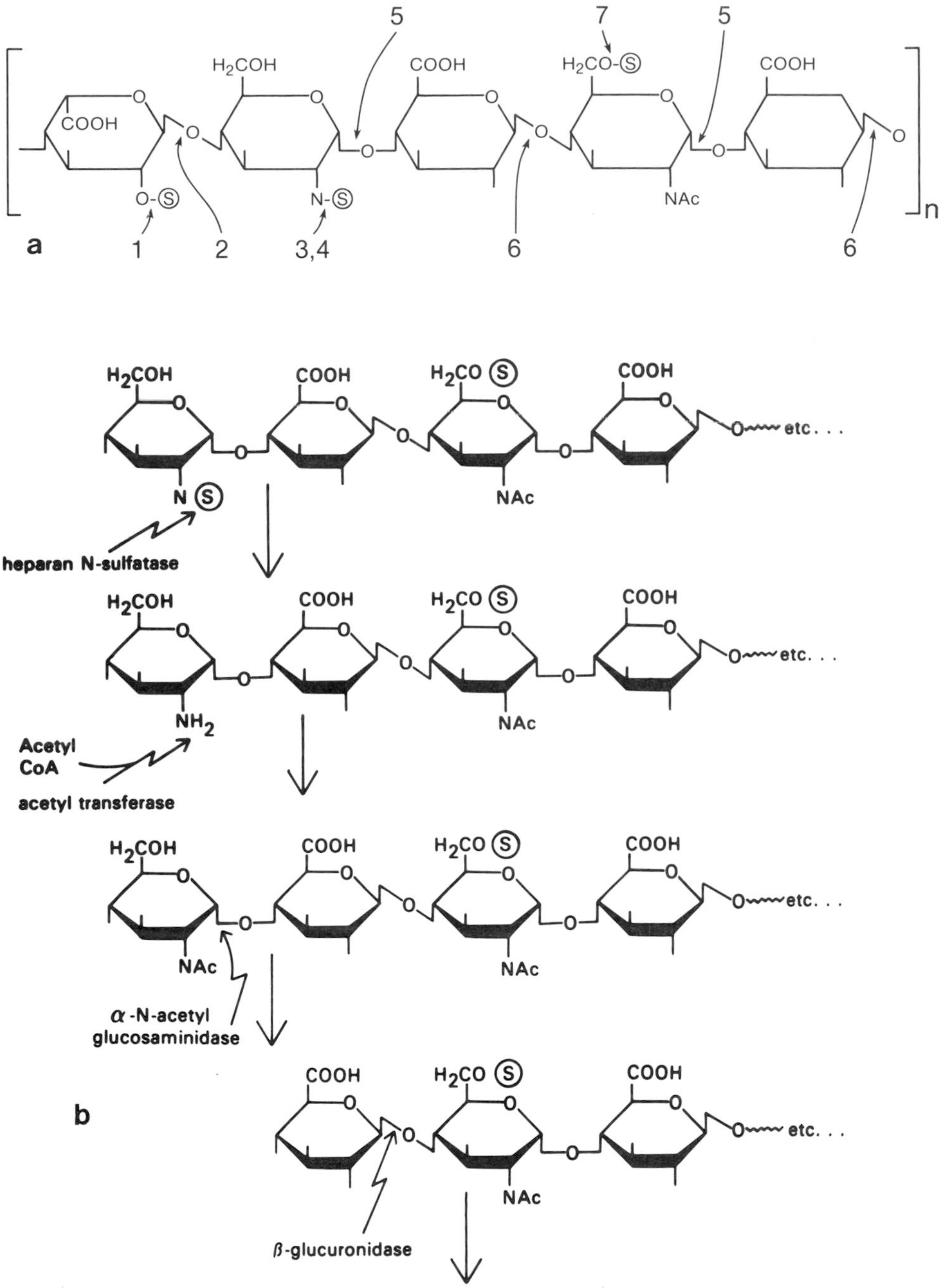

FIG. 6.9. (a) Heparan sulfate is degraded in the sequence indicated by the numbers. The enzyme activities involved in the numbered steps are as follows (some of these activities may be carried out by enzymes more usually known by other names): (1) iduronate sulfatase; (2) α-L-iduronidase; (3) heparan-N-sulfatase; (4) acetyl-CoA-α-glucosaminide-N-acetyl transferase; (5) N-acetyl-α-D-glucosaminidase; (6) β-glucuronidase; (7) N-acetylglucosamine 6-sulfatase. S, sulfate; NAc, N-acetyl. Courtesy of Dr. R. H. Glew (same reference as Fig. 6.8).

(b) Segment of the sequence in a for which acetylation is essential, as discussed in text. Courtesy of E. E. Neufeld.

and other diseases, hexosaminidase A is thought to be the most versatile of these enzyme with respect to range of natural substrates, and to be the hexosaminidase most important for normal ganglioside metabolism. Hexosaminidase B has a more restricted substrate range and a less certain cellular role. The best known relevant storage diseases involve hexosaminidase A deficiencies but in certain gangliosidoses, both A and B activities are deficient (Table 6.3). Because the latter diseases seem to result from single mutations, the observations helped generate the understanding that A and B share common subunits: The A enzyme is made both of α and β subunits (see Fig. 6.10) whereas the B form is constructed of subunits only of the β variety. When β subunits are defective, both enzyme species are affected. In fact, the absence of normal β subunits can be accompanied by the rapid degradation of α subunits, even if these are normal.

6.4.1.3. Activators; Stabilizers

For the storage diseases in which a single mutation produces deficiencies both in β-galactosidase activities and in neuraminidase (sialidase) activities, the defect seems to be the lack of a protein that activates the neuraminidase and stabilizes its activity. This protein also influences β-galactosidase because it engages the galactosidase in high-molecular-weight, multimeric complexes, which resist intralysosomal degradation more effectively than does noncomplexed enzyme. The complexes may associate with the lysosomal membrane.

The implication in this last example, that some lysosomal proteins have nonenzymatic roles, is borne out in other storage disorders as well. In several, lysosomal dysfunctions have been traced to the absence of specific "activator" proteins (Fig. 6.10) needed for effective enzymatic attacks upon particular species of lipids. One such activator is necessary for sulfatase attack on sulfatide lipids; its importance was realized from the discovery of sulfatide storage diseases that differ from classical metachromatic leukodystrophy (Table 6.3) in that aryl sulfatase A activity can still be detected but sulfatides accumulate nonetheless. An activator is also required for hexosaminidase A attack on gangliosides and glycolipids; hexosaminidase B cannot utilize this activator, which probably accounts for its relative inactivity against gangliosides. These two known activators (they may actually be the same protein, rather than two different species), and several other suspected ones, are glycoproteins of molecular weights between 20,000 and 30,000. Each acts on a selected group of lipids. Some of the activators form water-soluble complexes with lipids and some can act as *in vitro* lipid-exchange proteins, capable of "extracting" lipids from one membrane and carrying them to another (Fig. 6.10). It thus may be that "activation" is achieved by the proteins binding to lipid molecules present in membranes or micelles, and pulling the molecules at least part way out of the structure, thereby making them accessible to hydrolases.*

---

*Participation of activators in lysosome functions helps explain why, when purified and tested *in vitro*, several lipid-degrading enzymes are highly active against water-soluble artificial substrates but show very little activity against natural substrates of types the enzymes would encounter in lysosomes. Before activators were discovered, this seemingly paradoxical inactivity was overcome, *in vitro*, by including detergents in the incubation medium.

Also being toyed with is the idea that yet-to-be-identified molecules from the serum enter lysosomes endocytotically to aid digestion—by converting large lipid structures into smaller aggregates or, for trypanosomes, by activating lysosomal proteases.

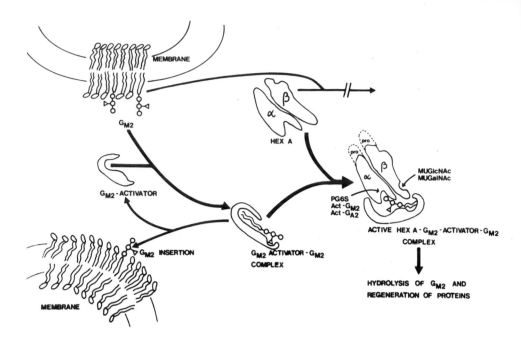

FIG. 6.10. Proposed cooperative phenomena in the lysosomal degradation of membrane-associated forms of the ganglioside G_{M2} An activator protein forms a water-soluble complex with the ganglioside, making the ganglioside accessible to cleavage by β-hexosaminidase A (HEX A). The hexosaminidase active sites that cleave G_{M2} are on the α subunits of the enzyme molecules. After hydrolysis of the substrate, the proteins are freed ("regenerated") to be reused. In the absence of the activator, the enzyme cannot attack the ganglioside (interrupted arrow). In the absence of the enzyme, the activator can transfer ganglioside molecules from one membrane to another ("insertion"). Precursor ("pro") forms of the hexosaminidase, which are present transiently during its genesis (Chapter 7), as well as the mature form of the enzyme, can hydrolyze the gangliosides in the complexes.

MUGlcNAc and GalNAc are positioned to suggest the likelihood that β subunits of the lysosomal hexosaminidases hydrolyze 4-methylumbelliferyl derivatives of the hexosamines more effectively than do α subunits. (4MU derivatives are artificial, fluorogenic substrates widely used in enzyme assays.) PG6S, Act-G_{M2}, and Act-G_{A2} are positioned to indicate that it is the α subunits that are principally involved in hydrolyzing p-nitrophenyl substrates and are required for effective hydrolysis of G_{M2} and G_{A2} (when these lipids are complexed with activator proteins).

From K. Sandhoff (1984) In *Molecular Basis of Lysosomal Storage Disease* (J. A. Barranger and R. O. Brady, eds.), Academic Press, New York.

A protein identified in Gaucher's disease spleens can modulate the lysosomal β-glucosidase ("glucocerebrosidase") activity that the cell uses to hydrolyze glucocerebrosides. The protein apparently complexes directly with the hydrolase and by so doing potentiates the activation of the enzyme by phospholipids.

6.4.1.4. Multiple Hydrolase Deficiencies; I-Cell Disorder

Fibroblasts from two of the "mucolipidoses"—I-cell disorder and "pseudo-Hurler's polydystrophy"—are characterized by simultaneous intracellular deficits of many lysosomal hydrolases. The "I" in I-cell refers to the abundance of inclusion bodies that accumulate in fibroblasts and other cell types. The disorders have their origins in mutations that divert newly synthesized hydrolases out of their normal paths to the lysosomes and into a secretory path, a discovery that set in motion the lines of research leading to the present understanding that M6P is a key signal in lysosome biogenesis (Sections 2.1.3.2 and 7.3). The present view is that, in mammals at least, the presence of M6P residues is the feature by which the cell is able to recognize many of the lysosomal hydrolases, to sort them from other proteins, and to direct them correctly to the lysosomes. The first evidence for this diversion was that I-cell patients show elevated levels of lysosomal hydrolases in the urine, serum, and other extracellular fluids and that cultured I-cell fibroblasts synthesize all the usual lysosomal enzymes but secrete a far higher proportion of their hydrolase production than do normal fibroblasts. "Cross-feeding" ("cross-correction") experiments were central to the subsequent analysis of this situation. It has been observed for several storage diseases that cocultivation of fibroblasts from different disorders—Hurler's and Hunter's syndromes, for example—could result in marked depletion of the abnormal intracellular stores. This led to determinations that the cells secrete and endocytose hydrolases at rates sufficient for the lysosomes of one strain of defective fibroblasts to acquire appreciable amounts of their missing enzyme from the secretions of the other strain. I-cell fibroblasts differ from those of other storage diseases in that they are poor donors in such experiments—their enzymes are not taken up efficiently by fibroblasts from other disorders—even though the I cells can readily endocytose hydrolases from the other cells.

These observations led, in turn, to the discovery that the efficient endocytosis of lysosomal hydrolases depends upon the receptors that recognize M6P residues. In broad outline, receptor-mediated endocytosis of lysosomal hydrolases resembles the processes described in Chapters 2 and 3 in that coated vesicles, endosomes, and dissociation at low pH are involved and the receptors are saturable and cycle between intracellular locales and the cell surface. Endocytotic recycling of the M6P receptors probably is constitutive, at least in some cell types. But unlike most endocytotic receptors whose intracellular pools are viewed as chiefly in a state of passage—immediate or delayed—back to the cell surface, the intracellular pools of receptors for M6P have a major intracellular source of ligand—the population of newly made hydrolases passing through the ER and Golgi apparatus (Chapter 7).

Some of the molecules of a given type of lysosomal hydrolase released by cultured fibroblasts are much more efficiently endocytosed than are other molecules of the same enzyme. These "high-uptake" forms are ones with M6P residues attached to their oligosaccharides. Experimenters obtain supplies of high-uptake hydrolases from culture supernatants or urine, or from the secretions of slime mold cells (see Section 7.3.3.4). I cells, however, do not make this sort of molecule; their hydrolases all lack M6P residues. The defect in I cells turned out to be in a phosphotransferase responsible for linking phosphates to the mannoses of newly made lysosomal enzymes (Section 7.4.2).

The defect in pseudo-Hurler's disorder has been studied less than that in I-cell disease, but in the terms discussed here the two diseases seem fundamentally similar.

Two additional observations on the I-cell disorder are particularly important for later concerns (Sections 7.3.3.5). (1) Not all lysosomal enzymes are equally affected: For example, levels of acid phosphatase and β-glucosidase activities detectable within the I cells are more nearly normal than those of other lysosomal enzymes. (2) Not all cells are affected to the same extent: Liver, brain, and other tissues obtained from I-cell patients on autopsy or biopsy have much more normal complements of intracellular acid hydrolases than do the cultured fibroblasts.

A multiple sulfatase deficiency disorder has been described in which both the lysosomal sulfatases and nonlysosomal sulfatases are abnormal. The molecular grounding of this situation is not known.

6.4.2. Etiology; Therapy

6.4.2.1. Which Cells Are Affected?

For several of the storage diseases, reasonable guesses can be made about the relations between the enzymatic deficits and the observed patterns of tissue and organ damage. Thus, one might anticipate that the spleen and liver would be affected strongly by Gaucher's disease and other deficiencies that diminish capacities to degrade glucosyl-ceramide glycolipids ("glucocerebrosides") and related molecules, because these molecules abound in the membranes of erythrocytes or leukocytes and the spleen and liver have large phagocyte populations that destroy blood cells. Fibroblasts, in contrast, are "spared" the full impact of Gaucher's disease probably in part because these cells do not confront large amounts of ceramide lipids. (Fibroblasts also are claimed to show capacities for handling excess glucocerebrosides effectively by routes that shunt these molecules into the genesis of plasma membrane lipids.)

It makes "sense" that sites of active heterophagy such as the kidney would often be drastically affected by storage disorders or that secondary deposition of lipids in the vascular system can be quite extensive: Combinations of impaired digestion and increased circulation of degradation intermediates should produce such effects because, for example, cells active in pinocytosis or phagocytosis will confront large amounts of storage disease products released to the bloodstream or urine by many other tissues.

Not surprisingly, the nervous system is a main site of abnormalities resulting from lesions in the lysosomal metabolism of the sulfatide lipids that abound in myelin, or of gangliosides and other lipids found in neuronal synaptic terminals and plasma membranes. Most likely retinal and central nervous tissues are so often grossly abnormal in storage diseases because these tissues maintain high rates of membrane turnover but discharge indigestible lysosomal residues at low rates (or do damage to their surroundings when discharge is too extensive). Consistent with these considerations are findings that long-lasting storage-disease-like lysosomal accumulations can be induced in cultured neurons by growing them for a while in media containing excesses of lipids. Limited defecatory capabilities may also help explain the pathology of cardiac muscle and other striated muscles in storage diseases, and of various cell types in disorders where hydrolase activities are diminished rather than completely absent.

6.4.2.2. Are Cells Disabled?

It is taken for granted that the presence of excessive lysosomal stores is bad for cells but translating this premise into testable terms has been troublesome ever since the first lysosomal storage disease was recognized as such: In Pompe's disease the cells accumulate autophagocytosed glycogen in their lysosomes. Lysosomal breakdown of glycogen, however, seems unlikely to be of quantitatively major import for cellular energy metabolism—well-known nonlysosomal mechanisms mobilize most cellular glycogen. And there are no obvious metabolic by-products of the lesion that should cause difficulties. Hence, by default, the pathology in Pompe's disease is assumed to stem somehow from the presence of too many large lysosomes in the cells.

A long list of disabilities might afflict cells stuffed with lysosomal stores: Autophagy and heterophagy might be thrown off kilter. Mechanical properties of the cells could be affected so that, for example, muscle contractions would be abnormal. Cell geometry might be changed leading to aberrant cell–cell interactions during development. Leakage of hydrolases within the cell might be enhanced. Diffusion in the cytoplasm, or the movement of organelles from one place to another might be impeded. Ionic and osmotic balances could be changed by the presence of large amounts of charged macromolecules. The molecules that accumulate might inhibit synthetic pathways, perturb regulatory enzymes such as kinases, or imbalance the production of membrane lipids or the assembly of membranes. Chain reactions of degradative disarray may be set off in which storage products like sulfated GAGs or cholesterol inhibit otherwise normal hydrolases. Rates of synthesis of hydrolases may be accelerated, or heterophagic cells induced to proliferate through mechanisms that "try" to compensate for the degradative deficits.

Much scarcer than these good ideas is concrete information about the capabilities of storage disease cells, even at the level of functions directly related to the lysosomes—endocytotic, heterophagic, autophagic, crinophagic, and recycling processes. A few findings suggest that cultured storage disease fibroblasts are not disastrously impaired in overall fluid phase endocytosis or in turnover of endogenous proteins though the cells do show the marked reductions in abilities to degrade specific endocytosed materials expected from their enzymes deficiencies. I cells have trouble in breaking down materials such as exogenous lipoproteins, but the wide-ranging lysosomal depletions in I-cell disorder produce fewer obvious quirks in heterophagy or intracellular turnover by cultured fibroblasts than might have been anticipated.

6.4.2.3. Organ and Organism

A few of the gross pathological effects of storage diseases have seemingly simple relations to the cellular defects. That the liver is enlarged, connective tissues are disarrayed, or the cornea is opaque, can be traced in part to the presence of the large deposits of undigested materials. Local circulation may be impeded by crowds of cells enlarged or distorted by their stores. The skeletal abnormalities often seen in storage diseases have been attributed both to mechanical distortion of forming bone by the abundance of engorged cells in the marrow, and to defects in remodeling during growth. The dramatic impacts of abnormal GAG metabolism can plausibly be tied to the many roles now suspected for extracellular matrices in development, maintenance of tissue architecture and so forth. Suboptimal functioning of macrophages and other phagocytes helps explain the lowered resistance to infection that characterizes several of the storage diseases.

In trying to move, however, from these broad considerations to more specific pictures of how each storage disease produces its particular cluster of pathological changes, it rapidly becomes obvious that the simplicity of the underlying cellular and molecular "causes" in storage diseases obviates depressingly little of the need to understand pathology in terms of processes integrated at the level of organ and organism: We must attend, that is, to long-term metabolic effects, imbalances in networks of cellular and tissue interactions, perturbed developmental sequences, and changes in cellular environments. Were the diseases simply manifestations of the cells being killed rapidly by their own lysosomal defects, most storage disorders would probably still be undetected, as affected embryos or fetuses would not survive. Culturing storage disease cells presumably would also be markedly more difficult than it is, though it may be that the media generally used to maintain mammalian cell cultures impose less of a load on the lysosomes than do the cells' environments *in situ*.

Too little is known about steady-state turnover, the importance of degradation in normal development, and the routes and mechanisms normally used to dispose of lysosomal residues to understand, for example, why defects in the catabolism of galactosyl-sulfatide lipids in metachromatic leukodystrophy or Krabbe's disease should eventuate in undermyelination or demyelination. Myelin normally turns over slowly, though some of its lipids may turn over fairly rapidly, especially during myelin formation or growth. Do degradation intermediates from this turnover pile up in the storage diseases and prove toxic to the cells (oligodendrocytes) that produce myelin in the central nervous system?

Part of the problem in dealing with such issues is that so many ramifying effects can be imagined for the enzymatic deficits that "initiate" storage diseases as almost to immobilize sorting through them for each particular disorder.

6.4.2.4. Detection and Therapy

Many storage diseases can be recognized prenatally though enzyme assays of cultured fetal cells obtained by amniocentesis or chorionic villus sampling. For Tay–Sachs disease and the X-linked disorders, among others, asymptomatic heterozygotes can also be identified biochemically, and screening of high-risk populations is increasingly routine. Diagnostic and screening programs are likely to intensify as appropriate DNA probes and monoclonal antibodies become available.

Hopes are high for rational therapy. The underlying genetic changes often are discrete and are increasingly finely characterized so that gene "replacement" may be feasible. The turnover of the hydrolases generally is not too fast, so that adequate enzyme levels might be maintained even with low rates of production of the corresponding mRNAs (see Section 7.5.2.1). Retroviral insertion of corrective genes is already being explored with cultured systems.

Because the diseases affect organelles relatively accessible to outside agents, effective intervention by direct manipulation of lysosomes and endosomes may prove feasible even before genetic therapy comes on line. Prospects for such intervention are somewhat better for disorders centered in peripheral, "reticuloendothelial" locations, especially Gaucher's disease and some of the mucopolysaccharidoses, than they are for diseases with heavy impact on the central nervous system—the blood–brain barrier is an obstacle to reaching nervous tissue. From cross-feeding studies in tissue culture like the experiments outlined in Section 6.4.1.4, it is established that lysosomes of storage disease cells can acquire hydrolases and activator or stabilizing proteins from their surroundings, in quantities sufficient to make a difference. One lively line of therapeutic endeavor therefore is attempting to supply enzymes either by infusing them as such into the bloodstream or by transplanting cells from which, it is hoped, there will be sufficient leakage to supply hydrolases to the recipient's cells. (Use of lymphocytes to target direct cell-to-cell transfers may also be possible; see Section 6.1.1.1.) Starting with early successes in using maltase to reduce the glycogen deposits in hepatocytes of patients with Pompe's disease, there has been a series of reports of improvements in the biochemical, histological, and clinical status of patients subjected to various enzyme replacement regimens. Dramatic breakthroughs have not been scored, but enough progress has been made to suggest that immunological consequences may not be limiting and to encourage efforts at producing vehicles for enzymes that target them to tissues of particular importance (see Section 4.8). More difficulty is foreseen in promoting uptake of enzymes by crucial tissues, such as cardiac muscle, that normally are not very active in endocytosis.

Another approach is the systematic draining of the stores that are readily accessible in the hopes of favorably altering balances so as to induce discharge of less accessible stores. Removal of circulating forms of stored lipids has been tried through plasmapheresis or infusion of appropriate enzymes into the circulation. Unfortunately, intracellular stores respond very slowly to such treatments so for the present at least, these approaches promise more for slowing the progress of disease than for reversing its effects. On the other hand, remember that cysteamine has proved useful in treating patients with cystinosis, presumably through its impact in intracellular deposits (Section 3.4.1.5), and that (limited) successes have been achieved in using low-molecular-weight chelating agents or penicillamine to reduce accumulations of metals acquired in excess from the environment or generated in disorders like Wilson's disease (Section 3.5.2.2).

Acknowledgments

I am particularly indebted for information, for comments, or for discussions over the years to: A. Allison (silica), M. Chalfie (*Caenorhabditis*), A. Chandler (lung), A. B. Fisher (lung), H. Gainer (processing of neurosecretions), R. H. Glew (storage diseases), S. Goldfischer (Wilson's disease; silica), S. Gordon (hydrolase secretion from macrophages), S. Hoffstein (enzyme release), L. Jacobson (*Caenorhabditis* cathepsin mutant), A. Kaplan (*Dictyostelium*), Y.-P. Loh (processing of neurosecretions), E. Neufeld (I cells; endocytosis of lysosomal enzymes), L. Orci (secretory cells), L. Rome (lysosomal acetylation), K. Sandhoff (storage diseases), W. Sly (M6P-dependent endocytosis), P. Stahl (endocytosis of hydrolases), J. Tager (storage diseases), P. Talbot (acrosomes), G. Vaes (osteoclasts), G. Weissmann (gout; hydrolase release; arthritis), Z. Werb (hydrolase secretion), H. Winkler (processing of adrenal secretions).

Further Reading

Barnes, D. M. (1987) Close encounters with an osteoclast, *Science* **236**:914–916.
Baron, R., Neff, L., Louvard, D., and Courtoy, P. J. (1985) Cell mediated extracellular acidification and bone resorption: Evidence for a low pH in resorbing lacunae and localization of a 100kD

lysosomal membrane protein in the osteoclast ruffled border, *J. Cell Biol.* **101:**2210–2222. (See also *J. Cell Biol.* **106:**1863–1872, 1988.)

Barranger, J. A., and Brady, R. O. (eds.) (1984) *Molecular Basis of Lysosomal Storage Disorders*, Academic Press, New York.

Baumbach, G. A., Saunders, D. T., Bazer, F. W., and Roberts, R. M. (1984) Uteroferrin has N-asparagine linked high mannose type oligosaccharides that contain mannose 6 phosphate, *Proc. Natl. Acad. Sci. USA* **81:**2985–2989.

Bowen, I. D., and Lockshin, R. A. (eds.) (1981) *Cell Death in Biology and Pathology*, Chapman & Hall, London.

Brands, R., Slot, J. W., and Geuze, H. J. (1982) Immunocytochemical localization of β-glucuronidase in the male rat preputial gland, *Eur. J. Cell Biol.* **27:**213–220.

Brown, J. A., Novack, E. K., and Swank, R. J. (1985) Effects of ammonia on processing and secretion of precursor and mature lysosomal enzymes from macrophages of normal and pale ear mice: Evidence for two distinct pathways, *J. Cell Biol.* **100:**1894–1904.

Callahan, J. W., and Lowden, J. A. (eds.) (1981) *Lysosomes and Lysosomal Storage Diseases*, Raven Press, New York.

Campbell, D. J. (1987) Circulating and tissue angiotensin systems, *J. Clin. Invest.* **79:**1–6.

Davidson, H. W., Rhodes, C. J., and Hutton, J. C. (1988) Intraorganellar calcium and pH control proinsulin cleavage in the pancreatic β-cell two distinct site-specific endopeptidases, *Nature* **333:**93–96.

Davies, I., and Sigee, D. C. (eds.) (1984) *Ageing and Cell Death*, Cambridge University Press, London.

Dean, M. F., and Diament, S. (1984) Exchange of lysosomal enzymes between cells, *Biochem. Soc. Trans.* **12:**524–526.

Dean, R. T., and Stahl, P. (eds.) (1985) *Developments in Cell Biology 1: Secretory Processes*, Butterworths, London.

de Hostos, E. L., Togasaki, R. K., and Grossman, R. (1988) Purification and biosynthesis of a derepressible aryl sulfatase from *Chlamydomonas reinhardtii*, *J. Cell Biol.* **106:** 21–28.

Dimond, R. L., Burns, R. A., and Jordan, K. B. (1981) Secretion of lysosomal enzymes in the cellular slime mold, Dictyostelium, *J. Biol. Chem.* **256:**6565–6572. (See also *J. Cell Biol.* **100:**1777–1787, 1985)

Docherty, K., Hutton, J. E., and Steiner, D. F. (1984) Cathepsin B-related proteases in the insulin secretory granule, *J. Biol. Chem.* **259:**6041–6044.

Faust, P. L., Chirgwin, J. M., and Kornfeld, S. (1987) Renin, a secretory glycoprotein. acquires phosphomannosyl residues, *J. Cell Biol.* **105:**1947–1956.

Gainer, H., Russell, J. T., and Loh, Y. P. (1985) The enzymology and intracellular organization of peptide precursor processing: The secretory vesicle hypothesis, *Neuroendocrinology* **40:**171–184.

Gal, S., and Gottesman, M. M. (1986) The major excreted protein of transformed fibroblasts is an activatable acid protease, *J. Biol. Chem.* **261:**1760–1765. (See also *J. Biol. Chem.* **263:**254, 1988 and *J. Cell Biol.* **106:**1879–1884, 1988.)

Ghosh, P. (1978) Lysosomal enzymes in joint disease, *Aust. N.Z. J. Med.* **8**(suppl. 1):12–15.

Glew, R. H., Basu, A., Prence, E. M., and Remaley, A. T. (1985) Biology of disease: Lysosomal storage disease, *Lab. Invest.* **53:**250–269.

Glynn, L. E. (ed.) (1981) *Tissue Repair and Regeneration*, North-Holland, Amsterdam.

Goldstein, P. (1987) Cytolytic T cell melodrama, *Nature* **327:**12.

Kane, A. B., Stanton, R. P., Raymond, E. G., Dobson, M. E., Knafele, M. E., and Farber, J. L. (1980) Dissociation of intracellular lysosomal rupture from the cell death caused by silica, *J. Cell Biol.* **87:**643–651.

Kornberg, S., and Sly, W. S. (1985) Lysosomal storage defects, *Hosp. Pract.* **20**(8):71–82.

Kuettner, K. E., Scheyerbach, R., and Hascall, V. C. (eds.) (1986) *Articular Cell Biochemistry*, Raven Press, New York.

Lew, P. D., Monod, A., Waldvogel, F. A., Dewald, B., Baggiolini, M., and Pozzan, T. (1986) Quantitative analysis of the cytosolic free calcium dependency of exocytosis from three subcellular compartments in rabbit and human neutrophils, *J. Cell Biol.* **102:**2197–2204.

Li, S-C., Sonnino, S., Tettamanti, G., and Li, Y-T. (1988) Characterization of a nonspecific activator protein for the enzymatic hydrolysis of glycolipids, *J. Biol. Chem.* **263:**6588–6591.

Neufeld, E. (1981) Recognition and processing of lysosomal enzymes in cultured fibroblasts, in *Lysosomes and Lysosomal Storage Diseases* (J. W. Callahan and J. A. Lowden, eds.), Raven Press, New York, pp. 115–130.

Orci, L. (1987) Proteolytic maturation of insulin is a post-Golgi event which occurs in acidifying, clathrin coated secretory vesicles, *Cell* **49**:865–868.

Paw, B. H., and Neufeld, E. F. (1988) Normal transcription of the β-hexosaminidase α-chain gene in the Ashkenazi Tay–Sachs mutation, *J. Biol. Chem.* **263**:3012–3018.

Scher, W. (1987) The role of extracellular proteases in cell proliferation and differentiation, *Lab. Invest.* **57**:607–623.

Silver, I. A., Murrib, R. J., and Etherington, D. J. (1988) Microelectrode studies on the acid microenvironment beneath adherent macrophages and osteoclasts, *Exp. Cell Res.* **175**:266–276.

Spring-Mills, E., and Hafez, E. S. E. (eds.) (1980) *Male Accessory Sex Glands*, Elsevier, Amsterdam.

Stanbury, J. B., Wyngaarden, J. B., Frederickson, D. S., Goldstein, J. L., and Brown, M. S. (eds.) (1983) *The Metabolic Basis of Inherited Disease* (5th ed.), McGraw-Hill, New York.

Tager, J. M. (1985) Biosynthesis and deficiency of lysosomal enzymes, *Trends Biochem. Sci.* **10**:324–326.

Takemura, R., and Werb, Z. (1984) Secretory products of macrophages and their physiological functions, *Am. J. Physiol.* **246**:C1–C9. [See also Werb, Z. (1987) In *Basic and Clinical Immunology* (6th ed.) (D. P. Sittes, J. D. Stobo, and J. V. Wells, eds.), Appleton & Lange, Los Angeles.]

Taugner, R., Buhrle, C. P., Nobiling, R., and Kirschke, H. (1985) Coexistence of renin and cathepsin B in epitheloid cell secretory granules, *Histochemistry* **83**:103–108. (See also *Cell Tissue Res.* **239**:575–587, 1985.)

Tiedke, A., Hunster, P., Florin-Christensen, J., and Florin-Christensen, M. (1988) Exocytosis, endocytosis and membrane cycling in *Tetrahymena thermophila*, *J. Cell Sci.* **89**:515–520.

Travis, J., and Salvsen, G. S. (1983) Human plasma proteinase inhibitors, *Annu. Rev. Biochem.* **52**:655–709.

Tsuchiya, M., Perez-Tamayo, R. Okazaki, I., and Maruyama, K. (eds.) (1982) *Collagen Degradation and Mammalian Collagenases*, Excerpta Medica, Amsterdam.

Vaes, G. (1980) Collagenase, lysosomes and osteoclastic bone resorption, in *Collagenases in Normal and Pathological Connective Tissues* (D. E. Wooley and J. M. Evanson, eds.), Wiley, New York, 185–207.

Van Golde, L. M. G., Batenberg, J. J., and Robertson, B. (1988) The pulmonary surfactant system: Biochemical aspects and functional significance, *Physiol. Rev.* **68**:374–455.

Wassarman, P. M. (1987) The biology and chemistry of fertilization, *Science* **235**:553–564.

Watts, R. W. E., and Gibbs, D. A. (1986) *Lysosomal Storage Diseases*, Taylor & Francis, London.

Wood, L., and Kaplan A. (1985) Transit of α-mannosidase during its maturation in *Dictyostelium discoideum*, *J. Cell Biol.* **101**:2063–2069.

Williams, M. C. (1984) Endocytosis in alveolar type II cells: Effects of change and size of tracer, *Proc. Natl. Acad. Sci. USA* **81**:6054–6058.

Wright, S. D., and Silverstein, S. C. (1984) Phagocytosing macrophages exclude proteins from the zone of contact with opsonized targets, *Nature* **309**:259–261.

Young, J. D. E., and Chon, Z. A. (1986) Cell mediated killing: A common mechanism? *Cell* **46**:641–642. (See also *J. Exp. Med.* **166**:1894, 1987.)

7

Genesis

This chapter deals with lysosome formation, especially with the synthesis and transport of lysosomal proteins. The central issues will be ones of processing, sorting, and packaging: What are the steps by which the proteins mature into the forms in which they are found in lysosomes? How is the intracellular transport of nascent lysosomal hydrolases organized so as to avoid their doing damage to other cellular components? How do lysosomal components come to be packaged separately from other proteins?

Once again, the literature forces overemphasis on proteins. There is not much to say, for example, about where the lipids in the membranes that bound lysosomes come from and whether the lipids are delivered only as parts of preassembled membranes or can also arrive as individual molecules. In what ways do the lysosomes acquire the high concentrations of certain lipid-related molecules claimed to be present in unusually high concentrations [e.g., dolichols, and certain phospholipids (Section 3.2.2.1)]? And how are the lipids in the membranes affected by acidification or membrane cycling or by the hydrolase activities going on nearby (Section 3.2.1)?

7.1. The Hydrolases Are Glycoproteins, Synthesized by the Rough ER and Golgi Apparatus

Most lysosomal hydrolases are relatively long-lived glycoproteins.* Estimated half lives for hydrolases in rodent liver or kidney, cultured mammalian fibroblasts, and macrophages, range from a half day to a few weeks for different enzymes. Half lives of a day to a week or so seem typical, though some enzymes, such as the sulfatases, are reportedly especially long-lived in some cell types, and published hydrolase half-lives for cultured fibroblasts are longer

---

*Reliable data on steady-state turnover have been hard to come by, partly because of problems in discriminating turnover of intrinsic lysosomal components from the lysosomal breakdown of autophagocytosed and heterophagocytosed proteins and partly because some of the newly made lysosomal hydrolase molecules are secreted rather than entering lysosomes. Additionally, it is difficult to incorporate high enough levels of radioactivity into the lysosomal hydrolases for precise turnover studies: In the cell types studied most (e.g., hepatocytes, fibroblasts), synthesis of hydrolases represents a very small proportion of the cells' overall protein synthetic operations and even for macrophages it probably amounts to no more than a few percent. Needless to add, the general problems in measuring turnover discussed in Chapter 5 apply to lysosomes as well, sometimes with considerable force because the hydrolases often are not very abundant.

FIG. 7.2. Structure, and degradation, of N-linked oligosaccharides of glycoproteins (see Section 7.1.1.1). The diagram illustrates a "complex" ("terminally glycosylated") oligosaccharide as a branched (in this case "biantennary"—two branches) structure attached at one end to an asparagine (Asn) in a polypeptide and terminating at the other end in sialic acids (SA). Gal, galactose; GlcNAc, N-acetylglucosamine; Man, mannose; Fuc, fucose. (The ± means that these residues are especially variable among different complex oligosaccharides, being present in many but not in many others.) The vertical arrows, and Greek letters followed by numbers, are the conventional indications of the locations and orientations of the bonds between adjacent units in the chains.

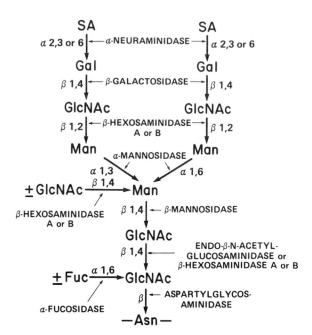

The "core" portion of this structure (the two GlcNAcs attached to the Asn and the three Mans attached to the GlcNAcs) is common to virtually all N-linked oligosaccharides in familiar mammalian glycoproteins; the remainder varies considerably. See also Fig. 7.11.

During lysosomal degradation, units are removed sequentially, starting at the sialic acid ends, by the exoglycosidases indicated in the diagram.

From Beaudet, A. L. (1983). In *Metabolic Basis of Inherited Disease* (5th ed.) (J. B. Stanbury, J. B. Wyngaarden, D. S. Frederickson, J. L. Goldstein, and M. S. Brown, eds.), McGraw-Hill, New York, p. 794.

For proteins that leave the ER, further processing is carried out by enzymes mostly located in the Golgi apparatus (Section 7.4). To produce the "complex" ("terminally glycosylated") oligosaccharides found on many secretory proteins and plasma membrane proteins, a mannosidase (α-mannosidase I) trims the chain down to five Mans and, after a GlcNAc is added by action of GlcNAc transferase I, a second mannosidase (α-mannosidase II) removes two more Mans. At this point, a "core" of two GlcNAcs and three Mans is what remains of the original oligosaccharide. Another GlcNAc transferase (II) now adds additional GlcNAcs to the oligosaccharide. Then, in the final steps, galactoses and sialic acids (and sometimes fucoses) are added by corresponding transferases.

Two features of the oligosaccharides on lysosomal hydrolases are of particular importance for later concerns. First, the collection of oligosaccharides found on the hydrolases produced by a given cell population often includes some chains of the high-mannose type, some complex chains, and some of a "hybrid" variety in which one branch ("antenna") of the chain contains sialic acids and galactose while another antenna of the same chain has only mannoses. In addition, lysosomal oligosaccharides are almost unique in their content of the mannose-6-phosphate (M6P) groups to which repeated allusion has already been made; these groups are added in the Golgi apparatus by a sequence of enzymatic steps (Section 7.4.2). In *Dictyostelium*, sulfate groups are present as well, probably having also been added by the Golgi apparatus (Section 7.3.3.4).

Lysosomal hydrolases like β-glucuronidase or cathepsin D, resemble secretory glycoproteins in that their exit from the ER is markedly delayed in cells exposed to the glucosidase inhibitor **deoxynojirimycin. Tunicamycin,** which prevents N-linked glycosylation, has variable effects on secretory proteins—affecting the transport of some relatively little and retarding movements of others. This agent sometimes decreases exit rates of cathepsin D and other hydrolases from the ER of cultured cells. More strikingly, owing to the absence of oligosaccharides to which M6Ps can be attached, tunicamycin-exposed cells show increases in the proportions of lysosomal enzymes missorted to the extracellular environment rather than being packaged into the lysosomes. Monensin, which inhibits the exit of many newly made secretory proteins from the Golgi apparatus, has complex effects on lysosomal enzymes. It not only alters Golgi transport but also affects endosomes and lysosomes directly (see Sections 3.1.1.1 and 3.3.4.2) and its influences on pH can result in inhibitions of some of the proteolytic processing steps by which lysosomal hydrolases mature (Section 7.4.3.1).

7.2. Whence the Humble Lysosome?

Everyone agrees that the sorting of lysosomal hydrolases from other ER-borne proteins and their packaging into membrane-delimited containers, depends crucially on the Golgi apparatus or structures closely associated with it. But agreement does not extend much beyond such generalities.

7.2.1. The Classical View: Primary Lysosomes Bud from the Golgi Apparatus

The specialization of mammalian phagocytes for lysosome-related functions made them favorable material for early work on hydrolase packaging. Some of the phagocytes even show cytochemically demonstrable acid hydrolases in their ER (Fig. 7.3). [That this is not seen in most other material was (and still is) assumed to reflect the relaxed pace of production of hydrolases in cells whose profession is not endocytotic.] Neutrophil granules arise at the Golgi apparatus (Fig. 2.2) and function by fusing with vacuoles bounded by membranes derived from the plasma membrane; these features are so strongly reminiscent of secretory cells that they have predisposed most observers to think of the neutrophil's primary lysosomes (azurophilic granules) almost as specialized secretion granules.

The neutrophil's sorting of different proteins into different packages is dramatic, in that at least two abundant types of cytoplasmic storage granules result (Section 2.1.2.1). One aspect of the sorting mechanism is temporal: Morphological studies and cytochemical demonstrations of acid phosphatase, alkaline phosphatase, aryl sulfatase, and peroxidase activities, indicate that differentiating neutrophils make most of their azurophilic granules before they begin to produce the nonlysosomal specific granules. But the neutrophils may also employ "spatial" sorting mechanisms—different routes to the final packages. Both varieties of granule form from Golgi sacs, in a manner calling to mind the formation of secretory bodies. However, as far as can be judged from conventional morphological landmarks—curvature of the stacked Golgi sacs, location of centrioles (Fig. 2.2), and the like—the face of the Golgi apparatus

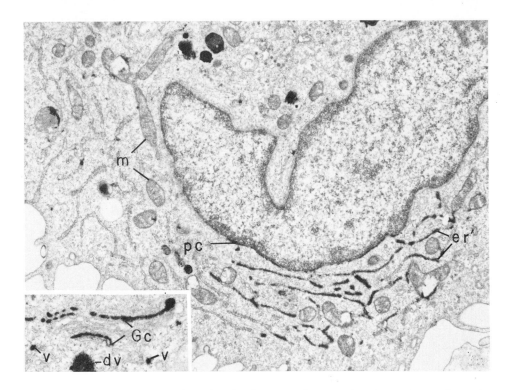

FIG. 7.3. Electron micrograph of a rabbit alveolar macrophage incubated, cytochemically, to demonstrate aryl sulfatase activity. Reaction product is seen in the rough ER (er') including the nuclear envelope (pc). Near the top of the picture, several lysosomal granules also show reaction. Mitochondria (m) do not. × 20,000. The insert shows sulfatase activity in the Golgi region of another macrophage: reaction product is seen in Golgi sacs (Gc), small vesicles (v), and a lysosomal digestive vacuole (dv). × 35,000 (approx.). From Nichols, B. A., Bainton, D. F., and Farquhar, M. G. (1971) *J. Cell Biol.* **50:**498.

from which the azurophilic granules form is opposite the one from which the specific granules form.* Unfortunately, the *cis–trans* organization of the neutrophil's Golgi apparatus has yet to be worked out. The face of the Golgi apparatus at which the azurophilic granules form somewhat resembles the *trans* face of the Golgi apparatus of secretory cells and the specific granules form from a "*cis*-like" face. But these judgments are essentially on morphological

---

* A generations-long debate about the functioning of the Golgi apparatus has left a confused terminology in its wake. The most characteristic morphological feature of the Golgi apparatus is the stack of closely associated Golgi sacs. This stack usually is polarized in its morphology and function. For instance, the sacs toward one side of the stack—currently called the "*trans* face" or *trans* aspect—often show a considerably more electron-dense content than do those toward the opposite side of the stack (the "*cis* face"). The majority of materials processed by the Golgi apparatus are assumed to pass from the ER to the *cis* face of the apparatus and then to move down the stack, often becoming increasingly concentrated as they approach the *trans* face. The electron density of the content of the *trans* sacs, when seen, is thought to reflect this "condensation" of the sacs' contents; but bear in mind that the appearance of the *trans* sacs, and of the rest of the Golgi apparatus, varies considerably from cell type to cell type and with different modes of tissue preparation. It is from the sacs at the *trans* face that secretory granules of gland cells and the vehicles that carry cell surface materials to the plasma membrane usually form.

grounds and certainly cannot be taken as reliable indications of the functional organization of the Golgi apparatus (e.g., of the distribution of the enzymes responsible for oligosaccharide processing; see footnote on p. 368). Conceivably, the polarity of its Golgi apparatus changes as the neutrophil switches from making one granule to making the other type. Indeed, because the contents of the specific granules are secreted more readily than are those of the azurophilic granules (Section 4.4.1.2), some observers argue that the specific granules more closely resemble secretory structures and therefore should be regarded as the *trans* Golgi product.

The monocytic precursors of macrophages resemble the neutrophils in that sacs at one face of the Golgi apparatus show acid phosphatase, aryl sulfatase (Fig. 7.3), and peroxidase activities and seem to package these enzymes into the monocytes' primary lysosomal granules (Section 2.1.3.1).

For most other cell types, a Golgi origin of primary lysosomes has been widely presumed even though primary lysosomes are not readily recognizable. This presumption, remember (see Section 2.1.3.1), rests upon generalization from the neutrophils and monocytes, and on findings of cytochemically demonstrable acid phosphatase in small (50- to 100-nm diameter) vesicles near the Golgi apparatus and in Golgi-associated sacs (Fig. 7.4). As always, there is the risk of overreliance on acid phosphatase. Occasionally, the acid phosphatase demonstrated cytochemically in Golgi sacs has been reported to differ subtly from the lysosomal enzyme, for example, in its relative preference for certain substrates. On the other hand, in some cytochemical preparations of macrophages and neurons, esterase or aryl sulfatase are seen in Golgi-associated vesicles and sacs; this demonstration has been accomplished only in a handful of cases but does provide some reassurance that Golgi-derived vesicles do carry lysosomal enzymes.

7.2.1.1. Coated Vesicles

Section 2.1.3.1 also mentioned that some of the presumptive Golgi-vesicle primary lysosomes are coated. This was first suspected from work in my laboratory and elsewhere showing acid phosphatase activity in coated vesicles of

---

Before "*cis*" and "*trans*" came into general use, the two Golgi faces were referred to by terms such as "convex and concave" (see below), "forming face and maturing face," or "entry face and exit face."

Rough rules of thumb have evolved to help microscopists discriminate between the *cis* and *trans* faces in electron microscope thin sections, though these rules cannot always be applied unambiguously: (1) The *cis* face of the stack tends to be convex in profile whereas the *trans* face is more often concave. (2) When, as is common, the cell's centrioles are located near the Golgi apparatus (Fig. 2.2), they frequently associate with the *trans* face. (3) The *trans* face often exhibits accumulations of newly formed secretory bodies. Nowadays, particular attention is paid to the *cis–trans* distribution of Golgi enzymes detectable immunocytochemically and in cell fractions. Current stereotypes of the Golgi apparatus locate the mannosidase I involved in oligosaccharide processing toward the *cis* side of the Golgi stack, mannosidase II and GlcNAc transferase I in "medial" sacs (ones in the middle of the stack), and the sialyl and galactosyl transferases near the *trans* face. Acid phosphatase activity usually is concentrated near the *trans* face (see Section 7.2.2). But in these regards too, problems of identification can arise: (1) The distribution of enzymes in the Golgi apparatus varies somewhat from cell type to cell type. (2) It is very difficult to separate Golgi apparatus from other cell structures, let alone to subfractionate the apparatus clearly into *cis* and *trans* components. (3) There is considerable uncertainty as to how the ER elements most directly involved in feeding materials into the Golgi apparatus at its *cis* face (the "transitional elements") contribute to the findings.

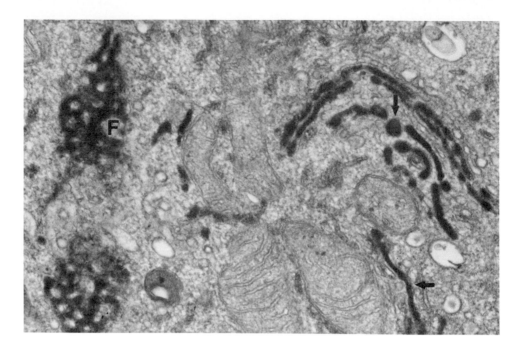

FIG. 7.4. Golgi region in an electron micrograph of an acid phosphatase preparation of a neuron from a cultured chick ganglion. Reaction product is readily seen in sacs, tubules, and vesicles toward the *trans* Golgi face (arrows indicate two of the many reactive structures). In face view (F) the sacs appear as "honeycombs" of tubular and saclike elements. From Teichberg, S., and Holtzman, E. (1973) *J. Cell Biol.* **57**:88. × 35,000.

neurons and several other cell types (see Fig. 2.15) and now is indicated by biochemical evidence as well (see Section 7.4.3.4). The coating is thought to include clathrin, from its appearance and from immunocytochemical demonstrations of clathrin in vesicles of appropriate size and distribution. Similar coating is present along regions of the sacs at the *trans* Golgi face from which the acid phosphatase-containing vesicles seem to arise (see below).

Because coated vesicles are so useful to the cell, it has proved tempting to go overboard and presume that all the vesicles that carry things from one place to another in the Golgi region are clathrin coated. Actually, however, in many cell types most of the clearly identifiable clathrin coating on Golgi sacs and vesicles, detectable by morphology or with antibodies, is toward the *trans* side; less has been detected along lateral aspects of the apparatus, or at the *cis* face (but see Section 7.3.2.3). At the lateral edges of the Golgi sacs and elsewhere in the Golgi region, one does encounter an abundant class of small vesicles whose membranes sometimes give the impression of being "thicker" and more electron dense than other cellular membranes. This thickening could be due to a coating of some sort, but it is not comprised of the distinctive radially arranged "spikes" that characterize clathrin-coated vesicles in thin section.

7.2.2. The Romantic Era: GERL Meets Man

7.2.2.1. GERL

Novikoff's pioneering cytochemical work was of major influence in establishing the likely involvement of Golgi structures in lysosome formation. This work led eventually to a unifying hypothesis that lysosomes derive from a set of sacs and tubules that, though juxtaposed to the *trans* face of the Golgi apparatus, constitute a system distinct from the apparatus itself. The chief grounds for this proposition were discoveries that some of the acid phosphatase-reactive sacs and tubules associated with the Golgi apparatus, though located near the *trans* Golgi face, show features different from the closely stacked Golgi sacs that constitute the morphological "core" of the apparatus (Fig. 7.4). For example, the most strongly acid phosphatase-reactive elements frequently are spaced a greater distance from the Golgi stack than most of the stacked sacs are spaced from one another. In addition, the reactive elements often have a striking and seemingly distinctive form, appearing as elaborately interconnected networks of tubules or as extensively fenestrated sacs (Fig. 7.4; see also Fig. 5.11 and note that such configurations provide a large surface area). They also show stretches ("rigid lamellae") that give the visual impression of being much stiffer than most cellular membrane systems. And, as noted already, they exhibit clathrin-coated zones.

In some cases the sacs or tubules in question and the structures that seem to be budding from them are the *only* Golgi-associated compartments other than recognizable lysosomes, that stain for acid phosphatase and for certain other enzymes. And sometimes they are unreactive, or very weakly reactive, for thiamine pyrophosphatase activity, in contrast to the strong reaction of their immediate neighbors in the Golgi stack. Because the presence of thiamine pyrophosphatase activity was, historically, a very important criterion for identifying and mapping the Golgi apparatus, the absence of this activity in the systems under consideration seemed to set these structures apart from typical Golgi elements.

The acronym GERL was coined to convey the idea that this set of structures constitutes a discrete system with distinctive cytochemical and morphological features and special functions. The acronym emphasizes the systems' relations to the Golgi apparatus (**G**) and the notion that it is a source of lysosomes (**L**); vesicular primary lysosomes were thought to bud from GERL, and some AVs to form through the envelopment of cytoplasm by GERL-derived sacs and tubules. The **ER** came from the belief that GERL is a part of the endoplasmic reticulum. This belief carried with it the possibility that lysosomal hydrolases could be transported within a specialized network directly from ER-bound ribosomes to the lysosomes, without necessarily passing through the Golgi stack. Some investigators even flirted with the heterodox idea that the ER regions involved might be specialized for hydrolase synthesis, which would step the problem of sorting lysosomal proteins from other ER products back to the question of how specific polysome species could concentrate along particular stretches of the ER, transiently or stably.

The GERL structures are among the systems at the *trans* Golgi face deemed most likely to be acidified. Although fledgling efforts to estimate pHs with microscopically visible weak bases (Section 3.1.2.1) suggest that the pH in the *trans* Golgi compartments is higher (ca. 6–6.5) than that in lysosomes, the acidification is another generic similarity between GERL and lysosomes.

7.2.2.2. The "*Trans* Golgi Network"

The concept of GERL focused intense attention on specialized features of the structures at the *trans* Golgi aspect. But the details of this concept have yet to be convincingly reconciled with several important observations, a disability shared in one form or another by all current views of how lysosomes arise.

That GERL is connected to the ER was concluded from observations of intimate appositions and other close associations between sacs near the *trans* Golgi face and smooth-surfaced sacs and tubules known to be ER because they are directly continuous with ribosome-studded cisternae. At first it was assumed that the continuities with the rough ER extended to the acid phosphatase-reactive sacs. But now it is appreciated that several different types of smooth-surfaced membrane systems coexist in the *trans* Golgi region and often are difficult to discriminate from one another morphologically. So far, unambiguous continuities of ribosome-studded ER specifically with the acid phosphatase-containing structures at the *trans* Golgi aspect have been observed so rarely as to leave most observers doubting the general existence of continuities between GERL and the ER.

Furthermore, by immunocytochemistry, lysosomal enzymes—proteases, glycosidases, and others—proved to be demonstrable not just in GERL, but also in many or all of the sacs of the Golgi stack. Often the immunocytochemical signals in *cis* and medial sacs are weak so that what stands out is the concentration of the enzymes in systems near the *trans* Golgi face (Fig. 7.5). But the findings do show that lysosomal hydrolases in the Golgi zone are not confined only to the ER and to GERL-like structures, implying, at the minimum, that some hydrolase molecules move through the Golgi stack. [Note that unlike

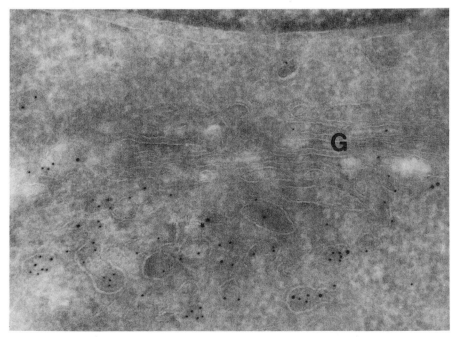

FIG. 7.5. Part of the Golgi region of an immunocytochemical preparation of an ultrathin cryosection (see footnote on p. 373) of a human hepatoma (HepG2) cell incubated, with gold procedures, to show the locations of the M6P receptor (215 kD) and simultaneously to demonstrate lysosomal enzymes (the antienzyme antibodies used were a "cocktail" directed against cathepsin D, α-glucosidase, and β-hexosaminidase). The large gold particles locate the lysosomal enzymes and the small ones locate the receptor; both sizes of gold are present in the same Golgi-associated compartments of these cells. Overall, few gold particles are present over the stacked Golgi sacs (G) but many are present over systems adjacent to the *trans* Golgi face (the *trans* Golgi network, or GERL; see text). From Geuze, H. J., Slot, J. W., Strous, G. J. A. M., Hasilik, A., and von Figura, K. (1985) *J. Cell Biol.* **101**:2253. × 100,000 (approx.).

enzyme cytochemical methods, which rely upon the presence of active enzymes, immunocytochemistry can detect "immature" or "precursor" molecules that are not yet enzymatically active. Several of the lysosomal hydrolases may traverse the Golgi apparatus in just such inactive forms (Section 7.4.1.1).]

In secretory cells, or in cells producing melanin granules, the acid phosphatase-containing sacs designated as GERL seem to be the same ones that package the cells' secretory or pigment granules (see Section 6.1.5). This was a confusing observation because it indicated that the functional distinctions between GERL and other systems of the Golgi region cannot be absolute—Golgi products do enter GERL. Others of the supposed distinctions between GERL and the Golgi apparatus also have proved not to be universal. For example, there are cell types and physiological or technical circumstances in which *trans* or even medial sacs that clearly are part of the Golgi stack, stain for acid phosphatase. And there are cases in which structures with the appearance and acid phosphatase reactivity of GERL also stain for thiamine pyrophosphatase.

These observations could mean simply that GERL exchanges materials extensively with the Golgi apparatus, perhaps via some of the numerous vesicles of the *trans* Golgi region. More often, it is argued that the structures designated as GERL really are part of the Golgi apparatus, being forms assumed by the *trans*-most Golgi elements. Proponents of this last viewpoint emphasize that the three-dimensional geometry of the Golgi apparatus and its associated systems is tortuous so that thin sections can convey misleading impressions. In addition, they suspect that the membrane systems in question are highly dynamic ones, whose morphology and local organization change rapidly as structures fuse with them or bud off and as the materials within them undergo transformations. At the same time, these investigators recognize that the sacs and tubules at the *trans* Golgi face do have distinctive features. To convey this recognition without implications of special relations to the ER, they substitute names like **trans Golgi network** (or *trans* Golgi reticulum) for "GERL."

Largely from immunocytochemical work,* the *trans* Golgi network (TGN) is considered to include regions where sialyl transferase is present along with acid phosphatase. (Galactosyl transferase is less readily evident; see also footnote on p. 376.) Lysosomal hydrolases, Golgi enzymes, secretory products such as the albumin made by hepatocytes, and membrane proteins—especially viral membrane proteins made in suitably infected cells (Fig. 7.6)—all are believed to occupy the same TGN. There may, however, be differences in the relative concentrations of different ones of these proteins in the different elements—

---

*These studies have gained much of their momentum from the recent reliance of immunocytochemists on electron microscopic techniques that use antibodies labeled with gold particles. One of the advantages of the immunogold procedures is that different-sized gold particles can be utilized to demonstrate different antigens simultaneously in the same tissue section (Fig. 7.5). This minimizes the potential for confusion arising when decisions as to whether different antigens occupy the same structure must be made by comparing different preparations each incubated to demonstrate a single antigen. Another advantage of the gold methods is that by counting the number of particles located over a region, information can be obtained about the relative number of antigen molecules present. (Such semiquantitative immunocytochemistry must be done very cautiously owing to the many factors that influence the binding of labeled antibodies to antigens within tissues subjected to the preparative procedures required for electron microscopy.)

The use of frozen thin sections (Figs. 7.5 and 7.6) has also been valuable for immunocytochemical work in part because the absence of embedding media facilitates penetration of detecting antibodies and because denaturations and other "inactivations" of antigens that can occur during the preparation of plastic-embedded thin-sectioned material are avoided.

sacs, tubules, attached vesicles, and the like—of the network. In functional terms: the TGN is regarded as a compartment where enzymatic machinery of the Golgi apparatus is present, and through which pass lysosomal enzymes, material destined for the cell surface, and products slated for secretion.

As in Fig. 7.6, the TGN in BHK tissue culture cells, expands considerably when the temperature is lowered to 20°C, a temperature that inhibits the production of the membrane-delimited carrier structures that transport materials away from the network, and also alters the interplay of endosomes and lysosomes (Section 2.4.7.1).

7.2.2.3. Endosomes and Lysosomes near the Golgi Apparatus; Hydrolase-Containing Sacs Far Away

The endocytotic bodies and secondary lysosomes present near the Golgi apparatus, often accumulate in the vicinity of the *trans* face. Endosomes and

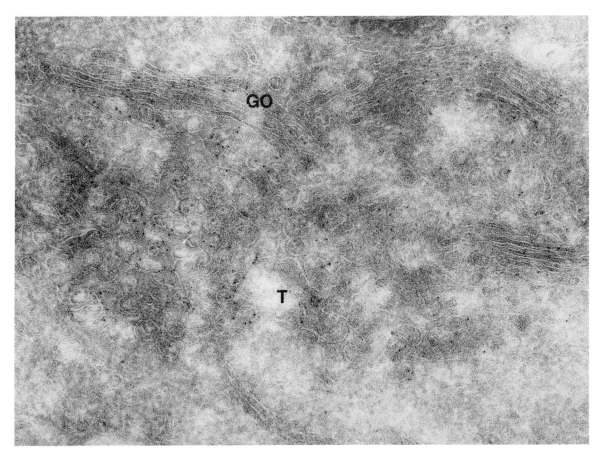

FIG. 7.6. Golgi region of a cultured BHK cell infected with vesicular stomatitis virus but incubated at 20°C, which inhibits passage of viral membrane proteins out of the Golgi apparatus. This is an immunocytochemical preparation (cryosectioning–gold procedure; see footnote on p. 373) demonstrating the localization of the G protein of the virus. Gold particles are seen over Golgi sacs (GO) and over a network of tubules and attached vesicles (T) at the *trans* Golgi face. Companion sections incubated to demonstrate acid phosphatase show reaction product in this *trans* Golgi network, producing patterns resembling the face views in Fig. 7.4. From Griffiths, G., Pfeiffer, S., Simons, K., and Matlin, K. (1985) *J. Cell Biol.* **101**:949. × 70,000 (approx.).

endocytotic tubules can look much like elements of the GERL–TGN systems and lysosomes also sometimes are interconnected in networks resembling these systems. What are the relations among these several classes of structures?

With occasional exceptions, the endocytotic tracers like HRP that are presumed to be taken up largely in the fluid phase, do not enter the networks or other structures regarded as integral parts of the GERL–TGN systems, even when these tracers accumulate in nearby endocytotic vesicles and tubules, endosomes, and secondary lysosomes. For the most part, therefore, the endocytotic and secondary lysosomal compartments near the Golgi apparatus probably are not structurally continuous with the major elements of the GERL–TGN systems. Nevertheless, endocytotic structures (and lysosomes?) do seem to communicate with the GERL–TGN compartments, perhaps via vesicles or other transport vehicles: Certain membrane-associated endocytotic tracers, such as conjugates of cholera toxin or ricin, accumulate in Golgi-associated sacs or tubules that seem to be GERL–TGN components (see Section 3.3.4.5). And the fact that glycoproteins can be reglycosylated during their cycling between the cell surface and the cell interior (Section 5.3.3.3) indicates that some membrane molecules interiorized by endocytosis gain access to compartments containing glycosylation enzymes such as sialyl transferase; almost by definition such compartments are thought of as parts of the GERL–TGN systems or of the Golgi apparatus proper (but see footnote on p. 376).

In cells retrieving membrane after extensive exocytosis, endocytosed tracers—especially membrane-associated ones—can pass into sacs throughout the Golgi stack (Section 5.3.3.2), showing that membrane recycling need not be confined to *trans* Golgi elements. A similar conclusion has been tentatively drawn from the finding, with erythroleukemia cells, that glycoproteins synthesized and transported to the cell surface in the presence of the mannosidase inhibitor **deoxymannojirimycin** can be made normal during their subsequent cycling into and out of the cell, once the inhibitor has been removed. As deoxymannojirimycin affects relatively early steps in Golgi-mediated glycosylation, these findings are taken to mean that recycling cell surface proteins can reach medial or even *cis*-ward Golgi structures.*

Another side to the complexity of the relations among lysosomes, endocytotic structures, and *trans* Golgi elements is that lysosome-related tubules, networks, or saclike structures similar to those present near the Golgi apparatus, often are present at a considerable distance from morphologically recognizable Golgi stacks. An especially striking example is encountered in the "beige mouse," a homologue of Chediak–Higashi disease (Section 4.4.2.4). Hepatocytes of these mice show an elaborate system of interconnected, acid phosphatase-containing tubules and sacs with modestly electron-dense content (Fig. 7.7); some of this network lies close to the stacked Golgi sacs but much of it extends into cytoplasmic regions far from recognizable Golgi structures. Other examples of extended systems are the tubular lysosomes seen in macrophages under some circumstances (Fig. 2.20a) or the elongate sacs, containing cytochemically demonstrable acid phosphatase and aryl sulfatase activity, demonstrable in the axons of injured or stressed neurons (Fig. 2.20b). These

---

*There are recent conflicting reports as to whether cells actually do "repair" glycoproteins made in the presence of deoxymannojirimycin. Involvement of *trans* Golgi systems in "repair" is more securely established than involvement of other Golgi structures.

various configurations have sometimes been regarded as hypertrophied variants or derivatives of GERL or of the Golgi apparatus and have been ascribed roles in transport of recently made hydrolases to lysosomes.* Many of them, however, are accessible to endocytotic tracers (see Fig. 2.20a) and may better be thought of as unusual secondary lysosomes than as transport pathways.

7.2.2.4. *Trans* Golgi Systems: A Summary

Despite the fragility of present evidence and assumptions, attractive compromise portraits can be cobbled that blur the disagreements about the structures of the *trans* Golgi region even if they do not completely resolve them. Popular current conceptions hold that the *trans* Golgi region contains, as structurally separate systems:

1. A distinctive set of saclike and tubular elements associated closely with the *trans* Golgi face and exhibiting clathrin-coated regions, whose derivatives transport materials processed by the Golgi apparatus to other compartments.
2. Secondary lysosomes and related structures, some of which interconnect with one another.
3. Endosomes.
4. Vesicular or tubular transport and storage vehicles of several sorts, some clathrin-coated, some not. Included are vehicles that move membranes from lysosomes and/or endosomes to Golgi compartments; ones carrying materials between Golgi elements; ones carrying products from the Golgi apparatus to secretion granules, to the plasma membrane, or to nascent lysosomes (see the discussion of intermediate compartments in Section 7.3.2.2); and ones shuttling membranes from endosomes or lysosomes to the cell surface.

The first of these systems—the sacs and tubules intimately associated with the *trans* Golgi face—is the one that has been referred to as GERL or TGN. It

---

*From time to time when smooth-surfaced systems, such as acid-phosphatase containing sacs or tubules, are seen extending considerable distances from the Golgi region toward other cytoplasmic zones, it is suggested that these structures actually are *trans* Golgi or GERL–TGN sacs that carry out Golgi functions at locales where no readily recognizable Golgi stacks are present. There has been a recent flurry of controversy in this connection, over reports that sialyl transferase and galactosyl transferase—two principal "terminal" transferases of the Golgi apparatus—may have somewhat different distributions in the Golgi zone (e.g., the galactosyltransferase supposedly is more concentrated in *trans*-ward sacs of the Golgi stack whereas the sialyl transferase is more evident in the GERL–TGN systems). Related reports have it that sialyl transferase-containing structures pervade an area of cytoplasm far greater than do other Golgi structures. These claims are based largely on immunocytochemistry. Presently unresolved disputes about them concern possibilities that different cell types differ dramatically in distribution of transferases, and technical matters such as the concentrations of antibodies used, the sensitivities of the methods, and the specificities of the antibodies. (E.g., some of the antitransferase antibodies are directed against carbohydrate components, raising possibilities that they bind to several proteins sharing similar oligosaccharides; there have, however, been reports of staining of structures far from the Golgi stacks, and even of the plasma membrane, with anti-sialyl transferase antibodies seemingly directed against the polypeptide part of the enzyme.)

Proposals that lysosomes contribute hydrolases to Golgi structures as well as receiving enzymes from the Golgi apparatus occasionally are made, but there is little supportive evidence.

contains acid phosphatase demonstrable cytochemically and has other lysosomal hydrolases demonstrable immunocytochemically, as well as secretory products and plasma membrane proteins. It probably is among the acidified Golgi-associated structures and possesses sialyl transferase. It is accessible to some membranes and receptors cycling from endocytotic structures but it tends to be inaccessible to the bulk contents of endocytotic structures.

This scheme, and its many relatives, accommodate the available evidence most comfortably, if one assumes that a given type of molecule can follow alternative pathways within the complex of Golgi-associated structures. The "choice" of path to be followed by a given molecule can be influenced by the molecule's nature, by its interactions with neighbors in the membranes or compartments it enters, by cytoskeletal elements, ligands, and pH, and by other factors. But there may also be "random" switching so that, for example, even when circumstances remain constant, a receptor or other membrane molecule would move along different paths in successive cycles of endocytosis or intracellular transport.

Glaring gaps? (1) The persistent uncertainties about how endocytotic structures and Golgi structures interact structurally and functionally. (2) The lack of

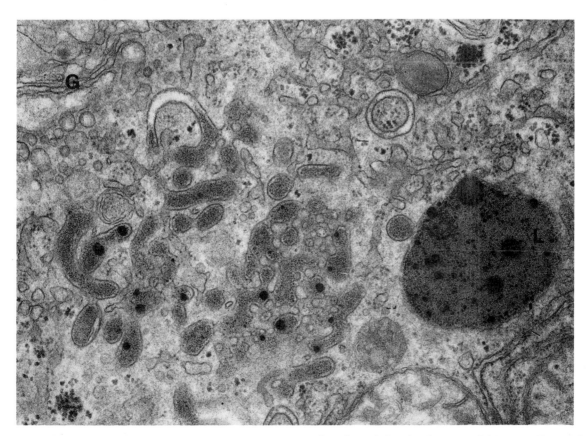

FIG. 7.7. Region near the Golgi apparatus (G) from a hepatocyte of the beige mouse (a homologue of the human Chediak–Higashi disorder; Section 4.4.2.4). L indicates a lysosome—from its appearance, probably a residual body. Much of the field is occupied by a network of membrane-delimited tubules with electron-dense content. Companion studies showed that the elements of this network react for acid phosphatase activity and that some are in direct continuity with recognizable lysosomes. From Essner, E., and Oliver, C. (1974) *Lab. Invest.* **30:**596. × 30,000 (approx.). (Copyright: U.S.–Canadian Academy of Pathologists.)

an explanation for the observed diversity of structural relations between ER and Golgi apparatus. (The existence of very close ER–Golgi appositions both at *cis* and at *trans* faces, and the existence—rarely seen though they may be—of direct ER to Golgi sac continuities, indicate that the two systems may communicate in ways additional to the shuttling of vesicles between the ER and the *cis* Golgi face.) (3) The inadequacy of information about how the clathrin coats and cytoskeletal arrays that are prominent in the *trans* Golgi region contribute to controlling the movements and membrane phenomena that occur there. (4) The incompleteness of knowledge about how vacuoles form in plants: Sections 5.1.6.1 and 5.4.2 summarize possible autophagic modes of origin, but there is a long history of proposals claiming that vacuoles arise by expansion or fusion of Golgi or ER derivatives. A fifth unknown—how different Golgi products entering common *trans* compartments eventually are sorted to different destinations—may be on its way to solution, as will next be discussed.

7.2.3. Secretion-Recapture: One Way in Which Receptors for M6P Might Function in Hydrolase Transport

The discovery that storage disease fibroblasts secrete hydrolases that can be taken up by receptor-mediated endocytosis, focused attention on M6P as a sorting signal almost unique to lysosomal enzymes (Section 6.4.1.4). Can it be concluded that lysosomes obtain their hydrolases from the Golgi region circuitously instead of by direct transport at the *trans* Golgi face? Perhaps the enzymes first move from the Golgi apparatus to the cell surface, either in their own containers or in consort with secretory material or plasma membrane components. At the cell surface, endocytotic retrieval, via high-affinity M6P receptors like those participating in the cross-feeding of tissue-cultured fibroblasts, could then pick up the hydrolases for transport to lysosomes. Sorting of the hydrolases from other proteins packaged by the ER and Golgi apparatus would, in effect, be postponed until after the proteins reach the cell surface where the hydrolases' unique possession of M6P residues would target them differentially to the lysosomes.

This hypothesis has a partial precedent in the mechanisms that operate in the thyroid gland (Section 4.2.4.1), and it gained significant currency for a while as it became clear that some cells in culture secrete surprisingly large proportions of their newly made hydrolases: With cultured rodent liver cells, 30% or even more of the newly made molecules of cathepsin B or β-glucuronidase can be detected in growth media within the first few hours after their synthesis. Also consistent with the hypothesis were the findings that the M6P receptor of cultured fibroblasts recycles between the cell surface and the interior much like a typical endocytotic receptor. Even when protein synthesis is inhibited, the cells can endocytose much larger quantities of hydrolases than would be possible were each M6P receptor present in the cells to be used only once. Later it was established that the half lives of the M6P receptors in cultured cell lines are on the order of 12–24 hr, ample time for many reuses. And recent investigations on CHO cells and others indicate that desialylated M6P receptors can acquire sialic acids relatively rapidly during recycling.* It

---

*Two related lines of experiments established this: (1) M6P receptors shorn of their sialic acids during their exposure at the cell surface, by neuraminidase added to the medium, can reacquire the sialic acids within a few hours. (Galactose residues are restored more slowly.) (2) The M6P

would appear therefore that M6P receptors from the cell surface are among the proteins that can experience effective passage through compartments with functional Golgi transferases before reemerging at the plasma membrane.

Despite all this, the secretion-recapture hypothesis for hydrolase packaging has been eclipsed of late by proposals that most of the sorting mediated by M6P and its receptors takes place entirely within the cell. This decline was precipitated by experiments showing that delivery of endogenous hydrolases to lysosomes in cultured fibroblasts continues largely unabated when the cells are grown, even for many days, in high concentrations (tens of mM) of free M6P. (Comparable studies suggest that the "endosomal" hydrolases considered in Section 2.4.4.1 are also largely unaffected by inclusion of M6P in the culture medium.) At such concentrations, free M6P can compete with "high-uptake" forms (Section 6.4.1.4) of lysosomal hydrolases for M6P receptors at the cell surface; therefore, the M6P should derail delivery via a mechanism dependent upon engagement of the lysosomal hydrolases with cell surface receptors.

An undertone of apprehension persists that these last experiments have been given too much weight: Perhaps hydrolases coming from inside the cell enjoy a special status during their tour to the cell surface and somehow are protected from competition with M6P. Could it be, for example, that enzymes already bound to the receptors by the time they reach the cell surface are not stripped off by M6P during the brief time—conceivably, less than a minute— they spend exposed to the extracellular medium? Maybe the affinity of the receptor for free M6P is not sufficiently high relative to the affinity for M6P-containing hydrolases or perhaps the M6P has only limited access to the cell surface domains at which hydrolase receptor complexes are delivered during their cycling into and out of the plasma membrane? The critics of secretion-recapture respond: But mutational analyses in yeast suggest that the vacuoles still get their hydrolases even if secretion is disrupted (Section 7.3.3.2). And if endogenous hydrolases moving to the cell surface are so tightly bound or protected, why should there be so much spontaneous release of newly made lysosomal enzymes from certain cultured cell lines to their surrounding media and why should it be so difficult to demonstrate plasma membrane involvement in hydrolase sorting, through cell fractionation? Obvious counterrejoinders are that cultured cells often are abnormal and that yeast are not animals and do not use M6P to target their hydrolases (Section 7.3.3.2), but these arguments simply open other Pandora's boxes that most of us hope stay closed.

7.3. M6P Receptors

If hydrolases are targeted to lysosomes by virtue of possessing M6P residues but if the M6P receptors at the cell surface do not mediate the targetting, then, it was reasoned, there must be **intracellular** M6P receptors to control sorting of the hydrolases. Such receptors were soon demonstrated in tissues and in cultured cells by cell fractionation and by finding great increases in the

---

receptors produced by cell lines genetically deficient in galactosyl transferase lack both galactoses and sialic acids but acquire the latter once galactoses are added by exogenous enzymes introduced in the culture medium.

"Reglycosylation" in such experiments seems to be possible for both classes of M6P receptors described below and is quite rapid compared with that for other membrane proteins.

binding of M6P ligands to cells and cell fractions after disruptive treatments that enable the ligands to cross membranes. The intracellular receptors are most abundant in "microsomal" cell fractions consisting chiefly of fragments of Golgi elements, ER, plasma membrane, and endosomes. They are several to many times more numerous than the receptors at the cell surface, but in affinities and other properties the surface and interior populations appear to be identical.

7.3.1. There Are at Least Two Classes of M6P Receptors: 215-kD ("Cation-Independent") and 46-kD ("Cation-Dependent")

The receptors to which we have been referring so far were the first of the M6P receptors to be discovered. They are transmembrane, N-glycosylated glycoproteins. From SDS–PAGE estimates, molecular weights appeared to be about 215 kD, so that the receptor has come to be called the 215-kD M6P receptor, although "cation-independent" M6P receptor is used by a growing number of authors (see Section 7.3.1.4). The single polypeptide chain shows extensive internal cysteine–cysteine disulfide bonding, reminiscent of the bonding within some of the receptors discussed in connection with receptor-mediated endocytosis (Section 2.3.2.9). The oligosaccharides include chains of the complex type, in which sialic acids and galactoses are present, along with some high-mannose chains. The binding sites for M6P-containing ligands are oriented so as to face the extracellular space at the cell surface, and the interior of intracellular compartments. The opposite, cytoplasmic "tail" of the receptor corresponds to the COOH terminus of the molecule.

cDNA sequence data have been obtained, first for the gene specifying the bovine receptor and then for the human form. In both species the deduced amino acid sequence is of almost 2500 amino acids and includes an amino terminal signal sequence 40–45 amino acids long; a presumed membrane spanning region ("transmembrane domain") of 23 amino acids; and a carboxyl-terminal cytoplasmic domain ("tail") of about 160 amino acids that contains clusters of acidic amino-acid residues (this last feature is one of several ways in which the M6P receptor's cytoplasmic domain resembles those of some of the endocytotic receptors considered earlier). Most of the mass of the protein is in the "extracytoplasmic" domain, which consists chiefly of a set of 15 contiguous similar sequences, each about 150 amino acids long and containing several cysteine residues. (These sequences in the human receptor resemble those of the bovine protein.) The extensive repetition of homologous sequences is somewhat puzzling in functional terms since the protein does not seem to have a corresponding number of high-affinity binding sites for M6P. (Perhaps, as some data suggest, when M6P-containing ligands bind to the receptor, there are conformational changes in the extracytoplasmic domain; these changes might reduce the affinities of the regions of the domains that were not involved in the initial binding; see also Section 7.6.)

The information just summarized predicts a molecular weight in excess of 270 kD (even more—perhaps 300 kD—when the effects of glycosylation are added). The discrepancy between this value and the apparent molecular weight of 215 kD obtained by SDS–PAGE studies on the protein could conceivably reflect extensive proteolytic processing of the M6P receptor during its genesis and maturation, but such processing has not been demonstrated.

Unexpected similarities have turned up between the amino acid sequence and other properties of the 215-kD M6P receptor and the corresponding features of a receptor that can bind to and can probably mediate endocytosis and degradation of a mitogenic growth factor—the insulin-like growth factor IGF-II. The amino acid sequences match so closely in fact as to suggest strongly that the same protein molecules can carry out both receptor functions (though it has not been ruled out that different populations of the molecules are responsible for the two roles). Efforts are underway to determine whether this situation has special functional significance (could it somehow tie lysosomal enzymes to growth control?*) or whether it instead represents evolution's economic utilization of one class of molecules for two unrelated purposes (could the endocytotic phase of the receptor's cycling have arisen partly to serve the cell's need to dispose of IGF-II molecules bound to its surface?) Binding of IGF-II to the receptor seems not to involve M6P and seems to take place at different sites on the receptor's extracellular domain than those used to bind M6P-containing ligands.

7.3.1.1. The Cell Surface and Interior Populations of 215-kD Receptors May Exchange Extensively with One Another

Cultured mammalian fibroblasts expose 15,000–30,000 215-kD M6P receptors per cell to the extracellular space at a given moment and contain 10^5 or more inside. The proportion of the receptors found inside the cell reaches 80–90% in some cells and circumstances. But from the studies summarized in Section 7.2.3, one would expect that many of the intracellular receptors cycle through the cell surface. Indeed, when living fibroblasts are exposed for several hours to antibodies directed against the M6P receptor, the cells show both a diminished capacity to endocytose M6P-bearing hydrolases and an impaired ability to sort new, endogenously synthesized hydrolases to the lysosomes— the hydrolases are secreted instead (Fig. 7.8). The interpretation is that the receptors mediating endocytosis and those transporting hydrolases to the lysosomes cycle through a common pool and that on successive cycles a given receptor molecule can be recruited to carry out either task. The antibodies used in the experiment are assumed to encounter the receptors at the cell surface and, by remaining bound until the receptors ultimately are degraded in lysosomes or other structures, to interfere with the cycling and functioning of the receptor molecules. As the antibodies seem eventually to affect virtually all the cell's M6P receptors, it is likely that over the course of a few hours each receptor passes to the cell surface at least once. It is conceivable, however, that the antibodies can enter the cells endocytotically, thereby gaining access to M6P receptors that do not themselves move to the cell surface. [Note also that some investigators report the turnover of the M6P receptor to be relatively insensitive both to the effects of the I cell disorder (Section 7.3.3.5) and to the usual lysosomal inhibitors, raising questions about the site of receptor degradation (see Section 7.3.2.2).]

*Is it simply coincidence that "proliferin," a prolactin-related protein secreted by cultured cells, can also bind to the 215-kD M6P receptor? (Proliferin probably does so by virtue of its content of phosphomannoses.) Remember also that some tumor cells secrete high levels of certain acid hydrolases (Section 6.2.3; tumor cells that have cell surface proteins closely akin to membrane proteins normally confined largely to the lysosomes have also been reported).

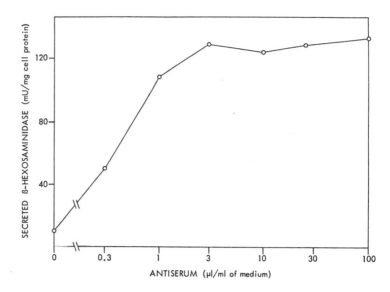

FIG. 7.8. When cultured human fibroblasts are exposed to antibodies against human liver 215-kD M6P receptors, they rapidly lose most of their capacity to endocytose lysosomal hydrolases, and increase their secretion of endogenous hydrolases. Shown is the secretion of β-hexosaminidase, detected with antihexosaminidase antibodies, during a 24-hr exposure to anti-M6P receptor antibodies. Most of the β-hexosaminidase released is in the form of "precursor" molecules (ones that have not undergone the processing that takes place in lysosomes; see Section 7.4.1). From von Figura, K., Gieselmann, V., and Hasilik, A. (1984) *EMBO J.* **3**:1281.

7.3.1.2. The Affinity of the 215-kD Receptor for M6P Is Diminished by Low pH

Like many of the receptors considered earlier, the M6P receptors normally deliver ligands to acidified compartments where the ligands dissociate. The receptors then depart from the compartments, to be reused. The affinity of the 215-kD M6P receptor for M6P-bearing hydrolases is known to be high at pH 6–7 and to drop off markedly at pHs much below 6, becoming low at pHs near 5. Correspondingly, hydrolase transport is perturbed by the usual "antacids": As Section 6.2.3.1 noted, fibroblasts or macrophages exposed to ammonium chloride or other weak bases, divert a considerably larger proportion than normal of their newly made hydrolases to the extracellular medium. The cells also become substantially less capable of endocytosing exogenous "high-uptake" forms of lysosomal enzymes. The explanation usually proffered is that the pH rise induced by the weak bases abolishes dissociation of endogenous M6P-bearing hydrolases from the M6P receptors so that the receptors eventually all become tied up in complexes with their ligands and no longer are available either for endocytosis or for intracellular sorting. There may be additional (osmotic?) effects of the bases on membrane cycling such that, for example, the total number of receptors at the cell surface diminishes somewhat (see also Section 3.3.4.3).

7.3.1.3. An Interim Model

This last explanation helped crystallize the prevailing conception of M6P receptor functioning: Newly made lysosomal hydrolases bearing M6Ps on their oligosaccharides encounter M6P receptors in the Golgi apparatus or in structures closely associated with the apparatus. The receptors bind to the hydrolases and segregate them to transport vehicles, which carry the enzymes to their sites of intracellular use. The transport vehicles are assumed to be vesicles that bud from the Golgi-associated compartments and fuse with "target compartments"—lysosomes or prelysosomal structures (see Section 7.3.2.2) or the extracellular digestive chambers of osteoclasts (Section 6.1.1). The target compartments are acidified so that, upon delivery, the hydrolases dissociate from the receptors. This dissociation frees the hydrolases to begin their degradative functions while the M6P receptors are enabled to leave the target compartments in vesicles or small tubules, eventually to return to the Golgi region.

In Golgi-associated structures, or at the target compartments, the intracellular cycle intersects with the cycling of M6P receptors to and from the cell surface and any given receptor molecule can choose either route—intracellular or to the cell surface—at the intersections.

The leakage of some newly made hydrolase molecules into secretory pathways under normal conditions is accounted for by the "accidental" failure of some of the enzymes to encounter M6P receptors correctly (perhaps because hydrolase molecules sometimes fail to "obtain" M6Ps in proper amounts or distributions). Or maybe some of the M6P receptors complexed with hydrolases move accidentally into structures destined for secretion.

One can imagine that M6P receptors maintain long-term interactions with one other or with other molecules, such that the membranes in which they travel maintain persistently high receptor concentrations. Alternatively, the receptors might reversibly congregate and disperse in the plane of the membrane during their cycling.

The limited information available on the sorting of nonhydrolase lysosomal components such as the activators and protective proteins described in Section 6.4.1.3 suggests that these proteins are handled like the hydrolases. For example, the I-cell mutation, or exposure of normal cells to ammonium chloride enhances secretion of some of these proteins. And when the proteins are placed in culture media, they can be delivered, endocytotically, to lysosomes.

Lysosomal membrane proteins are discussed in Section 7.4.3.7.

7.3.1.4. 46-kD M6P Receptors

The discovery of a second class of probable M6P receptors in a variety of cell types helped plug some holes in the picture just sketched. These receptors were sought under the impetus of discoveries that certain lines of cultured cells package lysosomal hydrolases fairly efficiently into lysosomes despite the absence of detectable 215-kD M6P receptors. Some cells that lack the 215-kD receptor—the P388D mouse macrophage-like cell line, for example—secrete as much as 60–70% of their hydrolases into surrounding media but others (e.g., MOPC cells) retain more than 50% of their enzyme production. At first, the packaging abilities of these "mutants" were attributed to hypothetical M6P-independent routes for hydrolase transport. Speculation that such routes exist still continues on other grounds (see Section 7.3.3.5). Now, however, several of the cell types that lack the 215-kD receptor (and various of those that have this receptor) are known to possess a protein with a molecular weight of about 46 kD, (as estimated by SDS–PAGE), which binds to phosphomannan columns and to mannose phosphate-containing proteins, with high affinity and specificity. This protein is thought to be a second type of M6P receptor.

Reliable data have only begun to accumulate on the distribution of this "46-kD M6P receptor" in the cell, on its recycling properties, and on other important attributes; even its receptor functions have not yet been rigorously demonstrated. Thus far, however, the essential properties of the 46-kD protein seem to be similar to those of the larger receptor; for instance, its affinity for ligands is high at pH 6–7 but diminishes markedly at pH 5.5, and its cycling through the cell involves stays both at the cell surface and in Golgi-associated locations. The most obvious known difference between the two receptor types, aside from their size, is that the smaller one requires divalent cations like Mn^{2+} for high-affinity binding of M6P ligands, whereas the larger receptor does not have this dependency. (For this reason some authors prefer to call the receptor the "cation dependent" M6PR rather than the "46-kD" receptor.) The 46-kD receptor also has a much lower affinity than does the 215-kD protein for oligosaccharides containing methyl diesters of M6P, like those found in *Dictyostelium* (see Section 7.3.3.4).

The protein-coding portion of the bovine 46-kD receptor gene has been cloned and shown to correspond most likely to a 279-amino-acid chain with an NH_2-terminal signal sequence followed by a 165-amino-acid extracytoplasmic domain, a hydrophobic membrane-spanning region of 25 amino acids, and a cytoplasmic COOH-terminal domain of 67 amino acids. The amino acid sequence of the extracytoplasmic domain resembles the 145-amino-acid sequences that repeat in the extracytoplasmic domain of the larger, cation-independent, 215-kD M6P receptor. The cytoplasmic tail, though not much like that of the larger receptor in overall amino acid sequence, does contain cysteines and clusters of acidic amino acids. The protein is relatively heavily N-glycosylated, probably carrying four or more oligosaccharides. Its functional form could well be dimeric or even more highly multimeric.

How are tasks apportioned or shared between the two receptor types in the cell types that possess both? Do both do exactly the same things? Does the smaller receptor function as a "safety device" ensuring that most of those hydrolase molecules that escape binding to the larger one are nonetheless correctly sorted? Or do the 46- and 215-kD receptors differ subtly in their affinities for different specific hydrolases or in the specific types of compartments they favor for acquiring or delivering hydrolases (see also Sections 7.3.2.3 and 7.4.3.5)? There is little basis yet for discussing these questions, except for rumors that the two types of receptors may not be identically distributed within cells, and reports that the 46-kD receptors at the cell surface bind hydrolases and mediate their endocytosis poorly as compared to the larger receptors. (Speculation has it that the 46-kD receptor changes its conformation when it cycles into the plasma membrane. In addition, initial findings suggest that the highest affinity of the receptor for hydrolases is at pHs similar to those in Golgi compartments—6 to 6.5—rather than at neutral pH; see also footnote on p. 391.)

Receptors additional to the 215- and 46-kD species are being looked for and preliminary leads suggest some may turn up. Especially if some of these prove to have very specialized functions or are expressed only in certain cell types, their existence could help clarify some of the mysteries I have left hanging.

7.3.2. Which Intracellular Compartments Take Part in the Sorting Mediated by M6P Receptors?

385

GENESIS

7.3.2.1. Golgi Sacs and Shuttling Vesicles

Immunocytochemical findings like those exemplified by Fig. 7.5 and 7.9 attest to the fact that the 215-kD receptors normally are concentrated in one or more sacs of the Golgi stack, in coated vesicles near the Golgi apparatus, and in large membrane-bounded vacuoles resembling endosomes or lysosomes. This has been shown for several cell types including those of the exocrine pancreas, hepatocytes, and several cultured cell lines. The receptor-containing vacuoles include structures in which Lucifer Yellow becomes concentrated upon its endocytosis.

According to some studies, when lysosomal hydrolases lacking M6P are produced—in I cells or because N-glycosylation has been suppressed by tunicamycin—the M6P receptors accumulate especially in the Golgi sacs and in nearby coated vesicles (Fig. 7.9). Contrastingly, exposure to weak bases leads to receptor depletion from Golgi-associated compartments and accumulation in the vacuoles; this redistribution is reversed by removal of the bases or by allowing the cells to endocytose M6P from the medium.* From such observations, the time it takes for a M6P receptor to acquire hydrolases, move away from the Golgi apparatus, and then cycle back, has been estimated at a bit more than an hour. That the loading of hydrolases into the vesicles and the movement from the Golgi apparatus to target compartments consumes most of this period is suggested by the fact that when weak bases are removed from the medium, only a few minutes is needed for receptors to reaccumulate at Golgi-associated sites.

Other studies, however, have given different results. For instance, in mammalian hepatoma cells, weak bases have been found either to produce slight increases in accumulation of M6P receptors and of lysosomal hydrolases in *trans* Golgi compartments or to have little effect on these compartments (the receptors were also noted in endosomal structures, some of which enlarge markedly in the presence of the bases).

According to some reports, treatment of fibroblasts with cycloheximide, which should suppress the production of new hydrolase molecules (i.e., of endogenous ligands for the M6P receptor), has little discernible short-run impact on the distribution of the 215-kD receptor, though after several hours, the effects begin to resemble those of tunicamycin described above. In other research, neither cycloheximide nor tunicamycin had much influence on the amounts of M6P receptor in endosomes, as evaluated by allowing the cells to take up HRP-conjugated ligands and then using the internalized HRP conjugates to iodinate M6P receptors. (This method is thought to tag selectively those M6P receptors present in compartments directly accessible to endocytosis; see Section 3.3.3.1.)

---

*Endocytosed M6P is presumed to effect this rescue by competing with the hydrolases for M6P receptors, thereby promoting dissociation of the hydrolases from the receptors even at relatively elevated pHs. But why would receptors bound to "free" M6P molecules return to the Golgi region whereas the same receptors seemingly are held back when they are bound to M6Ps that are residues in the oligosaccharides of hydrolase molecules? One guess is that the hydrolases, as multivalent ligands (Section 3.3.4.3), have a different impact on the mobility and distribution of the receptors in the plane of the membrane than do M6P molecules.

Recycling of the 215-kD receptor between the cell surface and intracellular compartments continues in fibroblasts exposed for prolonged periods to weak bases plus high concentrations of M6P-containing ligands, or to cycloheximide. (Recycling was monitored in these experiments, by immunological means and by establishing that over time, most of the cell's receptors were accessible to attack by proteases in the extracellular medium, although at any given moment only a relatively small proportion of the receptors is attacked.)

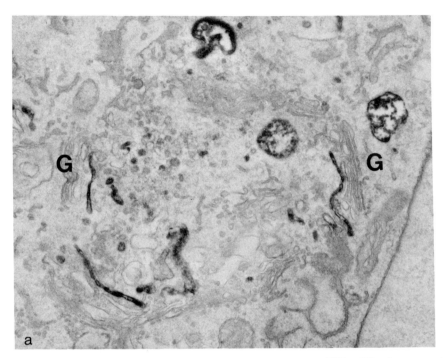

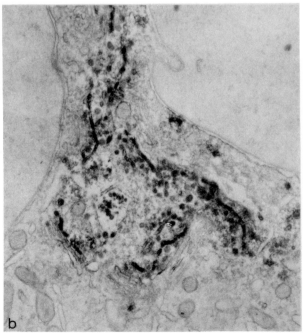

FIG. 7.9. Immunocytochemical preparations (HRP technique; Section 7.3.2.3) showing localizations of the 215-kD M6P receptor in Golgi regions of clone 9 cultured rat hepatocytes grown in ordinary medium **(a)** or in the presence of tunicamycin **(b)**. Receptors are demonstrated in sacs and tubules along one face of the Golgi stack (G), in small vesicles, some of which are attached to the Golgi systems, and in larger vacuolar structures (lysosomes or prelysosomes; see text). Many of the reactive vesicles are of the "coated" type. Judging from the positions of centrioles in the Golgi region of these cells, the reactive Golgi sacs are on the *cis* side of the Golgi apparatus. From Brown, W. J., Constantinescu, E., and Farquhar, M. G. (1984) *J. Cell Biol.* **99**:320. a, × 25,000 (approx.); b, × 15,000 (approx.).

Overall, these diverse observations agree in confirming both the role of the Golgi apparatus in hydrolase sorting and the use of coated vesicles to deliver hydrolases to other sites. And they strongly suggest that Golgi apparatus investment of M6P-receptor-rich membrane in delivery vesicles is recouped by the shuttling back and forth of the receptors between the Golgi apparatus and the sites of hydrolase delivery.

But how should the divergent findings with weak bases and inhibitors be interpreted? The reports that M6P receptors not occupied by hydrolases tend to remain in vesicles and sacs associated with the Golgi apparatus suggested that binding of M6P-bearing hydrolases to the receptors in some sense "triggers" or "permits" movement of receptor-enriched vesicles from the Golgi apparatus to target vacuoles. Dissociation of the hydrolases from the receptors seemingly promotes receptor movement back from the vacuoles to the Golgi apparatus. This pattern of behavior, however, apparently is not universal. Many of the other results seem better interpreted in terms of a constitutive cycling of the receptors that is not prevented when ligands are absent, or upon interference with dissociation of ligands from receptors. Models reconciling these findings propose that the cycling of the M6P receptor does have constitutive aspects, but that the steady-state distribution of receptors and the rates of movements from one place to another also are influenced by ligand occupancy, with influences varying in different cell types and experimental conditions.

7.3.2.2. Are Hydrolases Delivered to Lysosomes?

What has been said so far about the intracellular behavior of M6P receptors dovetails nicely with earlier conceptions of how lysosomes acquire their enzymes, with the proviso that "primary" lysosomes in most cells are now to be thought of more as reusable carriers than was hitherto the case (see Section 2.1.3.1). But there also are jarring observations. Especially surprising is that both cell fractionation and immunocytochemistry show the types of structures that can be identified readily as definitive secondary lysosomes to be poor in M6P receptors. Moreover, by immunocytochemistry, the intracellular distribution of M6P receptors in non-Golgi structures sometimes coincides better with that of endocytosed Lucifer Yellow or cationic ferritin than with that of the glycoproteins characteristic of the lysosomal membrane (Fig. 7.10), though these distributions all overlap with one another.

Could it be that M6P receptors depart so rapidly and so efficiently from secondary lysosomes to which they deliver hydrolases that they simply do not accumulate in the lysosomes? But even in weak-base-treated cells, in which M6P receptors are thought to spend more of their time at sites of hydrolase delivery (see Section 7.3.2.1), the distribution of M6P receptors does not coincide very well with that of lysosomal membrane protein. Present opinion, therefore, is shaded in favor of the idea that the M6P-receptor-rich carrier vesicles do not deliver lysosomal hydrolases predominantly to definitive secondary lysosomes but rather fuse preferentially with "intermediate" structures (see also Section 2.4.4). As they receive the hydrolases, these structures mature into the familiar types of secondary lysosomes but by the time they do mature, most of the M6P receptors acquired during hydrolase delivery have departed, eventually to recycle back to the Golgi apparatus.

The observations with Lucifer Yellow, and electron microscopic work with other endocytotic tracers, hint strongly that the intermediate structures include endosome-related structures, some of which are of the multivesicular type. Consistent with such endosomal involvement are the fact that several of the cell lines defective in acidification of endosomes (Section 3.1.5.4) also show impaired abilities to package hydrolases into mature lysosomes and the observations that endosomal cathepsin D behaves like a precursor of the lysosomal cathepsin (Section 2.4.4.1). The M6P receptors detectable immunocytochemically in endosome-like bodies show some tendency toward preferential lo-

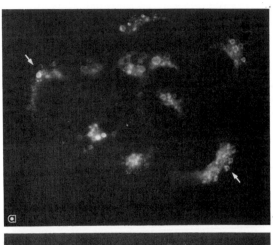

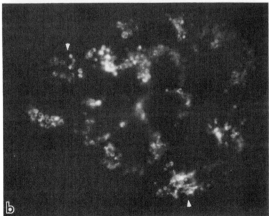

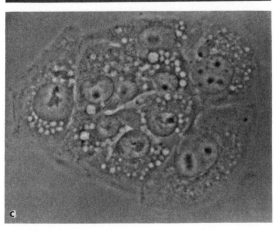

FIG. 7.10. The three panels show the same cells in a culture of rat clone 9 cultured hepatocytes treated with NH$_4$Cl and immunostained to localize 215-kD M6P receptors (panel **a**) and (panel **b**) a protein of the lysosomal membrane (LGP 120; Section 3.2.3.3; see also Lewis *et al.* (1985) *J. Cell Biol.* **100**:1839). **c** is a phase-contrast micrograph of the cells. The two antigens overlap in distribution but their localizations are not identical; the arrows and arrowheads indicate examples of the many bodies in which very little if any overlap is evident. Conclusions: The M6P receptor is not locked into the company of lysosomal membrane proteins, and definitive lysosomes can be poor in M6P receptors. From Brown, W. J., Goodhouse, J., and Farquhar, M. G. (1986) *J. Cell Biol.* **103**:1235.

calization in the tubular extensions, as described in Section 3.3.4.4, for recycling endocytotic receptors.

One group of investigators studying cultured mammalian cells (especially NRK cells) believes that these cells form intermediate bodies that are post-Golgi and probably post-*trans* Golgi network as well, but are distinctly preendosomal in the sense that endocytosed materials fail to enter them in cells maintained at 18° C. By immunocytochemistry this group has identified a set of bodies—possibly components of a larger complex compartment—with the appropriate characteristics, among the tubules and vacuoles near the Golgi apparatus. The bodies are membrane delimited and often have numerous internal, membrane-bounded tubules (see below): they contain M6P receptors, lysosomal membrane glycoproteins, and lysosomal hydrolases but seem not to acquire viral membrane proteins that are on their way to the cell surface (the viral proteins do accumulate in *trans* Golgi structures nearby). The bodies do not become labeled with endocytic tracers at 18°, but similar bodies do acquire the tracers at higher temperature. Their internal pH probably is in the acid range, judging from the fact that they accumulate DAMP (Section 3.1.2.1). The interpretation made is that these bodies are compartments that carry lysosomal components from the Golgi apparatus to endosomes and can be "caught" before making their delivery to endosomes because low temperature inhibits the fusions needed for delivery. Under normal circumstances these structures would presumably operate, in essence, as late endosomes: They would receive newly made lysosomal proteins from the *trans* Golgi network, fuse or otherwise mix content with early endosomes, and participate in the cycling of M6P receptors and perhaps of some endocytic receptors and other membrane components as well.

An unexplained finding is that some of the suspected intermediate bodies just described contain a goodly proportion of their immunocytochemically demonstrable M6P receptors on the membranes that, in thin section, appear to be within the interior of the bodies: Are some of the bodies autophagic? Are the interior membranes actually connected to the surface of the structure? Or do the interior receptors somehow move back and forth from the surface of the bodies? Other questions also need addressing: How do the intermediates form and how do they relate to the coated vesicles hitherto believed to be principal carriers of hydrolases? (E.g., do the proposed intermediates arise directly by transformation of *trans* Golgi systems or do they acquire their materials from vesicular carriers?) To what extent does the possible disordering of *trans* Golgi functions at low temperature (Section 7.2.2.2) contribute to the observations?

So far, these bodies are the most impressive candidates for special intermediates in the delivery of hydrolases to lysosomes, but they have been discovered too recently for final judgment about their importance or nature. On general grounds one would expect intermediates to be dynamic, heterogeneous, and relatively short-lived and therefore difficult to characterize. And what should we call them? At what points in their history are structures to be designated as transport vehicles, as true lysosomes, as new lysosomes, as dynamic forms of secondary lysosomes, as prelysosomes, as hydrolase-containing endosomes, or as nameless transients? The ambiguities encountered when Chapters 2 and 3 took up endosomal hydrolases and transformations return here in force. As often in the history of lysosomes, a lively dispute about nomenclature has begun, masking the conceptual crisis: (1) Information is still very sparse about the state of the hydrolases delivered to the intermediate bodies. Are the proteases present in active form or as inactive precursors (Section 7.4.1.1)? Is a body containing lysosomal enzymes some of which are active and some not, a lysosome? (2) Does the onset of degradation within nascent lysosomal digestive structures depend on changes in their membrane—the delivery or activation of proton pumps or the removal of other proteins (see Chapter 3)? (3) What are the relationships between hydrolase delivery via the supposed intermediates, and the fusions of "old" secondary lysosomes with newer bodies (Section 2.1.3.3)? The latter fusions have almost slipped from view in current schemes for hydrolase delivery to endocytotic structures, and their frequency and timing relative to the events considered here are unknown. Recall that very early endosomes seemingly do not like to fuse with old lysosomes (Section 2.4.4.2) but that newer and older bodies must eventually coalesce or transfer content in some fashion—through direct fusion (Section 2.1.3.3), via hydrolase-bearing bodies that bud from maturing residual bodies (Section 2.1.3.4), or through the movements of compartments involved in the membrane cycles discussed in Sections 2.4 and 3.3 (see also Section 3.5.2.4). Could the "intermediate bodies" include structures involved in such phenomena? (4) How does autophagy fit into the picture? Can the same transport vehicles deliver newly made hydrolases either to AVs or to heterophagic vacuoles?

7.3.2.3. *Cis* versus *Trans*

Currently, there is profound discord about the relationship of M6P-mediated hydrolase delivery to Golgi polarity. In several of the cell types studied immunocytochemically for the distribution of the 215-kD receptor, including the cultured hepatocyte line used for the work illustrated by Fig. 7.9, the receptors have been demonstrated only in compartments toward the *cis* face of the Golgi stack. Concentration of M6P receptors at the *cis* Golgi aspect could mean that lysosomal proteins are picked up by the receptors and moved away from the Golgi apparatus at a quite different place from the sites (the *trans* face) generally used to package proteins into secretory structures. This would be a simple way to achieve the differential packaging of lysosomal enzymes. In addition, proteins slated for lysosomes would avoid transit into the moderately acidified *trans* Golgi structures whose low pH might both induce premature dissociation of the hydrolases from the M6P receptor, and foster attack by the lysosomal enzymes on other materials in the same Golgi compartments.

At the same time, however, sorting of hydrolases to lysosomes from the *cis* Golgi face, would leave unexplained the *trans* localizations of acid hydrolases seen with acid phosphatase cytochemistry and by immunocytochemistry (Section 7.2). Furthermore, in some cell types the 215-kD M6P receptor has been detected toward the *trans* side of the Golgi apparatus, as well as at the *cis* side: The series of experiments illustrated in Fig. 7.9 demonstrated this location for cultured NRK cells and for osteoclasts. (The latter cell type may, of course, be a special case, given its extensive secretion of hydrolases; Section 6.1.1) Figure 7.5 illustrates the concentration of 215-kD receptors near the *trans* Golgi face, as has been demonstrated in a growing number of other cell types. (Preliminary data on the 46-kD receptor suggest a similar distribution.)

Investigators—probably the majority at present—who favor the *trans* Golgi face as the chief site from which hydrolases are directed to lysosomes are not much worried about problems that could be posed by Golgi acidification: The pH in the systems at the *trans* Golgi face—6 to 6.5—may not be low enough to have disastrous effects on ligand binding to the M6P receptor.* As for danger from the acid hydrolases to other material, the most dangerous enzymes—the cathepsins—seem likely to pass through the Golgi apparatus in precursor forms that may have very little activity at pH 6–6.5 (see also Section 7.4.1.1).

Conceivably, technical matters underlie the discrepancies between *cis* and *trans* demonstrations of the M6P receptors. For instance, the work in Fig. 7.9 was done with "preembedding," HRP procedures. (The cells are fixed, permeabilized with detergents to permit access of antibodies, and exposed to the antibodies, and only then embedded and thin-sectioned for electron microscopy; the locations of the antibodies are visualized through use of HRP–antibody conjugates whose sites are located by incubating the tissue in a cytochemical medium for demonstrating HRP activity.) The distributions emphasizing *trans* Golgi locations were obtained with quite different procedures like those used in Figs. 7.5 and 7.6: The cell interior is made accessible by cutting thin sections from frozen tissue that has not been embedded in plastic and antibodies are applied directly to the thin sections; the antibodies used are conjugated with gold or other electron-dense particles and observed in the electron microscope (see footnote on p. 373).

Enthusiasts for the latter methods assert that they are less likely to miss antigenic sites and are more reliable for semiquantitative studies of antigen distribution than are preimbedding, HRP procedures. But whether this accounts for the dissonance will not be known until there have been systematic comparisons of the two approaches using the same antibodies and the same cells. The few such efforts made thus far suggest that the different techniques may in fact give somewhat different results. On the other hand, cell fractionation studies in which the Golgi apparatus was separated on density gradients, into bands enriched in *cis* elements and ones rich in *trans* structures, seem to indicate that some cell types actually do concentrate their M6P receptors at the *cis* side of the apparatus whereas others have a more *trans*-oriented distribution.

actually takes place both directly from the *cis* Golgi face and after transit to the *trans* face. The relative importance of the two routes might vary not only with cell type, but also with physiological and or experimental situation and for different lysosomal proteins. The concentration of M6P receptors at the *cis* Golgi face, when it occurs, could reflect the dynamics of transport: The *cis* systems might, for example, function as staging areas where M6P receptors accumulate awaiting their charges of M6P-bearing ligands. Once charged, some of the receptors might move rapidly away to lysosome-related bodies while the remainder pass through the Golgi stacks to the *trans* Golgi face and then move away to delivery sites; the latter movement could sometimes be fast enough that, in the steady state, the receptors do not accumulate in *trans* compartments at levels high enough to be readily detectable by immunocytochemical procedures. (Or might ligand binding affect the immunocytochemical detectability

---

*Could it be that the 46-kD receptor, which binds M6P-containing hydrolases more avidly at pH 6–6.5 than at pH 7, picks up the occasional hydrolase molecules that escape from the 215-kD receptor in acidified Golgi compartments? (The 215-kD receptor, however, also retains high affinity for M6P-containing hydrolases at pH 6.3—some reports suggest the affinity is as high or higher than at near-neutral pH.) Or perhaps the escape of a few lysosomal enzyme molecules from M6P receptors during passage through *trans* Golgi structures explains why even under normal conditions some newly made acid hydrolases "leak" into secretory pathways (see Section 6.2.3).

of M6P receptors? Relevant changes are conceivable in accessibility of epitopes to immunocytochemical antibodies, in receptor conformation, or in the details of association of receptor with membranes.)

Other proposals are reminders that no conception failing to take into account the cell surface phase of the M6P receptors' life history can ultimately be satisfactory. If every M6P receptor does indeed spend part of its time at the cell surface, one root of the microscopists' confusion could be the superimposition of two distributions—that of receptors transporting hydrolases within the cell and that of receptors returning from the cell surface. It is not presently clear, for example, whether in each functional cycle, a given receptor normally makes a long circuit involving the Golgi apparatus, the cell surface, and the sites to which hydrolases are delivered or whether there are two or more alternative but interacting pathways for the receptors—one involving the cell surface and one entirely intracellular. Some investigators believe that the M6P receptors may actually resemble many other endocytotic receptors (Section 3.3.4.5) in visiting the Golgi apparatus only once in every few cycles.

7.3.3. What about Us? (More Paradoxes, Puzzles, and Conundrums)

The mechanisms discussed thus far probably apply to many hydrolases studied in mammalian cells. Nonmammalian material, however, affords clear cases in which sorting of hydrolases to the lysosomes does not depend on M6P. And even in mammalian cells some hydrolases show variant behavior not readily explained by the mechanisms considered up to now.

7.3.3.1. β-Glucuronidase

In certain mammalian cells, especially those of the liver and kidney of mice and rats, considerable activity of β-glucuronidase is demonstrable in the ER as well as in the lysosomes. In a number of other cases, seemingly anomalous ER locations of lysosomal components have eventually been explained on the basis of mutations or other ill-understood factors that delay the exit of newly synthesized proteins to the lysosomes. This sort of explanation has also been offered recently to account for the relatively high concentrations of M6P receptors detectable in the ER of cultured CHO cells. For β-glucuronidase, however, the ER form does not behave like a delayed precursor of the lysosomal form. The oligosaccharides of the ER form differ from those of the lysosomal enzyme and, in rats at least, there may be a few differences in amino acids that have yet to be explained. But as far as is known from the effects of mutations, the β-glucuronidase in the ER of mouse tissues is coded for by the same gene as the lysosomal enzyme; the divergence takes place subsequent to translation and leads to 30–40% or more of the molecules taking up relatively stable residence in the ER while the rest move to the lysosomes. The ER enzyme has a half-life of several days, a life span that compares respectably with other ER residents.

In mice, both the ER's β-glucuronidase and the lysosomal enzyme are tetramers of 75-kD subunits. But only in the ER are these tetramers associated intimately with the protein **egasyn.** When egasyn is missing, through mutation, glucuronidase does not accumulate in the ER (and the oligosaccharides of the lysosomal glucuronidase differ from those in wild-type mice). Egasyn is an enzyme of the esterase class and organophosphate esterase inhibitors can compete with β-glucuronidase for binding sites on egasyn. Only a proportion of the cell's egasyn is complexed with β-glucuronidase; much is present in other forms.

Though the significance of egasyn and the roles of the ER's β-glucuronidase are not yet understood, the "β-glucuronidase situation" affords an emphatic demonstration that the cell can handle particular lysosomal hydrolases differently from the others (see Section 7.5.1.2).

7.3.3.2. Yeast

In *Saccharomyces cerevisiae*, carboxypeptidase Y and proteases pass to the vacuole via the ER and Golgi apparatus. Because vacuolar enzymes show effects of some of the mutations that affect the production and processing of yeast secretory proteins, it is likely that vacuolar enzymes and secretory proteins share common pathways at early stages in their genesis and transport. Moreover, appreciable amounts of carboxypeptidase Y can be diverted to a secretory path—for example, by genetic manipulations that result in over-production of the enzyme.

But cells in which secretion is mutationally disabled can still package vacuolar enzymes correctly, probably because in these cells, and in wild-type cells as well, carboxypeptidase Y and other vacuolar components are efficiently sorted from secretory materials, at about the time they pass through the Golgi apparatus or soon thereafter. The carboxypeptidase is a glycoprotein, with mannose-rich oligosaccharides containing phosphomannosyl residues (many of which are diesterified to a second mannose). However, the sorting of the enzyme to the vacuole does not depend on M6P or even on the oligosaccharides: Carboxypeptidase Y is transported to the vacuole at more or less normal rates even in cells exposed to tunicamycin at concentrations that block glycosylation of nascent glycoproteins. What is more, among the nonlysosomal proteins normally secreted by yeast, several have phosphomannoses in their oligosaccharides (which often are much more extensive and richer in mannoses than is typical of mammalian glycoproteins). (See also Section 7.6.)

Rather than residing in the oligosaccharides, the carboxypeptidase sorting signal(s) is present in the polypeptide chain, toward its NH$_2$-terminal end: This has been shown by a variety of manipulations of the molecule, including the demonstration that when yeast cell genes are engineered so that a 50-amino-acid-long NH$_2$-terminal segment of carboxypeptidase Y becomes attached to molecules of the secretory protein **invertase,** the invertase molecules are redirected to the vacuole.

Findings similar to those on the sorting of carboxypeptide Y have begun to accumulate for other proteolytic enzymes of the yeast vacuole. It has also been reported that at least one of the vacuolar proteases is O-glycosylated.

7.3.3.3. Protein Bodies of Plants

Neither oligosaccharide-linked M6P residues nor M6P receptors have been demonstrated to play any role in transport of hydrolases or other materials to vacuoles in plants. The proteins stored in the protein bodies of seeds (Section 5.4.2) have been studied most thoroughly. Several of these proteins are not glycoproteins, at least as isolated in their mature forms. Those that are glycoproteins can be shipped to the protein bodies even in the presence of tunicamycin, demonstrating that N-linked oligosaccharides are not essential for their transport.

The glycoproteins destined for storage in the protein bodies normally are N-glycosylated by mechanisms fundamentally similar to those in animal cells, though the plant cell proteins often contain xylose, a sugar rare in animal cell glycoproteins, and there are other differences in detail. En route from the ER to the protein bodies, the proteins pass through the Golgi apparatus where some of their high-mannose oligosaccharides are fashioned into "complex" oligosaccharides by trimming off mannose residues and adding N-acetylglucosamines, xyloses, and fucoses. [Subsequent processing in the protein bodies (Section 5.4.2) sometimes includes removal of N-acetylglucosamines.]

When yeast are induced to make the plant vacuolar lectin, phytohemagglutinin, by transfection of genes, the yeast sort the protein to their vacuoles. It is hardly rash to conclude that yeast and plants may well use similar systems to package their lysosomal enzymes.*

7.3.3.4. Mannose Phosphate Diesters, Mannose-Linked Sulfates, and Other Modifications of Oligosaccharides

The slime mold *Dictyostelium discoideum* phosphorylates mannoses in the oligosaccharides of its lysosomal hydrolases but the phosphates are in diesters linked to methyl groups. The oligosaccharides also contain sulfate groups, some of which are mannose-6-sulfates (some mammalian lysosomal enzymes also carry sulfates). The receptors responsible for hydrolase sorting in *Dictyostelium* are still to be described. No convincing data have implicated either oligosaccharides or acid compartments in the process. The mammalian 215-kD receptor does have appreciable affinity for *Dictyostelium* hydrolases containing the methyl diesters, but neither this type of receptor nor any other M6P receptor has been found in *Dictyostelium*. At early stages in their transport to lysosomes, the enzymes in slime mold amebas seem to be closely associated with membranes. Normally, proteolysis loosens this association as the hydrolases move into the lysosomes (Section 7.4.1). When such proteolysis is inhibited, newly made lysosomal enzymes tend to be secreted rather than packaged into the lysosomes, suggesting that portions of the hydrolases' polypeptide chains might be directly involved in sorting.

The sorting signals for protozoan hydrolases are not known either. *Acanthamoeba* makes phosphomannosyl residues for its lysosomal enzymes, but how (or if) these function in hydrolase sorting has only begun to be investigated. The acid hydrolases of *Tetrahymena*, at least those that are released to the growth medium, contain neither phosphates nor sulfates. They are unusual in that the hydrolases are secreted with glucoses attached to the oligosaccharides.

In chick retina, and various other avian and mammalian tissues, M6Ps diesterified to glucoses have been identified in certain proteins, including hydrolases, bound to cell surfaces. A receptor-like protein, **ligatin**, is responsible

---

*The fact that the M6P system is not universally used for sorting lysosome-related hydrolases raises fascinating questions about the evolution of M6P receptors. Could it be, for example, that the receptors evolved initially as endocytotic receptors dedicated to recapturing hydrolases that leaked out of cells and would ultimately be wasted or that posed threats to neighboring cells? (See also the comments on the IGF-II receptor in Section 7.3.1.)

for the binding. This protein, and M6P-binding proteins on lymphocytes have been speculatively assigned functions in cellular recognition or adhesion phenomena but these assignments, and the identities of the proteins, deserve reexamination in light of the relations between insulin-like growth factor receptor and the M6P receptor outlined in Section 7.3.1.

7.3.3.5. I Cells

The most disconcerting observations for M6P extremists, if die-hard believers that M6P does it all still survive, are that many cell types in patients afflicted with I-cell disorder maintain intralysosomal hydrolase activities at levels reasonably near normal. And even cultured I-cell fibroblasts maintain acid phosphatase and glucocerebrosidase activities in their lysosomes at levels approaching 50% of normal. Is there some route or mechanism that is not dependent upon M6P but nevertheless can efficiently collect hydrolases into lysosomes of mammalian cells?

Lysosomes contain at least several dozen different types of proteins so that it would not be astonishing were several variations on transport themes to coexist. Possible traces of deviant routes and mechanisms abound. The behavior of cathepsin D in endosomes (Section 2.4.4.1) is a case in point. There also are claims that transport of cathepsin D to lysosomes in human fibroblasts is inhibited much more completely by ammonium ions than is transport of α-glucosidase. This could be due to differences in affinities of M6P receptors for the two enzymes, or it could mean that the glucosidase has access to transport routes or mechanisms not available to the protease or that cathepsin D's normally retarded passage from endosomes to lysosomes makes it unusually vulnerable to agents that perturb endosomal function. In their initial studies, investigators of the transport of glucocerebrosidase detected little if any phosphorylation during the life history of this enzyme and argued that glucocerebrosidase can be sorted to lysosomes through a still unknown M6P-independent route based upon the integration of the enzyme molecules directly into membranes or into stable complexes with membrane constituents. This idea has persisted, even though some researchers now believe that glucocerebrosidase actually does become phosphorylated. Glucocerebrosidase is not an integral membrane protein but it is unusually difficult to extract from disrupted lysosomes, arguing for some sort of special association with the lysosomal structure. (A possibility raised for the association of cathepsin D with endosomal membranes, which could apply to glucocerebrosidase as well, is that the attachment is to a lipid in the membrane rather than by insertion of part of the polypeptide chain into the bilayer.) There are yeast mutants that secrete several vacuolar enzymes but retain the vacuolar mannosidase, a membrane-associated enzyme. Remember, however, the evidence in Section 7.3.2 suggesting that the membranes transporting new hydrolases to the lysosomes ordinarily do not persist for long in the lysosome surface. Special features would have to be sought that would enable the membrane that transports enzymes like glucocerebrosidase either to stay in the lysosome surface (or its interior) or to transfer the enzyme to membrane already resident in the lysosome.

Could M6P-independent hydrolase transport depend on alternative receptors, such as the macrophage's mannose receptors, which are known to cycle between the cell surface and intracellular compartments and can bind lysosomal hydrolases? Or could the phosphates, sulfates, or other groups in the

premature or otherwise undesirable proteolytic attacks in the ER and Golgi compartments. Full activation of several of the mammalian cathepsins (and of yeast vacuolar proteases and peptidases) is delayed until about the time the enzymes enter the lysosomes. Activation depends on proteolysis, perhaps analogous to the activation of zymogens in the gut. Autoactivation *in vitro* is prompted by low pH.

The carboxypeptidase of yeast is activated by removal of an NH_2 terminal segment that also contains the sorting signal (Section 7.3.3.2) Yeast mutants are known in which carboxypeptidase Y and other vacuolar proteolytic hydrolases fail to undergo much activation owing to the inactivity of vacuolar proteinase A, but this proteinase by itself cannot activate the others *in vitro*, suggesting that some other vacuolar components collaborate with it.

Unlike the proteases, other lysosomal enzymes including glycosidases like the hexosaminidases, are capable of near-mature activity while still in precursor forms (though, of course, ambient conditions—pH, availability of activators, and so forth—may not favor such activity until the enzymes reach the lysosomes). Nevertheless, some of these enzymes are synthesized as precursors that mature into notably smaller polypeptide chains, suggesting that the precursor forms may be important for folding, stabilizing, or sorting the proteins during their early lives. In the case of the lysosomal enzymes of *Dictyostelium*, the proteolytic conversion of the precursors into smaller, more mature forms is accompanied by marked increases in the ease with which the enzymes can be solubilized from disrupted cell fractions. This has been taken to indicate that the precursors are tightly bound to membranes and that such binding facilitates transport and sorting to lysosomal or prelysosomal sites, where the hydrolases are freed from the membranes by proteolytic conversion.

7.4.2. Glycosylation and Phosphorylation

Phosphorylation of mannoses in newly synthesized lysosomal proteins takes place in two steps catalyzed by separate enzymes (Fig. 7.11). First the phosphate is attached to the mannose by a **phosphotransferase** (UDP-N-acetylglucosamine:lysosomal enzyme N-acetylglucosamine-1-phosphotransferase), which links an N-acetylglucosamine-1-phosphate via a phosphodiester bond to the hydroxyl on carbon-6 of a mannose in an oligosaccharide of the lysosomal hydrolase. Then a **phosphodiester glycosidase** (N-acetylglucosamine-1-phosphodiester α-N-acetylglucosaminidase) removes the N-acetylglucosamine, leaving the phosphate still linked to the mannose. Typically, several of the mannoses in a given hydrolase molecule become phosphorylated in this way. α-Glucosidase molecules, for example, acquire an average of three or four M6P residues; other hydrolases have even more.

7.4.2.1. Selectivity of Phosphorylation

Why it is that the lysosomal hydrolases are **selectively** phosphorylated in this way is obviously an issue of fundamental importance. Nothing special has been discovered about the oligosaccharides of the hydrolases that might permit the transferase to differentiate the lysosomal enzymes from other proteins (see next section). What is known about amino acid sequences of the hydrolases provides no satisfactory clues either—the primary sequences of the enzymes studied in detail (e.g., a number of the proteases, and β-glucuronidase) show no outstanding similarities to one another, or other regularities. The best present guess is that some distinctive conformational features common to most or all the hydrolases can be specified by more than one amino acid sequence.

Strong indirect evidence that the polypeptide portions of the hydrolases do contribute to the recognition phenomena underlying selective phosphorylation of the lysosomal oligosaccharides is that the phosphotransferase works very much better, *in vitro,* with lysosomal hydrolases such as hexosaminidases as substrates than with proper-looking oligosaccharides not linked to proteins. The enzyme does little with nonlysosomal glycoproteins such as thyroglobulin, immunoglobulins, or even with RNase B or other glycoproteins and glycopeptides that bear oligosaccharides not too different from those of the lysosomal hydrolases. Mature forms of the lysosomal enzymes often are good substrates

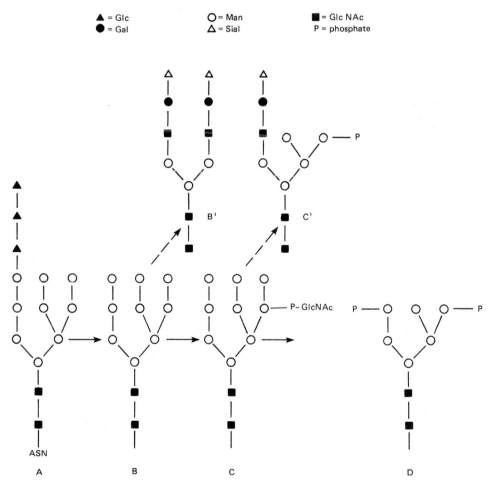

FIG. 7.11. N-Glycosylation and phosphorylation of lysosomal hydrolases. The high-mannose oligosaccharide attached initially in the ER **(A)** is trimmed **(B)**. Some molecules may then be converted to complex-type oligosaccharides without phosphorylation (B′). However, most are phosphorylated by formation of GlcNac-phosphodiesters **(C)** and subsequent removals of sugars. **D** represents a fully processed high-mannose form of the oligosaccharide, with two M6Ps. C′ is a "hybrid" structure with one "phosphomonoester" (M6P) and a complex type antenna on different branches of the same oligosaccharide. Based on Goldberg, D., Gabel, C., and Kornfeld, S. (1974) In *Lysosomes in Biology and Pathology* (J. T. Dingle, R. T. Dean, and W. Sly, eds.), Vol. 7, Elsevier, Amsterdam, p. 45.

but this no longer is true after treatments expected to disrupt their conformation, such as the proteolytic fragmentation of cathepsin D, or heat denaturation of uteroferrin. [In native form this last protein is a good substrate, lending additional weight to the idea that it might be a lysosomal enzyme or a close relative (Section 6.2.3).] The phosphotransferase is competitively inhibited by deglycosylated lysosomal hydrolases.

Genetic defects rendering the phosphotransferase catalytically nonfunctional account for the I-cell disorder. In some varieties of the pseudo-Hurler syndrome (Section 6.4.1.4), the transferase retains the capacity to transfer N-acetylglucosamine-phosphates to low-molecular-weight substrates such as methylmannoside but it cannot recognize lysosomal hydrolases as appropriate substrates. This could mean that the catalytic site of the enzyme is distinct from the site required for specific recognition of protein-linked oligosaccharides.

7.4.2.2. Phosphorylation Pattern and Receptor Affinity

Different molecules of a given species of hydrolase in a given cell type vary in the positions and numbers of the mannoses in their oligosaccharides, that undergo phosphorylation. As many as five of the mannoses in a hydrolase's oligosaccharide chain are potential locations for phosphorylation and either one or two phosphomannoses can be created in a given chain. Some of the phosphates remain "covered"—diesterified to N-acetylglucosamine—apparently because the glycosidase has failed to act on them.

Experiments analyzing the binding of test compounds to M6P receptors immobilized on columns and studies of receptor-mediated endocytosis of hydrolases by cells, show the apparent affinity of the receptors to be best when oligosaccharides have more than one M6P. The affinity is much lower when sialic acids are present on the same oligosaccharides as the M6Ps or when the M6Ps are "covered." These facts plus the finding that the M6P receptor's affinity is greater for M6P-bearing hydrolases than it is for free M6P, probably indicate that three-dimensional configurations of ligands and the number of sites of interaction of a given hydrolase molecule with the receptor are important in determining the strength of binding. Conceivably, the receptor interacts with the oligosaccharide or protein moieties of lysosomal hydrolases as well as with the phosphomannoses. (See Section 7.6.)

As suggested at several points above, differences in the phosphorylation patterns of different hydrolases, or in the affinity of the M6P receptors for the enzymes, could help account for variations in details of behavior of various hydrolases during transport and packaging. And some of the secretion of hydrolases observed with normal cells might reflect the stochastic production of molecules with oligosaccharides toward which the M6P receptor is relatively indifferent so that the molecules tend to slip past the sorting system. The extensive secretion of certain lysosome-related proteins by tumor cells, and the secretion of uteroferrin molecules by the porcine endometrium have been attributed to the secreted proteins' having oligosaccharides the M6P receptor does not much like, especially ones with phosphomannoses "covered" with N-acetylglucosamines.

7.4.3. Where and When

Yeast carboxypeptidase undergoes all the major steps in its processing and transport to the vacuole within a few minutes after translation (Table 7.1). Most cell types operate at slower paces. With *Dictyostelium* amebas, 10 min (for glucosidase) to 20 min (for mannosidase) or more elapses between translation of lysosomal hydrolase molecules and their arrival in quantity at the Golgi apparatus.* At this point, sulfation is initiated but an additional 10–20 min or more is needed for the new proteins to turn up in readily identifiable lysosomes. For cultured mammalian fibroblasts and other cell types of higher animals, the better part of an hour and sometimes much more (in some reports on α-glucosidase or hexosaminidases, up to several hours) is required for extensive transport of newly made hydrolases to definitive secondary lysosomes. In developing plant seeds, the half-time of movement of newly made storage proteins from the ER to the protein bodies is greater than an hour.

7.4.3.1. Proteolytic Processing

Most of the proteolytic "maturation" of new hydrolases subsequent to the removal of signal sequences is thought to occur in lysosomes. It begins at more or less the time the enzymes arrive in lysosomes and is blocked, both for endogenous enzymes and for endocytosed ones, by exposure to temperatures low enough to interfere with movement of materials into definitive lysosomes. Once they reach the lysosomes, many of the hydrolases continue to lose portions of their polypeptide chains over ensuing periods that stretch out for hours or even a day or two, as if the proteins are first attacked at particularly susceptible points and then are gradually nibbled down to a final, resistant form. The proteolytic processing of plant storage proteins that takes place in the protein bodies is a similarly leisurely affair.

Occasional claims that acid phosphatase, β-glucuronidase, or other hydrolases undergo prelysosomal proteolytic cleavages additional to removal of signal sequences, turn on observations of decreases in molecular weight detected at times before much entry of new proteins into the lysosomes is expected. Or sometimes, decreases are seen that are not much affected by manipulations of lysosomal pH or by other procedures expected to inhibit lysosomal digestion. Interpretations are controversial because it is difficult to determine precisely when the first molecules of new hydrolases reach lysosomal compartments and because processing of oligosaccharides and other nonproteolytic modifications can diminish the molecular sizes of glycoproteins. Indeed, several tentative claims of prelysosomal proteolytic processing of lysosomal hydrolases have eventually been withdrawn by their proponents upon careful scrutiny. But the possibility does remain open, for example, that the Golgi-associated enzymes responsible for processing secretory proteins also can initiate proteolytic modifications of such hydrolases as the lysosomal α-mannosidase of *Dictyostelium*. Or endosomal enzymes might initiate proteolyses of hydrolases delivered via the supposed intermediate compartments considered in Section 7.3.2.2. Perhaps egasyn's esterase activity is pressed into service to modify glucuronidase (see Section 7.3.3.1; little evidence has turned up for such action by egasyn).

Hydrolases assembled from products of more than one gene, such as certain hexosaminidases (Section 6.4.1.2), begin their assembly before proteolytic maturation; the subunits probably come together in the ER or Golgi apparatus.

---

*This timing, and the timing of arrival at lysosomes are evaluated from cell fractionation experiments. Ambiguities arise from difficulties in separating Golgi structures from other cell components in density gradients, and from the problems in deciding when a structure qualifies as a lysosome (Sections 1.2.1 and 2.4.4).

It is less clear where protective associations, such as those involving neuraminidase and β-galactosidase (Section 6.4.1.3), are initiated or where hydrolases come to associate with the lysosomal membrane, if they actually do so.

The hydrolases secreted by mammalian cells, especially when secretion is augmented (as in I-cell disease or on exposure to weak bases), include a high proportion of molecules that have not undergone proteolytic processing. This is a prime piece of evidence that secretion can occur by diverting newly made hydrolases before they reach the lysosomes (Section 6.2.3.1). When "mature" hydrolases are secreted, their origin is assumed to be the lysosomes.

7.4.3.2. Glycosylation; Phosphorylation

The "high-mannose" forms of oligosaccharides produced in the ER are most often detected by virtue of the fact that N-linked oligosaccharides containing five or more mannoses and lacking "terminal" units such as sialic acids or galactoses, are susceptible to *in vitro* removal by **endoglycosidase H.** When newly made glycoproteins show such susceptibility, it is concluded that the proteins have not been processed by medial or *trans* Golgi compartments. Conversely, acquisition of endoglycosidase H resistance testifies to passage of glycoproteins through *trans* Golgi compartments*: resistance is taken as a sign that the oligosaccharides have been converted to "complex" form (i.e., that they have been trimmed by mannosidases and have had groups such as galactoses, sialic acids, or fucoses added).

Lysosomal hydrolases seem to arrive at the Golgi apparatus with their oligosaccharides in a high-mannose form comparable to that characterizing immature secretory and plasma membrane glycoproteins. Because the enzymes responsible for subsequent steps in glycosylation are believed to be organized in a sequential *cis*-to-*trans* layout within the Golgi apparatus, one might hope to determine which Golgi compartments the hydrolases enter on their way to the lysosomes (see Section 7.3.2.3) by monitoring the state of the hydrolases' N-linked oligosaccharides. What gives this task a challenging twist is that the oligosaccharides of hydrolase populations, from many cell types, are a mixed lot—some being high-mannose forms, some complex, some hybrid, and some truncated stumps comprising only one or two residues. The proportions of chains containing residues added in post-*cis* Golgi compartments—galactoses, sialic acids, or fucoses—vary for different hydrolases in a given cell type, for given hydrolases in different cell types and under different experimental conditions. For example, the cathepsin D and β-glucuronidase molecules isolated from porcine kidney cells are relatively rich in high-mannose oligosaccharides whereas the glucocerebrosidase molecules isolated from the same cells have many chains terminating in sialic acids. (Glucocerebrosidase, of course, is one of the hydrolases that may have an "atypical" route to the lysosomes; see Section 7.3.3.5.)

---

*Diagnosis of Golgi processing from susceptibility to endoglycosidase H or other glycosidases must be confirmed by direct analyses of the oligosaccharides. It is suspected, for instance, that protection against endoglycosidase H can sometimes be afforded by changes in the oligosaccharides other than, or additional to, those just discussed, possibly including the sulfation of mannoses seen in the acid hydrolases in slime molds. (Sulfation, at least of tyrosines, seems to be a *trans* Golgi function.) Note also the comments below on the effects of M6P on oligosaccharide processing.

Phosphorylation of mannoses begins at about the time the new hydrolases reach the Golgi apparatus, which is congruent with biochemical data indicating that the phosphorylation system much prefers to work on high-mannose oligosaccharides before they have undergone trimming and processing of the sort carried out by the Golgi apparatus. Density gradient centrifugation indicates a Golgi localization of the transferase and glycosidase activities responsible for producing M6Ps. The enzymes probably occupy more than one sac in the Golgi stack: The two enzymes turn up in overlapping subfractions of Golgi-enriched cell fractions, but differences in their relative activities in the subfractions suggest the two may be slightly different in distribution within the apparatus. This suggestion is reinforced by the fact that temperatures low enough to inhibit passage of proteins through the Golgi stack can block the removal of N-acetylglucosamines from the mannose-linked phosphodiesters while having little effect on the creation of the phosphodiesters. The presently accepted view is that N-acetylglucosamine phosphates are added to the hydrolases' oligosaccharides at or near the *cis* Golgi aspect, with removal of N-acetylglucosamines commencing very soon afterward in the same compartment or in a Golgi compartment very close by. Still to be accounted for, however, are data indicating that the several phosphates on a given hydrolase molecule may be added asynchronously; addition of the first phosphate is not blocked by conditions (low temperature or inhibitors) thought to prevent passage of proteins from the ER to the Golgi apparatus sacs responsible for glycosylation.

Usually, M6Ps are found on endoglycosidase H-sensitive oligosaccharides or "antennae" (branches) of oligosaccharides. Other oligosaccharides along the same hydrolase molecule or even other antennae of the same oligosaccharide can terminate in sialic acid residues and correspondingly be endoglycosidase H resistant. Evidently, formation of M6Ps in an oligosaccharide inhibits further processing of that chain; enzymes such as the mannosidases seemingly cannot act on mannoses linked to phosphates or to N-acetylglucosamine-phosphates. A possible implication is that those hydrolase molecules that become particularly heavily phosphorylated can pass through the Golgi stack, including the *trans* sacs, without their oligosaccharides being modified into complex forms.

7.4.3.3. More on Sorting

These last considerations have complicated efforts to decide whether lysosomal hydrolases are sorted from *cis* or from *trans* Golgi compartments by looking at the oligosaccharides of the hydrolases. *Trans* Golgi involvement in processing of some of the molecules is indicated by the presence in lysosomes, of enzymes with oligosaccharides terminating in galactoses and sialic acids. But does the presence in the same lysosome population, of hydrolases with "unprocessed" oligosaccharides signify transport of these molecules directly from the *cis* Golgi face? It seems best to suspend judgment until more is known of the details of Golgi enzymology.

In any event, shortly after undergoing phosphorylation, hydrolases begin to bind to M6P receptors. Hydrolases so bound have been identified in Golgi-rich cell fractions. This is important, because it confirms the expectation that hydrolases complex with the receptors intracellularly. But no consensus prevails as to where in the cell complexing begins; the Golgi-rich fractions contain a mixture of structures, probably including some post-Golgi compartments and endocytotic structures along with Golgi elements. In certain cell types, large steady-state pools of phosphorylated hydrolases not bound to receptors seem to be present, along with the receptor-bound enzymes. Perhaps the dynamics of hydrolase transport through the Golgi apparatus and the details of the cycling of M6P receptors are such that some hydrolase molecules acquire M6Ps and meet M6P receptors in *cis* Golgi compartments while other molecules experience delays and do not bind to receptors until they enter medial or *trans* Golgi sacs.

In many situations when cells secrete large amounts of newly made hydrolases, the secreted molecules have notably higher proportions of complex oligosaccharides than do those retained within comparable cells; this is seen, for example, in comparisons of I-cell fibroblasts with normal fibroblasts or when cells are exposed to weak bases. It seems certain, therefore, that most of the lysosomal enzyme molecules released from the cell rather than being transported to the lysosomes are ones that have passed through *trans* Golgi compartments.*

7.4.3.4. Coated Vesicles?

Cell fractions rich in clathrin-coated structures ("coated vesicle" fractions) contain acid hydrolases (Fig. 7.12) and M6P receptors to which some of the hydrolases are bound. When the fractions were subfractionated to separate coated membranes of endocytotic origin from the rest, only 10% or so of the hydrolases remained with the endocytotic membranes. (Subfractionation was by agarose gel electrophoresis or through using endocytosed enzymes or endogenous ones, to generate products that affect the vesicles' buoyant density. The principle of the latter approach was as in Section 1.3.1 and Fig. 4.10 but acetylcholinesterase rather than peroxidase was employed to achieve the density shifts.) More difficult to evaluate is the extent to which the fractions contain fragments of the clathrin-coated stretches of sacs and tubules present near the *trans* Golgi face, along with structures that actually were vesicular *in situ*; in immunocytochemical studies, some of the coated regions of Golgi-associated sacs and tubules are found to be rich in M6P receptors and in hydrolases. The coated configurations seen along Golgi-associated sacs and tubules often look like nascent coated vesicles and may simply be vesicles caught as they were budding off, so they might not really be troublesome in the present context. But the controversies about three-dimensional morphology alluded to in Sections 1.5.3 and 2.3.1.3, and some of the friendly feuding between biochemists and microscopists, have made many of us wary about jumping to conclusions in such matters, particularly in public.

Most of the cathepsin molecules in the "coated vesicle" fractions are in precursor form insofar as proteolytic processing is concerned. Determinations of the oligosaccharide structure of these and other hydrolases (are they high-mannose or complex?) will help clarify some of the issues previously discussed, as will investigation of the pH in the "vesicles." (For example: Are the hydrolase-bearing structures among the acidified coated vesicles alluded to in Chapter 3? Do they carry proton pumps to their target compartments? Is their

---

*In contemporary "secretion-recapture" models for hydrolase transport (Section 7.2.3), the presence of hydrolases bearing complex oligosaccharides within lysosomes is taken as a sign that the enzymes have passed through the *trans* Golgi systems and then to the cell surface before being transported to the lysosomes.

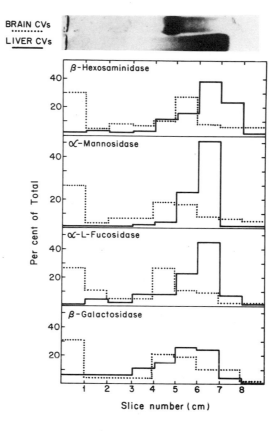

FIG. 7.12. Coated vesicles isolated from brain and liver contain acid hydrolases. In the final preparative step used to obtain the material illustrated, the vesicles were electrophoresed on agarose gels (top) and slices of the gels were analyzed for the presence of the enzymes. From Campbell, C. H., and Rome, L E. (1983) *J. Biol. Chem.* **258**:13347.

internal pH low enough that some of the hydrolases might dissociate from the M6P receptors and begin to process one another?)

Yeast seemingly do not require clathrin to transport hydrolases to their vacuoles; clathrin-deficient mutants still can package the enzymes.

7.4.3.5. Dephosphorylation; Deglycosylation

The enzymes normally present in lysosomes mostly lack phosphomannosyl groups whereas their precursors, obtained by using weak bases to promote secretion of prelysosomal forms, have these groups. This difference is assumed to be a consequence of the hydrolases' encounters with phosphatases during transport to the lysosomes or during the first hour or two after entry into the lysosomes. But if new hydrolases do move through the *trans* Golgi systems on their way to the lysosomes, an explanation is needed for the survival of M6P residues on the precursors despite the presence of acid phosphatase activity and somewhat acid conditions at the *trans* Golgi face. Is the acid phosphatase relatively inactive, because the pH is not very low? Or can the enzyme not attack the M6Ps because of its specificity or because the M6Ps are bound to M6P receptors?

It has been taken for granted that dephosphorylation on entry into lysosomes helps promote retention of the hydrolases in the lysosomes by rendering the enzymes unattractive to M6P receptors. On reflection, such a mechanism seems a bit redundant because the low intralysosomal pH should minimize complexing even of M6P-bearing hydrolases with receptors, especially given the scarcity of receptors in the membranes of definitive lysosomes. Still, the idea might eventually help explain, for example, why several of the cell lines lacking the 215-kD M6P receptor but possessing the 46-kD receptor, secrete hydrolases at elevated rates. The hydrolases obtainable from the lysosome-enriched fractions of these cells show an unexpected abundance of phosphomannosyl groups. Does this abundance imply that the 46-kD receptor targets hydrolases differentially to lysosomes or packaging machinery from which acid phosphatases that can attack M6Ps are absent? (The cells do contain ordinary acid phosphatases some of which should be able to dephosphorylate mannoses.)

As with the storage proteins of protein bodies (Section 5.4.2), once hydrolases reach the lysosomes they are subject to enzymatic deglycosylation, which can continue for many hours after they arrive. Correspondingly, the oligosaccharides of hydrolases extracted from the lysosomes of normal cells differ from those from storage disease cells deficient in lysosomal neuraminidase or galactosidase activities: These differences—in electrophoretic mobility, in binding to lectins, and in other properties—seem due to the "failure" of the storage disease cells to remove terminal residues from the oligosaccharides of lysosomal proteins.

7.4.3.6. Neutrophils

The cells on the path to normal neutrophil differentiation are very difficult to obtain in quantity, so that detailed biochemical analysis of the processing and packaging of lysosomal enzymes in this crucial cell type has lagged. Recourse has been had most often to cells such as the HL60 human "promyelocyte" line, which can differentiate either into macrophage-like cells or into neutrophil-like cells. With this cell line it has been found, for example, that the myeloperoxidase destined for azurophil-like granules is made as a large precursor (Table 7.1) that is processed proteolytically over a period of a day or two beginning at about the time the enzyme enters the granules. The HL60 cells do have M6P receptors and some of the new peroxidase molecules they make are phosphorylated, probably on their mannoses—these molecules have been detected by their ability to bind to endocytotic M6P receptors of cultured fibroblasts. Indications that the cells use an acidified compartment for sorting hydrolases to lysosomes include findings that they augment their secretion of cathepsins when exposed to weak bases.

Doubts that the orthodox scheme accounts for all of neutrophil packaging of hydrolases arise, however, from difficulties in demonstrating effects of ammonium chloride on the processing and packaging of peroxidase in HL60 cells (conflicting findings have been made) and from reports that I-cell disorder has relatively little effect on the hydrolase content of neutrophils.

7.4.3.7. Lysosomal Membrane Proteins

The ER and Golgi apparatus are presumed to be the ultimate source of the proton pumping machinery, transport carriers, and other proteins of the membranes bounding lysosomes and related compartments. This origin has been suggested, from microscopy, for systems such as plant cells in which Golgi derivatives appear to provide the membrane needed for rapidly enlarging vacuole surface area during cell elongation or protein body formation. In addition, fractionation studies suggest that the antigens of the lysosomal membrane discussed in Section 3.2.3.3 move to the lysosomes via the Golgi apparatus, with kinetics more or less similar to those of the hydrolases. These antigens turn over, in cultured cells such as the J774 macrophage line, with half lives of a half day to a day or two. As with the other membrane proteins considered in Chapter 5, the heterogeneity in half lives of different lysosomal membrane proteins remains unexplained. Quite likely the recycling phenomena discussed in Chapters 2 and 3 will provide part of the explanation.

Several of the lysosomal membrane proteins have been shown to be extensively sialylated. In their mature form at least, the proteins lack M6P residues or other phosphorylated units. Evidently then, their passage through Golgi systems involves a stay in *trans* sacs, and they are directed to the lysosomes by mechanisms that do not directly involve the M6P-based sorting system. This last point would explain why transport of new lysosomal membrane proteins to lysosomes is not blocked by tunicamycin (see Section 3.2.1 and 7.6).

7.5. Loose Ends

7.5.1. Heterogeneity and Specialization II

To what extent do the primary or secondary lysosomes of a given cell or those of different cell types, differ from one another in their enzymatic content or other functional capacities (see Section 2.1.2.2)? The answers are not clear, except for a few relatively special cases. The questions are of importance not only for understanding lysosome function but also for a comprehensive view of how lysosomes form. For instance, if small coated vesicles do serve as principal carriers of the hydrolases, then interesting problems of coordination and regulation of delivery arise. Lysosomes contain many dozens of different proteins and, given the diminutive size of Golgi-associated coated vesicles (mostly 100 nm or less in diameter), it is difficult to imagine that each coated transport vesicle could be made to carry precisely the same proportions of each of these proteins, to say nothing of the membrane components and other materials that must be provided. (Strictly speaking, it is not even ruled out yet that each vesicle transports only a single species of protein, though this seems very unlikely from what is known of other Golgi products, such as secretory structures; see also the footnote on p. 409.) Perhaps, over time, a given secondary lysosome usually receives what it needs by multiple deliveries both from the vesicles and from other lysosomes. Recall also the evidence in Section 2.4.4.1 that endosomes may receive or retain specific hydrolases differentially.

If, in fact, the delivery vehicles are not all the same in their enzyme content, then possibilities arise for regulated delivery yielding different enzyme contents at different target compartments (see Section 2.4.4.1). Examples discussed in Section 7.3.3.5 and elsewhere do suggest that certain lysosomal hydrolases may be quite atypical in their transport through the cell and their cellular locations.

7.5.1.1. Variation among Cells and among Organisms

In a few cases, differences among the lysosomes of different cell types can readily be correlated with function. Neutrophils are an extreme example, as

7.5.1.3. Secreted Hydrolases

The collections of lysosomal hydrolases secreted by cells can differ from the collections within the cells in terms of the relative levels of activity of different hydrolases. The apparent implication, as suggested for protozoa in Section 6.1.4.3, is that the hydrolase-bearing bodies that undergo exocytosis differ from the average lysosome retained by the cell. (It seems very unlikely that all such situations are ones in which the observers have been misled by the differential inactivation of released enzymes; see Sections 4.5.1.1 and 6.2.3.2.)

The secretion of lysosomal hydrolases can, however, involve at least three rather different phenomena with different ramifications for the issues under consideration:

1. When enzymes are released before they reach functional lysosomes—by "diversion" or "leakage" from the packaging route (Sections 6.2.3.1 and 7.2)—the relative activities of the secreted hydrolases will depend upon the rates and mechanisms of synthesis and transport of the different enzymes and also upon the extent of activation, which varies considerably for the various hydrolases (Section 7.4.1.1).

2. When hydrolases are transported to the cell surface from functioning lysosomes during recycling, the makeup of the population of secreted enzyme molecules will be determined by the access had by different hydrolases to the vehicles that mediate recycling. In *Acanthamoeba*, the ratios among hydrolase activities released from the cell as a consequence of endocytic recycling (Section 3.3) vary considerably with intralysosomal pH, probably because pH affects the association of the hydrolases in phagolysosomes with the structures that carry membrane back to the cell surface. Some of the "leakage" of hydrolases from mammalian cells has been explained by postulating that the mannose receptors (Sections 2.2.1.2 and 4.5.1), or other membrane molecules with significant affinity for lysosomal enzymes, pick up the enzymes as the receptors cycle transiently through the lysosomes or endosomes; if so, the composition of the mixture of hydrolases transported to the exterior would depend on the mobilities of the enzymes in the lysosomes or endosomes and on the enzymes' affinities for the cycling receptors.

3. When hydrolases are defecated from the cell by direct exocytosis of secondary lysosomes, the relative activities of the different enzymes will depend upon the prior life histories of the defecated particles (e.g., upon the rates at which enzymes in lysosomes are inactivated spontaneously, or are degraded by the proteases in the lysosomes, or are affected by contact with the materials undergoing digestion).

Is defecation selective for those lysosomes that need to be taken out of intracellular circulation because they have finished all the work they are expected to do, or are enfeebled by a surfeit of indigestible contents or a plethora of inactive hydrolases (Section 3.5.2.4)? In *Paramecium*, older food vacuoles—mainly residual bodies—are much more likely to undergo exocytosis at the cytoproct than are younger forms, perhaps in part because motility mechanisms and cell geometry are aligned so as to minimize premature arrival of bodies at the cytoproct. For *Tetrahymena*, reports have been somewhat contradictory. In some studies, phagocytosed particles such as newly phagocytosed latex seem to remain in the cell for the better part of an hour before extensive defecation begins. Once particle release is under way, it seems to take place with "first-order" kinetics (see footnote on p. 218). But when *Tetrahymena* egests radioactively labeled dimethylbenzanthracene particles taken up endocytotically, similar percentages of radioactivity are released per minute from cells that were first fed the radioactive particles and then fed nonradioactive dimethylbenzanthracene, as from cells given the nonradioactive particles before the radioactive ones. If phagolysosomes must pass through an orderly preegestion period before becoming eligible for release, one might have expected preferential release of the particles taken up earliest and therefore a difference dependent upon the order of presentation of the labeled and unlabeled particles.

Of course, for the many cell types where extensive fusion of secondary lysosomes with newly forming digestive bodies is assumed to occur, notions of "older" and "newer" would apply clumsily to much of the lysosomal population (see footnote on p. 298). Data on egestion of old lysosomal content from such cells are inconsistent and ambiguous. For example, metals or detergents such as Triton WR-1339 (Section 1.3.1) loaded into hepatic lysosomes *in vivo* are cleared from rat livers in a matter of days, probably by defecation. But these same materials are retained by the livers of Chinese hamsters for months.

In addition, as Section 6.4.2.2 outlined, even for heavily loaded storage disease cells or cells full of lipofuscin, critical evidence is lacking for "demobilization" of the supposedly overloaded organelles: they still do have at least some functional hydrolases and may still be able to fuse with newer digestive structures.

7.5.2. Regulation

One reason why heterogeneity is worth careful study is for the light to be shed on the regulation of hydrolase production and of lysosome functioning. Anecdotal information on these topics abounds, but coherent analysis does not (a situation likely to change soon, given the rapid progress in cloning lysosome-related genes and jockeying them around).

Though the genes for lysosomal hydrolases are scattered through the genome, their expression must somehow be coordinated or balanced. At a minimum, the dozens of enzymes and other proteins must normally be produced and transported in proportions appropriate for lysosomal functioning. The relevant controls cannot be solely matters of intrinsic transcription and translation rates: particular enzymes do vary differentially in different cells and conditions. For mice, the variations in levels of hydrolases among different strains, among the tissues of a given strain and at different developmental stages, suggest that genes additional to the structural genes for hydrolases are important in regulating the timing of hydrolase production and the steady-state enzyme levels.

Developmental and physiological regulation of hydrolase synthesis or processing and of lysosome behavior is easily illustrated. Redistributions of endosomes or lysosomes to the Golgi region or to the nuclear region reputedly are evoked by hormones or serum factors. Macrophages alter rates of hydrolase production considerably when they pass from one state to another (see Section 4.4.1; e.g. activation induces considerable synthesis of the enzymes). Under certain conditions, lysosomes in macrophages assume the form of elongate tubules stretched out along microtubules (Fig. 2.20) whereas in other situations they are roughly spherical bodies each much smaller than one of the tubules.

Seedlings turn on the digestion of the contents of their protein bodies when they begin to sprout. When grown in rich media, the amebas of slime molds like *Dictyostelium discoideum* store most of their hydrolases intracellularly. But when starved—a condition that can induce the familiar developmental program of aggregation and differentiation—the cells secrete much of their enzyme content and also seem to undertake considerable autophagy (see Section 5.1.6). 3T3 cells in culture augment their hydrolase content when the cultures stop growing, and they increase their rates of protein turnover (Section 5.5.1.1).

7.5.2.1. Correlations of Hydrolase Amounts with Degradative Activities; . . . with mRNA Levels; Hydrolase Turnover

Macrophages and *Tetrahymena* are among several cell types in which phagocytosis of digestible materials, but not of indigestible bodies, leads to increases in lysosomal hydrolases. In cultured fibroblasts, however, prolonged endocytosis of sucrose can induce hydrolase production, despite the indigestibility of this disaccharide. In *Tetrahymena* and some other microorganisms, entry into stationary phase is accompanied by increases in hydrolase activity associated, perhaps, with the enhanced levels of secretion or autophagy that often are seen. Yeast also augment their vacuolar hydrolase content in stationary phase or in nitrogen starvation. Rates of hydrolase synthesis by *Dictyostelium* vary substantially at different points in the developmental cycle. Hydrolase levels rise in tissues of multicellular organisms, such as muscle, on exposure to agents like thyroid hormones or to other physiological conditions promoting cytoplasmic breakdown. But autophagy in protozoa and in cells of "higher" organisms can also be induced without a short-term increase in acid hydrolases.

In various of the storage diseases, increased levels of hydrolase activities other than the missing enzymes are noted, as if the tissues were "trying" to compensate for their deficiencies.

Little is known of the mechanisms whereby changes in hydrolase synthesis are controlled, though altered levels of mRNA have been detected in a few of the situations.* The steady-state abundance of mRNAs for given lysosomal enzymes in mouse organs like kidney or liver has been estimated, provisionally, as on the order of one or a few molecules per cell so that small changes could have large effects. Increases in the relevant mRNA take place, for instance, during the massive induction of β-glucuronidase by androgens in mouse kidney. (This induction is particularly interesting in that it is selective for the glucuronidase—female mice can show changes of 100-fold in levels of glucuronidase while producing only 2- or 3-fold more of other enzymes such as β-galactosidase.) mRNAs for certain of the lysosome-related proteins secreted by transformed cells increase under the impact of growth factors that enhance such secretion, and HL60 promyelocytes seem to reduce their content of mRNAs for myeloperoxidase as they differentiate into cells that no longer produce peroxidase.

---

*More is known of the regulatory mechanisms for the molecules that collaborate with the lysosomes, such as the endocytic receptors. Changes in mRNA levels underlie some of the long-term alterations in abundance of LDL receptors, and Sections 4.5.2 and 5.2.1.2 have described other devices, such as "down-regulation," that affect the numbers, functioning, or intracellular distributions of various receptors.

That not all of the differences in hydrolase amounts among mouse strains correlate with differences in pertinent mRNAs is one of many bits of indirect evidence for regulation at levels additional to transcription. The quantities of each type of lysosomal enzyme present at a given moment presumably are determined principally by balances among the synthesis and ultimate degradation of the enzyme molecules, but are influenced also by the efficiency with which the molecules are captured by the packaging machinery rather than dribbling out of the cell, and by the production of other proteins that affect the fate of particular hydrolases [egasyn, for example (Section 7.3.3.1), or stabilizer proteins (Section 6.4.1.5)].

A mouse kidney or liver cell has a few thousand molecules of each lysosomal hydrolase. The turnover rates of different hydrolases differ considerably (Section 7.1). In the absence of embarrassing evidence to the contrary, lysosomes are assumed to turn themselves over, in the sense of degrading their own enzymes. Most likely, the rates of degradation of a given hydrolase are governed largely by the intrinsic susceptibilities of the molecules to attack. These susceptibilities probably are not much regulated physiologically but it is conceivable that some of the molecules or structures entering lysosomes by autophagy or heterophagy affect the stability of the hydrolases.

7.5.2.2. Controls of Heterophagy, Autophagy, Movement, and Fusion

A few of the more obvious external conditions that regulate heterophagy and autophagy have been identified. Phagocytosis can be induced by antibody-coated particles (Section 2.2.1.2). Autophagy responds to hormones and to circulating levels of nutrients (Section 5.1.5.1). A few clues to the integration of such effects and to internal cellular controls have turned up: With macrophages for instance, provision of proteins for heterophagic digestion reduces turnover of endogenous proteins, whereas deprivation of amino acids can induce endocytosis. Seemingly, this cell type can "evaluate" the adequacy of its internal amino acid pools and adjust both heterophagy and autophagy accordingly. Amebas, after feeding for a while, reduce their rates of endocytosis of proteins, as if they too control heterophagy in response to internal metabolite pools.

At least some phagocytes, mammalian and protozoan, regulate the extent and duration of phagocytosis, slowing uptake when they become stuffed with particles, and resuming once their cytoplasm becomes partly cleared by digestion and defecation. The cells' membrane economies and the devices that keep the size of lysosomes within bounds in normal cells (Section 2.4.1) may help govern this behavior. For example, when macrophages engulf large amounts of indigestible materials rapidly, they seem eventually to reach a point where they have to synthesize new plasma membrane components before resuming phagocytosis, as if the cells had to pause in order to replace the membrane tied up in replete lysosomes (see Section 3.3.2.1). Macrophages also sometimes slow rates of pinocytosis when the presence of indigestible solutes like sucrose has induced the cells to form too many large endocytotic vacuoles. In *Acanthamoeba* the surface area of the large intracellular vacuoles in which endocytosed materials accumulate seems to be kept roughly constant as the cells shift from pinocytosis, in which the vacuoles do not contain large particles, to phagocytosis, in which particles are present. During division of cultured mammalian cells, endocytotic rates decline owing, most likely, to the reorganizations of the cytoskeleton and of the cell surface that characterize dividing cells.

Calcitonin induces a redistribution of a lysosome-related membrane antigen from the osteoclast cell surface to multivesicular structures in the cell, and also reduces levels of acid hydrolases demonstrable cytochemically in the ER. Do both effects depend on the same calcitonin-induced signal or are there several messengers? As in this example, there is a mass of circumstantial evidence implicating calcium ions and other inorganic ions, arachidonic acid metabolites, phosphorylations of proteins, cyclic nucleotides, cytoskeletal structures, soluble cytoplasmic proteins, and so forth in the control of lysosomal behavior (Section 2.3.2.10, 2.4.4, 2.4.6, and 4.4.1), which needs to be sifted and codified into testable propositions about regulatory mechanisms. A daunting variety of agents affect lysosome-related phenomena, and the effects themselves are diverse, and sometimes ill-defined. Moreover, present views on the roles of regulatory ions and molecules, fusogens, and the like are seriously skewed by considerations of experimental ease—it is much simpler, for instance, to study fusions of lysosomes with the cell surface than fusions within the cell but extrapolating from the one situation to the other could prove misleading (see Section 2.4).

More generally, the state of understanding of organelle movements and membrane fusions is still so primitive as to preclude incisive discussion of many central questions. For example, does the distribution of lysosomes to daughter cells at cell division—which usually is roughly evenhanded—require special controls or orienting devices? Lysosomes and related endocytotic structures often seem to be clustered near the asters of the spindles of animal cells but it is not known whether this grouping arises chiefly as a consequence of the exclusion of organelles from the mitotic apparatus itself and from other regions of the cell or chiefly because the astral microtubules actively direct the lysosomes. In yeast containing a single large vacuole, the vacuole had been thought to fragment as part of the preparations for cell division. More recent data suggest instead that the vacuole of the parent cell can remain intact and can generate the vacuole of the daughter cell by some sort of budding or via vesicular intermediates, at least under some growth conditions.

Are the movements and positions of lysosomes in non-dividing cells aimed chiefly at promoting proper fusions (Section 2.4) or do they also help distribute digestion products efficiently within the cell?

Fortunately, biology is becoming accustomed to coping with such problems by finding or creating special opportunities. Mutations are already known that affect lysosomal fusions or size controls—examples are the Chediak–Higashi disorder in humans (Section 4.4.2.4), and the variety of mouse pigment mutations, some of which mimic the Chediak–Higashi syndrome and many of which produce decreased rates of hydrolase secretion. Cell-free reconstruction of the behavior of lysosomes or endosomes is well underway (Section 2.4.6). Few of the existing gaps seem unbridgeable and several of the questions left open in this book are likely to be resolved by the time the book is published.

7.6. Bulletins from the Front

The following information became available after the book had entered into the editing and production process:

1. "Cation-independent" and "cation-dependent" have largely replaced

"215-kD" and "46-kD" in referring to the larger and smaller M6P receptors (see Section 7.3.1).

2. Suspicion is growing that the lysosomal–endosomal proton pump contains proteins more closely related in properties (and evolution?) to components of the H^+-ATPases of archaebacteria than to the systems of mitochondria, chloroplasts, and eubacteria.

3. Initial truncation experiments suggest that regions of the cytoplasmic domain of cation-independent (215-kD) M6P receptor (Section 7.3.1) are needed for the internalization of hydrolases from the culture medium, for delivery of the enzymes to lysosomes, and for correct cycling of the receptor.

4. One model under discussion for the functioning of the M6P receptors proposes that a molecule of the cation-independent (215-kD) receptor folds so that it binds to two hydrolase-associated M6Ps, whereas the cation-dependent (46-kD) receptor molecules function in pairs (or larger combinations) that together bind to two or more M6Ps (see Section 7.4.2.2).

5. Disagreement has arisen as to whether tunicamycin invariably reduces the half-life of lysosomal membrane proteins drastically (see Sections 3.2.1 and 7.3.4.7).

6. L cells engineered to overproduce transferrin receptors show increased numbers of coated pits, suggesting that the receptors actively influence the formation of the pits rather than simply being recruited by pits that form independently of the receptors.

7. Although many newly made acid phosphatase molecules normally do exhibit M6Ps, when a cloned human gene for acid phosphatase is transfected into BHK cells, acid phosphastase molecules are produced that associate closely (as transmembrane proteins?) with membranes and move to the lysosomes, seemingly without the involvement of M6Ps.

8. Thus far, M6P residues have not been found in appreciable numbers in the oligosaccharides of M6P receptor molecules, but during its biogenesis, the large form of the receptor can acquire phosphates on serine residues.

9. Initial proposals for the mechanisms of binding of M6P receptors to their ligands envisage involvement of ionic interactions between basic amino acids in the receptors and phosphates in the ligands, and also hydrogen bonding between the receptor proteins and the saccharide (and polypeptide?) moieties of the ligands.

10. Studies on *Saccharomyces* mutants defective in vacuole acidification (detectable by alterations in the accumulation of quinacrine), and on the effects of bafilomycin and weak bases on yeast, indicate that in yeast cells that are unable to maintain a low vacuolar pH, much of the carboxypeptidase and proteases is secreted rather than delivered to the vacuoles. The cells do not show similar abnormalities in the handling of membrane-associated vacuole proteins such as α-mannosidase. Acidification of Golgi structures was not studied. In these investigations, and in work on other mutants in which the vacuoles (normally present as one to four large structures occupying about 20% of the cell's volume) are replaced by smaller vesicles or other structures, there are signs that, as in plants, the vacuole participates centrally in regulating the yeast cell's responses to altered external pH and to osmotic stress.

11. Contrary to the information presented on page 64, recent findings (*J. Cell Biol.* **107**:801) suggest that binding of ligands is not necessary to bring about clustering of EGF receptors, at least in fibroblasts. Correspondingly, models of receptor down-regulation (see page 268) are being prepared in which ligands alter rates of receptor internalization and affect the intracellular behavior of receptors without the involvement of ligand-driven clustering phenomena (and sometimes without the obligatory involvement of coated vesicles).

12. The rapid apparent release of a proportion of endocytosed traces (described on pages 131–132) may include the seepage of material trapped in the extracellular spaces as well as genuine regurgitation of endocytosed molecules (*J. Cell Physiol.* **136**:389).

13. A recent report indicates that cathepsin D that is engineered to remain in the ER nonetheless undergoes the initial steps of mannose phosphorylation (*EMBO J* **7**:9–13).

14. Since DNA can transform prokaryotes, DNA molecules can evidently enter cells nonendocytotically, but the routes and mechanisms of DNA entry into eukaryotic cells are not well understood.

Acknowledgments

J. Bergmann, W. J. Brown, C. Gabel, H. Geuze, G. Griffiths, and S. Kornfeld read large parts or all of this chapter and made very useful suggestions as well as calling my attention to very important unpublished material. B. Storrie, who read most of the manuscript of the book (see Preface), was particularly helpful with this chapter. A. Hasilik and K. von Figura patiently corrected some of my misapprehensions about work of their laboratory and others, and generously added unpublished details. In addition, I am grateful for specific information provided by: D. Bainton (neutrophils), W. J. Brown (M6P receptors), R. O. Elferink (hydrolase genesis; α-glucosidase). A. Erickson (early stages in hydrolase genesis), M. G. Farquhar (neutrophils, Golgi apparatus), R. E. Fine (coated vesicles), C. Gabel and D. Lazzarino (phosphorylation of hydrolases), D. E. Goldberg (hydrolases as glycoproteins), G. Griffiths (Golgi organization; intermediate compartments), A. Hasilik (hydrolase transport), A. Kaplan (*Dictyostelium*), S. Kornfeld (hydrolases as glycoproteins; M6P receptors), P. Lobel (M6P receptors), I. Mellman (intermediate compartments), E. Neufeld (M6P receptors; secretion recapture), K. Paigen (molecular biology of mouse lysosomal enzymes), L. Rome (coated vesicles), M. G. Rosenfeld (lysosomal membrane proteins), J. Roth (Golgi organization), J. Slot (immunocytochemistry of Golgi apparatus and lysosomes), W. Sly (M6P receptors), P. Stahl (receptors in macrophages), R. Swank (mouse mutants; β-glucuronidase), K. von Figura (hydrolase transport).

Further Reading

Banta, L. M., Robinson, J. S., Klionsky, D. J., and Emr, S. D. (1988) Organelle assembly in yeast: Characterization of yeast mutants defective in vacuole biogenesis and protein sorting, *J. Cell Biol.* **107**:1369–1384.

Barriocanal, J. G., Bonifacino, J. S., Yuan, L., and Sandoval, I. N. (1986) Biosynthesis, glycosylation, movement through the Golgi apparatus by an N-linked carbohydrate independent mechanism of three lysosomal integral membrane proteins, *J. Biol. Chem.* **261**:16755–16763.

Bowles, D. J., and Pappin, D. J. (1988) Traffic and assembly of concanavalin A, *Trends Biochem. Sci.* **13**:60–64.

Braulke, T., Gartung, G., Hartung, A., and Von Figura, K. (1987) Is movement of mannose 6-phosphate receptor triggered by binding of lysosomal enzymes? *J. Cell Biol.* **104**:1735–1742.

Brown, W. J., and Farquhar, M. G. (1987) The distribution of 215 kD mannose-6-phosphate receptors within cis (heavy) and trans (light) Golgi subfractions varies in different cell types, *Proc. Natl. Acad. Sci. USA* **84**:9001–9005.

Brown, W. J., Goodhouse, J., and Farquhar, M. G. (1986) M6P receptors for lysosomal enzymes cycle between the Golgi complex and endosomes, *J. Cell Biol.* **103**:1235–1247.

Cardelli, J. A., Golumbeski, G. S., and Dimond, R. L. (1986) Lysosomal enzymes in *Dictyostelium discoideum* are transported to lysosomes at distinctly different rates, *J. Cell Biol.* **102**:1264–1270.

Cladaras, M. H., and Kaplan, A. (1984) Maturation of alpha-mannosidase in *Dictyostelium discoideum*, *J. Biol. Chem.* **259**:14165–14169.

Dahms, N. M., Lobel, P., Breitmeyher, J., Chirgwin, J. M., and Kornfeld, S. (1987) 46kD cation-dependent mannose-6-phosphate receptor: Cloning, expression, and homology to the 215 kd cation-independent mannose-6-phosphate receptor, *Cell* **50**:181–192.

Duncan, J. R., and Kornfeld, S. (1988) Intracellular movement of two manose-6-phosphate receptors. Return to the Golgi apparatus, *J. Cell Biol.* **106**:617–628.

Dunphy, W. G., and Rothman, J. E. (1985) Compartmental organization of the Golgi apparatus, *Cell* **42**:13–21.

Erickson, A. H., and Blobel, G. (1983) Carboxyl-terminal proteolytic processing during biosynthesis of the lysosomal enzymes beta-glucuronidase and cathepsin D, *Biochemistry* **22**:5201–5205.

Erickson, A. H., Ginns, E. I., and Barranger, J. A. (1985) Biosynthesis of the lysosomal enzyme, glucocerebrosidase, *J. Biol. Chem.* **260**:14319–14324. (See also *J. Biol. Chem.* **261**:50–53, 1986.)

Farquhar, M. G., and Palade, G. E. (1981) The Golgi apparatus (complex) 1954–1981: From artifact to center stage, *J. Cell Biol.* **91**:77S–103S.

Faye, L., and Chrispeels, M. J. (1987) Transport and processing of the glycosylated precursor of concanavalin A in jack bean, *Planta* **170**:217–241.

Freeze, H. H., Mierendorf, R. C., Wunderlich, R., and Dimond, R. L. (1984) Sulfated oligosaccharides block antibodies to many *Dictyostelium discoideum* acid hydrolases, *J. Biol. Chem.* **259**:10641–10643. (See also *J. Biol. Chem.* **261**:135–141, 1986.)

Gabel, C. A., and Foster, S. A. (1986) Mannose-6-phosphate receptor mediated endocytosis of acid hydrolases: Internalization of beta-glucuronidase is accompanied by a limited dephosphorylation, *J. Cell Biol.* **103**:1817–1827.

Geuze, H. J., Slot, J. W., and Schwartz, A. L. (1987) Membranes of sorting organelles display lateral heterogeneity in receptor distribution, *J. Cell Biol.* **104**:1715–1723.

Goldfischer, S. (1982) The internal reticular apparatus of Camillo Golgi, *J. Histochem. Cytochem.* **30**:717–733.

Green, S. A., Zimmer, K. P., Griffiths, G., and Mellman, I. (1987) Kinetics of intracellular transport and sorting of lysosomal membrane and plasma membrane proteins, *J. Cell Biol.* **105**:1227–1240.

Griffiths, G., and Simons, K. (1986) The trans Golgi network: Sorting at the exit site of the Golgi apparatus, *Science* **234**:438–443.

Griffiths, G., Hoflack, B., Simons, K., Mellman, I., and Kornfeld, S. (1988) The mannose 6 phosphate receptor and the biogenesis of lysosomes, *Cell* **53**:329–341.

Hansen, W., and Walter, P. (1988) Preprocarboxypeptidase Y and a truncated form of pre-invertase but not full length pre-invertase can be posttranslationally translocated across microsomal vesicle membranes from *Saccharomyces cerevisiae*, *J. Cell Biol.* **106**:1075–1082.

Hoflack, B., and Kornfeld, S. (1985) Purification and characterization of a cation dependent mannose 6 phosphate receptor from murine P338D macrophages and human liver, *J. Biol. Chem.* **260**:12008–12014.

Jessup, W., Bodmer, J. L., Dean, R. T., Greenaway, V. A., and Leoni, P. (1984) Intracellular turnover and secretion of lysosomal enzymes, *Biochem. Soc. Trans.* **12**:529–531.

Johnson, L. M., Bankates, V. A., and Emr, S. D. (1987) Distinct sequence determinants direct intracellular sorting and modification of a yeast vacuolar protease, *Cell* **48**:875–885. (See also *Mol. Cell Biol.* **8**:2105–2116, 1988.)

Kelly, R. B. (1985) Pathways of protein secretion, *Science* **230**:25–32. (See also *Nature* **326**:14–15, 1987.)

Kornfeld, R., and Kornfeld, S. (1985) Assembly of asparagine-linked oligosaccharides, *Annu. Rev. Biochem.* **54**:631–664.

Lazzarino, D. A., and Gabel, C. A. (1988) Biosynthesis of the mannose-6-phosphate recognition marker in transport-impaired mouse lymphocyte cells: Demonstration of a two-step phosphorylation, *J. Biol. CHem.* **263**:10118–10126.

Lee, S-J., and Nathans, D. (1988) Proliferin secreted by cultured cells binds to mannose 6 phosphate receptors, *J. Biol. Chem.* **263**:3521–3527.

Lemansky, P., Hasilik, A., Von Figura, K., Helmy, S., Fishman, T., Fine, R. E., Kederska, N. L., and Rome, L. H. (1987) Lysosomal enzyme precursors in coated vesicles derived from exocytic and endocytic pathways, *J. Cell Biol.* **104**:1743–1748.

Lobel, P., Dahms, N. M., Breitmeyer, J., Chirgwin, J. M., and Kornfeld, S. (1987) Cloning of the bovine 215-kDa cation-independent mannose-6-phosphate receptor, *Proc. Natl. Acad. Sci. USA* **84**:2233–2237. (See also *J. Biol. Chem.* **263**:2563, 1988.)

Morgan, D. O., Edman, J. C., Standring, D. N., Fried, V. A., Smith, M. C., Roth, R. A., and Rutter, W. J. (1987) Insulin-like growth factor II receptor as a multifunctional binding protein, *Nature* **329**:301–307. (See also Tong *et al.*, *J. Biol. Chem.* **263**:2585, 1988.)

Murphy, R. F. (1988) Processing of endocytosed material, *Adv. Cell Biol.* **2**:(in press).

Neefjes, J. J., Verkeck, J. M. H., Broxterman, H. J. G., van der Marrel, G. A., van Bloom, J. H., and Ploegh, H. L. (1988) Recycling glycoproteins do not return to the *cis*-Golgi, *J. Cell Biol.* **107**:79–88. (See also Snider and Rogers in Further Reading, Chapter 5.)

North, M. J. (1985) Cysteine proteinases of cellular slime molds, *Biochem. Soc. Trans.* **13**:288–290.

Novick, P. (1985) Intracellular transport mutants of yeast, *Trends Biochem. Sci.* **10**:432–434.

Novikoff, P. M., Novikoff, A. B., Quintana, N., and Hauw, J.-J. (1971) Golgi apparatus, GERL and lysosomes of neurons in rat dorsal root ganglion studied by thin section and thick section cytochemistry, *J. Cell Biol.* **50**:859–886.

Oliver, C., and Hand, A. R. (1983) Enzyme modulation of the Golgi apparatus and GERL, *J. Histochem. Cytochem.* **31**:1041–1048. (See also *J. Histochem. Cytochem.* **31**:222–223, 1983).

Oshima, A., Nolan, C. M., Kyle, J. W., Grubb, J. M., and Sly W. S. (1988) The human cation-independent mannose 6-phosphate receptor. Cloning and sequence of the full length cDNA and expression of a functional receptor in L-cells, *J. Biol. Chem.* **263**:2553–2562.

Paigen, K. (1981) Genetic regulation of lysosomal enzymes, in *Lysosomes and Lysosomal Storage Diseases* (J. W. Callahan and J. A. Lowden, eds.), Raven Press, New York, pp. 1–17.

Pertoft, H., Warmegard, B., and Hook, M. (1978) Heterogeneity of lysosomes originating from rat liver parenchymal cells, *Biochem. J.* **174**:309–317.

Pfeffer, S. R. (1987) The endosomal concentration of mannose 6 phosphate receptors is unchanged in the absence of ligand synthesis, *J. Cell Biol.* **105**:229–234.

Pfeffer, S. R. (1988) Mannose-6-phosphate receptors and their role in targeting proteins to lysosomes, *J. Memb. Biol.* **103**:7–16.

Rosenfeld, M. G., Krebich, G., Popov, D., Kato, K., and Sabatini, D. D. (1982) Biosynthesis of lysosomal hydrolases: Their synthesis on bound polysomes and the role of carbohydrate and posttranslational processing in determining their subcellular distribution, *J. Cell Biol.* **93**:135–143.

Roth, J., Taatjes, D. J., Lucocq, J. M., Weinstein, J., and Paulson, J. C. (1985) Demonstration of an extensive transtubular network continuous with the Golgi apparatus stack that may function in glycosylation, *Cell* **43**:287–295.

Rothman, J. H., and Stevens, T. H. (1986) Protein sorting in yeast: Mutants defective in vacuole biogenesis mislocalize proteins into the late secretory pathway, *Cell* **47**:1041–1051.

Sahagian, G. G. (1984) The mannose 6 phosphate receptor: Function, biosynthesis and translocation, *Biol. Cell.* **51**:207–214.

Stein, M., Zijerhand-Bleckmolen, J. E., Geuze, M., Hasilik, A., and Von Figura, K. (1987) Mr 46,000 M6P receptor: Its role in targetting of lysosomal enzymes, *EMBO J.* **6**:2677–2682. (See also *Hoppe-Seyler Zeit Physiol. Chem.* **368**:927, 1988.)

Stoorvogel, W., Geuze, H. J., Griffith, J. M., and Strous, G. J. (1988) The pathways of endocytosed transferrin and secretory protein are connected in the trans-Golgi reticulum, *J. Cell Biol.* **106**:1821–1830.

Storrie, B. (1987) Assembly of lysosomes: Perspectives from comparative molecular biology, *Int. Rev. Cytol.* (in press).

Tartakoff, A. M. (1983) Mutations that influence the sorting pathway in animal cells, *Biochem. J.* **216**:1–9.

Valls, L. A., Hunter, C. P., Rothman, J. H., and Stevens, T. H. (1987) Protein sorting in yeast: The localization determinant of yeast vacuolar CPY resides in the propeptide, *Cell* **48**:887–897.

Van Dongen, J. M., Barneveld, R. A., Geuze, H. J., and Galjaard, H. (1984) Immunocytochemistry of lysosomal hydrolases and their precursor forms in normal and mutant human cells, *Histochem. J.* **16**:941–954. (See also *J. Histochem. Cytochem.* **31**:1049–1056, 1983.)

Von Figura, K., and Hasilik, A. (1986) Lysosomal enzymes and their receptors, *Annu. Rev. Biochem.* **59**:167–193.

Waheed, A., Gottschalk, S., Hille, A., Krentler, C., Pahlmann, R., Braulke, T., Hauser, H., Geuze, H., and Von Figura, K. (1988) Human lysosomal acid phosphatase is transported as a transmembrane protein in transformed baby hamster kidney cells, *EMBO J.* **7**:2351–2358.

Weisman, L. S., Bacallao, R., and Wickner, W. (1987) Multiple methods of visualizing the yeast vacuole permit evaluation of its morphology and inheritance during the cell cycle, *J. Cell Biol.* **105**:1539–1547.

Index

The use of boldface indicates sites of important definitions or of principal discussions for items with many entries. Page numbers in italics indicate figures or tables. Question marks (?) following an entry indicate unresolved issues. Items prefixed by Greek letters are indexed according to the first letters of the English segments of their names.